Geochemistry and Genesis of Hydrothermal Ore Deposits

Geochemistry and Genesis of Hydrothermal Ore Deposits

Guest Editors

Yuichi Morishita
Napoleon Q. Hammond

MDPI
Basel • Beijing • Wuhan • Barcelona • Belgrade • Novi Sad • Cluj • Manchester

Guest Editors

Yuichi Morishita
Center for Integrated Research and Education of Natural Hazards
Shizuoka University
Shizuoka
Japan

Napoleon Q. Hammond
Department of Geology and Mining
University of Limpopo
Sovenga
South Africa

Editorial Office

MDPI AG
Grosspeteranlage 5
4052 Basel, Switzerland

This is a reprint of the Special Issue, published open access by the journal *Minerals* (ISSN 2075-163X), freely accessible at: https://www.mdpi.com/journal/minerals/special_issues/hydrothermal_ore_deposits.

For citation purposes, cite each article independently as indicated on the article page online and as indicated below:

Lastname, A.A.; Lastname, B.B. Article Title. *Journal Name* **Year**, *Volume Number*, Page Range.

ISBN 978-3-7258-3828-8 (Hbk)
ISBN 978-3-7258-3827-1 (PDF)
https://doi.org/10.3390/books978-3-7258-3827-1

Contents

About the Editors

Yuichi Morishita

Yuichi Morishita is a visiting professor at Shizuoka University. He received his Ph.D. in Geology from the University of Tokyo in 1987. He worked for the Geological Survey of Japan (GSJ) for many years before moving to Shizuoka University. He was a postdoctoral fellow at the Department of Geological Sciences, Brown University, where he conducted experimental mineral diffusion studies. His research has focused on magmatic–hydrothermal ore deposits using multidisciplinary approaches, including field observations (visiting more than 50 ore deposits), oxygen and carbon isotope geochemistry, and microanalysis. He was the leader of the Magma-Hydrothermal Research Group, the SIMS (Secondary Ion Mass Spectrometry) laboratory, and the stable isotope laboratory of the GSJ. He served as the President of the Society of Resource Geology for 2 years.

Napoleon Q. Hammond

Napoleon Quaye Hammond is a professor at the University of Limpopo. He holds a PhD from Rhodes University in South Africa, an MSc from the University of Tokyo in Japan, and a BSc Honours degree from the University of Ghana. He has been involved in ore deposit research for the past 25 years in a variety of research groups at different institutions in Japan and South Africa. His key research interests are genetic aspects of ore deposits including orogenic gold mineralization in Precambrian settings, platinum-group element (PGE) mineralization in magmatic deposits, the application of fluid inclusion and stable isotope geochemistry to fluid evolution in hydrothermal systems, and the geo-metallurgical investigation of refractory ores.

Preface

Hydrothermal ore deposits are formed via the circulation of ore-forming hydrothermal fluids from a variety of sources within the crust, including fluids of mixed origin. These deposits have been associated with different tectonic processes and geodynamic settings, and are hosted in a wide variety of lithologies. These deposits may form near the surface, such as epithermal deposits, or at deeper levels, such as hypothermal deposits including porphyry deposits. Their formation is engendered by the various physicochemical conditions encountered by the related ore-forming hydrothermal fluids along their flow paths. Furthermore, several research studies in seismology and volcanology have clearly evidenced the role of hydrothermal fluids in current earthquake occurrence and volcanic eruptions.

The involvement of mixed hydrothermal fluids in other types of ore deposits has been established. For example, magmatic fluids have been found to potentially evolve into hydrothermal fluid that is mixed with meteoric water in many geological studies. The involvement of hydrothermal fluids in non-hydrothermal-related deposits has been established worldwide; this includes, for example, paleo-placer sedimentary ore deposits such as the uranium–gold deposits found in the vast Witwatersrand goldfields in South Africa.

In recent years, significant strides have been made regarding the application of modern analytical techniques to advance the understanding of these deposits. This Special Issue aims to elucidate the role and importance of hydrothermal fluid in the formation of hydrothermal ore deposits in a broad sense. This Special Issue contains research papers that apply analytical techniques, including geothermometry and geobarometry, geochemical techniques, and geological observations, to address the source and evolution of hydrothermal fluid. With this, we may be able to disclose the genesis of ore deposits, as well as their relevance to the exploration of these deposits.

Yuichi Morishita and Napoleon Q. Hammond
Guest Editors

minerals

MDPI

Article

The Magmatic-Hydrothermal Ore-Forming Processes of the Tonggou Cu-Zn Deposit, NW China: Constraints from Magnetite Chemistry and Fluid Inclusions

Chang-Cheng Han [1], Xue-Bing Zhang [1,2,*], Shi-Shan Wu [1] and Ying-Ting Liu [1]

1 College of Geology and Mining Engineering, Xinjiang University, Urumqi 830017, China; hanchangcheng@xju.edu.cn (C.-C.H.); xjuwss@163.com (S.-S.W.); liu_y_t@stu.xju.edu.cn (Y.-T.L.)
2 Xinjiang Key Laboratory for Geodynamic Processes and Metallogenic Prognosis of the Central Asian Orogenic Belt, Xinjiang University, Urumqi 830047, China
* Correspondence: zhangxb@xju.edu.cn

Abstract: The Tonggou deposit is a porphyry Cu and vein-type Cu-Zn mineralization system located in the Bogda Orogenic Belt, north of Eastern Tianshan. Systematic fluid inclusion analyses were performed on granular quartz from the magnetite–quartz stage and pyrite–chalcopyrite–quartz stage from the porphyry Cu mineralization. During the early stage of porphyry Cu mineralization, the ore-forming fluids were at high temperatures (450–501 °C) and high salinity (51.2–55.2 wt.% NaCl equiv.) H_2O-NaCl hydrothermal fluids with fluid boiling. These fluids evolved to high temperature (412–450 °C) and intermediate to low salinity (8.3–14.2 wt.% NaCl equiv.) H_2O-NaCl hydrothermal fluids during the pyrite–chalcopyrite–quartz stage. In addition, magnetite from the Tonggou deposit was studied as a marker for the ore-forming process evolution of porphyry and vein-type mineralization. Sampled magnetite can be divided into Mag_I (allotriomorphic magnetite from altered granodiorite), Mag_{II} (magnetite from altered granodiorite found in veinlets or as granular aggregates), Mag_{III} (from the magnetite–quartz stage of porphyry mineralization), and Mag_{IV} (from the polymetallic sulfide–epidote–quartz stage of vein-type mineralization). Magnetite LA-ICP-MS data indicate a hydrothermal origin. The contents of Ti, Si, Al, and Ta are controlled by temperature, and these elements gradually decrease from Mag_I to Mag_{IV}. Moreover, fO_2 has considerable influence on the substitution of Sn, V, and Mn in magnetite, and the contents of these elements generally decrease from Mag_I to Mag_{III}—increasing only in Mag_{IV}. Indeed, high fO_2 in the polymetallic sulfide–epidote–quartz stage (Mag_{IV}) of vein-type mineralization is shown by the presence of a replacement texture in ilmenite grains within hydrothermal magnetite. On the other hand, magnetite samples from the Tonggou deposit have relatively low Ti + V contents compared to other porphyry Cu deposits—plotting in the skarn field of the (Ti + V) vs. (Ca + Al + Mn) diagram—and shows negative correlations in the (Ti + V) vs. Sn diagram. These data indicate that the porphyry Cu mineralization at Tonggou formed at relatively lower fO_2 conditions than the Tonggou vein-type mineralization and other typical porphyry Cu deposits. Finally, porphyry and vein-type mineralization at Tonggou are both sourced from the porphyry system, as a result of ore-forming fluid transfer to a different location.

Keywords: magnetite; trace element composition; fluid inclusion; Tonggou deposit; Eastern Tianshan

Citation: Han, C.-C.; Zhang, X.-B.; Wu, S.-S.; Liu, Y.-T. The Magmatic-Hydrothermal Ore-Forming Processes of the Tonggou Cu-Zn Deposit, NW China: Constraints from Magnetite Chemistry and Fluid Inclusions. *Minerals* **2022**, *12*, 485. https://doi.org/10.3390/min12040485

Academic Editor: António Manuel Nunes Mateus

Received: 25 March 2022
Accepted: 13 April 2022
Published: 15 April 2022

1. Introduction

Magnetite is a common mineral in intermediate to acidic rocks and metallic deposits, which is stable at a wide range of temperatures [1–4]. Indeed, magnetite may display different chemical compositions depending on its formation temperature, mineralization style, and genetic processes involved in its formation. For instance, magnetite crystallized from silicate and sulfide melts at high temperatures is expected to be enriched in Ti, Si, Al, Ba, Ta, Sc, and Y [3]. Thus, due to its wide temperature stability across different ore-forming processes, magnetite trace element analysis by LA-ICP-MS has increasingly been used to discuss the ore-forming conditions of different deposits [1,4–6].

Based on elemental concentrations and ratios in magnetite, several studies have established geochemical discrimination diagrams for various types of deposits [4,7,8]. On the other hand, some research efforts have been devoted to distinguishing magnetite from igneous, hydrothermal, and sedimentary origins on the basis of elemental variations [1,9,10]. Therefore, some studies have focused on the trace-element geochemistry of magnetite to fingerprint the evolution of the ore-forming hydrothermal fluids and reconstruct the mineralization processes [1,2,11]. However, compositional variations in magnetite from porphyry Cu systems dependent on mineralization style (e.g., porphyry and vein types) are still poorly understood, as are their specific implications for ore-forming conditions and ore-forming fluid evolution. In addition, the study of fluid inclusions is a primary method for getting the information about the ore-forming fluids (e.g., temperatures, salinities, and pressures), which is the key to revealing the ore genesis and ore-forming progress of the hydrothermal deposits [12].

The Tonggou Cu-Zn deposit is located in the Eastern Tianshan section of the Central Asian Orogenic Belt (CAOB)—one of the world's largest Phanerozoic subduction-accretionary orogens (Figure 1a) [13–15]. This area hosts a number of porphyry Cu deposits, for instance, Tuwu-Yandong, Fuxing, and Sanchakou, located in the Jueluotage Belt [16–18]. Tonggou porphyry Cu and vein-type Cu-Zn mineralization both formed in Late Carboniferous (302–303 Ma; [19,20]), which provide an ideal target to study the chemical variations among different types and generations of magnetite.

This paper presents new magnetite compositional data, acquired using LA-ICP-MS, coupled with a petrographic investigation—i.e., detailed classification and texture analysis from different mineralization stages of the Tonggou deposit. We provide new insight into ore-forming conditions as well as genetic relationships between different mineralization types in porphyry Cu systems. In addition, we acquired fluid inclusion data from porphyry mineralization to constrain the evolution of the ore-forming fluids. The present study could potentially be applied to other similar vein mineralization in porphyry Cu systems.

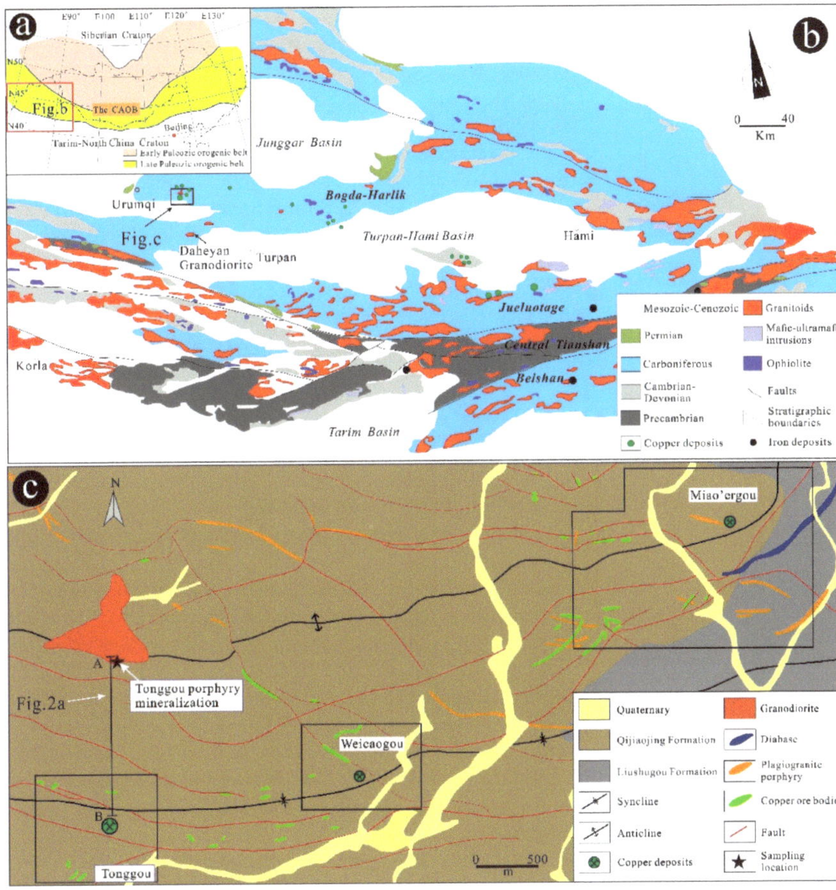

Figure 1. (**a**) Schematic geologic map of the Central Asian Orogenic Belt (CAOB); (**b**) geological map of Eastern Tianshan [15]; (**c**) geological map of the mineralized district in the Tonggou-Miao'ergou area (modified after [20]).

2. Regional Geological Setting

2.1. Regional Geology

The Eastern Tianshan area is located in the southern part of the CAOB, bordered by the Junggar Block to the north and the Precambrian Tarim Block to the south (Figure 1b). Eastern Tianshan can be divided from north to south into the Bogda-Harlik Orogenic Belt, the Jueluotage Belt, the Central Tianshan massif, and the Beishan terrane [21]. This area witnessed the evolution and closure of the Paleo-Asian Ocean [22–24], and the formation of a number of economic Au, Fe, porphyry Cu-Mo, magmatic Cu-Ni, and VMS Cu-polymetallic deposits [25–28]. Porphyry Cu-Mo mineralization in Eastern Tianshan is mainly distributed along the Jueluotage Belt, while the Tonggou porphyry deposit is located in the Bogda Orogenic Belt, with coexisting vein mineralization.

The Bogda Orogenic Belt, situated north of Eastern Tianshan, consists of Devonian, Carboniferous, Permian, and Jurassic strata [29]. The Devonian strata mainly consist of marine and terrigenous sediments, tuffaceous sandstone, and volcanic rocks. The Carboniferous strata show fault contacts with the Devonian rocks and are divided into the Lower Carboniferous Qijiaojing Formation and the Upper Carboniferous Liushugou and Qijiagou Formations, respectively. The Lower Carboniferous Formation mainly consists of marine volcanic ignimbrites, tuffaceous sandstone, and bimodal volcanic lavas, while the Upper Carboniferous Formation is dominated by marine (pillow) basaltic lava and felsic ignimbrites, with minor sandstone and siltstone. The Lower and Upper Carboniferous Formations are separated by regional faults. The Permian strata unconformably overlie the Carboniferous rocks and are mainly composed of terrestrial conglomerate, sandstone, and siliceous mudstone, intercalated with bimodal volcanic lavas. In the southeastern region of the study area, Jurassic clastic sediments unconformably overlie the Permian strata [19,29].

Meanwhile, Late Carboniferous to Early Permian intermediate-to-felsic intrusions occur sporadically in the Bogda Orogenic Belt [30–33]. These intermediate-to-felsic intrusions are associated with a series of Cu polymetallic vein deposits [19], such as the Tonggou, Weicaogou, and Miao'ergou deposits (Figure 1c). In addition, porphyry Cu mineralization has been discovered at the top of the Tonggou granodiorite. Moreover, Re-Os and U-Pb isotopic ages indicate that both porphyry and vein Cu polymetallic mineralization at Tonggou formed in the Late Carboniferous (303 Ma, [19,20]).

2.2. Deposit Geology

The Tonggou deposit is located in the western part of the Bogda Orogenic Belt. The exposed strata mainly correspond to the Late Carboniferous Qijiaojing Formation—consisting of argillite, siltstone, and fine sandstone.

Porphyry Cu mineralization occurs at the top (or contact zone) of the Tonggou granodiorite (Figure 2a). The strata in Tonggou porphyry mineralization belong to the First Segment of Devonian Qijiaojing Formation and consist of metamorphic fine sandstone and siltstone. Metallic minerals are mainly hematite, magnetite, pyrite, and chalcopyrite (Figure 3a–g), with lesser amounts of galena. Porphyry Cu mineralization mainly occurs as disseminated, veinlet, and stockwork veins, accompanied by potassic and propylitic alteration. Widespread gangue minerals contain abundant K-feldspar, chlorite, epidote, quartz, calcite, and lesser anhydrite. Magnetite is widely present in the porphyry Cu mineralization and is partially replaced by hematite (Figures 3e and 4). Based on mineral assemblages and crosscutting relationships, hydrothermal mineralization at Tonggou can be classified into two stages: the magnetite–quartz stage (stage I) and the pyrite–chalcopyrite–quartz stage (stage II).

Vein-type Cu-Zn mineralization occurs 1.5 km south of the porphyry Cu mineralization and is mainly hosted by Late Carboniferous Qijiaojing Formation argillite, siltstone, and fine sandstone. The 36 economic ore bodies discovered are mainly distributed in the E-W trend faults, which are widely developed and act as the ore-controlling structures in the Tonggou deposit. For vein-type Cu-Zn mineralization, the shapes and sizes of the ore bodies are controlled by faults, mainly occur as disseminated, veinlet disseminated, lenticular, and vein-shaped [19]. Metallic minerals are mainly magnetite, pyrite, chalcopyrite, and sphalerite, with lesser amounts of hematite, ilmenite, galena, and bornite (Figures 3h,i and 4). Gangue minerals, including quartz, calcite, chlorite, and epidote, occur widely in the Tonggou deposit and are similar to the propylitic alteration of porphyry mineralization [34]. The hydrothermal mineralization at Tonggou can be divided into three stages: the pyrite–quartz stage (stage 1), the polymetallic sulfide–epidote–quartz stage (stage 2), and the pyrite–calcite stage (stage 3).

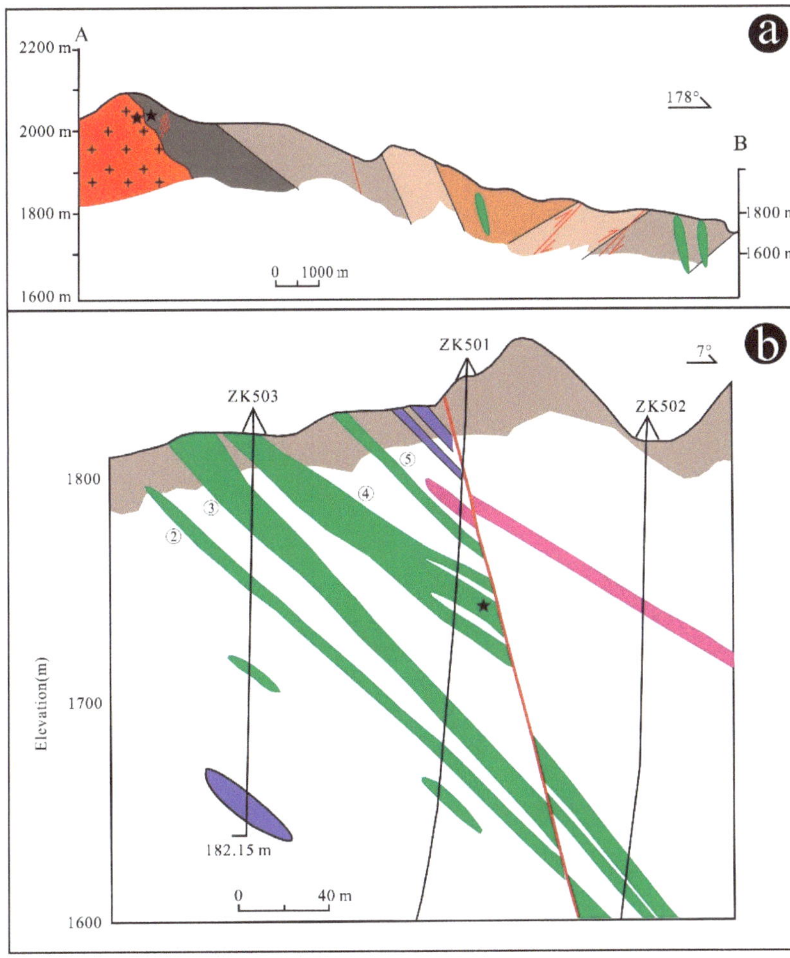

Figure 2. (**a**) Cross-section of A–B of the Tonggou deposit; (**b**) cross-section along exploration line 5 of the Tonggou deposit (modified after [32]).

Figure 3. Representative photographs and photomicrographs of magnetite samples at the Tonggou porphyry and vein mineralization. (**a**) A Mag_I sample from altered granodiorite; (**b**) reflected-light photomicrograph of Mag_I; (**c**) A Mag_{II} sample accompanied with pyrite from altered granodiorite; (**d**) reflected-light photomicrograph of veinlet-disseminated Mag_{II}; (**e**) reflected-light photomicrograph of euhedral Mag_{II}; (**f**) A Mag_{III} sample from magnetite–quartz stage; (**g**) reflected-light photomicrograph of Mag_{III}; (**h**) the specimen of Mag_{IV} representing polymetallic sulfide–epidote–quartz stage; (**i**) reflected-light photomicrograph of Mag_{IV}. Ccp—chalcopyrite; Hem—hematite; Mag—magnetite; Py—pyrite; Qz—quartz.

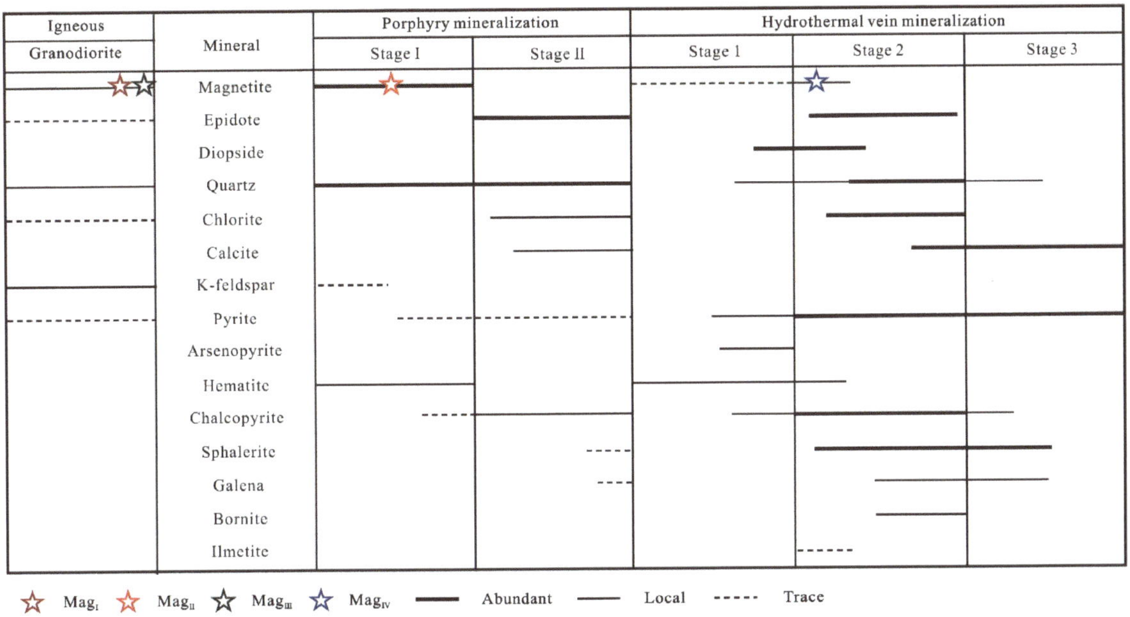

Figure 4. Mineral paragenesis at the Tonggou deposit, showing hydrothermal magnetite.

3. Samples and Analytical Methods

3.1. Sample Preparation

Quartz samples were selected for fluid inclusion analyses, specifically granular quartz from magnetite–quartz stage (stage I) and stage- II pyrite–chalcopyrite–quartz veins (Figure 5a–c). In addition, representative samples were collected from the altered granodiorite and wall rocks of the different mineralization stages. Magnetite samples include two from granodiorite (Mag$_I$ and Mag$_{II}$), one from stage- I porphyry Cu mineralization (Mag$_{III}$), and another from stage 2 of vein-type Cu-Zn mineralization (Mag$_{IV}$). Mag$_I$ from altered granodiorite is allotriomorphic and occurs as isolated or granular aggregates (Figure 3a,b). Mag$_{II}$ from altered granodiorite is subidiomorphic and occurs as veinlet or granular aggregates accompanied by pyrite (Figure 3c–e). Mag$_{III}$ from stage I is allotriomorphic and occurs alongside quartz (Figure 3f,g). Mag$_{IV}$ (stage 2) is allotriomorphic and is partially replaced by chalcopyrite (Figure 3h,i). In addition, some Mag$_{IV}$ samples exhibit ilmenite replacement, which indicates that magnetite from stage 2 formed at high oxygen fugacity conditions.

3.2. Analytical Methods

Fluid inclusion petrography and microthermometric analyses were performed at the Xinjiang Key Laboratory for Geodynamic Processes and Metallogenic Prognosis of the Central Asian Orogenic Belt, Xinjiang University, China. Microthermometric measurements were performed using a Linkam THMS-600 heating-freezing stage mounted on a Carl Zeiss Axiolab microscope with 10× and 50× ultra-long working distance objectives. The heating–freezing rate was generally from 0.2 to 5 °C/min but was reduced to less than 0.2 °C/min near phase transformations. Uncertainties of freezing and heating measurements were ± 0.1 °C from −120 to 31 °C, ± 1 °C from 31 to 300 °C, and ± 2 °C above 300 °C. Homogenization temperature data for fluid inclusions were calculated using the software MacFlincor [35].

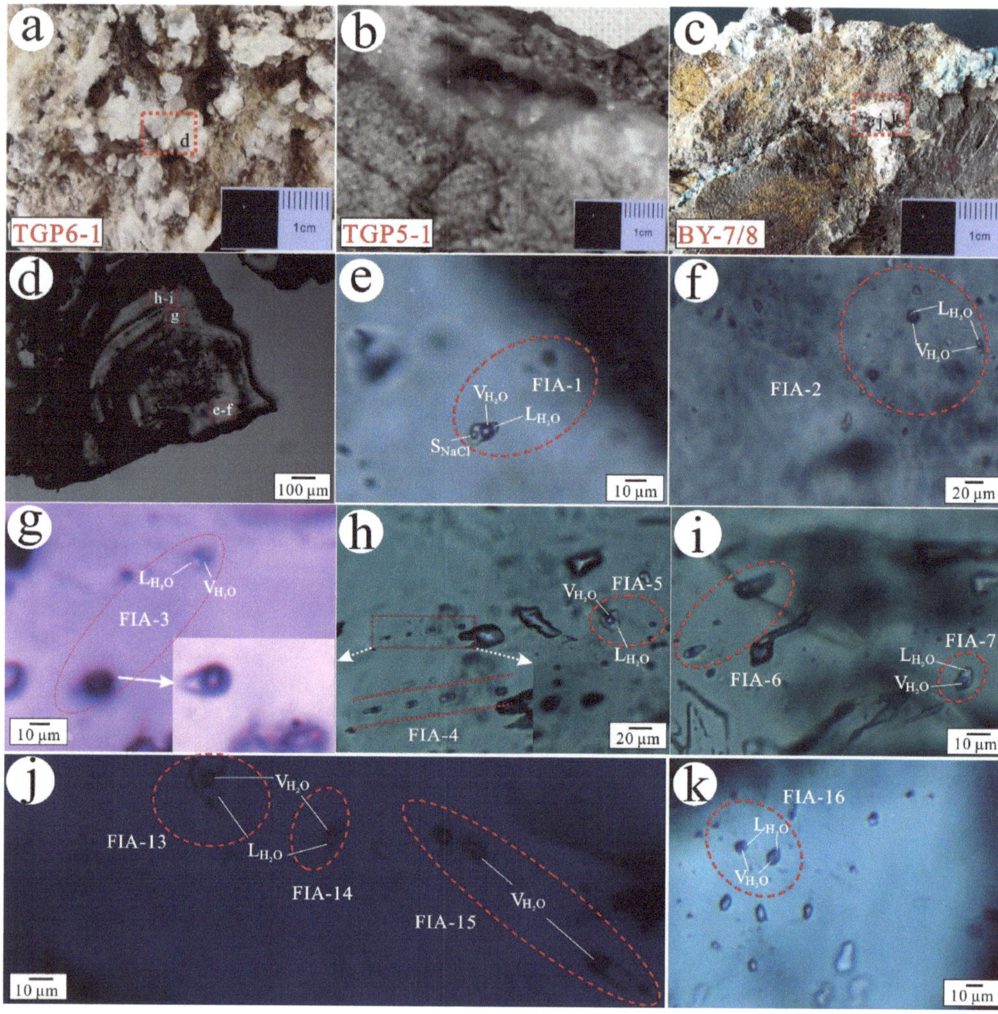

Figure 5. Specimen photographs and photomicrographs of FIAs in quartz from the porphyry Cu mineralization of Tonggou deposit. (**a**) Granular quartz sample from magnetite–quartz stage; (**b**) granular quartz vein; (**c**) chalcopyrite–quartz vein; (**d**) granular quartz sample from magnetite–quartz stage show growth bands; (**e**) Boiling FIA (S- and V-type FIs) in the core of sample TGP6-1; (**f**) L-type FIA and V-type FIA in the cire of sample TGP6-1; (**g**) L-type FIA and V-type FIA in the transition of sample; (**h**) L-type FIA and V-type FIA in the rim of sample TGP6-1; (**i**) L-type FIA in the rim of sample TGP6-1; (**j**,**k**) L-type FIA and V-type FIA in sample BY-7; L (liquid phase), V (vapor phase), S (daughter mineral).

Energy dispersive spectrometer (EDS) and back-scattered electron (BSE) analyses were performed on magnetite in the Beijing Kerongen Science and Technology Ltd., Beijing, China, prior to LA-ICP-MS analysis. EDS analyses were carried out to obtain Fe, O, Al, Mg, Si, and Ti concentrations, while BSE images were obtained for inclusion identification in magnetite (Figure 6).

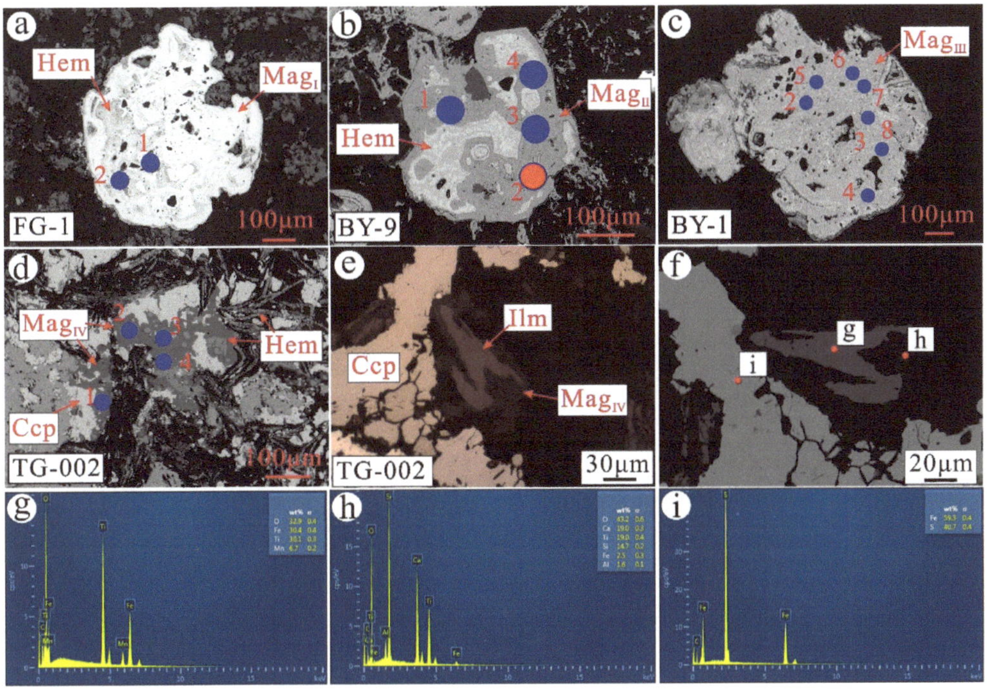

Figure 6. (**a**) BSE image of Mag_{I} from sample FG-1; (**b**) BSE image of Mag_{II} from sample BY-9; (**c**) BSE image of Mag_{III} from sample BY-1; (**d**) BSE image of Mag_{IV} from sample TG-002; (**e**) replacement texture of Mag_{IV} in reflected-light photomicrograph from sample TG-002; (**f**) replacement texture of Mag_{IV} in BSE image from sample TG-002; (**g**) EDS analysis on ilmenite; (**h**) EDS analysis on magnetite; (**i**) EDS analysis on chalcopyrite.

Trace element analyses of magnetite were performed with LA-ICP-MS at Wuhan Sample Solution Analytical Technology Co., Ltd., Wuhan, China. Detailed operating conditions for the laser ablation system, the ICP-MS instrument, and data reduction are as described by Zong et al. [36]. Laser sampling was performed using a GeolasPro laser ablation system that consists of a COMPexPro 102 ArF excimer laser (wavelength of 193 nm and maximum energy of 200 mJ) and a MicroLas optical system. An Agilent 7900 ICP-MS instrument was used to acquire ion-signal intensities. Helium was applied as a carrier gas. Argon was used as the make-up gas and mixed with the carrier gas via a T-connector before entering the ICP-MS. A "wire" signal smoothing device is included in this laser ablation system [37]. Laser spot size and frequency were set to 44 μm and 5 Hz, respectively. Trace element compositions of magnetite were calibrated against various reference materials (BHVO-2G, BCR-2G, and BIR-1G) without using an internal standard [38]. Each analysis incorporated a background acquisition of approximately 20–30 s followed by 50 s of data acquisition from the sample. The Excel-based software ICPMSDataCal was used to perform off-line selection and integration of background and analyzed signals, time-drift correction, and quantitative calibration for trace element analysis [38].

4. Results

4.1. Petrography of Fluid Inclusions

The standard for distinguishing different generations of FIs within hydrothermal quartz was proposed by Goldstein and Reynolds [39]. FIs along growth bands are considered as separate FIA. Primary FIs occur as isolated inclusions or random groups, while pseudosecondary inclusions are found in infilled internal cracks. Based on fluid inclusion petrography, three types of fluid inclusions were identified in our samples according to their phase compositions at room temperature (25 °C) and the phase transitions observed during heating and cooling. The three types of inclusions include liquid-rich aqueous inclusions (L), vapor-rich aqueous inclusions (V), and NaCl daughter mineral-bearing three-phase inclusions (S) (Figure 5d–k).

L-type FIs generally consist of a liquid phase and a minor vapor phase at room temperature, with 10 to 30 vol.% of vapor bubble (Figure 5f–k). With maximum dimensions of 20 μm, these FIs occur randomly in quartz and are present in granular quartz and stage 2 of porphyry Cu mineralization. VL-type FIs consist of a liquid with a vapor bubble—which accounts for 60–90% of the FI volume (Figure 5f–h,j,k). This type of FI is typically elliptical and 10–15 μm in size. These FIs were trapped in granular quartz and stage 2 mineralization and occurred randomly in quartz. S-type FIs are comprised of halite, liquid phase, and vapor bubble that account for 15–20% of the total volume at room temperature and are generally less than 15 μm in diameter (Figure 5e). This type of FI is only present in the core of granular quartz and is accompanied by minor V-type FIs.

4.2. Fluid Inclusion Microthermometry

In total, 19 FIAs and 131 FIs were chosen for microthermometric analysis. Detailed data are listed in Table 1. The homogenization temperature vs. salinity diagram of the different FIAs is presented in Figure 7.

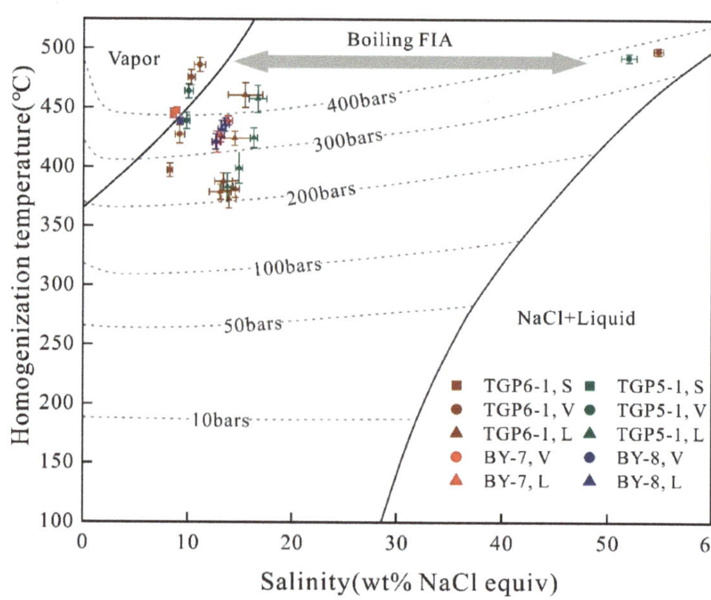

Figure 7. Homogenization temperature vs. salinity in the NaCl-H_2O system for FIAs from porphyry Cu mineralization in the Tonggou deposit. Isobars were calculated using the equations of Driesner and Heinrich [40].

Table 1. Microthermometric data from FIAs in quartz from porphyry Cu mineralization of the Tonggou deposit.

Sample	Host Mineral	FIA	FI Type	Number	Gas–Liquid Ratio (%)	Tm (Ice) (°C)	Tm (NaCl) (°C)	Salinity (wt% NaCl Equiv.)	Th (°C)	Note
TGP6-1	Quartz from granular quartz	FIA-1	V	3	80	From −7.9 to −7.0		10.5–11.6 (avg.11.1)	480–492	core
			S	2	20		458–465	54.3–55.2	495–501	
		FIA-2	V	4	70	From −7.1 to −6.6		9.9–10.6	469–482	
			L	6	15	From −13.1 to −9.8		13.8–17.1	450–471	
		FIA-3	V	3	70	From −6.3 to −5.6		8.7–9.6	420–435	transition
			L	8	20	From −11.9 to −9.1		13.0–15.9	418–430	
		FIA-4	V	2	90	From −5.5 to −5.1		8.0–8.5	391–403	rim
			L	7	25	From −10.2 to −8.7		12.5–14.2	380–395	
		FIA-5	L	5	30	From −10.1 to −8.2		12.0–14.1	372–385	
		FIA-6	L	6	20	From −10.6 to −9.3		13.2–14.6	365–380	
		FIA-7	L	5	15	From −10.9 to −10.1		14.1–14.9	374–388	
TGP5-1	Quartz from granular quartz vein	FIA-8	S	3	20		439–445	51.2–52.7	489–496	core
		FIA-9	V	3	80	From −6.9 to −6.3		9.6–10.4	458–470	
			L	7	20	From −13.5 to −11.8		15.8–17.5	446–469	
		FIA-10	V	2	70	From −6.7 to −6.2		9.5–10.1	432–446	transition
			L	6	15	From −12.6 to −11.9		15.9–16.6	416–433	
		FIA-11	L	7	20	From −11.2 to −10.5		14.5–15.2	386–412	rim
		FIA-12	L	5	30	From −10.1 to −9.3		13.2–14.3	371–395	rim
BY-7	Quartz from pyrite–chalcopyrite–quartz vein	FIA-13	L	6	10	From −9.2 to −8.5		12.3–13.1	412–430	
		FIA-14	L	6	20	From −9.5 to −8.8		12.7–13.4	421–432	
		FIA-15	V	3	80	From −5.7 to −5.3		8.3–8.8	441–449	
			L	8	30	From −10.3 to −9.5		13.4–14.2	435–443	
BY-7	Quartz from pyrite–chalcopyrite–quartz vein	FIA-16	V	2	60	From −5.9 to −5.6		8.7–9.1	445–450	
BY-8	Quartz from pyrite–chalcopyrite–quartz vein	FIA-17	V	3	80	From −6.2 to −5.8		8.9–9.5	436–441	
			L	6	20	From −9.9 to −9.2		13.1–13.9	430–441	
		FIA-18	L	7	15	From −9.6 to −8.9		12.8–13.6	425–439	
		FIA-19	L	6	10	From −9.1 to −8.4		12.2–13.0	415–428	

Granular quartz from porphyry mineralization displays a clearly zonal structure. Grain cores contain abundant L-, V-, and S-type FIs. Final ice-melting temperatures of L-type FIs range from −13.5 °C to −9.8 °C, corresponding to salinities of 13.8–17.5 wt.% NaCl equiv. Total homogenization of L-type FIs to the liquid phase occurred at temperatures of 446–471 °C. Final ice-melting temperatures of V-type FIs range from −7.9 to −6.3 °C, corresponding to salinities of 9.6–11.6 wt.% NaCl equiv. Total homogenization of V-type FIs to the vapor phase occurred at temperatures of 458–492 °C. For S-type FIs, the following homogenization mechanism was observed: halite crystals dissolved first, and the disappearance of the vapor phase occurred later. S-type FIs finally homogenized to a single liquid phase at temperatures of 489–501 °C. Halite crystals within S-type FIs dissolved at temperatures of 439–465 °C, corresponding to salinities of 51.2–55.2 wt.% NaCl equiv.

The granular quartz transition area contains abundant L- and V-type FIs. L-type FIs all homogenize to the liquid phase at temperatures of 416–433 °C. Final ice-melting temperatures of L-type FIs range from −12.6 °C to −9.1 °C, corresponding to salinities of 13.0–16.6 wt.% NaCl equiv. Final ice-melting temperatures of V-type FIs range from −6.7 to −5.6 °C, corresponding to salinities of 8.7–10.1 wt.% NaCl equiv. Total homogenization of V-type FIs to the vapor phase occurred at temperatures of 420–446 °C.

Granular quartz rims also contain abundant L- and V-type FIs. Final ice-melting temperatures of L-type FIs range from −8.2 °C to −11.2 °C. Their salinities are estimated to range between 12.0 and 15.2 wt.% NaCl equiv., and these FIs homogenized to a liquid phase at temperatures ranging from 365 °C to 412 °C. On the other hand, V-type FIs all homogenized to the liquid phase at temperatures of 391–403 °C. Final ice-melting temperatures of L-type FIs range from −5.5 °C to −5.1 °C, corresponding to salinities of 8.0–8.5 wt.% NaCl equiv.

Abundant L-and V-type FIs occur in quartz from stage 2. Final ice-melting temperatures of L-type FIs range from −10.3 °C to −8.4 °C, and their salinities are estimated between 12.2 and 14.2 wt.% NaCl equiv. L-type FIs are homogenized to a liquid phase at temperatures of 412 °C to 443 °C. V-type FIs all homogenize to the liquid phase at temperatures of 436–450 °C. Final ice-melting temperatures of L-type FIs range from −6.2 °C to −5.3 °C, corresponding to salinities of 8.3–9.5 wt.% NaCl equiv.

4.3. Magnetite Chemistry

LA-ICP-MS spot analyses were conducted on four magnetite samples, with two, eight, seven, and eight spots on Mag_{I}, Mag_{II}, Mag_{III}, and Mag_{IV}, respectively. Representative bulk trace-element compositions (Si, Al, Ti, Mn, Ca, P, Mg, V, Cr, Co, Ni, Cu, Zn, Ga, Zr, Ba, As, Ge, Sn, Sc, Sr, Y, Nb, Mo, W, Hf, Ta, and Bi) are summarized in Table 2. In addition, EDS analyses were conducted to acquire elemental concentrations of ilmenite crystals in magnetite, which show a phenomenon of alteration of the ilmenite (Figure 6e–i).

Table 2. LA-ICP-MS trace element data (ppm) of magnetite from Tonggou deposit.

Sample	Generation	Mg	Al	Si	Ca	Sc	Ti	V	Cr	Mn	Co	Ni	Cu	Zn	Ga	Ge	Sn	W	Pb	Bi
FG1-01	1	1007	10,806	28,084	5225	23.4	723	234	34	471	401.5	470.6	7	590	12.3	6.7	0.9	0.6	8.7	35.7
FG1-02	1	915	8736	18,636	5153	45.3	872	310	136	429	232.4	633.2	6	655	26.2	6.4	0.9	0.9	4.9	3.4
BY-9(1)-01	2	1956	20,850	33,635	4226	37.6	222	356	879	294	7.6	491.1	86	3908	19.1	5.7	0.4	3.4	42.6	4.7
BY-9(1)-02	2	2151	3452	17,063	1316	3.6	19	240	138	534	95.1	732.4	17	1229	2.62	4.9	0.0	0.8	11.9	0.7
BY-9(1)-03	2	1404	20,099	33,527	4561	31.7	319	383	1215	545	7.1	466.8	80	3841	18.9	4.7	1.1	2.8	45.1	2.5
BY-9(1)-04	2	1531	20,763	32,390	4406	34.2	516	342	862	620	7.4	497.8	82	3740	18.2	3.8	0.0	3.0	33.9	2.1
BY-9-01	2	1685	19,564	32,353	4604	34.1	352	347	1008	581	14.9	516.3	75	3232	17.0	5.4	0.6	2.3	28.9	1.9
BY-9-02	2	1817	10,291	26,291	2649	17.5	287	295	685	945	62.1	688.5	47	1909	9.34	5.2	0.0	1.7	23.9	0.7
BY-9-03	2	1864	15,439	31,205	3675	29.9	655	359	381	884	15.8	457.4	59	2629	14.5	6.3	0.2	1.8	38.3	2.9
BY-9-04	2	2002	22,472	40,869	4933	29.8	247	412	1844	617	16.4	606.1	108	3795	27.0	3.6	0.5	3.3	32.0	2.2
BY-1-02	3	623	6285	29,965	777	0.5	294	2	17	518	463.3	761.1	2	261	0.0	2.0	0.5	0.0	2.3	6.4
BY-1-03	3	954	3605	21,831	444	0.6	626	3	13	497	426.6	706.5	2	251	0.2	3.6	1.5	0.1	2.2	4.2
BY-1-04	3	437	1559	21,602	357	0.1	69	1	12	451	552.6	1199.7	5	240	0.0	2.5	0.4	0.0	0.9	1.2
BY1-05	3	585	5391	30,415	668	0.4	280	2	15	367	525.3	885.7	3	206	0.4	3.7	0.4	0.0	1.5	1.4
BY1-06	3	611	3775	23,137	386	0.6	354	2	11	448	451.7	757.2	2	260	0.4	2.6	0.0	0.0	1.6	3.1
BY1-07	3	909	6470	32,784	672	1.0	421	5	19	521	360.1	599.1	1	251	1.0	2.6	0.6	0.1	3.2	7.6
BY1-08	3	618	3371	22,733	548	0.6	380	2	13	371	497.4	826.2	3	254	0.4	5.2	0.2	0.0	1.7	4.2
TG002-2-01	4	2642	5245	15,177	362	4.5	59	30	11	612	2.3	3.9	22,282	481	4.6	41.0	6.2	3023.3	2.1	13.4
TG002-2-02	4	980	2928	9821	249	2.9	18	14	2	498	0.5	0.6	2010	163	2.2	39.0	5.4	3414.5	1.3	4.7
TG002-2-03	4	560	2514	9170	430	2.7	18	24	2	606	4.4	0.3	16,829	217	0.9	32.9	9.4	4653.6	2.6	28.3
TG002-2-04	4	907	1512	15,492	1270	2.5	25	18	8	447	1.9	7.9	10,096	182	1.1	61.8	9.1	5735.6	1.8	15.0
TG002-01	4	400	2351	14,150	561	2.6	41	28	15	487	0.3	0.5	517	151	1.7	76.5	16.2	4616.2	1.7	3.7
TG002-02	4	520	2756	9581	188	3.3	44	24	1	583	1.1	2.9	111	55	1.7	44.7	14.9	4204.6	0.5	10.4
TG002-03	4	482	2332	9911	413	2.6	12	15	27	447	0.4	0.0	58,878	208	1.6	45.3	6.1	3376.0	2.1	14.5

5. Discussion

5.1. Controlling Factors on Magnetite Compositions

Trace elements in magnetite always depend on the formation temperatures of the rock or ore, as well as oxygen fugacity (fO_2), fluid/melt composition, host rock composition, and coexisting minerals during mineral deposit formation [2,4,41]. Titanium is favored in high-temperature ulvospinel–magnetite solid solutions and is incorporated into magnetite by coupling substitution with Fe^{2+} for Fe^{3+} [3,42]). Ti concentrations gradually drop from Mag_I to Mag_{IV}, which is consistent with a higher formation temperature for Mag_{III} (~500 °C, as described in Section 4.2.) than for Mag_{IV} (~400 °C; [19]). These results are coherent with the affinity of magnetite for Ti at higher temperatures. Furthermore, Si, Al, and Ta in magnetite are found to decrease gradually from Mag_I to Mag_{IV}, which implies a positive correlation of these elements with Ti (Figure 8).

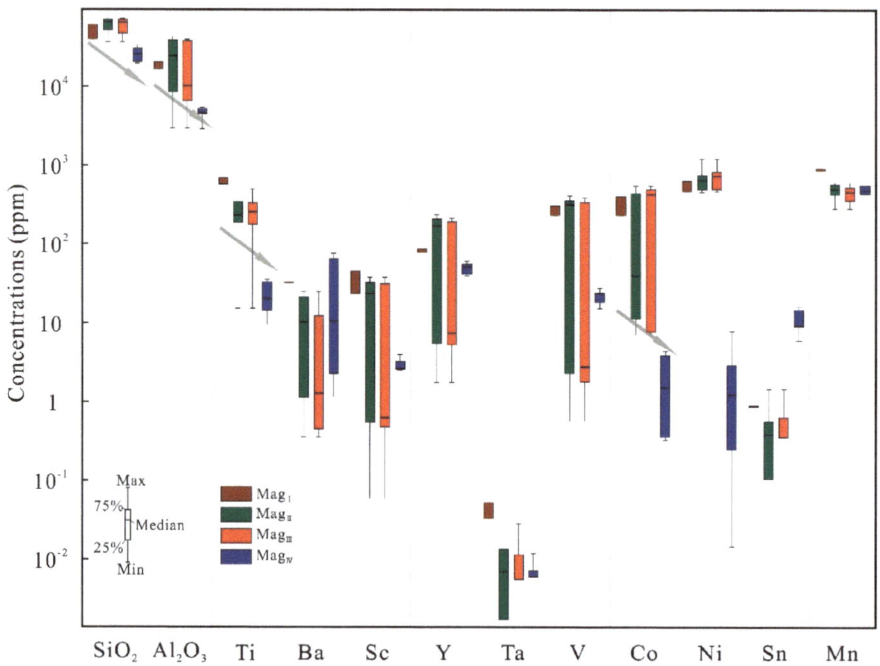

Figure 8. Comparative box plot of SiO_2, Al_2O_3, Ti, Ba, Sc, Y, Ta, V, Co, Ni, Sn, and Mn concentrations of the different magnetite generations.

On the other hand, some trace element concentrations (e.g., Sn, V, Mn) are controlled by oxygen fugacity since these elements have several valence states. For instance, in magnetite, Fe^{3+} is substituted by Sn^{4+} more readily than Sn^{2+} [41]. This is consistent with the fact that Mag_{IV} contains higher ratios of Sn than Mag_{III} (Figure 8). In addition, V and Mn generally decrease from Mag_I to Mag_{III}, following the same trend as Ti. However, V and Mn do not decrease with Ti in Mag_{IV} (Figure 8). This can be explained by the appearance of ilmenite replacing Mag_{IV} (Figure 3i), which indicates that high fO_2 is characteristic of stage 2 (polymetallic sulfide–epidote–quartz stage). Thus, Ti in magnetite can be considered a temperature indicator, while Sn, V, and Mn are dependent on both temperature and fO_2 [3,10]

In addition, considering the different partition coefficients of trace elements between magnetite and coexisting minerals, the compositions of certain trace elements in magnetite may also be affected by the co-precipitation of other minerals [1]. For instance, chalcophile elements (Co, Ni, Cu, Zn, and Bi) partition more readily into sulfides than magnetite [2]. Indeed, there is a decreasing trend in Co, Ni, and Zn contents from Mag_{III} to Mag_{IV} (Figure 8). However, Cu and Bi are not depleted in Mag_{IV} as expected (Figure 8), despite the coexistence of sulfides with Mag_{IV}. It may be a reasonable explanation that Mag_{IV} precipitated earlier than the sulfides, which is supported by the occurrence of secondary chalcopyrite replacing magnetite (Figure 6e). Similar cases have been described in the Yuleken porphyry Cu deposit in East Junggar [4].

5.2. Evolution of Ore-Forming Fluids

FIs identified in the cores of granular quartz grains are LV-, VL-, and S-type FIs with high temperatures (450–501 °C) and high salinity (51.2–55.2 wt.% NaCl equiv.). In addition, fluid boiling occurred in the core of granular quartz, as demonstrated by coexisting vapor-rich (VL-type) and high-salinity (S-type) FIs, which homogenized at similar temperatures, suggesting they were trapped during phase separation [43]. In view of the fact that the homogenization temperature and salinity of boiling FIAs are extremely high (~500 °C), they most likely originated from phase separation of magmatic water due to decompression. In addition, homogenization temperature and salinity decrease gradually from core to rim in granular quartz, which indicates an increasing contribution of meteoric water in ore-forming fluids during the later stage.

On the other hand, FIs from porphyry Cu mineralization (stage 2) have similar homogenization temperatures and salinities to vein-type Cu mineralization in the Tonggou deposit (this study and [19]). In addition, they both formed in Late Carboniferous Porphyry Cu Systems (303 Ma; [19,20]) and occur in adjacent geographical locations. Thus, we consider that the porphyry and vein-type Cu mineralization belong to the same porphyry Cu system.

5.3. Metallogenetic Processes

Based on magnetite chemistry, some researchers have proposed discrimination diagrams for distinguishing magnetite of different origins or from various deposit types [1,3,4,7,44]. Tonggou magnetite of different origins was mostly plotted within the magmatic range in the Ni/Cr vs. Ti diagram (Figure 9a). Due to the higher solubility of Ni in hydrothermal fluids with respect to Cr, hydrothermal magnetite has relatively low Ni/Cr ratios [1].

In the (Ti + V) vs. (Ca + Al + Mn) diagram proposed by Dupuis and Beaudoin [7], magnetite from Tonggou plots close to the skarn field (Figure 9b), indicating relatively low Ti + V contents compared to other porphyry Cu deposits. As discussed above, the temperature is a controlling factor for trace element incorporation and correlates positively with Ti, with the early stage of porphyry Cu mineralization showing the highest formation temperatures (~500 °C) and Ti contents of the Tonggou deposit. By observing our data plotted in the (Ti + V) vs. (Ca + Al + Mn) diagram (Figure 9c) by Nadoll et al. [2], it is evident that magnetite from Tonggou is high-temperature magnetite (generally >300 °C). However, V concentrations in Tonggou magnetite are influenced by both temperature and fO_2 and are lower than in other porphyry Cu deposits (e.g., Yuleken Cu deposit; [4]). In addition, zircons from Tonggou granodiorite have high Ce/Ce_N^* values (227–234) and Ce^{4+}/Ce^{3+} ratios (159–286; [20]). Furthermore, Sn is characterized by negative correlations with Ti + V, as seen in the (Ti + V) vs. Sn diagram (Figure 9d). It is worth noting that porphyry Cu mineralization at Tonggou likely formed at relatively high fO_2 conditions, though lower than other porphyry Cu deposits (e.g., Yuleken) and Tonggou vein-type mineralization.

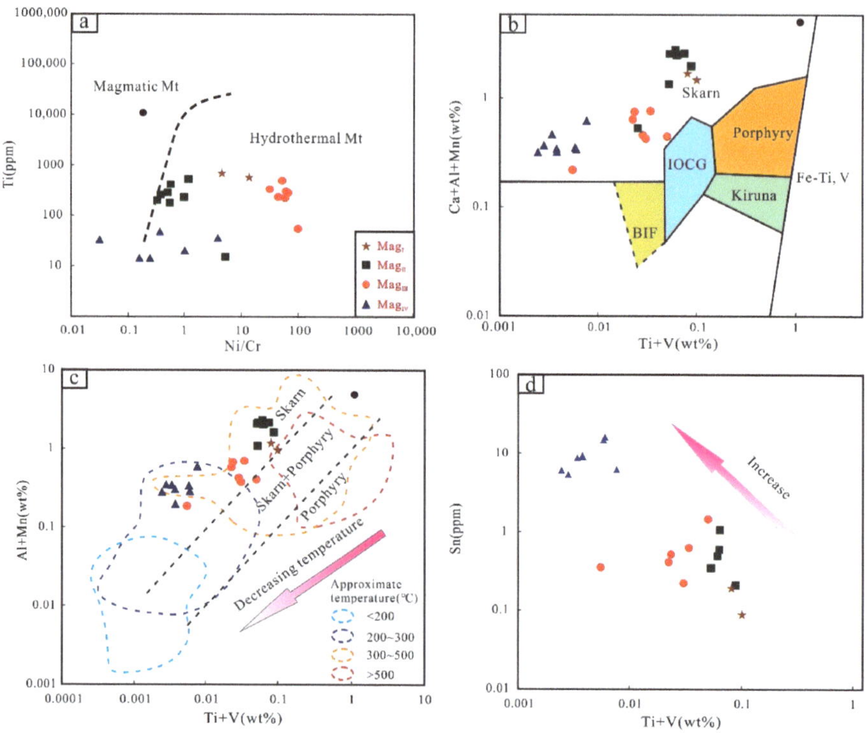

Figure 9. (**a**) Ti vs. Ni/Cr ratio in magnetite, where the boundary is defined by Dare et al. [1]. (**b**) (Ca + Al + Mn) vs. (Ti + V) discrimination diagram for magnetite from the Tonggou deposit, where the fields for Kiruna, IOCG, PCD, skarn, Fe-Ti-V, and BIF deposits were defined by Dupuis and Beaudoin [7]. (**c**) (Al + Mn) vs. (Ti + V) diagram for magnetite (modified after [1]) from the Tonggou deposit. (**d**) Sn vs. (Ti + V) diagram for magnetite from the Tonggou deposit.

Thus, although both porphyry and vein-type mineralization are sourced from the Tonggou porphyry system, the vein-type mineralization may have formed at a higher fO_2 than the porphyry mineralization. Generally, large and intermediate porphyry Cu deposits are associated with intrusive rocks derived from oxidized magmas [45]. This is consistent with the fact that the vein Cu-Zn mineralization at the Tonggou deposit is economically significant, while the porphyry Cu mineralization is less economically relevant.

6. Conclusions

Magnetite in the Tonggou deposit can be divided into allotriomorphic Mag_I from altered granodiorite, subidiomorphic Mag_{II} from altered granodiorite, allotriomorphic Mag_{III} from the magnetite–quartz stage of porphyry mineralization, and allotriomorphic Mag_{IV} from the polymetallic sulfide–epidote–quartz stage of vein-type mineralization.

LA-ICP-MS analyses reveal that trace elements in magnetite of the Tonggou deposit vary systematically from porphyry to vein-type mineralization. In addition, the content of Ti, Si, Al, and Ta in magnetite is positively correlated with high temperatures, while fO_2 has a considerable influence on Sn, V, and Mn.

Ore-forming fluids of the Tonggou porphyry mineralization were high temperature and high-salinity H_2O-NaCl hydrothermal fluids in the early stage, which derived from a magmatic fluid.

The occurrence of ilmenite replacing Mag_{IV} in the polymetallic sulfide–epidote–quartz stage is characterized by high fO_2 and intermediate to high temperatures. Porphyry and vein-type mineralization are sourced from the Tonggou porphyry system and are the result of the transfer of ore-forming fluids to a different location.

Author Contributions: Conceptualization, C.-C.H.; Methodology, X.-B.Z.; Software, S.-S.W. and Y.-T.L.; Validation, C.-C.H.; Formal Analysis, X.-B.Z.; Investigation, C.-C.H. and X.-B.Z.; Resources, X.-B.Z.; Data Curation, C.-C.H.; Writing—Original Draft Preparation, X.-B.Z.; Writing—Review and Editing, C.-C.H. and X.-B.Z.; Visualization, X.-B.Z. and S.-S.W.; Supervision, X.-B.Z.; Project Administration, X.-B.Z.; Funding Acquisition, X.-B.Z. All authors have read and agreed to the published version of the manuscript.

Funding: This research was supported by the following funding agency: Tianshan Young Project of Xinjiang, China (Grant No. 2019Q070); the Higher Educational Science and Research Program of Xinjiang, China (Grant No. XJEDU2021Y015); and Students Research Training Program Fund (Grant No. S202010755022).

Data Availability Statement: Not applicable.

Acknowledgments: We are most grateful to the staff of Beijing Kerongen Science and Technology Ltd. for EDS and BSE analysis on magnetite. In addition, we thank the staff of Wuhan Sample Solution Analytical Technology Co., Ltd. for LA-ICP-MS magnetite trace element analysis. We thank the editor and anonymous reviewers for their constructive comments that led to the significant improvement of the manuscript.

Conflicts of Interest: The authors declare no conflict of interest.

References

1. Dare, S.A.S.; Barnes, S.J.; Beaudoin, G.; Méric, J.; Boutroy, E.; Potvin-Doucet, C. Trace elements in magnetite as petrogenetic indicators. Mineral. *Deposita* **2014**, *49*, 785–796. [CrossRef]
2. Nadoll, P.; Angerer, T.; Mauk, J.L.; French, D.; Walshe, J. The chemistry of hydrothermal magnetite: A review. *Ore Geol. Rev.* **2014**, *61*, 1–32. [CrossRef]
3. Canil, D.; Grondahl, C.; Lacourse, T.; Pisiak, L.K. Trace elements in magnetite from porphyry Cu-Mo-Au deposits in British Columbia, Canada. *Ore Geol. Rev.* **2016**, *72*, 1116–1128. [CrossRef]
4. Wu, C.; Chen, H.-Y.; Hong, W.; Li, D.-F.; Liang, P.; Fang, J.; Zhang, L.-J.; Lai, C. Magnetite chemistry and implications for the magmatic-hydrothermal oreforming process: An example from the Devonian Yuleken porphyry Cu system, NW China. *Chem. Geol.* **2019**, *522*, 1–15. [CrossRef]
5. Huang, X.-W.; Zhou, M.-F.; Qiu, Y.-Z.; Qi, L. In-situ LA-ICP-MS trace elemental analyses of magnetite: The Bayan Obo Fe-REE-Nb deposit, North China. *Ore Geol. Rev.* **2015**, *65*, 884–899. [CrossRef]
6. Li, D.-F.; Chen, H.-Y.; Hollings, P.; Zhang, L.; Sun, X.-M.; Zheng, Y.; Xia, X.-P.; Xiao, B.; Wang, C.-M.; Fang, J. Trace element geochemistry of magnetite: Implications for ore genesis of the Talate skarn Pb-Zn (-Fe) deposit, Altay, NW China. *Ore Geol. Rev.* **2018**, *100*, 471–482. [CrossRef]
7. Dupuis, C.; Beaudoin, G. Discriminant diagrams for iron oxide trace element fingerprinting of mineral deposit types. *Mineral. Deposita* **2011**, *46*, 319–335. [CrossRef]
8. Pisiak, L.K.; Canil, D.; Lacourse, T.; Plouffe, A.; Ferbey, T. Magnetite as an indicator mineral in the exploration of porphyry deposits: A case study in till near the Mount Polley Cu-Au deposit, British Columbia, Canada. *Econ. Geol.* **2017**, *112*, 919–940. [CrossRef]
9. Chen, W.-T.; Zhou, M.-F.; Gao, J.-F.; Hu, R.-Z. Geochemistry of magnetite from Proterozoic Fe-Cu deposits in the Kangdian metallogenic province, SW China. *Mineral. Deposita* **2015**, *50*, 795–809. [CrossRef]
10. Chen, W.-T.; Zhou, M.-F.; Li, X.-C.; Gao, J.-F.; Hou, K.-J. In-situ LA-ICP-MS trace elemental analyses of magnetite: Cu-(Au, Fe) deposits in the Khetri copper belt in Rajasthan Province, NW India. *Ore Geol. Rev.* **2015**, *65*, 929–939. [CrossRef]
11. Liu, Y.; Fan, Y.; Zhou, T.; Xiao, X.; White, N.-C.; Thompsom, J.; Hong, H.; Zhang, L. Geochemical characteristics of magnetite in Longqiao skarn iron deposit in the Middle-Lower Yangtze Metallogenic Belt, Eastern China. *Mineral. Deposita* **2019**, *54*, 1229–1242. [CrossRef]
12. Ni, P.; Chi, Z.; Pan, J.-Y.; Wang, G.-G.; Chen, H.; Ding, J.-Y. The Characteristics of Ore-Forming Fluids and Mineralization Mechanism in Hydrothermal Deposits: A Case Study of Some Typical Deposits in China. *Bull. Mineral. Petrol. Geochem.* **2018**, *37*, 369–395.
13. Sengör, A.M.C.; Natal'In, B.A.; Burtman, V.S. Evolution of the Altaid tectonic collage and Palaeozoic crustal growth in Eurasia. *Nature* **1993**, *364*, 299–307. [CrossRef]
14. Jahn, B.M.; Wu, F.; Chen, B. Granitoids of the Central Asian Orogenic Belt and continental growth in the Phanerozoic. *Earth Env. Sci. Trans. R. Soc.* **2000**, *350*, 181–193.
15. Wang, J.-B.; Wang, Y.-W.; He, Z.-J. Ore deposits as a guide to the tectonic evolution in the East Tianshan Mountains, NW China. *Geol. China* **2006**, *33*, 461–469.

16. Wang, Y.-H.; Zhang, F.-F.; Liu, J.-J. Genesis of the Fuxing porphyry Cu deposit in Eastern Tianshan, China: Evidence from fluid inclusions and C-H-O-S-Pb isotope systematics. *Ore Geol. Rev.* **2016**, *79*, 46–61. [CrossRef]
17. Wang, Y.-F.; Chen, H.-Y.; Baker, M.J.; Han, J.-S.; Xiao, B.; Yang, J.-T.; Fred, J. Multiple mineralization events of the Paleozoic Tuwu porphyry copper deposit, Eastern Tianshan: Evidence from geology, fluid inclusions, sulfur isotopes, and geochronology. *Mineral. Deposita* **2019**, *54*, 1053–1076. [CrossRef]
18. Zhang, F.-F.; Wang, Y.-H.; Liu, J.-J.; Xue, C.-J.; Wang, J.-P.; Zhang, W.; Li, Y.-Y. Paleozoic Magmatism and Mineralization Potential of the Sanchakou Copper Deposit, Eastern Tianshan, Northwest China: Insights from Geochronology, Mineral Chemistry, and Isotopes. *Econ. Geol.* **2022**, *117*, 165–194. [CrossRef]
19. Zhang, X.-B.; Chen, C.; Xia, F.; Gao, L.-L.; Quan, H.-Y. Metallogenesis and hydrothermal evolution of the Tonggou Cu deposit in the Eastern Tianshan: Evidence from fluid inclusions, H-O-S isotopes, and Re-Os geochronology. *Geosci. Front.* **2019**, *10*, 2301–2312. [CrossRef]
20. Zhang, X.-B.; Chai, F.-M.; Chen, C.; Quan, H.-Y.; Wang, K.-Y.; Li, S.-D.; Wu, S.-S. Using whole rock and zircon geochemistry to assess porphyry copper potential of the Tonggou copper deposit, Eastern Tianshan. *Minerals* **2020**, *10*, 584. [CrossRef]
21. Han, C.-M.; Xiao, W.J.; Wan, B.; Ao, S.-J.; Zhang, J.-E.; Song, D.-F.; Zhang, Z.-Y.; Wang, Z.-M. Late Palaeozoic-Mesozoic endogenetic metallogenic series and geodynamic evolution in the East Tianshan Mountains. *Acta Petrol. Sin.* **2018**, *34*, 1914–1932.
22. Xiao, W.-J.; Zhang, L.-C.; Qin, K.-Z.; Sun, S.; Li, J.-L. Paleozoic accretionary and collisional tectonics of the eastern Tianshan (China): Implications for the continental growth of Central Asia. *Am. J. Sci.* **2004**, *304*, 370–395. [CrossRef]
23. Wu, F.-Y.; Sun, D.-Y.; Ge, W.-C.; Zhang, Y.-B.; Grant, M.L.; Wilde, S.A.; Jahn, B.M. Geochronology of the Phanerozoic granitoids in northeastern China. *J. Asian Earth Sci.* **2011**, *41*, 1–30. [CrossRef]
24. Xu, B.; Charvet, J.; Chen, Y.; Zhao, P.; Shi, G.-Z. Middle Paleozoic convergent orogenic belts in western Inner Mongolia (China): Framework, kinematics, geochronology and implications for tectonic evolution of the Central Asian Orogenic Belt. *Gondwana Res.* **2012**, *23*, 1342–1364. [CrossRef]
25. Mao, J.-W.; Goldfarb, R.T.; Wang, Y.-T.; Hart, C.J.; Wang, Z.-L.; Yang, J.-M. Late Paleozoic base and precious metal deposits, east Tianshan, Xinjiang, China: Characteristics and geodynamic setting. *Episodes* **2005**, *28*, 23–36. [CrossRef] [PubMed]
26. Xiao, B.; Chen, H.-Y.; Hollings, P.; Han, J.-S.; Wang, Y.-F.; Yang, J.-T.; Cai, K.-D. Magmatic evolution of the Tuwu-Yandong porphyry Cu belt, NW China: Constraints from geochronology, geochemistry and Sr-Nd-Hf isotopes. *Gondwana Res.* **2017**, *43*, 74–91. [CrossRef]
27. Zheng, J.-H.; Chai, F.-M.; Feng, W.-Y.; Yang, F.-Q.; Shen, P. Geochemistry and chronology of the early paleozoic diorites and granites in the Huangtupo volcanogenic massive sulfide (VMS) deposit, Eastern Tianshan, NW China: Implications for petrogenesis and geodynamic setting. *Lithos* **2018**, *302–303*, 455–466. [CrossRef]
28. Zheng, J.-H.; Mao, J.-W.; Yang, F.-Q.; Chai, F.-M.; Shen, P. Petrological and geochemical features of the early Paleozoic granitic gneisses and iron ores in the Tianhu iron deposit, Eastern Tianshan, NW China: Implications for ore genesis. *Lithos* **2017**, *286–287*, 426–439. [CrossRef]
29. BGMRXUAR (Bureau of Geology and Mineral Resources of Xinjiang Uygur Autonomous Region). *Regional Geology of Xinjiang Autonomous Region*; Geological Memoirs, No. 32, Map Scale 1:500,000; Geological Publishing House: Beijing, China, 1993; pp. 1–841.
30. Gu, L.-X.; Hu, S.-X.; Yu, C.-S.; Li, H.-Y.; Xiao, X.-J.; Yan, Z.-F. Carboniferous volcanites in the Bogda Mountains of eastern Tianshan: Their tectonic implications. *Acta Petrol. Sin.* **2000**, *16*, 305–316.
31. Lei, W.-S.; Xu, P.; Guo, J.-X.; Xiao, L.; Li, X.-C.; Li, Y. LA-ICP-MS zircon U-Pb dating, geological and geochemical features of Sujishan gabbro pluton, eastern Bogda Mountains, and their tectonic significances. *Geol. Rev.* **2016**, *62*, 317–330.
32. Li, P.; Zhu, Z.-X.; Chen, C.; Liu, X.; Chen, B.-X.; Jin, L.-Y.; Huang, L.-M. Geological characteristics and fluid inclusions studies of Tonggou copper deposit of Dabancheng area, Xinjiang. *Xinjiang Geol.* **2015**, *33*, 51–55.
33. Zhang, X.-B.; Chai, F.-M.; Chen, C.; Quan, H.-Y.; Gong, X.-P. Geochronology, geochemistry and tectonic implications of late Carboniferous Daheyan intrusions from the Bogda Mountains, Eastern Tianshan. *Geol. Mag.* **2020**, *157*, 289–306. [CrossRef]
34. Sillitoe, R.H. Porphyry copper systems. *Econ. Geol.* **2010**, *105*, 3–41. [CrossRef]
35. Brown, P.E.; Hagemann, S.G. MacFlincor and its application to fluids in Archean lode-gold deposits. Geochim. *Cosmochim. Acta* **1995**, *59*, 3943–3952. [CrossRef]
36. Zong, K.-Q.; Chen, J.-Y.; Hu, Z.-C.; Liu, Y.-S.; Li, M.; Fan, H.-H.; Meng, Y.-N. In-situ U-Pb dating of uraninite by fs-LA-ICP-MS. *Sci. China Earth Sci.* **2015**, *58*, 1731–1740. [CrossRef]
37. Hu, Z.-C.; Zhang, W.; Liu, Y.-S.; Gao, S.; Li, M.; Zong, K.-Q.; Chen, H.-H.; Hu, S.-H. "Wave" signal-smoothing and mercury-removing device for laser ablation quadrupole and multiple collector ICPMS analysis: Application to lead isotope analysis. *Anal. Chem.* **2015**, *87*, 1152–1157. [CrossRef]
38. Liu, Y.-S.; Hu, Z.-C.; Gao, S.; Günther, D.; Xu, J.; Gao, C.-G.; Chen, H.-H. In situ analysis of major and trace elements of anhydrous minerals by LA-ICP-MS without applying an internal standard. *Chem. Geol.* **2008**, *257*, 34–43. [CrossRef]
39. Goldstein, R.H.; Reynolds, T.J. Systematics of fluid inclusions in diagenetic minerals. Soc. Sediment. *Geol. Sepm Short Course.* **1994**, *31*, 1–199.
40. Driesner, T.; Heinrich, C.A. The system H_2O-NaCl. Part I: Correlation formulae for phase relations in temperature-pressure-composition space from 0 to 1000 °C, 0 to 5000 bar, and 0 to 1 XNaCl. *Geochim. et Cosmochim. Acta* **2007**, *71*, 4880–4901.

41. Carew, M.J. Controls on Cu-Au Mineralization and Fe Oxide Metasomatism in the Eastern Fold Belt, N.W. Ph.D. Thesis, James Cook University, Douglas, QLD, Australia, 2004.
42. Buddington, A.F.; Lindsley, D.H. Iron-titanium oxide minerals and synthetic equivalents. *J. Petrol.* **1964**, *5*, 310–357. [CrossRef]
43. Ramboz, C.; Pichavant, M.; Weisbrod, A. Fluid immiscibility in natural processes: Use and misuse of fluid inclusion data: II. Interpretation of fluid inclusion data in terms of immiscibility. *Chem. Geol.* **1982**, *37*, 29–48. [CrossRef]
44. Zhang, Y.; Hollings, P.; Shao, Y.-J.; Li, D.-F.; Chen, H.-Y.; Li, H.-B. Magnetite texture and trace-element geochemistry fingerprint of pulsed mineralization in the Xinqiao Cu-Fe-Au deposit, Eastern China. *Am. Mineral.* **2020**, *105*, 1715–1723. [CrossRef]
45. Shen, P.; Hattori, K.; Pan, H.-D.; Jackson, S.; Seitmuratova, E. Oxidation condition and metal fertility of granitic magmas: Zircon trace-element data from porphyry Cu deposits in the Central Asian orogenic belt. *Econ. Geol.* **2015**, *110*, 1861–1878. [CrossRef]

minerals MDPI

Article

Genesis of the Halılar Metasediment-Hosted Cu-Pb (±Zn) Mineralization, NW Turkey: Evidence from Mineralogy, Alteration, and Sulfur Isotope Geochemistry

Demet Kiran Yildirim

Department of Geological Engineering, Faculty of Mines, Istanbul Technical University, 34469 Istanbul, Turkey; kirand@itu.edu.tr

Abstract: This study contributes to our understanding of the evolution of Halılar Cu-Pb (±Zn) mineralization (NW Turkey) based on mineralogical and geochemical results and sulfur isotope data. The study area represents local Cu-Pb with some Zn brecciated-stockwork vein type mineralization along the NE–SW fault gouge zone at the lower boundary of the Sakarkaya and Düztarla granitoid rocks. Two main zones, consisting of sericite–quartz–chlorite ± kaolinite ± pyrite (i.e., zone-1) and calcite–epidote–albite ± chlorite ± sericite (i.e., zone-2), were observed within the central ore mineral zone at the mining site. Different mineralization assemblages were recorded; the main ore mineral contains chalcopyrite, galena, pyrite, and sphalerite within alteration zone-1, and the oxidation/supergene mineralization includes covellite and goethite. The mass balance calculations show that the samples of zone-1 show an increase in SiO_2, Fe_2O_3, K_2O, and LOI along with Ag, As, Cu, Mo, Pb, S, Sb, and Zn, reflecting high pyritization with sericitization and silicification. On the other hand, the samples from zone-2 are rich in CaO; Na_2O; P_2O_5; TiO_2; LOI; and carbon-reflecting calcite, epidote, and albite alterations. A uniform magmatic sulfur source of Halılar sulfides is suggested by their mean $\delta^{34}S$ value of −1.62‰. Furthermore, the primary metal source is metasediments and intrusive Düztarla granitoid magmatism. These observations suggest that the Halılar metasediment-hosted Cu-Pb (±Zn) mineralization was formed by epigenetic hydrothermal processes after sedimentation/diagenesis and metamorphism.

Keywords: mass-balance calculations; $\delta^{34}S$ data; Cu-Pb (±Zn) mineralization; Halılar; NW Turkey

Citation: Yildirim, D.K. Genesis of the Halılar Metasediment-Hosted Cu-Pb (±Zn) Mineralization, NW Turkey: Evidence from Mineralogy, Alteration, and Sulfur Isotope Geochemistry. *Minerals* **2022**, *12*, 991. https://doi.org/10.3390/min12080991

Academic Editor: Yuichi Morishita

Received: 6 July 2022
Accepted: 31 July 2022
Published: 4 August 2022

1. Introduction

Studies of hydrothermal alteration are important in the exploration of copper deposits in order to determine the processes of ore formation, as well as to identify potential ore zones [1]. Spectroscopic methods, geophysics, or multispectral remote sensing techniques are used in mapping alteration zones, as well as in identifying their mineral assemblages [1–9]. In addition, the geochemical changes from host rock to alteration zones provide alteration type and its degree, as well as the genesis and evolution of the hydrothermal system [5,10–19]. Hydrothermal alteration processes are responsible for mineralogical and chemical changes in the rock-forming minerals as a result of interactions between the hydrothermal fluids and host rocks along fracture zones and grain boundaries [1,2,20,21]. Schwartz [22] stated that the alteration generally depends on: (1) temperature, pressure, and chemical composition of the fluid; (2) the chemical and physical nature of the wall rocks; and (3) the water–rock ratio. The mechanism and types of mineral deposits are assigned by the nature of the alteration assemblages and the different hydrothermal systems. In addition, the mineral assemblages of the altered rocks are important to help identify the alteration types (e.g., phyllic alteration refers to assemblages of quartz + sericite + pyrite minerals; potassic alteration: orthoclase + biotite + sericite; propylitic alteration: epidote + chlorite + albite) [23]. Gifkins et al. [24] defined different types of mineral deposits by their alteration type and mineralogy, such as porphyry Cu deposits having potassic, phyllic, argillic, and propylitic alterations, while the low-sulfidation, epithermal, geothermal, VHMS, and sediment-hosted massive sulphide deposits having sericitic (or phyllic) and propylitic (or saussuritization) alterations.

In Turkey, mineralization in the structural zone of the Anatolian tectonic belt represents part of the Tethyan–Eurasian metallogenic belt (TEMB), which formed during the Mesozoic and Early Cenozoic [25]. This mineralization was controlled by extensional events that formed after the Neo-Tethys closure. It is associated with calc–alkaline magmatic activity during the Oligocene–Miocene/Pliocene within the post-collision continent-continent environments and led to the formation of Pb-Zn, Sb, As, and Au-Cu deposits [25].

The study area (Halılar area) is located about 25–30 km northeast of Edremit in Balıkesir Province (Biga Peninsula, Turkey) (Figure 1). Halılar Cu-Pb ($\pm$Zn) mineralization occurs in a vein-type deposit that formed in the volcanogenic metasediments of the Sakarkaya Formation. It is associated with the NE–SW fault gouge zone along with the lower boundary of the Bağcağız Formation and the Düztarla granitoid intrusion.

Although geological and geochemical studies of the Halılar area have been published [26], the genesis of base-metal Cu-Pb (-Zn) mineralization in this area remains ambiguous, as it has not been studied in detail. Therefore, this study focuses on mineralization in the Halılar area by reporting new data obtained from mineralogical, petrographical, and geochemical investigations of the mineralization and altered host rock. Using mass balance calculations, enrichment and/or depletion in the chemical components of the different alteration zones associated with this mineralization were calculated on the basis of their mass/volume changes (gain and loss). Sulfur isotope data from the sulfide minerals, including pyrite, chalcopyrite, and galena, were collected to understand the sulfur source(s), as well as to determine the $\delta^{34}S_{H2S}$ values of the hydrothermal fluid that caused the Halılar Cu-Pb-($\pm$Zn) mineralization.

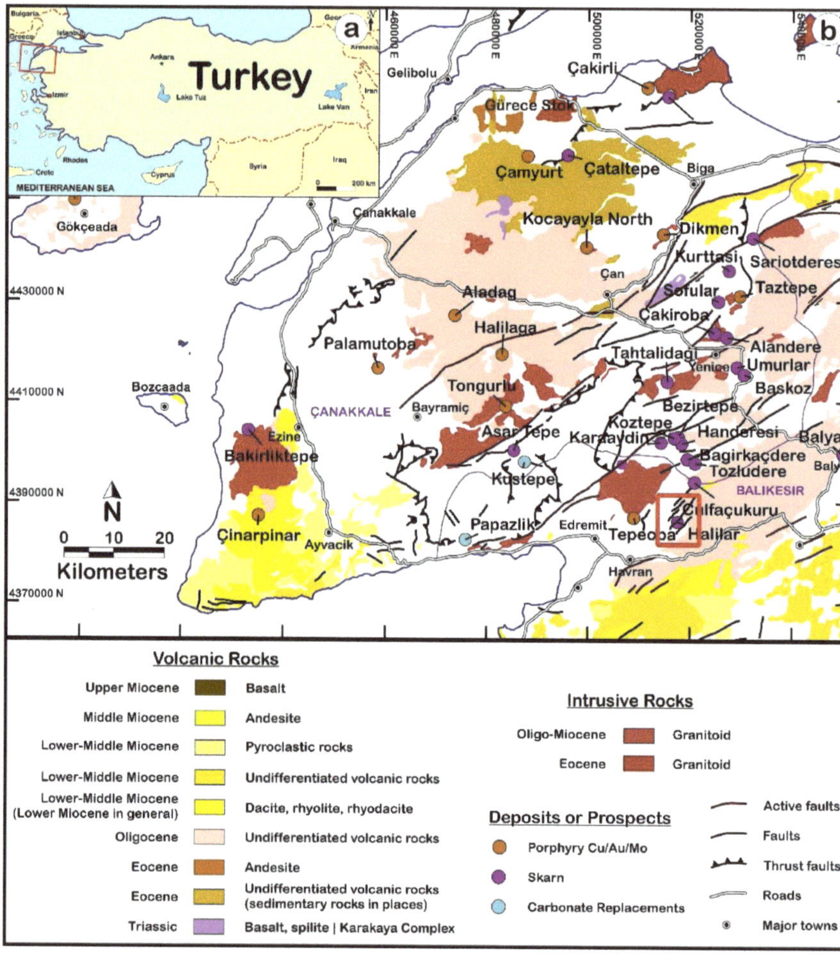

Figure 1. (**a**) Geological map of Turkey; (**b**) study area around Halılar in the Northeast of Edremit in Balıkesir after Yigit [27].

2. Geological Setting

The Halılar area contains two groups: the clastic Halılar Group, which is slightly metamorphosed and overlain by the pre-Late Triassic age or Permian limestone [28], and the Bilecik group. These two groups are in contact with the intrusive rocks to the N and NW of Halılar village (Figure 2). The Halılar Group consists of two formations: the Bağcağız and Sakarkaya Formations; the Bilecik Group is represented by two formations: the Taşçıbayırı Formation and Günören Limestone (Figure 2). The granitoid rocks intruded the Sakarkaya and Bağcağız Formations of the Halılar Group in the study area (Figure 2).

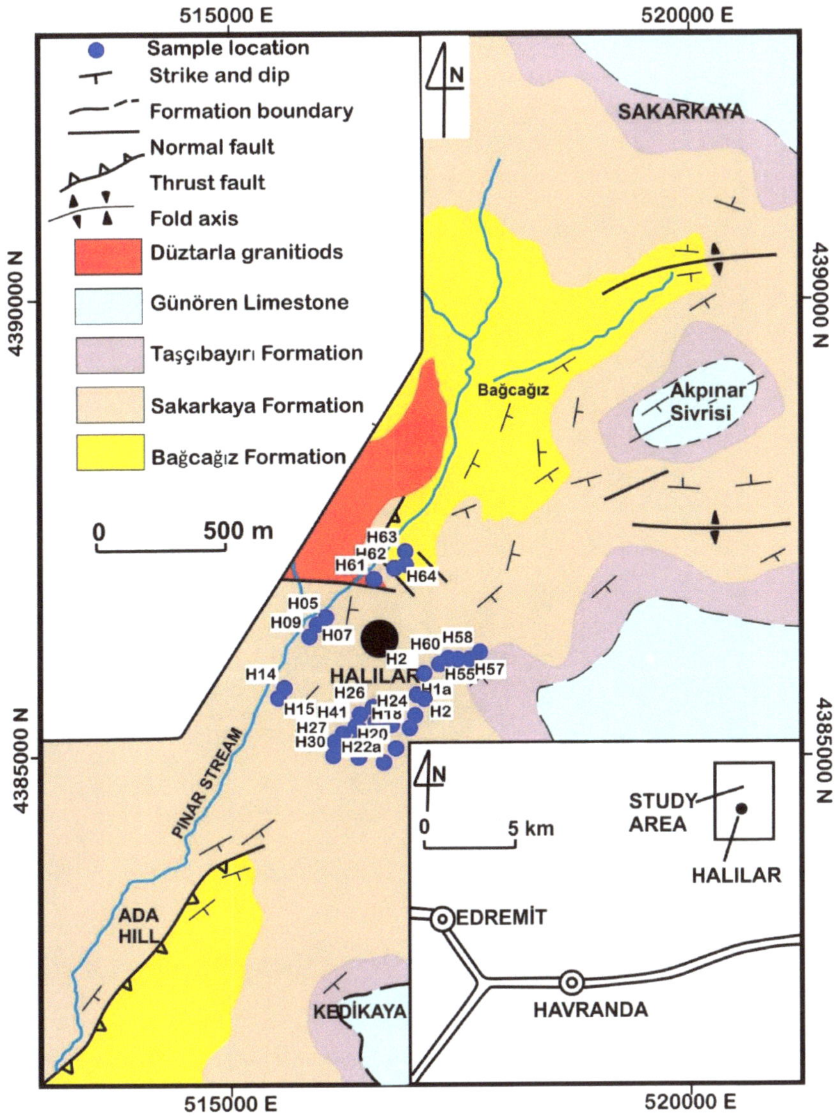

Figure 2. Geologic map of the Halılar area after Altiner et al. [26].

The Halılar Group was classified by Krushensky, Akcay, and Karaege [28] into the Bağcağız Formation (sandstone and shale) and the Sakarkaya Formation (sandstone and conglomerate). The Bağcağız Formation (sample IDs: H63 and H64) was intruded by the Düztarla granitoid at its lower boundary (Figure 2). It has dark siltstone at its upper boundary, which is overlain by the sandstone of the Sakarkaya Formation. This formation also has sandstone and siltstone alternations from bottom to top, consisting of dark-grayish-colored siltstones and silty shales with yellowish-colored, medium-bedded sandstones from the Lower Triassic to Middle Jurassic. The Bağcağız Formation is represented by carbonaceous dark metasiltstone and rhyolitic metatuffs (Figure 3a). The rhyolitic metatuffs are fine-grained light gray to yellowish rocks (Figure 3a) microscopically consisting of microperthite and quartz crystals embedded in a finer-grained tuffaceous matrix of kaolinitized and carbonatized feldspar, quartz, and Fe-oxide (Figure 3b).

Figure 3. (**a**) Rhyolitic metatuffs of the Bağcağız Formation; (**b**) XPL photomicrograph of the mineral composition of rhyolitic metatuffs; (**c**) yellowish-colored metasandstone of the Sakarkaya Formation; (**d**,**e**) XPL photomicrograph of the poorly sorted quartz, feldspar, and mica grains bounded by iron oxide in the subarkosic-to-quartz arenitic of the metasandstone; (**f**) general view of the Bilecik Limestone; (**g**) XPL photomicrograph of the calcite with feldspar, mica, and volcanic rock fragments in the sandy limestone of the Taşçıbayırı Formation; (**h**) XPL photomicrograph of the calcite and dolomite with Fe-oxide minerals in Günören dolomitic limestone; (**i**) granodiorite of Düztarla intrusive rocks; (**j**) XPL photomicrograph of the oligoclase, quartz, and microperthite with subordinate amount of biotite and Fe-oxide minerals in granodiorite; (**k**) granite from the Düztarla intrusion invaded into the Bağcağız Formation; (**l**) XPL photomicrograph of the mineral composition of the Düztarla granite intrusion. Abbreviations: biotite (bt), calcite (cal), dolomite (dol), K-feldspar (kfs), kaolinite (kln), muscovite (ms), opaque (opq), plagioclase (pl), quartz (qz), and sericite (ser).

The Sakarkaya Formation (sample IDs: H05, H07, H09, H14, H15, H18, H20, H22a, H55, and H60) outcrops approximately 500 m south of Sakarkaya Hill and 1.5–2 km north and northeast of Halılar village (Figure 2). It is represented by fine-grained, yellowish-colored metasandstone (Figure 3c). It has a sharp contact with the dark metasiltstones of the Bağcağız Formation. The unit rests with a distinct contact on the Bositra-bearing dark silty shale of the Bağcağız Formation [26]. The metasandstone ranges from subarkosic to wackes in composition and consists of poorly sorted quartz, sericitized and kaolinitized feldspar, and mica grains cemented by iron oxide (Figure 3d,e). These components are embedded in altered feldspar and silicified fine-grained matrix (Figure 3d,e). The upper portion of the formation is represented by cross-stratified beds.

The Bilecik Group is part of the Callovian–Hauterivian (Middle Jurassic–Lower Cretaceous) stratigraphy in NW Anatolia known as the Bilecik Limestone (Figure 3f–h). It has been divided into the two formations; Taşçıbayırı and Günören Limestone formations. The Taşçıbayırı Formation (sample IDs: H56, H57, and H58) underlies the Günören Limestone (sample ID: H59); they contain sandy limestone and dolomitic limestone, respectively. The sandy limestone of the Taşçıbayırı Formation is composed of calcite with feldspar, mica, and volcanic rock fragments (Figure 3g). The volcanic rock fragments are composed of broken and/or eroded volcanic rocks consisting of quartz and feldspar (Figure 3g), while the Günören dolomitic limestone consists of calcite and dolomite with Fe-oxide minerals (Figure 3h).

The Düztarla granitoid rocks (sample IDs: H61 and H62) reflect Upper Oligocene–Lower Miocene post-collisional magmatic activity in the study area (Figure 2), differentiated into granodiorite and granitic rocks (Figure 3i–l). The granodiorite consists of plagioclase (oligoclase in composition; 35–50 vol.%), quartz (20–35 vol.%), and microperthite (8–17 vol.%) with subordinate amounts of biotite and Fe-oxide minerals (Figure 3j). The plagioclase is slightly affected by sericite and kaolinite alteration (Figure 3j). The granite is composed of microperthite (30–45 vol.%), quartz (20–35 vol.%), and plagioclase (albite in composition; 20–30 vol.%) with muscovite and Fe-oxide minerals (Figure 3l).

The Halılar area has a well-described Upper Triassic–Liassic continuous succession (Figures 1 and 2). The tectonic sedimentary rocks formed at the Sakarya divergent margin, which evolved in the Late Triassic–Aptian interval [29,30]. As a result of the diachronic closure of the Tethys basin in western Anatolian, the Upper Triassic black shales were deposited in the Lias in the Karakaya euxinic basin throughout the Edremit region. This shale and the Hettangian arkosic sandstones were later intruded by the Düztarla granodioritic–granitic body due to the southward subduction of the Paleo-Tethys [29].

3. Sampling and Analytical Methods

A total of 45 host rocks, altered rocks, and mineralized samples were collected from the study area. Thin sections and a subset of polished sections were examined optically using transmitted and reflected light microscopes. Whole-rock major, trace, and rare earth element analyses were conducted at the Geochemistry Research Laboratories of Istanbul Technical University (ITU/JAL). The samples were grounded using a tungsten carbide milling device. Major elements were analyzed using a BRUKER S8 TIGER model X-ray fluorescence spectrometer (XRF) (Östliche Rheinbrückenstraße 49, 76187 Karlsruhe, Germany) with a wavelength range from 0.01–12 nm. Trace elements were analyzed by inductively coupled plasma-mass spectrometry (ICP-MS) using an ELAN DRC-e Perkin Elmer model (PerkinElmer, Waltham, MA, US). Approximately 100 mg of powdered sample was digested in two steps. The first step was completed with 6 mL of 37% HCl, 2 mL of 65% HNO_3, and 1 mL of 38%–40% HF in a pressure- and temperature-controlled Teflon beaker using a Berghoff Microwave™ at an average temperature of 180 °C. The second step was completed with the addition of 6 mL of 5% boric acid solution. The remaining solution sample was analyzed by ICP-MS. The altered rocks were also analyzed for mineralogy using a BRUKER X-ray diffractometer (XRD) (Östliche Rheinbrückenstraße 49, 76187 Karlsruhe, Germany). Calculation of the normative mineral abundances from the major element analyses and rare earth element diagrams were created using Igpet 2.3 [31]. The GEOISO-Windows of Coelho [32] were used to determine the absolute mobility of the elements using equations from Gresens [33] and isocon diagrams from Grant [34,35].

Sulfide minerals for sulfur isotope analysis were separated from slightly crushed (200 mesh) lode samples (>95 % pure pyrite, chalcopyrite, and galena). They were washed and handpicked under a binocular microscope. These analyses were carried out at the Geochron Laboratory (USA) using EA-IRMS (Elemental Analysis-Isotope Ratio Mass Spectrometry) techniques. All stable isotope data are reported in the delta (δ) notation, relative to Vienna-Canyon Diablo Troilite (V-CDT) for sulfur isotopes with 0.5‰ (1 σ) analytical uncertainty.

4. Halılar Cu-Pb (±Zn) Mineralization

The Halılar base metal mineralization represents Cu-Pb with some Zn brecciated-stockwork-veining-type mineralization. The mineralization is restricted to a fault gouge zone directed NE–SW, as well as along the lower boundary of the Sakarkaya and Düztarla granitoid rocks (Figure 2). It is also closely associated with intense hydrothermal alteration within the breccia and quartz stockwork veining (Figure 4a–d). Based on the field investigation and petrographic and mineralogical (XRD) data, the mineralized quartz veins and brecciated ore bodies are accompanied by two types of hydrothermal alteration zone with gradational boundaries: zone-1 (sericite–quartz–chlorite ± kaolinite ± pyrite) and zone-2 (calcite–epidote–albite ± chlorite ± sericite).

Figure 4. (**a**,**b**) Brecciation and quartz stockwork veining along with intense hydrothermal alteration; (**c**) base -metal-bearing brecciated zone; (**d**) pervasive chlorite and epidote alteration in the host rock; (**e**–**h**) XPL and PPL photomicrographs of alteration zone-1 containing quartz, wollastonite, kaolinite, and andradite with calcite in a mineralized and brecciated quartz vein; (**i**–**l**) XPL and PPL photomicrographs of alteration zone-2 with a high amount of sericite, quartz, chlorite, kaolinite, and opaque minerals; (**m**–**p**) XPL and PPL photomicrographs of the calcite, epidote, albite, and quartz with chlorite in alteration zone-2. Abbreviations: calcite (cal), chlorite (chl), epidote (ep), kaolinite (kln), opaque (opq), quartz (qz), sericite (ser), and wollastonite (wo).

The ore zone is represented by mineralized and brecciated quartz stockwork veining (Figure 4a–c). It has high amounts of Cu (9.9 %), Pb (11.3 %), and Zn (0.29 %) mineralization, with high amounts of chalcopyrite and galena with sphalerite and pyrite (Figure 4a–c). It contains quartz with a subordinate amount of wollastonite, kaolinite, andradite, and calcite (Figures 4e–h and 5 and Appendix A). These calc–silicate assemblages refer to the skarn that resulted from the metasomatism of sandy limestone in the Taşçıbayırı Formation in association with andradite (Figures 4e–h and 5 and Appendix A). The XRD data show quartz (low), wollastonite (1A, manganoan), kaolinite (1A), microcline, calcite, chalcopyrite, andradite, anglesite, and cubanite (high) with smaller amounts of pyrite, sphalerite (ferrous), galena, and quartz (high) (Figure 5 and Appendix A).

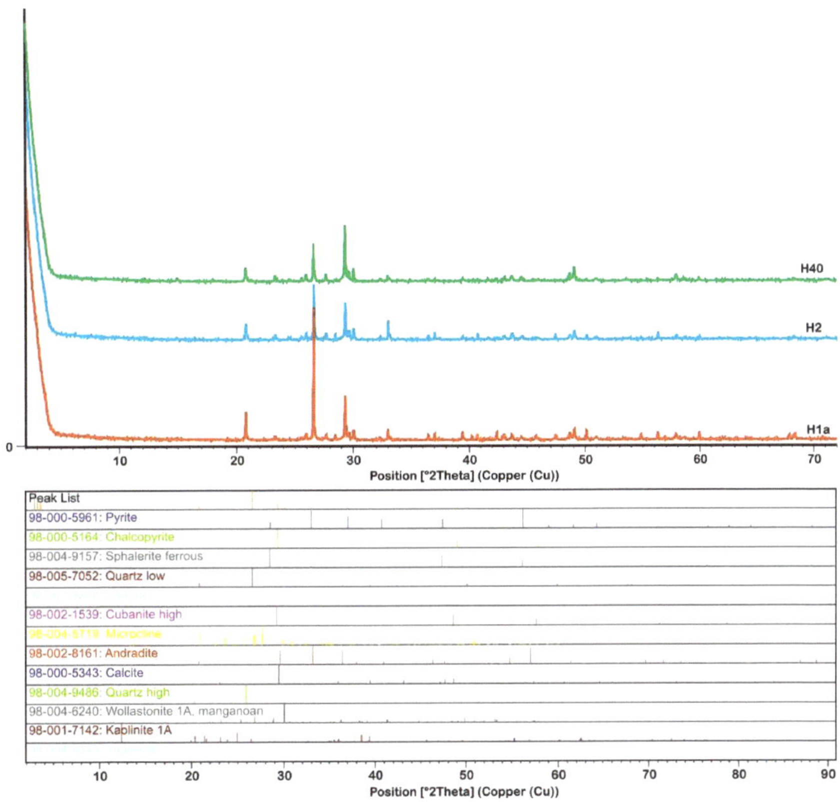

Figure 5. XRD patterns of the representative samples from the ore zone, having quartz; wollastonite; kaolinite; microcline; calcite; chalcopyrite; andradite; anglesite; and cubanite with pyrite, sphalerite, galena.

4.1. Hydrothermal Alteration Types

The hydrothermal alterations associated with Cu-Pb-Zn mineralization include extensive silicification (Figure 4e–h), sulfidation (Figure 4e–h), and carbonatization (Figure 4k–n), with some chloritization (Figure 4i–p), sericitization (Figure 4i–p), and calc–silicate alteration (Figure 4e–p) distributed in two alteration zones around the mineralized orebodies.

Alteration zone-1 (sericite–quartz–chlorite ± kaolinite ± pyrite) forms the main alteration zone developed outwards from the ore zone and has high amounts of sericite and quartz, with lesser amounts of chlorite, kaolinite, and pyrite (Figures 4i–l and 6 and Appendix B). It is characterized by the preferential replacement of the original K-feldspar and/or plagioclase–biotite by sericite/muscovite–kaolinite. XRD studies reveal a paragenesis of quartz (low), kaolinite (1A), clinochlore (1MIa), and sericite (2M1) with a subordinate amount of chamosite (1MIIb), pyrite, and chalcopyrite (Figure 6, Appendix B).

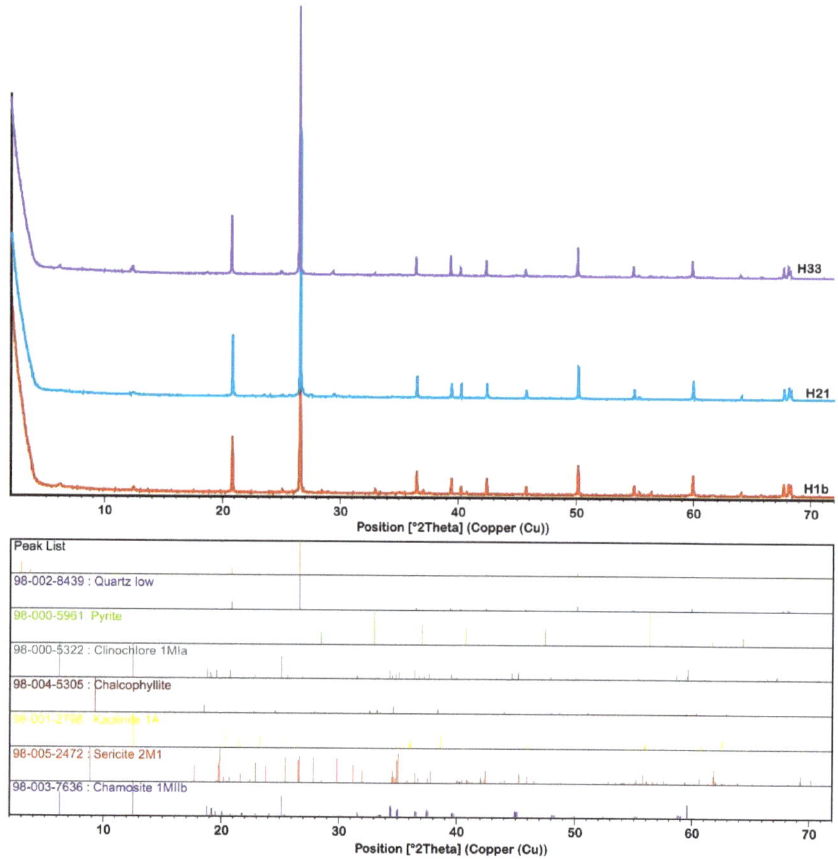

Figure 6. XRD patterns of representative samples from alteration zone-1 (sericite–quartz–chlorite ± kaolinite ± pyrite).

Alteration zone-2 (calcite–epidote–albite ± chlorite ± sericite) represents the distal zone and has higher amounts of calcite, epidote, and albite, with a subordinate amount of chlorite (Figures 4m–p and 7 and Appendix C). The sulfide minerals are less abundant in this zone (Figure 4d). The XRD data reveal that this alteration zone consists of quartz (low), albite (low), muscovite (2M1), clinochlore (IIb-4), microcline, sericite (2M1), and calcite, with lesser amounts of orthoclase, chamosite (1MIIb), and epidote (Figure 7 and Appendix C).

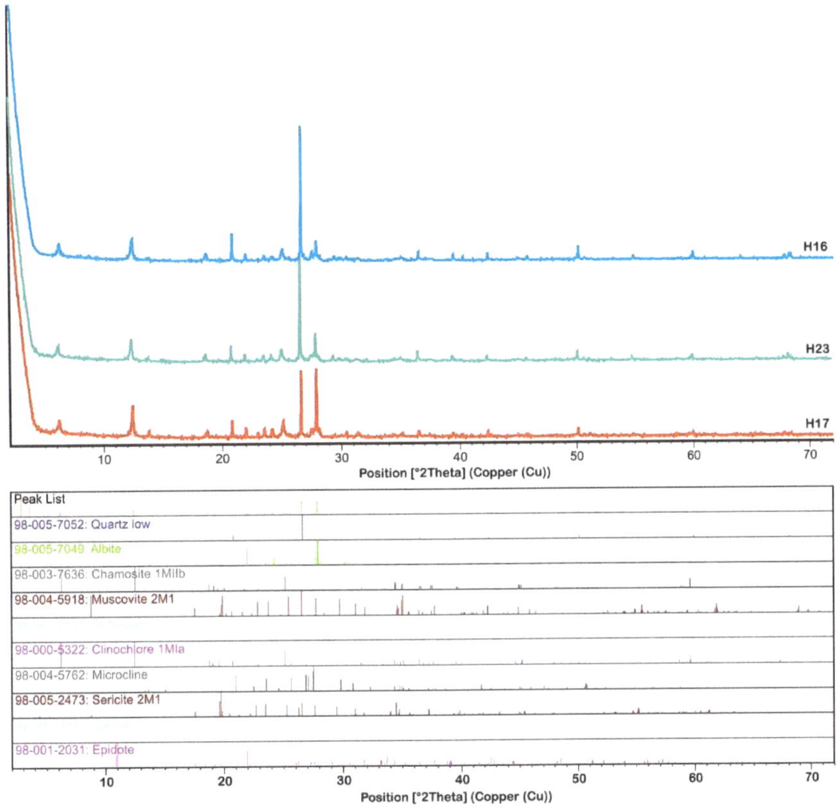

Figure 7. XRD patterns of representative samples from alteration zone-2 (calcite–epidote–albite ± chlorite ± sericite).

4.2. Ore Mineralogy

The ore mineral assemblage includes chalcopyrite, galena, pyrite, and sphalerite with covellite and goethite in abundant gangue minerals such as quartz, sericite, chlorite, and calcite forming along the quartz stockwork veins as well as in the brecciated ore zones (Figures 4–7 and Appendices A–C). Chalcopyrite and galena are the most common sulfide minerals in the ore bodies, occurring as yellow and whitish gray in color and with a subhedral granular texture (up to 2 mm), respectively (Figure 8a,b). Pyrite is either associated with or occurs as inclusions in chalcopyrite (Figure 8c,d). Sphalerite is characterized by dark gray coloring associated with chalcopyrite and pyrite, forming exsolution textures produced by chalcopyrite (Figure 8b,c,e). These minerals were developed in the main ore mineralization phase (Figure 9). On the other hand, the oxidation and supergene mineralization events represent the second phase of mineralization, including covellite and goethite formed after chalcopyrite and pyrite, respectively (Figures 8 and 9).

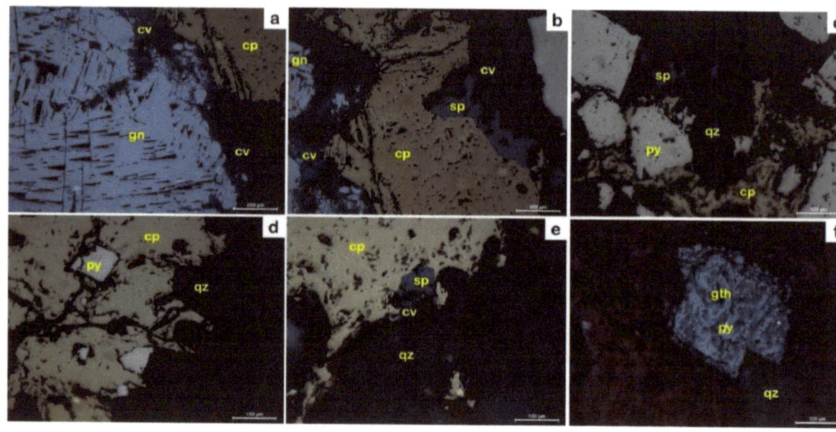

Figure 8. PPL photomicrographs of reflected light microscopy. (**a**) Galena with triangular cleavage pits associated with chalcopyrite replaced by covellite. (**b**) Chalcopyrite with sphalerite, galena, pyrite, and covellite. (**c–e**) Pyrite associated with chalcopyrite and sphalerite in a mineralized quartz vein. (**f**) Goethite, the main alteration product after pyrite. Abbreviations: chalcopyrite (cp), covellite (cv), galena (gn), goethite (gth), pyrite (py), quartz (qz), and sphalerite (sp).

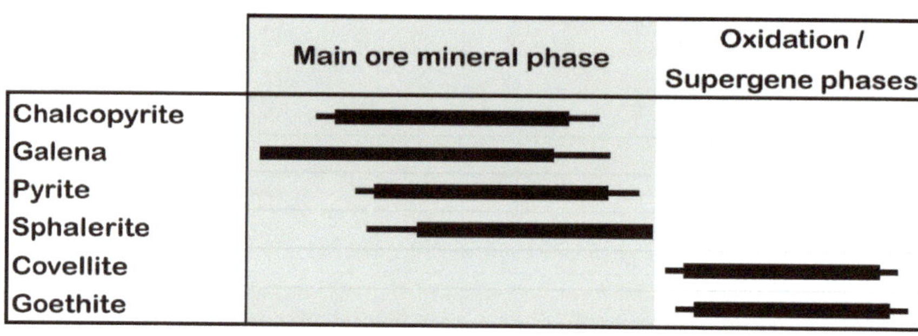

Figure 9. Paragenetic sequence of mineralization phases in the Halılar area.

5. Geochemical Characteristics

5.1. Geochemistry of the Least-Altered Metasediments

Ten representative samples collected from the least-altered metasediments of the Sakarkaya Formation were analyzed for major, trace, and rare-earth element contents (Table 1). Samples from the metasandstones are classified as mainly wackes and, rarely, Fe-sand and Fe-shale based on the geochemical classification of the terrigenous sandstones and shales by Herron [36] (Figure 10a). The samples have SiO_2/Al_2O_3 ratios ranging from 2.7 to 5.5 with an average of 4.3, which are similar to upper continental crust (UCC) [37] (~4.3 SiO_2/Al_2O_3 ratio), suggesting that they were sourced from the crustal felsic rocks. It also appears in Figure 10b,c that the Sakarkaya metasediments have acidic/intermediate characteristics, which lie mostly in the field of the metavolcanic tuffs, metagreywackes, and arkosic sands [38] according to their low K/Rb ratios (mean = 312.8). In the F1-F2 classification diagram (Figure 10d), the metasediments are mostly comparable with the compositional characteristics of the P4-quartoze sedimentary provenance that form within the passive and active continental margins (Figure 10e) due to recycling from old sedimentary rocks derived from highly weathered felsic terrains. The metasandstones have low total rare earth element contents (∑REE) (up to 145.14 ppm with an average of 88.96 ppm), ∑REE/∑HREE = 6.59–10.43 ppm, $(La/Yb)_N$ = 5.38–14.29 ppm, and positive Eu anomaly

(Eu/Eu* = 0.68–1.27 ppm) that are similar to the upper continental crust (UCC) of Taylor and McLennan [37] (Figure 10f).

Table 1. The major oxides and trace and rare-earth elements (REE) of metasediments in the Sakarkaya Formation.

Major Oxide (%)	H05	H07	H09	H-14	H-15	H-18	H-20	H-22a	H55	H60
SiO_2	59.2	62.3	67.6	68.2	50.2	64.0	58.9	63.5	65.2	61.0
Al_2O_3	14.0	13.7	16.4	16.2	18.4	13.6	14.6	11.5	13.6	15.1
Fe_2O_3	6.8	5.0	3.5	3.4	10.9	5.0	6.9	11.3	4.9	5.8
MgO	5.6	4.0	1.5	1.5	9.2	4.0	5.7	4.8	4.2	5.5
CaO	1.7	0.2	0.5	0.3	0.6	2.5	1.7	0.1	2.6	0.3
Na_2O	2.1	0.9	1.8	1.8	3.3	2.5	2.1	0.8	2.5	2.0
K_2O	4.7	3.9	5.5	5.8	1.0	4.8	4.8	1.6	4.9	4.8
TiO_2	0.6	0.6	0.5	0.5	0.7	0.5	0.7	0.3	0.8	0.4
P_2O_5	0.1	0.1	0.1	0.1	0.2	0.1	0.2	0.1	0.1	0.2
MnO	0.1	0.2	0.1	0.1	0.3	0.1	0.2	0.3	0.1	0.1
Cr_2O_3	0.0	0.1	0.0	0.0	0.1	0.0	0.0	0.0	0.0	0.0
LOI	3.2	3.9	3.6	1.8	5.0	2.6	3.8	4.0	3.5	3.4
Trace and Rare Earth Elements (ppm)										
Ag	6.7	10.7	6.0	1.7	11.7	9.7	2.3	7.0	4.6	6.8
As	19.7	17.9	222.7	33.5	24.6	14.9	21.0	424.3	222.0	18.8
Au	0.0	0.0	0.0	0.1	0.0	0.0	0.0	0.1	0.0	0.1
Ba	465.2	336.5	823.3	773.2	157.2	515.7	1130.9	381.5	756.2	423.4
Be	1.7	2.6	4.3	1.7	1.7	3.5	5.1	1.5	3.3	1.6
C	2520.7	1234.1	1284.5	545.6	708.1	2621.5	2419.9	48.2	1877.4	1259.3
Cd	0.4	0.8	1.0	0.4	0.4	1.2	0.9	1.3	1.1	0.9
Co	31.9	22.3	23.8	16.3	43.7	20.1	24.4	23.3	27.6	27.1
Cs	5.0	4.1	3.6	6.6	3.7	3.9	6.1	2.2	4.6	3.9
Cu	40.8	42.9	27.8	16.8	64.8	20.9	34.6	125.7	80.2	83.3
Ga	32.8	26.4	36.1	39.2	26.4	26.4	45.8	26.0	35.9	29.4
Hf	1.6	1.2	1.1	0.6	1.3	1.4	1.8	0.6	1.4	1.2
In	0.0	0.0	0.0	0.0	0.1	0.0	0.1	0.0	0.0	0.0
Ir	0.0	0.0	0.0	0.0	0.0	0.0	0.0	0.0	0.0	0.0
Li	67.7	83.5	63.7	16.2	119.3	47.8	79.6	104.8	92.2	86.2
Mo	2.5	2.5	2.5	1.0	2.0			3.0	2.8	2.6
Ni	42.7	37.7	37.9	32.0	53.4	22.0	53.7	45.1	49.4	43.9
Pb	232.5	272.3	306.4	33.8	101.4	54.1	61.5	911.8	356.9	416.0
Pd	1.1	1.6	1.7	0.8	1.5	1.7	1.7	0.7	1.2	0.9
Pt	0.0	0.0	0.0	0.0	0.0	0.0	0.1	0.0	0.0	0.0
Rb	122.5	92.3	107.4	125.3	46.6	149.1	182.9	62.1	99.8	103.6
Rh	0.0	0.0	0.0	0.0	0.0	0.0	0.0	0.0	0.0	0.0
Ru	0.0	0.0	0.0	0.0	0.0	0.0	0.0	0.0	0.0	0.0
S	265.1	167.7	540.9	342.1	188.1	147.4	934.3	871.9	265.1	167.7
Sb	1.5	1.7	1.6	2.0	1.3	1.1	2.1	5.3	1.8	1.6
Sn	1.8	1.4	1.6	3.3	2.3	3.6	2.5	1.1	1.5	2.2
Sr	272.1	208.3	153.9	60.3	122.8	163.3	380.8	35.7	240.2	181.1
Te	0.0	0.0	0.0	0.0	0.1	0.0	0.0	0.1	0.0	0.0
Tl	0.7	0.9	1.8	1.0	0.4	1.5	2.2	0.9	1.5	0.8
U										
Zn	420.0	500.9	460.5	197.4	72.4	270.4	258.1	581.9	480.7	470.6
Se	0.3	−0.2	3.2	0.2	1.8	-1.2	0.8	5.6	3.0	1.2
Sc	152.8	165.4	138.8	129.5	176.1	154.6	123.0	144.8	133.9	148.8
Y	18.9	11.9	12.0	7.3	8.7	19.1	18.8	5.1	15.4	12.0
La	16.0	22.5	20.4	9.8	12.0	26.1	29.6	5.7	17.1	14.7
Ce	34.6	39.7	56.3	43.0	26.1	53.2	59.3	10.6	35.0	22.6
Pr	4.0	2.6	3.3	2.3	2.6	5.8	6.8	1.2	2.9	3.1
Nd	13.6	19.4	17.7	8.5	9.9	22.5	26.0	4.6	14.7	12.6
Sm	5.3	3.5	3.3	2.3	2.2	4.9	5.8	1.3	4.4	3.4
Eu	1.4	1.3	1.5	0.9	0.6	1.6	1.5	0.4	1.0	1.1

Table 1. *Cont.*

Major Oxide (%)	H05	H07	H09	H-14	H-15	H-18	H-20	H-22a	H55	H60
Gd	3.4	4.0	4.1	2.6	2.9	6.0	5.8	1.3	4.5	3.5
Tb	0.2	0.3	0.3	0.3	0.3	0.6	0.7	0.2	0.2	0.2
Dy	1.7	3.3	3.3	1.7	2.0	3.5	3.9	1.1	1.1	3.7
Ho	0.3	0.4	0.8	0.3	0.4	0.7	0.8	0.2	0.2	0.8
Er	1.0	1.2	2.1	1.0	1.2	2.1	2.4	0.7	2.3	0.7
Tm	0.1	0.2	0.3	0.1	0.2	0.3	0.3	0.1	0.3	0.1
Yb	1.0	1.1	1.9	1.0	1.1	1.8	2.2	0.6	2.1	0.7
Lu	0.2	0.2	0.3	0.2	0.2	0.3	0.3	0.1	0.3	0.1
Th	7.6	12.0	12.3	6.8	3.9	12.6	12.7	2.1	12.0	6.9
Parameters										
K/Rb	317.8	350.0	428.3	382.1	169.1	266.2	216.7	209.3	406.0	382.8
F1	−8.6	−8.1	−5.2	−5.6	−3.8	−7.0	−8.5	−2.8	−7.8	−9.0
F2	−3.1	−4.6	2.7	3.0	−12.1	0.1	−3.2	−10.8	0.2	−3.4
K wt%	3.9	3.2	4.6	4.8	0.8	4.0	4.0	1.3	4.1	4.0
log SiO_2/Al_2O_3	0.6	0.7	0.6	0.6	0.4	0.7	0.6	0.7	0.7	0.6
log Fe_2O_3/K_2O	0.2	0.1	−0.2	−0.2	1.1	0.0	0.2	0.9	0.0	0.1
Σ REE	82.8	99.6	115.4	74.0	61.6	129.3	145.1	28.1	86.3	67.2
Σ REE/Σ HREE	10.4	9.4	8.9	10.3	7.5	8.5	8.9	6.6	7.7	6.9
Eu/Eu*	1.0	1.1	1.3	1.1	0.8	0.9	0.8	1.0	0.7	1.0
$(La/Yb)_N$	11.0	14.1	7.2	6.7	7.5	9.4	9.0	6.0	5.4	14.3

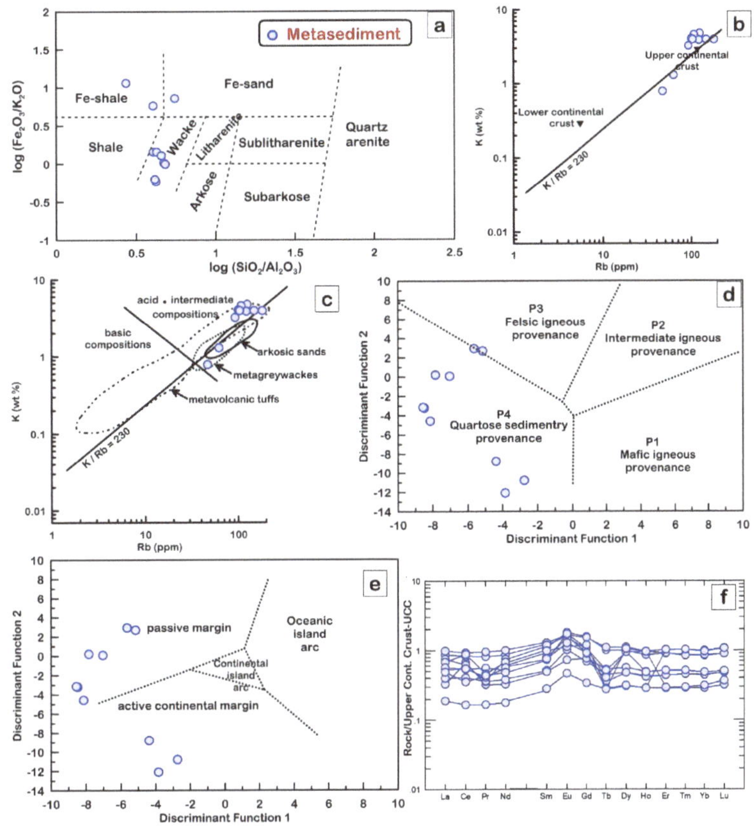

Figure 10. Geochemical classification diagrams of metasediments: (**a**) log(Fe_2O_3/K_2O)−log(SiO_2/Al_2O_3) diagram after Herron [36]. (**b**,**c**) K − Rb diagrams after Floyd and Leveridge [38]. Fields of unmetamorphosed arkosic sands after van de Kamp et al. [39], low−grade metagreywackes after Condie et al. [40] and Caby et al. [41], and higher-grade metavolcanic tuffs after van de Kamp [42]; (**d**) plot of samples in discriminant functions F1 vs. F2 (provenance fields are after Roser and Korsch [43]; (**e**) plot of discriminant scores along Function 1 vs. 2 after Bhatia [44]; (**f**) upper continental crust (UCC)−normalized REE patterns [37].

5.2. Alteration Geochemistry

Two main alteration zones surround the Cu-Pb±Zn-bearing ore mineralization in the Halılar area. These are represented by zone-1 (sericite–quartz–chlorite ± kaolinite ± pyrite) and zone-2 (calcite–epidote–albite ± chlorite ± sericite), and they were analyzed for major, trace, and REEs (Table 2). Based on the alteration index (AI) [45] and advanced argillic alteration index (AAAI) of Williams and Davidson [46], samples from each zone show opposite alteration trends (Figure 11a). The ore zone and alteration zone-1 fall along the trend of silicification/potassic alteration, while alteration zone-2 falls along the carbonation/chloritization alteration trend (Figure 11a). Based on the alteration boxplot relationship between the chlorite–carbonate–pyrite index (CCPI) of Large et al. [47] and the AI of Ishikawa et al. [45], the samples of the ore zone and zone-1 are clustered in the field of strongly altered rock, having chlorite–sericite–pyrite alteration while the ore zone is affected by extensive pyritization (Figure 11b). On the other hand, zone-2, within the carbonate-altered host rock field, shows Mn–carbonate–sericite–chlorite alteration (Figure 11b).

Table 2. Major oxides and trace and rare-earth elements (REE) of the ore zone and alteration zones 1 and 2 in the Halılar area.

	Ore Zone				Zone 1 (Sericite-Quartz-Chlorite ± Kaolinite ± Pyrite)													
Major Oxides (%)	H1-a	H2	H-25	H-40	H1-b	H3-1	H3-2	H5	H-6	H-7	H-8	H-9	H-11	H-12	H-13	H-21	H-26	H-28
SiO_2	39.5	21.3	38.6	15.2	75.7	88.6	91.2	80.2	79.1	71.3	87.0	78.5	66.3	88.2	57.9	83.0	67.4	89.7
Al_2O_3	2.1	3.5	0.8	1.3	4.8	2.3	1.5	6.5	7.0	6.0	4.0	2.3	4.3	4.0	16.6	6.0	9.1	2.8
Fe_2O_3	11.7	15.9	14.2	18.6	7.3	2.6	2.5	7.1	3.5	10.5	5.0	6.5	7.3	2.8	9.1	1.8	4.6	2.8
MgO	0.0	0.7	0.3	0.1	1.7	0.6	0.3	1.9	1.1	0.2	1.9	0.1	1.4	1.0	3.3	0.9	1.4	1.2
CaO	0.0	0.0	0.0	0.0	0.1	0.0	0.0	0.1	0.1	0.0	0.0	0.0	0.0	0.1	0.1	2.8	2.7	0.9
Na_2O	0.0	0.0	0.0	0.0		0.0	0.0	0.1	0.0	0.0				0.0	0.1	0.1	0.1	0.0
K_2O	0.8	0.8	0.0	0.2	0.5	0.6	0.3	2.0	2.4	3.7	0.1	1.2	0.7	0.8	6.5	2.3	5.0	0.3
TiO_2	0.0	0.0	0.0	0.0	0.2	0.0	0.0	0.1	0.2	0.1	0.0	0.1	0.0	0.1	0.6	0.1	0.4	0.0
P_2O_5	0.0	0.0	0.0	0.0	0.0	0.0	0.0	0.1	0.0	0.0	0.0	0.0		0.1	0.1	0.0	0.1	
MnO	0.0	0.0	0.0	0.0	0.1	0.0	0.0	0.1	0.0		0.1		0.1	0.1	0.2	0.1	0.2	0.1
Cr_2O_3	0.0	0.0	0.0	0.0	0.0			0.0	0.0						0.0	0.0	0.0	0.0
LOI	9.7	14.6	17.1	13.9	4.2	1.7	1.4	1.3	3.1	3.9	1.4	3.9	6.2	1.3	4.5	2.5	3.2	1.1
Trace and Rare Earth Elements (ppm)																		
Ag	328.7	677.8	34.0	781.3	3.3	18.8	3.3	4.7	6.2	66.2	3.3	112.5	24.1	27.3	9.9	1.3	14.4	2.6
As	188.9	236.0	8084.9	203.5	203.6	41.9	55.9	95.6	78.9	126.1	145.3	475.8	545.0	23.9	48.3	1377.2	68.1	40.6
Au	0.5	0.5	1.0	0.6	0.1	0.1	0.1	0.0	0.1	0.1	0.0	0.0	0.0	0.0	0.0	0.0	0.0	0.0
Ba	161.6	227.2	143.2	101.4	77.9	81.4	93.7	235.6	350.5	796.8	14.7	484.6	118.9	85.8	1422.3	684.6	1181.0	191.2
Be	0.6	0.8	2.6	1.3	0.6	0.4	0.6	0.7	1.3	0.7	0.4	0.4	0.5	0.6	2.0	2.5	0.7	2.3
C	62.2	267.0		308.3	106.8	94.6	36.2	353.6	97.4	805.2	3.4	246.0		107.4	230.5	4018.6	5636.7	1006.9
Cd	26.8	38.9	1.5	8.7	0.4	0.7	−0.2	−0.1	0.4	0.5	0.2	0.8	0.9	1.1	0.3	0.8	139.4	0.4
Co	24.6	28.4	609.6	18.5	32.3	77.1	45.3	23.7	28.3	20.0	31.1	67.6	131.7	32.2	8.0	60.2	25.1	53.3
Cs	1.2	1.3	0.8	1.1	2.4	1.7	1.1	1.7	3.2	3.0	0.2	1.5	0.7	2.3	5.3	3.3	3.8	2.2
Cu	678,54.1	126,125.6	10,936.9	194,721.6	133.6	4204.9	2824.8	439.9	948.7	1664.1	545.8	4871.0	24,584.5	1340.1	1610.0	30.9	453.6	1677.1
Ga	5.9	9.8	8.4	5.7	7.7	4.1	5.5	14.3	11.7	20.2	4.2	14.3	8.9	5.7	46.3	22.2	34.7	12.0
Hf	0.1	0.3	0.0	0.1	0.2	0.1	0.0	0.2	0.3	0.5	0.1	0.2	0.1	0.1	1.6	0.2	1.3	0.1
In	0.3	0.3	0.1	0.1	0.2	0.1	0.0	0.0	0.1	0.0	0.0	0.1	0.3	0.0	0.0	0.0	0.2	0.0
Ir	0.0	0.1	0.0	0.0	0.0	0.1	0.0	0.0	0.0	0.1	0.0	0.0	0.0	0.0	0.0	0.0	0.0	0.0
Li	66.2	34.1	28.3	54.0	38.6	39.9	39.0	51.3	64.2	35.1	70.2	41.5	43.0	60.5	64.6	79.9	42.4	65.4
Mo	33.0		15.0	137.0	11.0	24.0	21.0	22.0	69.0	48.0	24.0	84.0	37.0		45.0		4.0	9.0
Ni	8.7	12.2	7.7	7.7	11.4	8.8	19.6	9.7	5.0	2.3	12.8	6.8	32.1	1.6	15.8	17.5	27.1	7.5
Pb	81,009.8	191,128.0	5958.0	175,698.2	54.4	415.0	90.6	430.8	1427.6	1242.6	372.6	698.9	108.3	7088.1	1459.1	155.5	3864.8	75.8
Pd	0.1	0.2	0.1	0.1	0.2	0.2	0.1	0.4	0.3	0.8	0.2	0.3	0.1	0.1	2.1	0.2	1.9	0.1
Pt	0.1	0.1	0.1	0.0	0.0	0.0	0.0	0.0	0.1	0.0	0.0	0.0	0.2	0.1	0.0	0.1	0.0	0.1
Rb	27.5	33.5	4.7	10.1	17.1	21.7	10.6	47.8	67.4	95.1	1.8	36.1	18.5	24.8	225.5	85.3	169.7	17.3
Rh	1.9	2.7	0.3	3.4	0.0	0.1	0.1	0.0	0.1	0.1	0.0	0.1	0.5	0.1	0.1	0.0	0.1	0.1
Ru	0.1	0.1	0.0	0.0	0.0	0.0	0.0	0.0	0.0	0.0	0.0	0.0	0.0	0.0	0.0	0.0	0.1	0.0
S	162,930.0	211,150.0	256,310.0	203,370.0	34,415.0	11,038.0	6804.3	824.7	15,500.0	24,925.0	580.8	35,579.0	59,632.0	3205.2	759.1	1139.3	4052.6	3609.0
Sb	23.2	15.5	12.5	27.7	4.5	6.7	5.4	7.8	6.2	12.5	5.4	13.0	8.0	5.6	5.1	21.2	5.3	5.4
Sn	0.5	0.8	0.6	0.5	1.9	0.6	0.9	1.0	1.1	0.9	0.4	1.2	0.6	1.0	2.9	1.0	1.7	0.5
Sr	14.7	85.1	224.1	89.1	5.7	6.8	8.1	16.5	21.6	35.7	2.8	15.9	7.9	7.3	86.1	214.7	85.7	238.4
Te	16.9	35.8	2.5	27.9	0.5	0.9	0.3	0.7	0.9	4.7	0.4	2.8	1.1	1.4	0.4	0.1	0.1	0.0
Tl	1.5	2.8	0.1	2.3	0.2	0.4	0.2	0.7	1.1	1.6	0.0	0.7	0.2	0.3	2.8	1.2	2.2	0.2
Zn	4328.1	5932.9	356.5	1038.0	63.3	20.9	35.5	64.9	32.6	89.8	116.6	37.3	88.7	141.9	69.7	211.4	25,117.3	270.7
Se	192.8	377.5	54.4	472.5	3.4	5.0	6.5	6.8	4.4	19.8	6.7	29.9	7.7	4.7	2.9	6.0	3.1	2.1
Sc	53.4	31.5	54.4	31.0	141.8	176.2	146.9	111.3	157.7	167.0	141.9	172.2	137.9	157.7	178.7	118.7	149.7	135.3

	Zone 2 (Calcite-Epidote-Albite ± Chlorite ± Sericite)								
	H4	H-10	H-16	H-17	H-19	H-23	H-27	H-30	H-41
SiO_2	61.3	60.4	61.6	54.2	57.3	57.4	63.7	37.8	75.0
Al_2O_3	19.2	15.8	13.6	16.5	15.2	15.3	17.9	12.4	12.7
Fe_2O_3	5.2	7.1	6.6	11.9	9.1	7.4	4.7	7.4	0.7
MgO	1.2	4.9	5.0	4.6	5.9	5.7	2.0	5.5	0.2
CaO	2.8	0.7	2.8	2.1	0.4	2.8	0.3	19.3	0.8
Na_2O	1.9	3.0	2.3	3.4	2.1	2.6	2.1	1.8	3.3
K_2O	3.4	3.4	3.1	2.1	4.5	3.1	5.7	2.5	5.0
TiO_2	0.6	0.8	0.8	1.1	0.8	0.8	0.5	0.6	0.0
P_2O_5	0.1	0.2	0.2	0.3	0.2	0.2	0.2	0.1	0.0
MnO	0.2	0.2	0.2	0.2	0.3	0.2	0.1	0.2	0.0
Cr_2O_3	0.0	0.0	0.0	0.0	0.0	0.0	0.0	0.1	0.0
LOI	4.0	3.2	3.6	3.2	3.4	4.2	2.4	11.8	1.5
Ag	1.6	11.7	2.9	8.5	11.0	1.6	2.6	0.4	13.6
As	36.0	14.3	28.8	14.5	41.1	23.9	23.7	36.0	27.8
Au	0.1	0.1	0.0	0.0	0.1	0.0	0.1	0.0	0.1
Ba	303.9	846.4	676.1	451.0	1135.4	621.4	951.5	1145.4	461.1
Be	1.7	3.2	5.1	1.8	2.6	4.6	2.6	1.5	4.3
C	4242.9	917.1	3032.0	1153.2	212.1	3317.8	963.9	26,344.0	793.0
Cd	−0.2	0.8	3.5	0.2	0.2	0.3	0.2	0.8	0.6
Co	25.5	17.4	33.1	21.1	26.5	25.4	12.2	23.0	34.3
Cs	10.9	4.5	11.5	6.7	5.4	9.6	6.9	8.6	3.7
Cu	1.1	58.4	43.2	20.2	41.3	16.5	28.9	19.2	24.9
Ga	24.2	37.2	36.6	31.9	48.3	36.9	39.1	48.9	21.0
Hf	2.6	1.8	1.2	1.0	1.4	0.8	0.4	0.8	2.5
In	0.1	0.1	0.1	0.1	0.0	0.0	0.0	0.0	0.0
Ir	0.0	0.1	0.0	0.0	0.0	0.0	0.0	0.0	0.0
Li	32.8	64.7	73.9	70.0	84.6	76.6	25.5	61.3	29.6
Mo	5.0	3.0		1.0	6.0	2.0	4.0	2.0	4.0
Ni	30.7	45.3	70.5	23.8	58.7	32.3	12.9	37.9	8.5
Pb	9.4	78.2	45.9	28.0	51.1	38.2	43.5	24.9	112.5
Pd	4.0	2.6	1.1	1.0	1.9	0.6	0.6	0.9	3.0
Pt	0.0	0.0	0.1	0.0	0.0	0.0	0.1	0.0	0.0
Rb	110.3	131.7	139.4	85.2	186.5	143.4	148.8	98.2	130.6
Rh	0.0	0.1	0.1	0.1	0.0	0.1	0.1	0.0	0.0
Ru	0.0	0.0	0.0	0.0	0.0	0.0	0.0	0.0	0.0
S	529.9	327.8	392.5	482.6	3094.1	1553.3	497.5	2868.3	2847.1
Sb	2.1	2.0	1.4	2.7	4.1	1.0	2.3	1.8	3.2
Sn	2.4	3.3	2.3	2.8	7.2	2.3	7.5	1.6	0.6
Sr	127.7	232.5	408.8	320.7	197.7	329.9	114.8	289.8	274.6
Te	0.0	0.1	0.0	0.0	0.1	0.0	0.0	0.0	0.1
Tl	1.3	1.4	1.3	0.8	2.1	1.3	1.3	1.2	1.3
Zn	57.7	134.2	122.0	74.9	102.4	97.4	3.2	65.3	225.6
Se	1.7	4.1	1.5	0.8	3.6	7.7	−1.1	2.8	4.3
Sc	126.6	163.1	120.7	125.2	119.5	147.6	143.5	100.8	159.6

Table 2. *Cont.*

	Ore Zone			Zone 1 (Sericite-Quartz-Chlorite ± Kaolinite ± Pyrite)															Zone 2 (Calcite-Epidote-Albite ± Chlorite ± Sericite)									
Major Oxides (%)	H1-a	H2	H-25	H-40	H1-b	H3-1	H3-2	H5	H-6	H-7	H-8	H-9	H-11	H-12	H-13	H-21	H-26	H-28		H4	H-10	H-16	H-17	H-19	H-23	H-27	H-30	H-41
Y	0.2	1.2	0.4	0.4	2.5	0.6	0.4	1.3	2.5	2.9	1.2	1.0	1.5	1.4	13.7	4.3	12.1	1.2	Y	15.3	15.7	16.6	14.3	10.8	22.2	13.0	9.5	10.7
La	0.5	2.3	0.2	1.3	6.7	1.4	0.5	1.3	5.0	3.9	0.9	1.9	0.5	1.3	21.9	6.3	11.3	0.9	La	40.5	17.4	26.3	16.7	15.3	35.0	21.6	10.1	22.5
Ce	1.0	3.7	0.4	2.3	13.3	2.7	0.9	1.8	7.2	7.9	1.8	3.5	0.7	2.0	42.6	11.5	22.4	1.7	Ce	73.9	30.0	51.2	36.7	40.8	70.2	32.7	19.9	46.9
Pr	0.1	0.4	0.1	0.3	1.5	0.3	0.1	0.3	0.9	0.9	0.2	0.4	0.1	0.3	5.1	1.3	2.6	0.2	Pr	7.8	4.5	6.1	4.2	3.7	8.1	5.0	2.3	4.9
Nd	0.3	1.3	0.2	1.0	5.4	1.1	0.4	1.1	3.2	3.2	0.9	1.4	0.4	1.3	19.5	5.0	10.2	0.9	Nd	27.5	17.1	23.4	16.8	14.3	30.9	18.2	8.9	16.7
Sm	0.2	0.4	0.1	0.2	1.1	0.3	0.1	0.4	0.8	1.4	0.2	0.7	0.2	0.4	5.3	1.3	2.9	0.2	Sm	4.8	4.4	5.0	3.8	3.6	6.5	4.1	2.8	3.2
Eu	0.1	0.1	0.0	0.0	0.3	0.1	0.1	0.2	0.3	0.7	0.1	0.3	0.1	0.1	1.7	0.4	0.9	0.1	Eu	1.0	1.3	1.2	1.5	0.9	1.6	1.2	1.2	0.5
Gd	0.1	0.4	0.1	0.2	1.2	0.3	0.1	0.3	0.7	0.9	0.2	0.4	0.2	0.4	4.8	1.2	2.9	0.2	Gd	6.0	4.7	5.3	4.6	4.0	6.9	3.6	2.4	3.8
Tb	0.0	0.0	0.0	0.0	0.1	0.0	0.0	0.0	0.1	0.1	0.0	0.0	0.0	0.1	0.5	0.1	0.4	0.0	Tb	0.6	0.6	0.6	0.5	0.4	0.8	0.5	0.3	0.3
Dy	0.0	0.3	0.1	0.1	0.6	0.1	0.1	0.3	0.5	0.6	0.2	0.2	0.3	0.3	2.9	0.8	2.1	0.2	Dy	2.9	3.3	3.6	2.9	2.3	4.6	2.8	2.0	1.8
Ho	0.0	0.1	0.0	0.0	0.1	0.0	0.0	0.1	0.1	0.1	0.1	0.1	0.1	0.1	0.6	0.2	0.4	0.1	Ho	0.5	0.6	0.7	0.6	0.5	0.9	0.5	0.4	0.4
Er	0.0	0.2	0.0	0.1	0.3	0.1	0.1	0.2	0.3	0.4	0.1	0.1	0.2	0.2	1.7	0.5	1.3	0.1	Er	1.7	1.9	2.0	1.6	1.3	2.7	1.7	1.2	1.3
Tm	0.0	0.0	0.0	0.0	0.0	0.0	0.0	0.0	0.0	0.1	0.0	0.0	0.0	0.0	0.2	0.1	0.2	0.0	Tm	0.3	0.3	0.3	0.2	0.2	0.4	0.2	0.2	0.2
Yb	0.0	0.1	0.0	0.0	0.3	0.1	0.0	0.2	0.3	0.4	0.1	0.1	0.1	0.2	1.5	0.5	1.2	0.1	Yb	1.8	1.6	1.7	1.3	1.2	2.4	1.6	1.1	1.6
Lu	0.0	0.0	0.0	0.0	0.0	0.0	0.0	0.0	0.0	0.1	0.0	0.0	0.0	0.0	0.3	0.1	0.2	0.0	Lu	0.3	0.3	0.3	0.2	0.2	0.4	0.2	0.2	0.3
Th	0.3	1.7	0.2	0.4	2.4	1.0	0.2	0.9	1.5	2.6	0.5	1.1	0.2	0.6	8.9	1.8	5.7	0.3	Th	14.6	10.5	9.7	6.4	9.8	13.8	12.1	7.4	16.3
Parameters																			*Parameters*									
CCPI	93.2	94.6	99.7	99.0	94.1	82.0	88.1	79.7	63.5	71.8	97.9	83.4	91.8	80.2	63.4	51.6	51.7	92.3	CCPI	53.0	63.8	66.6	73.5	68.3	68.4	44.5	73.7	8.9
Ishikawa AI	98.8	99.0	95.2	96.2	97.3	95.6	93.0	94.6	97.5	98.1	98.2	97.9	98.0	95.5	98.2	53.3	68.7	61.9	Ishikawa AI	50.0	69.7	61.3	55.0	80.3	61.9	76.0	27.5	55.8
AAAI	98.6	76.0	92.5	93.1	81.0	92.9	95.9	79.2	86.7	96.7	82.0	98.8	82.4	89.1	62.7	68.7	61.4	80.7	AAAI	51.2	41.4	38.0	34.8	40.3	33.9	59.1	12.4	63.7

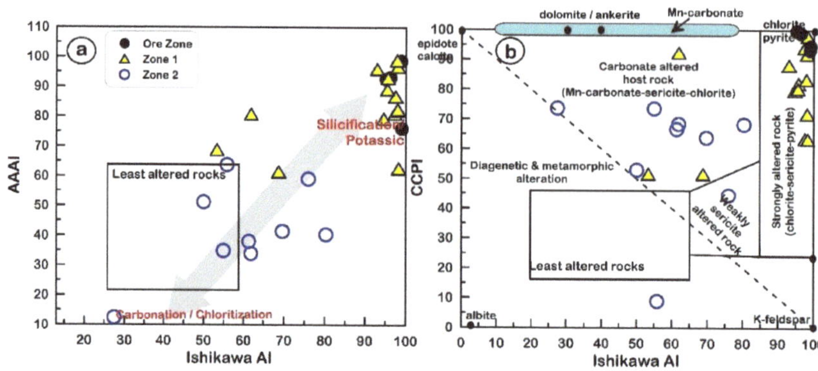

Figure 11. (**a**) AI [45] vs. AAAI [46]; (**b**) AI [45] vs. CCPI [47] of the studied alteration samples from the Halılar area.

5.3. Mass Balance Calculations

The behavior of different elements, excluding immobile ones, is changeable during hydrothermal alteration processes depending on their volume changes and their mass transfer [48,49]. Gresens [33] and Grant [34,35] used mass-balance calculations to quantify hydrothermal alteration effects on the host rock within the mineralized regions and to determine the relative gain and loss of the various major and trace elements during hydrothermal alteration.

Based on the trace element geochemical analyses, the ore zone and alteration zone-1 have high amounts of Cu and Pb, with an average of 9.9% and 11.3%, respectively, for the ore zone, and 0.32% and 0.12%, respectively, for zone-1 (Table 2). They are classified as a Cu-Pb type (Figure 12), which refers to the high concentrations of chalcopyrite and galena. Alteration zone-2 represents the Cu-Pb-Zn type (Figure 12), having low Cu, Pb, and Zn contents, with averages of 28.18ppm, 47.95ppm, and 98.07ppm, respectively.

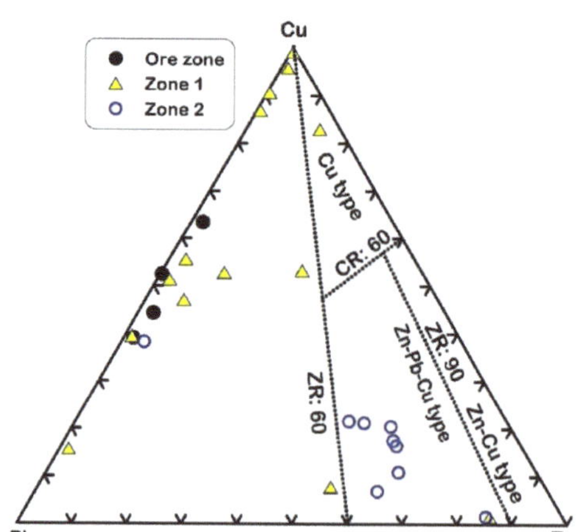

Figure 12. Metal content classification diagram after Large [50]. The ratios ZR = (100 Zn/(Zn + Pb) and CR = 100 Cu/(Cu + Zn) are based on the average ore grades in mass percent after Large [50].

Al_2O_3 and TiO_2 are immobile in all alteration zones during hydrothermal alteration; therefore, they were selected to assess the chemical changes due to the process of hydrothermal alteration by using the GEOISO-Windows software developed by Coelho [32]. The results of these calculations are illustrated through the isocon diagrams of Grant [34] and show the different patterns of major and trace element gains and losses (Figures 13 and 14 and Table 3). The samples from zone-1 are rich in SiO_2, Fe_2O_3, K_2O, and LOI, with lesser increases in the amount of CaO, P_2O_5, and MnO (Figure 14a). Gains in Ag, As, Cu, Mo, Pb, S, Sb, and Zn are also recognized within this alteration zone (Figure 14b). This zone is characterized by higher amounts of sulfur and iron, with variable copper, lead, and zinc contents reflecting high pyritization, with the main base metals providing higher mass (MC = 170.42) and volume change (VC = 182.1) (Table 3). SiO_2 and K_2O increases reflect high silicification and sericitization, which are comparable with the petrographic and mineralogical (XRD) data. In zone-2, CaO, Na_2O, P_2O_5, TiO_2, LOI, and carbon are enriched, reflecting calcite, epidote, and albite alterations (Figure 14c). The loss of Cu, Pb, and Zn is observed in this zone, providing lower MC (−3.18) and VC (−1.80) values (Figure 14d and Table 3).

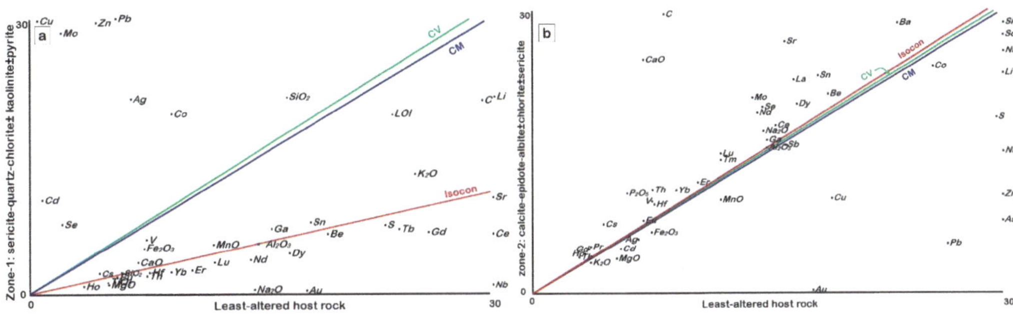

Figure 13. Isocon diagram comparing the mean composition of least-altered samples and altered samples from (**a**) alteration zone-1 (sericite–quartz–chlorite ± kaolinite ± pyrite) and (**b**) alteration zone-2 (calcite–epidote–albite ± chlorite ± sericite).

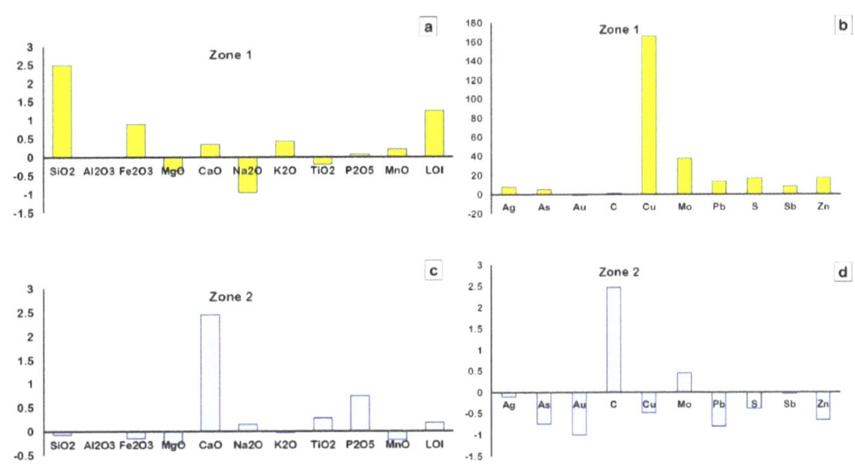

Figure 14. Gain/loss of major oxides (wt. %) (**a**,**c**) and trace elements (ppm) (**b**,**d**) in the alteration zones during hydrothermal alteration based on the mean data of the representative least−altered samples as a reference for calculations.

Table 3. Element/oxide mass changes in relation to the original whole-rock mass ((Mfi-Moi)/Mo) and in relation to the original element/oxide mass in the original rock ((Mfi-Moi)/Moi).

	Zone 1		**Zone 2**	
	(Mfi-Moi)/Mo	**(Mfi-Moi)/Moi**	**(Mfi-Moi)/Mo**	**(Mfi-Moi)/Moi**
SiO_2	152.3	2.5	−4.1	−0.1
Al_2O_3	0.0	0.0	0.0	0.0
Fe_2O_3	6.7	0.9	−1.1	−0.1
MgO	−1.8	−0.4	−1.3	−0.3
CaO	0.4	0.4	2.5	2.5
Na_2O	−2.0	−0.9	0.3	0.1
K_2O	1.5	0.4	−0.1	0.0
TiO_2	−0.1	−0.2	0.1	0.3
P_2O_5	0.0	0.1	0.1	0.7
MnO	0.0	0.2	0.0	−0.2
LOI	4.3	1.3	0.6	0.2
Ag	51.0	7.8	−0.7	−0.1
As	538.8	5.2	−77.2	−0.7
Au	−44.3	−1.0	−44.4	−1.0
Ba	532.3	0.9	117.4	0.2
Be	−0.1	0.0	0.2	0.1
C	1192.7	0.9	3139.3	2.5
Cd	27.3	34.2	−0.1	−0.2
Co	97.2	3.8	−2.1	−0.1
Cs	1.7	0.4	2.8	0.6
Cu	8702.9	165.5	−25.3	−0.5
Ga	8.1	0.2	2.1	0.1
Hf	−0.2	−0.1	0.2	0.2
Li	68.6	0.9	−17.7	−0.2
Mo	74.9	37.4	0.9	0.5
Nb	−186.5	−0.9	−101.0	−0.5
Ni	−6.9	−0.2	−6.8	−0.2
Pb	3144.6	13.5	−186.1	−0.8
S	36,833.0	16.8	−842.1	−0.4
Sb	19.3	8.4	−0.1	0.0
Se	18.9	9.0	0.6	0.3
Sn	0.4	0.2	0.6	0.2
Sr	−7.1	0.0	94.4	0.6
V	8.9	1.2	2.1	0.3
Zn	4815.7	17.4	−181.1	−0.7
Zr	−15.5	−0.2	76.2	0.8
Sc	258.7	1.8	−15.8	−0.1
Y	−2.8	−0.2	2.0	0.2
La	−4.3	−0.3	5.5	0.3
Ce	−15.3	−0.4	4.8	0.1
Pr	−0.9	−0.3	1.3	0.3
Nd	−3.9	−0.3	4.4	0.3
Sm	−0.4	−0.1	0.8	0.2
Eu	0.0	0.0	0.1	0.1
Gd	−1.1	−0.3	0.7	0.2
Tb	−0.1	−0.2	0.1	0.3
Dy	−0.6	−0.3	0.4	0.2
Ho	−0.1	−0.2	0.1	0.1
Er	−0.4	−0.3	0.2	0.1
Tm	0.0	−0.2	0.0	0.2
Yb	−0.3	−0.2	0.2	0.2
Lu	0.0	−0.2	0.0	0.2
Th	−2.2	−0.3	3.2	0.4
Mass change	**MC= 170.4**		**MC = -3.2**	
Volume change	**VC = 182.1**		**VC = -1.8**	

6. Sulfur Isotope ($\delta^{34}S$)

$\delta^{34}S$ isotopic data from the sulfide-bearing ore deposits were obtained to determine the source of the sulfur and the origin of the sulfur-bearing fluids [51]. The $\delta^{34}S$ isotope values of ten pyrite, chalcopyrite, and galena samples collected from the highly altered and mineralized altered metasediments host rocks are in the range of −1.1 to −0.1‰V_{CDT} (n = 3), −2.7 to −0.5 ‰V_{CDT} (n = 3), and −3.5 to −2.1‰V_{CDT} (n = 4), respectively (Table 4). Pyrites from a quartz vein have an average $\delta^{34}S$ of 0.4‰VCDT (Table 4 and Figure 15a). By assuming the H_2S as the sulfur species in solution, and based on the fractionation equations of Czamanske and Rye [52] and Ohmoto and Rye [51], the $\delta^{34}S_{H2S}$ values of the fluid have a narrow range of −2.54 to −0.08 ‰ V_{CDT} (Table 4 and Figure 15b).

Table 4. Sulfur isotope values of sulfides from the Halılar area.

Sample ID	Mineral	Host	Measured $\delta^{34}S$ (‰)	at T°C [(1)]	$\delta^{34}S_{H2S}$ Fluid (‰) [(2)]	Error range	
H25	pyrite	Mineralized altered rock	−0.1	285.86	−1.4	±	0.3
H2	pyrite	Qz-veined altered rock	−1.1	285.86	−0.9	±	0.3
H1-a	pyrite	Qz-veined altered rock	−0.7	285.86	−2.0	±	0.3
H11	pyrite	Qz-vein	0.4	285.86	−0.9	±	0.3
H25	chalcopyrite	Mineralized altered rock	−0.5	285.86	−0.3	±	0.3
H2	chalcopyrite	Qz-veined altered rock	−2.7	285.86	−2.5	±	0.3
H40	chalcopyrite	Mineralized altered rock	−1.1	285.86	−0.9	±	0.3
H1-a	galena	Qz-veined altered rock	−3	285.86	−1.0	±	0.2
H2	galena	Qz-veined altered rock	−3.4	285.86	−1.4	±	0.2
H25	galena	Mineralized altered rock	−2.1	285.86	−0.1	±	0.2
H40	galena	Mineralized altered rock	−3.5	285.86	−1.5	±	0.2

[(1)] Based on the fluid inclusion thermometry (unpublished data), [(2)] Calculated by using the sulfur isotope fractionation equations in Czamanske and Rye [52] and Ohmoto and Rye [51].

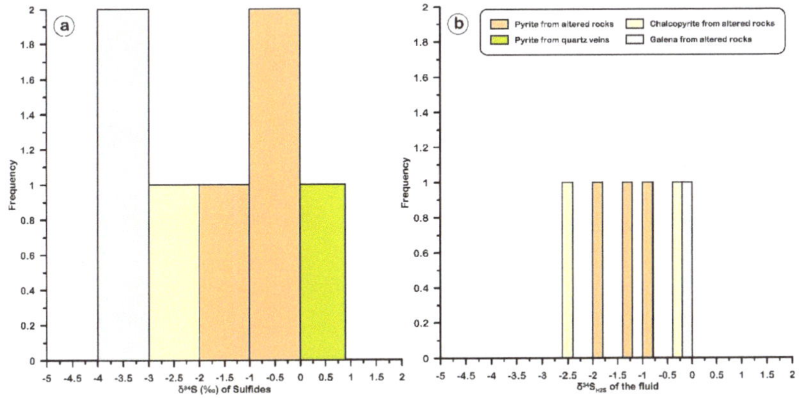

Figure 15. Histograms of (**a**) the $\delta^{34}S$ isotopic compositions for sulfide minerals (pyrite, chalcopyrite, and galena) and (**b**) the $\delta^{34}S_{H2S}$ of the fluid that formed the sulfides in the Halılar area.

7. Discussion

7.1. Sources of Sulfur

There are many sulfur sources with distinct $\delta^{34}S$ values: (1) the mantle source has a 0 ± 3‰ $\delta^{34}S$ value [53]; (2) the magmatic source, in which the sulfur resulted from desulfidation and/or dissolution or from magmatic sulfides, has 0 to +9‰ $\delta^{34}S$ [54]; (3) the seawater sources have a mean value of +20 ‰ $\delta^{34}S$; and (4) the strongly reduced sulfur source in the sedimentary rocks has very negative $\delta^{34}S$ values [55].

In the Halılar area, the mean $\delta^{34}S$ value of the sulfides is close to −1.62‰, suggesting a uniform magmatic sulfur source in which the sulfur originates either from the leaching and remobilization of the old magmatic sulfide or from the mantle source (Figure 16). Furthermore, the $\delta^{34}S$ values of the studied sulfide minerals decrease from pyrite (−1.1 to 0.4 ‰V_{CDT}) and chalcopyrite (−2.7 to −0.5 ‰V_{CDT}) to galena (−3.5 to −2.1‰V_{CDT}) (Figure 17), which is compatible with the suggested trend of differentiation of Ohmoto and Rye [51]. Thus, the ore-bearing fluid appears to have a magmatic (mantle) source [51] with magmatic–hydrothermal signatures [56] (Figure 18).

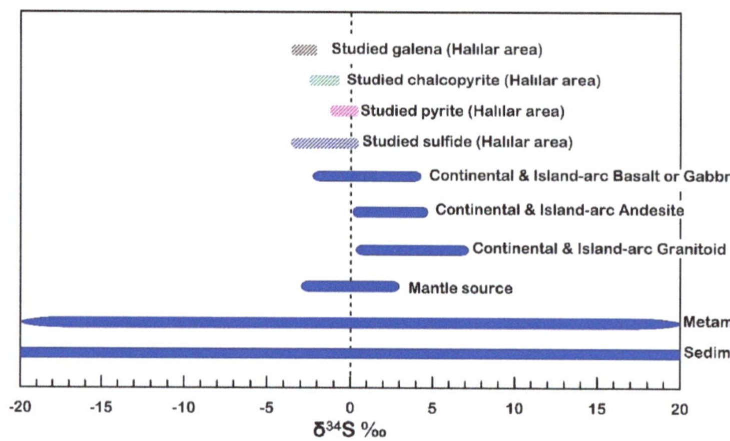

Figure 16. $\delta^{34}S$ values of sulfides from the mantle, the continental igneous setting, and metamorphic and sedimentary sources. Mantle source [53], island arc basalts and gabbros [57,58], andesites [59,60], granitoids [61–63], and metamorphic and sedimentary sources [64].

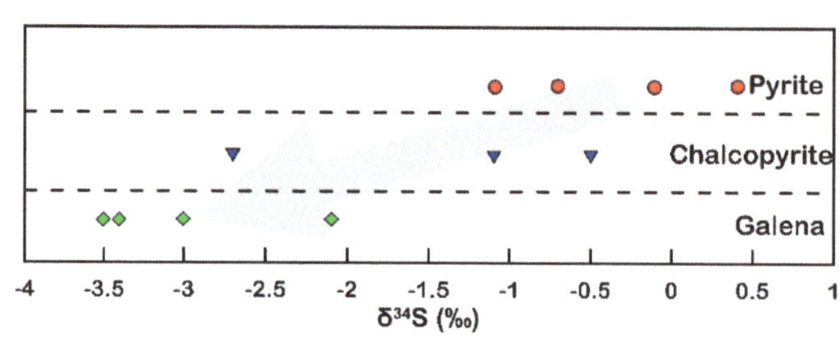

Figure 17. Distribution of the $\delta^{34}S$ values in the studied sulfide minerals from the Halılar area.

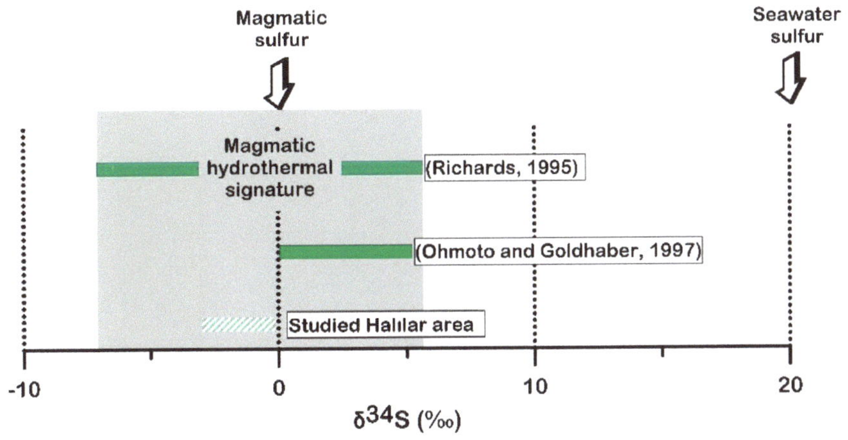

Figure 18. $\delta^{34}S$ values from the Halılar area compared with the magmatic–hydrothermal deposits [51,56].

7.2. Metal Source

The metasediments of the Sakarkaya Formation that host the Halılar Cu-Pb (±Zn) mineralization are slightly enriched in metallic elements (average of Ag = 6.7 ppm, As = 101.9 ppm, Au = 0.04 ppm, Cu = 53.8 ppm, Mo = 1.8 ppm, Pb = 274.7 ppm, S = 389.0 ppm, Sb = 2.0 ppm, and Zn = 371.3 ppm) relative to the average UCC (Table 5). Moreover, the contents of the metallic elements in the Düztarla granitoid rocks also show higher values than typical UCC (mean values of Ag = 1.14 ppm, As = 84.05 ppm, Au = 0.36 ppm, Cu = 368.91 ppm, Mo = 324.68 ppm, Pb = 49.52 ppm, S = 1396.7 ppm, Sb = 2.34 ppm, and Zn = 414.69 ppm).

Table 5. Upper continental crust (UCC)–normalized elements of metasediments in Sakarkaya Formation and Düztarla granitoid rocks.

		Ag	As	Au	Cu	Mo	Pb	S	Sb	Zn
UCC		50 *	1.5 *	1.8 *	25 *	1.5 *	17 *	621 **	0.2 *	71 *
Metasediments normalized to UCC (ppm)	H-14	0.0	22.3	0.0	0.7	0.7	2.0	0.6	9.8	2.8
	H-15	0.2	16.4	0.0	2.6	1.3	6.0	0.3	6.4	1.0
	H-18	0.2	9.9	0.0	0.8	0.0	3.2	0.2	5.4	3.8
	H-20	0.0	14.0	0.0	1.4	0.0	3.6	1.5	10.7	3.6
	H-22a	0.1	282.9	0.0	5.0	2.0	53.6	1.4	26.3	8.2
	H05	0.1	13.2	0.0	1.6	1.7	13.7	0.4	7.6	5.9
	H07	0.2	12.0	0.0	1.7	1.7	16.0	0.3	8.6	7.1
	H09	0.1	148.4	0.0	1.1	1.7	18.0	0.9	8.1	6.5
	H55	0.1	148.0	0.0	3.2	1.8	21.0	0.4	9.2	6.8
	H60	0.1	12.6	0.0	3.3	1.8	24.5	0.3	8.1	6.6
Granitoid normalized to UCC		0.0	56.0	0.2	14.8	216.5	2.9	2.2	11.7	5.8

* UCC after Taylor and McLennan [37]; ** UCC after Rudnick et al. [65].

When the sulfur is normalized to the UCC of Rudnick et al. [65], it is highly rich in granitoid rocks, but not in metasediments (Figure 19a,b). Thus, the primary metal suppliers appear to be the metasediments and intrusive Düztarla granitoid magmatism; together, they account for the metals in the Halılar brecciated-stockwork-type mineralization. Based on the geologic features and mode of occurrences, the Halılar metasediment-hosted Cu-Pb (±Zn) mineralization appears to be formed by epigenetic hydrothermal processes after sedimentation/diagenesis and metamorphism.

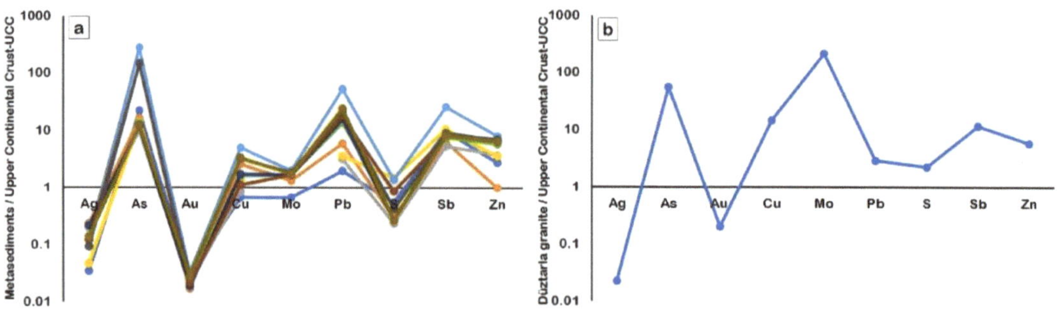

Figure 19. Upper continental crust (UCC)–normalized elements diagram of the metasediments (**a**) and Düztarla granitoid (**b**). After Taylor and McLennan [37] and Rudnick et al. [65].

8. Conclusions

The Halılar area contains two groups: the clastic Halılar Group that overlies the metamorphics of the pre-Late Triassic age or Permian limestones and the Bilecik Group. The Halılar Group consists of the Bağcağız and Sakarkaya Formations, and the Bilecik Group is represented by two formations, including the Taşçıbayırı Formation and the Günören Limestone. The Sakarkaya and Bağcağız Formations were later intruded by Oligo–Miocene Düztarla granitoid rocks.

The Halılar base metal mineralization consists mainly of Cu-Pb sulfide with some Zn sulfide in the brecciated stockworks and veins. This type of vein mineralization is restricted to a fault gouge zone directed NE–SW and along the lower contact of the Sakarkaya and Düztarla granitic rocks. Two types of hydrothermal alteration zones with gradual boundaries can be observed in the main ore zone. These include zone-1 (sericite–quartz–chlorite ± kaolinite ± pyrite) and zone-2 (calcite–epidote–albite ± chlorite ± sericite). The main ore mineral assemblage consists of chalcopyrite, galena, pyrite, and sphalerite in an abundant amount of gangue minerals such as quartz, sericite, chlorite, and calcite forming along the quartz stockwork veins, as well as in the brecciated ore zones. The other oxidation and supergene mineralization includes covellite and goethite formed after chalcopyrite and pyrite, respectively.

The least-altered Sakarkaya metasediments are classified mainly as wackes and, rarely, Fe-sand and Fe-shale, which are relatively similar in chemical composition to the upper continental crust (UCC). They are sourced from the crustal felsic rocks and a quartzose sedimentary provenance formed within the passive and active continental margins. Mass-balance calculations reveal that the samples of zone-1 are enriched in SiO_2, Fe_2O_3, K_2O, and LOI, with Ag, As, Cu, Mo, Pb, S, Sb, and Zn reflecting a high degree of pyritization with sericitization and silicification. On the other hand, the samples of zone-2 show an increase in CaO; Na_2O; P_2O_5; TiO_2; LOI; and carbon-reflecting calcite, epidote, and albite alterations.

The mean $\delta^{34}S$ value of the sulfides in the Halılar area is close to −1.62‰, suggesting a uniform magmatic sulfur source in which the sulfur originates either from leaching and remobilization from the old magmatic sulfide or from the mantle source. There is also a sulfur isotope having a differentiation trend from pyrite to galena. The ore-bearing fluid has $\delta^{34}S$ values of H_2S, ranging from −2.54 to −0.08 ‰, typical of a magmatic–hydrothermal signature [47].

Based on the normalization of the metallic elements in the Sakarkaya metasediments and Düztarla granitoid rocks to the UCC [38] and [65], these metasediments and granitoid rocks represent the primary source of metals forming the Halılar brecciated-stockwork-veining-type mineralization. Overall, the geologic features and the mode of occurrences of the Halılar metasediment-hosted Cu-Pb (±Zn) mineralization suggest that they were formed by epigenetic hydrothermal processes after sedimentation/diagenesis and metamorphism.

Funding: This work was supported by the BAP Project (No. 37872) of Istanbul Technical University (ITU, Turkey).

Data Availability Statement: The data presented in this study are available in this article and its Appendices A–C.

Acknowledgments: The author would like to acknowledge the assistance of members of the Geochemistry Research Laboratories at Istanbul Technical University (Turkey). Great appreciation goes to M. Kumral (ITU, Turkey) and A. Abdelnasser (Benha University) for their support during all stages of work on the article. Z. Doner (ITU, Turkey) and A. Unal (ITU, Turkey) are also thanked for their help during fieldwork. The help of F. Yavuz (ITU, Turkey) and G. Ustunisik (South Dakota School of Mines, USA) were highly appreciated during the review of the article. The editor and two anonymous reviewers are thanked for carefully reading the manuscript and for their constructive comments.

Conflicts of Interest: The author declares no conflict of interest.

Appendix A

Table A1. XRD analyses of representative samples from the ore zone.

No.	Ref. Code	Compound Name	Chemical Formula	SemiQuant [%]				Average
				H1a	H2	H40	H25	
1	98-005-7052	Quartz low	O2 Si1	35	16	13	51	28.75
2	98-004-6240	Wollastonite 1A, manganoan	Ca2.88 Mn0.12 O9 Si3	8	11	25	19	15.75
3	98-001-7142	Kaolinite 1A	H4 Al2 O9 Si2	10	13	16		9.75
4	98-004-5719	Microcline	Al1 K1 O8 Si3	18	9	1		7.00
5	98-000-5164	Chalcopyrite	Cu1 Fe1 S2	6	6	13	1	6.50
6	98-000-5343	Calcite	C1 Ca1 O3	3	17	5		6.25
7	98-002-8161	Andradite	Ca3 Fe2 O12 Si3	5	5	7		4.25
8	98-004-5343	Anglesite	O4 Pb1 S1	4	4	5	3	4.00
9	98-000-5961	Pyrite	Fe1 S2	4	4	2	6	4.00
10	98-002-1539	Cubanite high	Cu0.3333 Fe0.6667 S1	1	4	7		3.00
11	98-004-9157	Sphalerite ferrous	Fe0.215 S1 Zn0.785	1	2	1	4	2.00
12	98-008-5367	Albite	Al1 Na1 O8 Si3	1	5	1		1.75
13	98-007-8263	Biotite 1M	H1.47 Al1.92 F1.98 Fe2.59 K2 Mg3.15 Mn0.09 O21.47 Si5.98 Ti0.27				6	1.50
14	98-007-9292	Galena	Pb1 S1	1	1	2	1	1.25
15	98-004-9486	Quartz high	O2 Si1	2	1	1		1.00
16	98-002-2578	Halite	Br0.8947 Cl0.1053 Na1	1	2	1		1.00
17	98-001-7351	Goethite	H1 Fe1 O2				4	1.00
18	98-006-2562	Chalcocite high	Cu2 S1				3	0.75
19	98-005-2707	Clinochlore IIb-2	H2 Al2 Mg5 O15 Si3				2	0.50
20	98-002-1503	Barite high	Ba1 O4 S1				1	0.25

Appendix B

Table A2. XRD analyses of representative samples from alteration zone-1 (sericite–quartz–chlorite ± kaolinite ± pyrite).

No.	Ref. Code	Compound Name	Chemical Formula	SemiQuant [%]																Average
				H1b	H3.1	H3.2	H5	H6	H7	H8	H9	H11	H12	H13	H21	H26	H28	H31	H33	
1	98-002-8439	Quartz low	O2 Si1	48	66	83	51	49	56	53	84	64	65	15	58	38	84	63	46	57.69
2	98-001-1941	Muscovite 2M1	H2 Al3 K1 O12 Si3				40	25		32		12	4	2	10	31			25	11.31
3	98-001-2798	Kaolinite 1A	H4 Al2 O9 Si2	9	18	4		15	15	4	4	4	14		4				7	6.13
4	98-001-7363	Microcline	Al1 K1 O8 Si3		6	5	2	6			6	5	4	17	7	10	6		5	4.94
5	98-000-5322	Clinochlore 1MIa	H8 Al3.3 Fe1.65 Mg2.5 O18 Si2.2	4	4	2	2			5		4	4	4	4	5	3	14	14	4.31
6	98-005-2472	Sericite 2M1	H2 Al2.75 Ca0.011 Fe0.032 K0.727 Mg0.022 Na0.17 O12 Si3.128 Ti0.02	36										22	10	1				4.31
7	98-001-7679	Orthoclase	Al1 K0.94 Na0.06 O8 Si3				4		14				3	16		1				2.38
8	98-003-7636	Chamosite 1MIIb	H16 Al5.024 Fe4.964 Mg5.036 O36 Si5.7							4			4	7		5	4	7		1.94
9	98-003-4846	Albite	Al1 Ge2 Na1 O8 Si1			1		3		1				8	3	1				1.06
10	98-002-3975	Calcite	C1 Ca1 O3			2			1		2	6		1	2	1	1			1.00
11	98-002-3340	Pyrite	Fe1 S2	1	1	1		1	2		1	2	1	1	1	1		1	1	0.94
12	98-007-8271	Biotite 1M	H1.47 Al1.92 F1.98 Fe2.59 K2 Mg3.15 Mn0.09 O21.47 Si5.98 Ti0.27						7					3		1		4		0.94
13	98-000-5164	Chalcopyrite	Cu1 Fe1 S2		1							2	1	1	1	1				0.44
14	98-006-1164	Andradite	Ca3 Fe2 O12 Si3															6		0.38
15	98-008-5911	Dolomite	C2 Ca1 Mg1 O6						2		1			1		2				0.38
16	98-004-5305	Chalcophyllite	H48 Al1 As2 Cu9 O44 S1.5	2													1	1	1	0.31
17	98-004-1521	Magnetite	Fe3 O4					1	1					1		1		1		0.31
18	98-005-1007	Sphalerite	Fe0.372 S1 Zn0.628						1		1							1	1	0.25
19	98-001-7247	Melanterite	H14 Fe1 O11 S1		2					1										0.19
20	98-002-1503	Barite high	Ba1 O4 S1			1	1													0.13
21	98-001-7371	Galena	Pb1 S1		1															0.06
22	98-005-6956	Phengite 3T	H1.2 Al1.4 F0.8 Fe0.04 K1 Mg0.75 O11.2 Si3.81		1															0.06
23	98-002-8127	Ankerite	C2 Ca0.997 Fe0.676 Mg0.273 Mn0.054 O6						1											0.06
24	98-002-2570	Halite, bromian	Br0.1018 Cl0.8982 Na1															1		0.06

Appendix C

Table A3. XRD analyses of representative samples from alteration zone-2 (calcite–epidote–albite ± chlorite ± sericite).

No.	Ref. Code	Compound Name	Chemical Formula	SemiQuant [%]											Average
				H4	H10	H16	H17	H19	H23	H27	H29	H30	H32	H41	
1	98-004-6436	Albite low	Al1.005 Na0.986 O8 Si2.995	29	26	14	31	16	19	11	16	16	29	16	20.3
2	98-005-4829	Quartz low	O2 Si1	26	7	21	10	12	18	17	8	8	16	38	16.5
3	98-004-5914	Muscovite 2M1	H2 Al2.9 K1 O12 Si3.1	31	13	2	9	13	17	4	12	1	2	5	9.9
4	98-003-4690	Clinochlore IIb-4	H8 Al1.7 Fe0.33 Mg4.95 O18 Si3.02	12	10	14	12	11	7	5	17	14	5		9.7
5	98-001-7363	Microcline	Al1 K1 O8 Si3		10	9	8	11	9	7	7	7	16	12	8.7
6	98-005-2473	Sericite 2M1	Al2.724 Ca0.011 Fe0.032 K0.776 Mg0.022 Na0.181 O11 Si3.148 Ti0.02			8	6		9	28	8		22		7.4
7	98-001-8053	Anorthite	Al2 Ca1 O8 Si2					15		10	20	20		12	7.0
8	98-001-2962	Kaolinite 1A	H4 Al2 O9 Si2		6	15				5	12	14	8	3	5.7
9	98-005-7051	Illite 2M1	H3 Al4 K1 O12 Si2		20			18							3.5
10	98-004-5295	Chlorite IIb+4, chro-mian	H8 Al1.75 Cr0.25 Mg5 O18 Si3			14			18						2.9
11	98-006-0995	Calcite	C1 Ca1 O3			1	1		2		6	13	2		2.3
12	98-001-1876	Lepidolite 6M	H1 Al1 F1 K1 Mg3 O11 Si3							13				9	2.0
13	98-001-7534	Orthoclase	Al1 K1 O8 Si3		9		7								1.5
14	98-003-7636	Chamosite 1MIIb	H16 Al5.024 Fe4.964 Mg5.036 O36 Si5.7				9								0.8
15	98-002-2747	Titanite	Ca1 O5 Si1 Ti1											5	0.5
16	98-001-2031	Epidote	H1 Al2 Ca2 Fe1 O13 Si3				6								0.5
17	98-005-0411	Dolomite	C2 Ca1 Mg1 O6	1				1							0.2
18	98-001-2313	Phlogopite 3T	H1 Al1 F1 K1 Mg3 O11 Si3	2											0.2
19	98-000-6129	Bornite	Cu4.98 Fe1.02 S4		1										0.1
20	98-007-0040	Hematite	Fe2 O3					1							0.1
21	98-004-1521	Magnetite	Fe3 O4					1							0.1
22	98-003-8836	Phengite 3T	H2 Al1.848 K1 Mg0.58 O12 Si3.572						1						0.1
23	98-001-7325	Gypsum	H4 Ca1 O6 S1								1				0.1
24	98-002-8130	Ankerite	C2 Ca0.997 Fe0.676 Mg0.273 Mn0.054 O6											1	0.1

References

1. Rose, A.W.; Burt, D.M. Hydrothermal alteration. In *Geochemistry of Hydrothermal Ore Deposits*, 3rd ed.; Barnes, H.L., Ed.; John Wiley & Sons: New York, NY, USA, 1979; pp. 173–227.
2. Meyer, C.; Hemley, J.J. Wall rock alteration. *Geochem. Hydrothermal Ore Depos.* **1967**, *1*, 166–235.
3. Creasey, S.C. Some phase relations in the hydrothermally altered rocks of porphyry copper deposits. *Econ. Geol.* **1959**, *54*, 351–373. [CrossRef]
4. Lowell, J.D.; Guilbert, J.M. Lateral and Vertical Alteration-Mineralization Zoning in Porphyry Ore Deposits. *Econ. Geol.* **1970**, *65*, 373–408. [CrossRef]

5. Beane, R.; Bodnar, R. Hydrothermal fluids and hydrothermal alteration in porphyry copper deposits: Arizona Geological Society Digest. *Porphyry Copp. Depos. Am. Cordill.* **1995**, *20*, 83–93.
6. Sabins, F.F. Remote sensing for mineral exploration. *Ore Geol. Rev.* **1999**, *14*, 157–183. [CrossRef]
7. Watanabe, Y.; Hedenquist, J.W. Mineralogic and stable isotope zonation at the surface over the El Salvador porphyry copper deposit, Chile. *Econ. Geol.* **2001**, *96*, 1775–1797. [CrossRef]
8. Seedorff, E.; Dilles, J.H.; Proffett, J.M.; Einaudi, M.T.; Zurcher, L.; Stavast, W.J.; Johnson, D.A.; Barton, M.D. Porphyry deposits: Characteristics and origin of hypogene features. *Econ. Geol.* **2005**. [CrossRef]
9. Sillitoe, R.H. Porphyry copper systems. *Econ. Geol.* **2010**, *105*, 3–41. [CrossRef]
10. Tayor, H. The application of oxygen and hydrogen isotope studies to problems of hydrothermal alteration and ore deposit. *Econ. Geol.* **1974**, *69*, 843–883.
11. Gustafson, L.B.; Hunt, J.P. The porphyry copper deposit at El Salvador, Chile. *Econ. Geol.* **1975**, *70*, 857–912. [CrossRef]
12. Ohmoto, H. Stable isotope geochemistry of ore deposits. *Rev. Mineral. Geochem.* **1986**, *16*, 491–559.
13. Bowman, J.; Parry, W.; Kropp, W.; Kruer, S. Chemical and isotopic evolution of hydrothermal solutions at Bingham, Utah. *Econ. Geol.* **1987**, *82*, 395–428. [CrossRef]
14. Norman, D.; Parry, W.; Bowman, J.R. Petrology and geochemistry of propylitic alteration at Southwest Tintic, Utah. *Econ. Geol.* **1991**, *86*, 13–28. [CrossRef]
15. Dilles, J.H.; Solomon, G.C.; Taylor, H.P.; Einaudi, M.T. Oxygen and hydrogen isotope characteristics of hydrothermal alteration at the Ann-Mason porphyry copper deposit, Yerington, Nevada. *Econ. Geol.* **1992**, *87*, 44–63. [CrossRef]
16. Clark, A. Are outsize porphyry copper deposits either anatomically or environmentally distinctive? In *Giant Ore Deposits*; Society of Economic Geologists: Littleton, CO, USA, 1993.
17. Zaluski, G.; Nesbitt, B.; Muehlenbachs, K. Hydrothermal alteration and stable isotope systematics of the Babine porphyry Cu deposits, British Columbia; implications for fluid evolution of porphyry systems. *Econ. Geol.* **1994**, *89*, 1518–1541. [CrossRef]
18. Wilson, A.J.; Cooke, D.R.; Harper, B.J.; Deyell, C.L. Sulfur isotopic zonation in the Cadia district, southeastern Australia: Exploration significance and implications for the genesis of alkalic porphyry gold-Copper deposits. *Miner. Depos.* **2007**, *42*, 465–487. [CrossRef]
19. Urqueta, E.; Kyser, T.K.; Clark, A.H.; Stanley, C.R.; Oates, C.J. Lithogeochemistry of the Collahuasi porphyry Cu–Mo and epithermal Cu–Ag (–Au) cluster, northern Chile: Pearce element ratio vectors to ore. *Geochem. Explor. Environ. Anal.* **2009**, *9*, 9–17. [CrossRef]
20. Lentz, D.R. Exchange reactions in hydrothermally altered rocks: Examples from biotite-bearing assemblages. Alteration and alteration processes associated with ore-forming systems. *Geol. Assoc. Can. Short Course Notes* **1994**, *11*, 69–99.
21. Reed, M.H. Hydrothermal alteration and its relationship to ore fluid composition. *Geochem. Hydrothermal Ore Depos.* **1997**, *3*, 303–365.
22. Schwartz, G.M. Hydrothermal alteration. *Econ. Geol.* **1959**, *54*, 161–183. [CrossRef]
23. Pirajno, F. *Hydrothermal Processes and Mineral Systems*, 1st ed.; Springer: Heidelberg, Germany, 2009; p. 1250.
24. Gifkins, C.; Herrmann, W.; Large, R.R. *Altered Volcanic Rocks: A Guide to Description and Interpretation*; Centre for Ore Deposit Research, University of Tasmania: Hobart, Australia, 2005.
25. Janković, S. The copper deposits and geotectonic setting of the Thethyan Eurasian metallogenic belt. *Miner. Depos.* **1977**, *12*, 37–47. [CrossRef]
26. Altiner, D.; Kocyigit, A.; Farinacci, A.; Nicosia, U.; Conti, M. Jurassic-Lower Cretaceous stratigraphy and paleogeographic evolution of the southern part of north-western Anatolia (Turkey). *Geol. Romana* **1991**, *27*, 13–80.
27. Yigit, O. Mineral deposits of Turkey in relation to Tethyan metallogeny: Implications for future mineral exploration. *Econ. Geol.* **2009**, *104*, 19–51. [CrossRef]
28. Krushensky, R.; Akcay, Y.; Karaege, E. *Geology of the Karalar-Yesiler Area, Northwest Anatolia, Turkey*; US Geological Survey: Reston, VA, USA, 1974; pp. 1258–2331.
29. Şengör, A.C.; Yilmaz, Y. Tethyan evolution of Turkey: A plate tectonic approach. *Tectonophysics* **1981**, *75*, 181–241. [CrossRef]
30. Şengör, A.; Yılmaz, Y.; Sungurlu, O. Tectonics of the Mediterranean Cimmerides: Nature and evolution of the western termination of Palaeo-Tethys. *Geol. Soc. Lond. Spec. Publ.* **1984**, *17*, 77–112. [CrossRef]
31. Carr, M. *Igpet 2007 for Windows XP or Vista*, Terra Softa Inc.: Newark, NJ, USA, 2007.
32. Coelho, J. GEOISO—A Windows™ program to calculate and plot mass balances and volume changes occurring in a wide variety of geologic processes. *Comput. Geosci.* **2006**, *32*, 1523–1528. [CrossRef]
33. Gresens, R.L. Composition-Volume relationships of metasomatism. *Chem. Geol.* **1967**, *2*, 47–65. [CrossRef]
34. Grant, J.A. The isocon diagram; a simple solution to Gresens' equation for metasomatic alteration. *Econ. Geol.* **1986**, *81*, 1976–1982. [CrossRef]
35. Grant, J.A. Isocon analysis: A brief review of the method and applications. *Phys. Chem. Earth Parts A/B/C* **2005**, *30*, 997–1004. [CrossRef]
36. Herron, M.M. Geochemical classification of terrigenous sands and shales from core or log data. *J. Sediment. Res.* **1988**, *58*, 820–829.
37. Taylor, S.R.; McLennan, S.M. *The Continental Crust: Its Composition and Evolution*; Blackwell Scientific Publications: Hoboken, NJ, USA, 1985.

38. Floyd, P.; Leveridge, B. Tectonic environment of the Devonian Gramscatho basin, south Cornwall: Framework mode and geochemical evidence from turbiditic sandstones. *J. Geol. Soc.* **1987**, *144*, 531–542. [CrossRef]
39. van de Kamp, P.C.; Leake, B.E.; Senior, A. The petrography and geochemistry of some Californian arkoses with application to identifying gneisses of metasedimentary origin. *J. Geol.* **1976**, *84*, 195–212. [CrossRef]
40. Condie, K.C.; Macke, J.E.; Reimer, T.O. Petrology and geochemistry of early Precambrian graywackes from the Fig Tree Group, South Africa. *Geol. Soc. Am. Bull.* **1970**, *81*, 2759–2776. [CrossRef]
41. Caby, R.; Dostal, J.; Dupuy, C. Upper Proterozoic volcanic graywackes from northwestern Hoggar (Algeria)—Geology and geochemistry. *Precambrian Res.* **1977**, *5*, 283–297. [CrossRef]
42. van de Kamp, P.C. Geochemistry and origin of metasediments in the Haliburton-Madoc area, southeastern Ontario. *Can. J. Earth Sci.* **1968**, *5*, 1337–1372. [CrossRef]
43. Roser, B.; Korsch, R. Determination of tectonic setting of sandstone-Mudstone suites using content and ratio. *J. Geol.* **1986**, *94*, 635–650. [CrossRef]
44. Bhatia, M.R. Plate tectonics and geochemical composition of sandstones. *J. Geol.* **1983**, *91*, 611–627. [CrossRef]
45. Ishikawa, Y.; Sawaguchi, T.; Iwaya, S.; Horiuchi, M. Delineation of prospecting targets for Kuroko deposits based on modes of volcanism of underlying dacite and alteration halos. *Min. Geol.* **1976**, *26*, 105–117.
46. Williams, N.C.; Davidson, G.J. Possible submarine advanced argillic alteration at the Basin Lake prospect, Western Tasmania, Australia. *Econ. Geol.* **2004**, *99*, 987–1002. [CrossRef]
47. Large, R.R.; Gemmell, J.B.; Paulick, H.; Huston, D.L. The alteration box plot: A simple approach to understanding the relationship between alteration mineralogy and lithogeochemistry associated with volcanic-hosted massive sulfide deposits. *Econ. Geol.* **2001**, *96*, 957–971. [CrossRef]
48. Sánchez-España, J.; Velasco, F.; Yusta, I. Hydrothermal alteration of felsic volcanic rocks associated with massive sulphide deposition in the northern Iberian Pyrite Belt (SW Spain). *Appl. Geochem.* **2000**, *15*, 1265–1290. [CrossRef]
49. Kumral, M.; Abdelnasser, A.; Budakoglu, M. Geochemistry of Hydrothermal Alteration Associated with Cenozoic Intrusion-Hosted Cu-Pb-Zn Mineralization at Tavşanlı Area, Kütahya, NW Turkey. *Minerals* **2016**, *6*, 13. [CrossRef]
50. Large, R.R. Australian volcanic-Hosted massive sulfide deposits; features, styles, and genetic models. *Econ. Geol.* **1992**, *87*, 471–510. [CrossRef]
51. Ohmoto, H.; Rye, R. Isotopes of sulfur and carbon. *Geochem. Hydrothermal Ore Depos.* **1979**, 509–567.
52. Czamanske, G.K.; Rye, R.O. Experimentally determined sulfur isotope fractionations between sphalerite and galena in the temperature range 600 degrees to 275 degrees C. *Econ. Geol.* **1974**, *69*, 17–25. [CrossRef]
53. Chaussidon, M.; Lorand, J.-P. Sulphur isotope composition of orogenic spinel lherzolite massifs from Ariege (North-Eastern Pyrenees, France): An ion microprobe study. *Geochim. Et Cosmochim. Acta* **1990**, *54*, 2835–2846. [CrossRef]
54. McCuaig, T.C.; Kerrich, R. P—T—t—deformation—fluid characteristics of lode gold deposits: Evidence from alteration systematics. *Ore Geol. Rev.* **1998**, *12*, 381–453. [CrossRef]
55. Rollinson, H. *Using Geochemical Data*; Cambridge University Press: Cambridge, UK, 1993; p. 352.
56. Richards, J.P. Alkalic-Type epithermal gold deposits—A review. In *Magmas, Fluids, and Ore Deposits, Mineralogical Association of Canada Short Course*; Mineralogical Association of Canada: Quebec, QC, Canada, 1995; Volume 23, pp. 367–400.
57. Ueda, A.; Sakai, H. Sulfur isotope study of Quaternary volcanic rocks from the Japanese Islands Arc. *Geochim. Et Cosmochim. Acta* **1984**, *48*, 1837–1848. [CrossRef]
58. Chaussidon, M.; Albarede, F.; Sheppard, M. Sulphur isotope heterogeneity in the mantle from ion microprobe measurements of sulphide inclusions in diamonds. *Nature* **1987**, *330*, 242–244. [CrossRef]
59. Rye, R.; Luhr, J.; Wasserman, M. Sulfur and oxygen isotopic systematics of the 1982 eruptions of El Chichón Volcano, Chiapas, Mexico. *J. Volcanol. Geotherm. Res.* **1984**, *23*, 109–123. [CrossRef]
60. Luhr, J.F.; Logan, M.A.V. Sulfur isotope systematics of the 1982 El Chichón trachyandesite: An ion microprobe study. *Geochim. Et Cosmochim. Acta* **2002**, *66*, 3303–3316. [CrossRef]
61. Sasaki, A.; Ishihara, S. Sulfur isotopic composition of the magnetite-series and ilmenite-series granitoids in Japan. *Contrib. Mineral. Petrol.* **1979**, *68*, 107–115. [CrossRef]
62. Ishihara, S.; Sasaki, A. Sulfur isotopic ratios of the magnetite-series and ilmenite-series granitoids of the Sierra Nevada batholith—a reconnaissance study. *Geology* **1989**, *17*, 788–791. [CrossRef]
63. Santosh, M.; Masuda, H. Reconnaissance oxygen and sulfur isotopic mapping of Pan-African alkali granites and syenites in the southern Indian Shield. *Geochem. J.* **1991**, *25*, 173–185. [CrossRef]
64. Hoefs, J. Isotope fractionation processes of selected elements. In *Stable Isotope Geochemistry*; Springer: Berlin/Heidelberg, Germany, 2015; pp. 47–190.
65. Rudnick, R.; Gao, S.; Holland, H.; Turekian, K. Composition of the continental crust. *Crust* **2003**, *3*, 1–64.

minerals

MDPI

Article

Mineralogical and Fluid Inclusion Evidence for Reworking of Au Mineralization by Ag-Sb-Base Metal-Rich Fluids from the Bytíz Deposit, Příbram Uranium and Base-Metal Ore District, Czech Republic

Jiří Sejkora [1,*], Zdeněk Dolníček [1], Jiří Zachariáš [2], Jana Ulmanová [1], Vladimír Šrein [3] and Pavel Škácha [1,4]

1 Department of Mineralogy and Petrology, National Museum, Cirkusová 1740, 193 00 Prague, Czech Republic
2 Institute of Geochemistry, Mineralogy and Mineral Resources, Faculty of Science, Charles University, Albertov 6, 128 43 Prague, Czech Republic
3 Vladimír Šrein, Mříčná 154, 512 04 Mříčná, Czech Republic
4 Mining muzeum Příbram, Hynka Kličky Place 293, 261 01 Příbram VI, Czech Republic
* Correspondence: jiri.sejkora@nm.cz

Abstract: This mineralogical and fluid inclusion study was conducted on an Au-bearing quartz–sulfide vein encountered in the deep parts of the Bytíz deposit in the Příbram uranium and base-metal district, Bohemian Massif, Czech Republic. The samples were taken where the Au-bearing vein is crosscut by the common base-metal Zn-Pb ore vein Bt23C. The early mineralization of the Au-bearing vein is composed mainly of quartz (Q-1 to Q-3), illite–muscovite, Fe-Mg chlorite, arsenopyrite, and Au-Ag alloys, showing a wide range of compositions (4–69 at. % Ag) and a decrease in Au/(Au + Ag) ratios during vein evolution. Younger hydrothermal processes led to the crystallization of nests and veinlets composed of late quartz (Q-4), carbonates (siderite, dolomite–ankerite and calcite), base-metal sulfides (galena, sphalerite, chalcopyrite, and tetrahedrite), a suite of Ag and Bi-tellurides, and acanthite. The input of Sb is manifested by the partial to complete replacement of some gold grains by aurostibite and an unnamed (Ag,Au)-Sb oxide with a composition close to $AuSbO_3$. The fluid inclusion study, combined with chlorite thermometry and arsenopyrite thermometry, showed that the early mineralization crystallized from progressively cooled (from 300 to 400 °C down to ca. 180 °C), diluted (1.2–7.0 wt. % NaCl eq.) aqueous solutions. The late portion of the mineralization formed from aqueous fluids with highly variable salinity (0.2–23.4 wt. % NaCl eq.) and homogenization temperatures decreasing from ca. 250 °C to < 50 °C, which compare well with the base-metal mineralization of the vein Bt23C and other base-metal veins of the Příbram ore area. Our study illustrates the nature and intensity of the processes of the reworking of the early gold mineralization mediated by the younger Ag,Sb-rich base-metal fluids, giving rise to Příbram's typical late-Variscan vein Zn-Pb mineralization.

Keywords: vein gold deposits; metasomatism; alteration; mobilization; Bi-tellurides; Zn-Pb mineralization; fluid inclusions; Příbram uranium and base-metal district

Citation: Sejkora, J.; Dolníček, Z.; Zachariáš, J.; Ulmanová, J.; Šrein, V.; Škácha, P. Mineralogical and Fluid Inclusion Evidence for Reworking of Au Mineralization by Ag-Sb-Base Metal-Rich Fluids from the Bytíz Deposit, Příbram Uranium and Base-Metal Ore District, Czech Republic. *Minerals* **2022**, *12*, 1539. https://doi.org/10.3390/min12121539

Academic Editor: Yuichi Morishita

Received: 31 October 2022
Accepted: 27 November 2022
Published: 29 November 2022

1. Introduction

The Příbram uranium and base-metal ore district is remarkable for numerous reasons: it is one of the most significant ore districts in the Czech Republic, with more than 2500 hydrothermal ore veins (1641 with uranium mineralization, 35 with Pb-Zn, and finally 19 with monometallic silver mineralizations), with an exposed vertical extent of more than 1800 m from 26 main shafts—forming altogether 23 km of vertical shafts, more than 2100 km of horizontal adits, and 300 km of chutes in an area of 57.6 km^2 [1]. The total production of 48,432 t of pure U metal represented 49% of the Czechoslovak production in the period from 1947 to 1991. The parallel mining of base metals and silver from veins produced more than 6100 tons of Pb, 2400 tons of Zn, and 28 of tons Ag [2]. These data make the

Příbram uranium district comparable with the most important ones on a worldwide scale. It is also marked by wide mineralogical diversity (more than are 300 species known so far—https://www.mindat.org/loc-779.html, accessed on 15 October 2022), reflecting the occurrence of various types of mineralization—from the earliest gold-bearing veins and the main uranium and base-metal veins, with locally rich selenide mineralization as well as Ag, Ag-Sb, Sb, or As bonanzas, to the post-ore zeolite mineralizations. These remarkable facts render detailed studies of this ore district scientifically relevant as they contribute to the understanding of the origin and subsequent overprint of a complex, polyphase, hydrothermal system.

The ore veins of the Příbram ore area originated during multiple mineralizing stages. A common feature is the repeated dissolution of some vein minerals and/or their replacement by younger hydrothermal mineral phases, even leading to the completely re-worked vein fabric as well as the mineral assemblage of the original vein. Kutina [3] described the formation of fine-grained "hard ores", dominated by a mosaic of quartz and sulfides, which originated by the stepwise replacement of the old siderite vein. He documented up to three cycles of replacement operating in the same portion of the vein. Widespread features of vein dissolution and replacement, especially in the case of the siderite, dolomite-ankerite, galena, and Pb-Sb sulfosalts, were also found during recent investigations of the Pb-Zn ore veins from the Příbram uranium district [4–6]. The reason for these phenomena was identified in the temporal but repeated activity of corrosive fluids [5].

It is not known whether the mentioned hypogene corrosive processes took place only in the base-metal veins of the Příbram ore area or whether they could also influence the other types of older hydrothermal mineralizations occurring in this area. In this paper, we present a detailed description of the ore textures and paragenetic sequence of the unique occurrence of Au(-Bi) and Ag-Sb mineralizations hosted in the same vein structure of the Bytíz deposit and discuss the observed mineralogical variations as well as the redistribution processes, using microscopy, electron microprobe, and fluid inclusion analyses.

2. Geological Setting

Two main ore districts are known in the Příbram ore area (central Bohemia, Czech Republic), including the Březové Hory base-metal district and the Příbram uranium and base-metal district. The Březové Hory base-metal district, located in the western part of the area, was known from at least the 13th century, but the main mining activity took place from 1810 to 1978. A total of 22 million tons of extracted ore contained 3837 tons of Ag, 517,961 tons of Pb, and 70,300 tons of Sb [1]. The complex Příbram uranium and base-metal district mined between 1947 and 1991 represents the largest accumulation of vein-type hydrothermal U ores in the Czech Republic (see above) and is comparable to world-class deposits of this type.

Both of the main ore districts are tied to the southeastern contact of the Barrandian sedimentary rocks of the Neoproterozoic and Lower Cambrian age with the western margin of the granitoids of the Central Bohemian Plutonic Complex. The Příbram uranium district is located tightly in the exocontact of the Central Bohemian Plutonic Complex in the metasedimentary rocks of the Teplá-Barrandian Unit. The uranium hydrothermal veins are located in a 1–2 km wide and 25 km long zone of tectonically deformed Neoproterozoic rocks (Figure 1), including the Příbram antiform [2]. A predominant part of the uranium mineralization (96%) occurred in veins located in the Dobříš Group of the Neoproterozoic age (Figure 2) with a predominance of claystones, siltstones, and sandstones. The underlying Davle Group, composed of tuffs and tuffites, was encountered by two structural boreholes and by the deep levels of the mines in the Bytíz area.

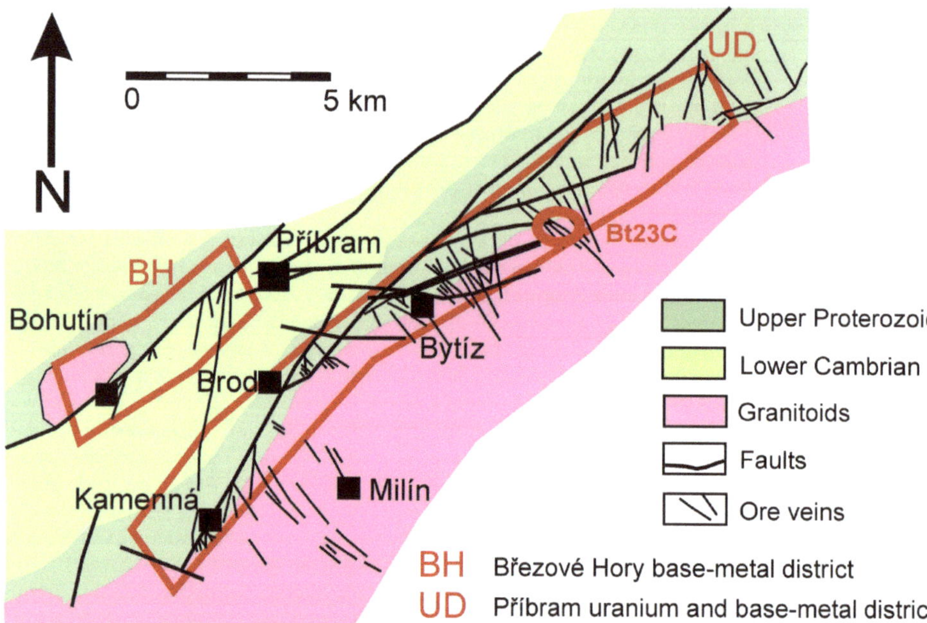

Figure 1. Geological position of the Příbram ore area and studied vein Bt23C, modified from Ettler et al. [1].

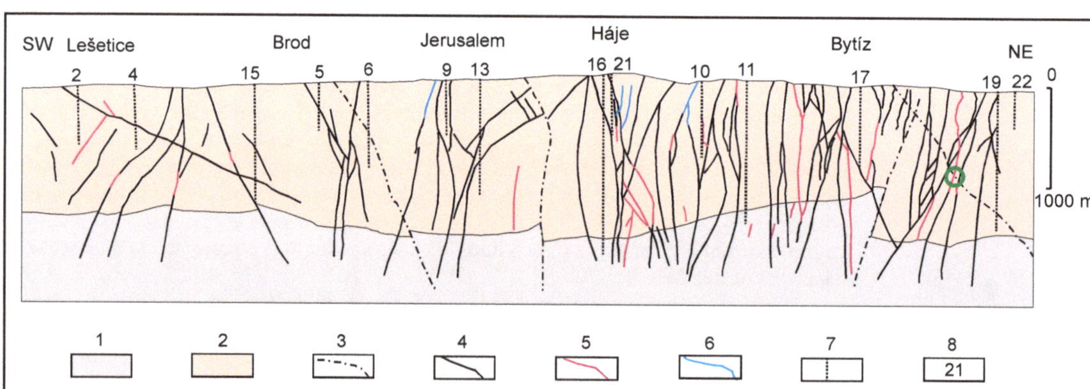

Figure 2. The SW–NE section of the Příbram uranium ore district (modified from Komínek [7]); localization of described ore stope is marked by green circle; 1—tuffs and tuffites of the Davle Group (Neoproterozoic); 2—claystones, siltstones, and sandstones of the Dobříš Group (Neoproterozoic); 3—faults; 4—hydrothermal veins with uranium mineralization; 5—hydrothermal veins with base-metal mineralization; 6—hydrothermal veins with monometallic Ag mineralization; 7—shafts; 8— shaft numbers.

The Lower Cambrian sediments, including the conglomerates overlying the Dobříš Group, were penetrated by numerous mines. This concerns the upper parts of the central and northeastern parts of the ore district and (at greater depth) the northern part of the vein Bt4. All the rock types are affected by the contact metamorphism caused by the granitoids of the Central Bohemian Plutonic Complex [7]. The granitoids are represented here by the so-called marginal type and the Blatná type granitoids. The mineralized veins wedge out when reaching the plutonic rocks.

The gold-bearing ore mineralization represents the earliest and the relatively widespread type of hydrothermal mineralization in this ore district. Its economic importance, however, was marginal to submarginal only. The wavy quartz to quartz-sulfide veins with a thickness in the range cm–dm form coulisse-like arranged systems of NW–SE to E–W directions with a variable dip (40°–75°) from SW to S or NE to N in the rock of the Barrandian Neoproterozoic and in the granitoids of the NW margin of the Central Bohemian Plutonic Complex. In the central part of this ore district, the gold-bearing mineralization was verified by mining works down to the depth of about 1400 m. Some ore veins are locally characterized by high metal content (about 100 ppm Au) and variable mineral associations containing Bi, Te, Ag, Sb, Pb, Cu, Fe, As, and S [8–10].

The most economically important uranium and base-metal mineralization originated during four main mineralization stages: (I) siderite–sulfidic, (II) calcite, (III) calcite–uraninite, and IV) calcite–sulfidic [7]. The oldest siderite–sulfidic stage (I) is developed on a smaller scale in comparison with that of the neighboring Březové Hory district. The younger calcite stages, characterized by notably lower formation temperatures, are more abundant; the calcite generations were used in separating the mineralization stages. For the calcite stage, (II) the pre-ore calcite DK and the calcite K1 are characteristic. In the calcite–uraninite stage (III), carrying the main part of the economic uranium mineralization (uraninite, coffinite, and U-bearing anthraxolite), are the present calcite types K2–K4. In the last calcite–sulfidic stage (IV), the post-ore calcite K5 appears. An extremely diverse (more than 25 mineral species) selenide mineralization appears locally around the boundary of stages III and IV [11]. The origin of the Ag, Ag-Sb, Sb, and As-Sb bonanza-type accumulations [12–14] is probably connected with the calcite–sulfidic stage IV. The age of the uranium mineralization, obtained by the U-Pb radiometric dating of two uraninite samples, is Permian, 275 ± 4 and 278 ± 4 Ma [15].

The youngest post-ore zeolite mineralization is bound to the independent calcite veins or occurs in drusy cavities of uranium-bearing veins. It is represented by occurrences of harmotome, laumontite, stilbite, natrolite, thomsonite, etc. [9,16], with local occurrences whewellite [17,18].

Our research concentrated on the Au-bearing quartz–sulfide vein of the U and Pb-Zn deposit of Bytíz (the vein cluster Bt17B-Bt22). This vein was found during the mining of the base-metal ores above the 20th level (at a depth of 900–950 m) of mine No. 19 (Bytíz). The NW–SE trending vein had a wavy character with a thickness ranging between 3 and 25 cm and was located in close proximity to the younger base-metal vein Bt23C. According to chip sampling at the length of 18 m, the Au-bearing vein may be characterized by variable contents of Au from 0.43 to 111 ppm (Figure 3). The vein filling is represented by grey-white to grey massive quartz with abundant arsenopyrite and younger tiny veins of carbonates and at some places with fine-grained gold (0.X mm) and grey aggregates of associated minerals with the size of about 1 mm.

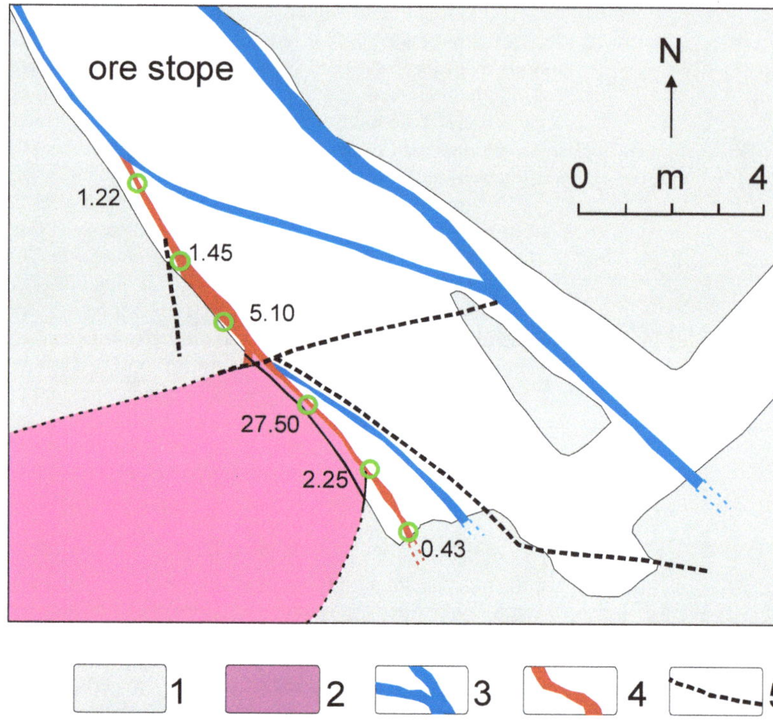

Figure 3. The geological situation of the described ore stope, modified according to the documentation of the Správa uranových ložisek, o.z., Příbram. 1—metamorphosed sediments of the Dobříš Group; 2—granodiorite porphyrite; 3—hydrothermal vein with base-metal mineralization; 4—hydrothermal vein with Au-bearing mineralization with determined Au contents (ppm); 5—small dislocations.

3. Analytical Methods

The studied samples were prepared for optical investigation and measurements of the chemical composition using standard diamond polishing techniques. The ore minerals and their relationships and textures were first studied in reflected light using a Nikon Eclipse ME600 polarizing microscope (National Museum, Prague, Czech Republic).

The chemical analyses of the Au, Ag, Pb, Zn, Sb, Cu, Bi, and Hg in the bulk ore were performed using atomic absorption spectroscopy (AAS) in the lab of GEMATEST Ltd. in Černošice.

The chemical analyses of minerals were performed using a Cameca SX-100 electron microprobe (National Museum, Prague, Czech Republic), operating in the wave-dispersive (WDS) mode. For most of the ***sulfide minerals, tellurides,*** and ***native metals,*** the following conditions, standards, and X-ray lines were used to minimize line overlaps: 25 kV, 20 nA, 0.7 μm wide beam; Ag (AgLα), apatite (CaKα, PKα), Au (AuMα), baryte (BaLα), Bi_2Se_3 (BiMβ), CdTe (CdLα), Co (CoKα), $CuFeS_2$ (CuKα, SKα), FeS_2 (FeKα), GaAs (GaLα), Ge (GeLα), HgTe (HgLα), InAs (InLα), Mn (MnKα), NaCl (ClKα), NiAs (AsLβ), Ni (NiKα), PbS (PbMα), PbSe (SeLβ), PbTe (TeLα), sanidine (KKα), Sb_2S_3 (SbLα), Sn (SnLα), and ZnS (ZnKα). Acanthite was measured at 4 nA and with a 5 μm wide beam. Part of the arsenopyrite analyses was taken with a special setup to increase the sensitivity to Au. These data were collected at 60 nA, with a 5 μm wide beam, and with a long counting time for the Au peak (120 s). With the exception of Au, whose detection limit was about 330 ppm, the contents of As, Co, Cu, Fe, Mn, Ni, Pb, S, and Sb were determined. For ***gangue minerals,*** 15 kV, 5–10 nA, and 2–5 μm wide beams were used. In feldspar, the contents of Al, Ba, Ca, Cs, Cu, F, Fe, Mg, Mn, K, Na, P, Pb, Rb, Si, Sr, and Zn; in TiO_2 mineral, Al, As, Ca, Cr, Fe,

Mg, Mn, Mo, Na, Nb, P, Pb, S, Sc, Si, Sn, Ta, Ti, U, V, W, Y, and Zr; in chlorite and white mica, Al Ba, Ca, Cl, Co, Cr, Cs, Cu, Fe, K, Mg, Mn, Na, Ni, P, Pb, Rb, Si, Ti, V, and Zn; in monazite, Al, As, Ba, Bi, Ca, Ce, Cl, Cu, Dy, Eu, Er, F, Fe, Gd, Hf, Ho, K, La, Lu, Mg, Mn, Na, Nd, P, Pb, Pr, S, Sc, Si, Sm, Sr, Tb, Th, Ti, Tm, U, V, Y, Yb, and Zr; and in carbonates, Ba, Ca, Cu, Fe, Mg, Mn, Na, P, Pb, S, Si, Sr, and Zn were measured, respectively. The following standards and lines were used: albite (Na*Kα*), almandine (Al*Kα*, Fe*Kα*), apatite (P*Kα*), baryte (Ba*Lα*), Bi (Bi*Mα*), celestite (S*Kα*, Sr*Lβ*), $CePO_4$ (Ce*Lα*), chalcopyrite (Cu*Kα*), clinoclase (As*Lα*), Cr_2O_3 (Cr*Kα*), $CrTa_2O_6$ (Ta*Lα*), Cs-glass (Cs*Lα*), diopside (Mg*Kα*), $DyPO_4$ (Dy*Lβ*), $ErPO_4$ (Er*Lα*), $EuPO_4$ (Eu*Lα*), $GdPO_4$ (Gd*Lα*), halite (Cl*Kα*), hematite (Fe*Kα*), Hf (Hf*Mα*), $HoPO_4$ (Ho*Lβ*), $LaPO_4$ (La*Lα*), LiF (F*Kα*), $LuPO_4$ (Lu*Lα*), Nb (Nb*Lα*), $NdPO_4$ (Nd*Lβ*), Ni (Ni*Kα*), $PrPO_4$ (Pr*Lβ*), Rb-Ge-glass (Rb*Lα*), rhodonite (Mn*Kα*), sanidine (K*Kα*, Si*Kα*, Al*Kα*), $ScVO_4$ (Sc*Kα*), scheelite (W*Lα*), $SmPO_4$ (Sm*Lα*), Sn (Sn*Lα*), stibnite (Sb*Lα*), $TbPO_4$ (Tb*Lα*), Th (Th*Mα*), TiO_2 (Ti*Kα*), $TmPO_4$ (Tm*Lα*), UO_2 (U*Mα*), V (V*Kα*), vanadinite (Pb*Mα*), wollastonite (Ca*Kα*, Si*Kα*), wulfenite (Mo*Lα*), $YbPO_4$ (Yb*Lα*), YVO_4 (Y*Lα*), zinkite (Zn*Kα*), and zircon (Zr*Lα*). The peak counting times were between 10 and 30 s; those of both backgrounds were half of the peak counting time. Raw intensities were converted to the concentrations of elements using the automatic "PAP" matrix-correction procedure [19]. The amount of oxygen in the oxygen-bearing minerals was calculated by stoichiometry. The contents of the above-listed elements, which are not included in the tables, were analyzed quantitatively, but their contents were below the detection limit (ca. 0.03–0.05 wt. % for most elements, around 0.2 wt. % for Ho-Lu). The collected data were corrected for the overlaps P vs. Ca, La vs. Dy, Eu vs. Dy, F vs. Ce, Bi vs. Ce, and Th vs. U.

The contents of REE were normalized on C1 chondrite with the data given by Anders and Grevesse [20]. The numerical values of the Ce and Eu anomalies were calculated using the equations given by McLennan [21].

The fluid inclusions in the quartz and carbonates were studied in standard doubly polished ~100 μm thick plates mounted by the Canada balsam. The fluid inclusion petrography was conducted using a Leica DMPL polarizing microscope (magnification up to 1000x) (Leica Microsystems, Wetzlar, Germany) and fluid inclusion microthermometry using a Linkam THMSG 600 heating–freezing stage (Linkam Scientific Instruments, Surrey, United Kingdom) mounted on an Olympus BX-50 microscope (Olympus Co., Tokyo, Japan) with 20× and 50× long-working-distance objectives. The stage was calibrated using synthetic and natural standards at −56.6 °C (CO_2), 0 °C (H_2O), +31.1 °C (CO_2), and +307 °C ($NaNO_3$). The precision of the measurements was ± 0.1 °C in the temperature span between 0 °C and +50 °C. Where possible, the microthermometric data were collected for homogeneous fluid inclusion associations only, such as individual growth zones, three-dimensional clusters, or well-defined fluid inclusion trails. The measured phase transitions included the freezing temperature (T_f), the temperature of the initial melting (T_i), the melting temperature of the last ice crystal (Tm-ice), and the homogenization temperature (Th). The degree of fill (F) is estimated as the L/(V+L) ratio at room temperature. The salinities were calculated as wt. % NaCl eq. using the equation of Bodnar [22]. Isochores were calculated in the ISOC software [23], with calibration by Knight and Bodnar [24].

4. Results

4.1. Bulk Chemical Composition of the Studied Ore

The samples for this study were prepared from the richest Au-mineralized quartz-sulfide gangue with the following bulk contents—Au: 183–232; Ag: up to 41.9; Pb: 70.5–643; Zn: 297–719; Sb: 37.2–270; Cu: up to 93.4; and Bi and Hg below 10 (all contents in ppm, three samples).

4.2. Mineral Description

Quartz is the predominating mineral of the studied gold-bearing vein. Four generations of quartz (Q-1, Q-2, Q-3, and Q-4) were distinguished. The Q-1 constitutes the bulk of the gangue. It is poorly transparent due to the great amount of submicroscopic

(<2 μm) fluid inclusions and ductile to semi-ductile deformation overprint. The deformation resulted in the formation of quartz porphyroclasts floating in a fine-grained matrix. Locally, the relics of less-deformed grains and grain clusters can be also identified. While the quartz deformed largely in a ductile manner, the arsenopyrite deformed mainly in a brittle manner. We suggest that both minerals crystallized coevally. The Q-1 is crosscut by numerous quartz (Q-2, Q-3) and carbonate veinlets, typically 20–70 μm (and, exceptionally, up to 1 mm) thick (Figure 4A–F). Some of these veinlets host Au-Bi-Te-(S) phases. They are well transparent and free of ductile deformation. Complex crosscutting relationships confirm several episodes of tectonic reactivation. Quartz Q-3 crosscuts Q-2 and is even more transparent. The undeformed crystals of late quartz Q-4 (Figure 4F) were also found. These comprise euhedral crystals growing into a cavity in the Q-1 or growing on the edge of the dolomite veinlets crosscutting the Q-1.

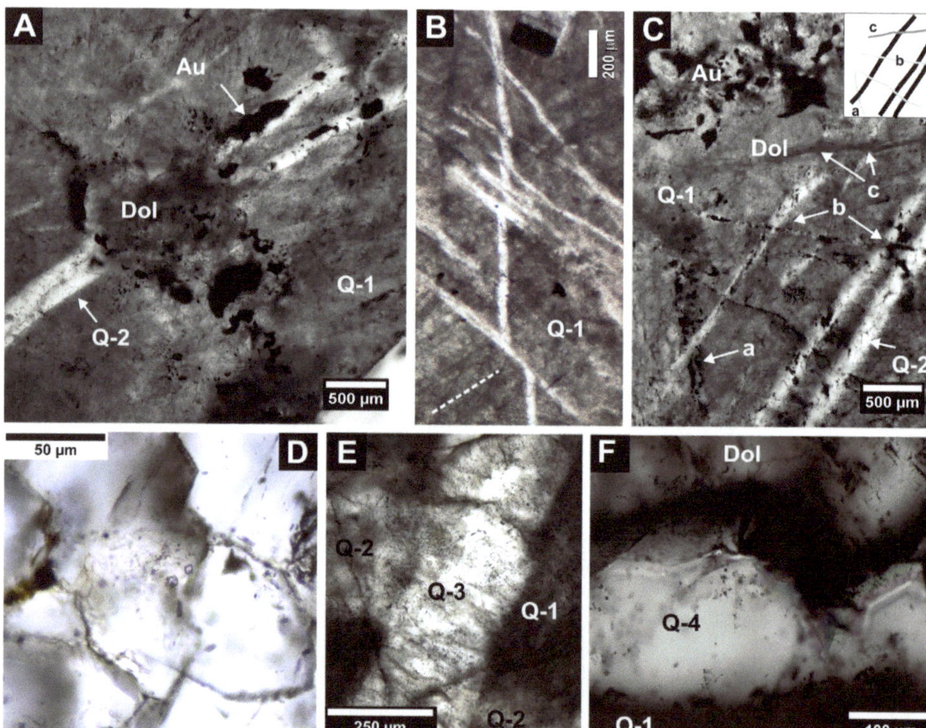

Figure 4. Microphotographs of gold-bearing quartz gangue: (**A**–**C**) Poorly transparent quartz (Q-1) crosscut by carbonate (Dol) and quartz (Q-2) veinlets. The black opaque phases largely correspond to gold and Bi-Te phases. Note that some quartz-2 veinlets host the gold (Au). (**C**) Complex crosscutting relationships between the quartz-2 and dolomite (Dol) veinlets. (**D**) Detailed image of a veinlet crosscutting oldest quartz (Q-1) composed of subhedral crystals of quartz-2. (**E**) Crosscutting relationship between the quartz Q-1, Q-2, and Q-3. (**F**) Euhedral crystals of the latest quartz (Q-4) on the edge of dolomite–ankerite (Dol) veinlet.

Fine-grained **white mica** is the main constituent of the early arsenopyrite-rich vein (Figure 5a). It is concentrated especially in ore-rich portions of the vein, whereas in the quartz-rich, arsenopyrite-poor domains it can even be completely missing. The tiny flakes of white mica often show hypo-parallel arrangement and belong to the oldest compounds of the vein. It is especially spatially related with the aggregates of chlorite, quartz

(Q-2, -3), siderite, and dolomite, whereas it is missing in the younger carbonate–sulfide–telluride veinlets.

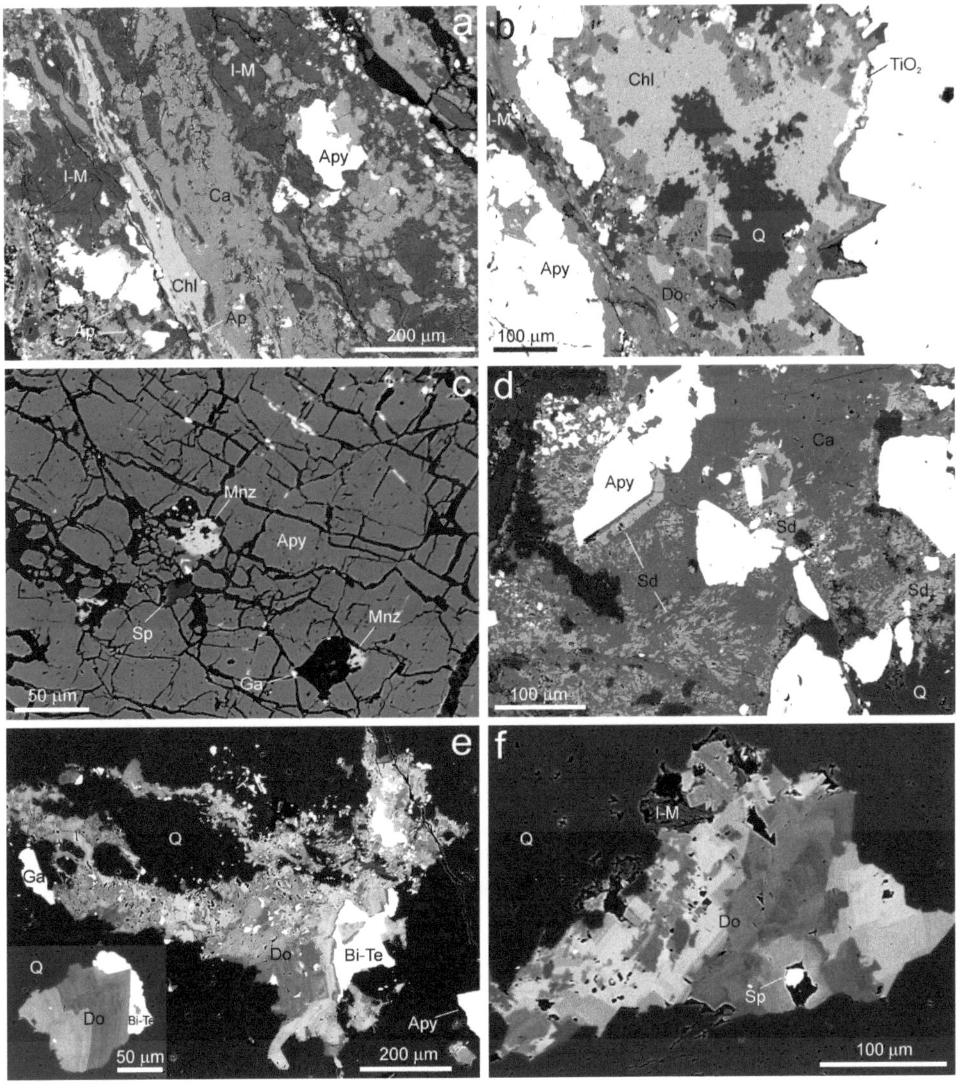

Figure 5. Gangue minerals of the studied Au-bearing vein in BSE images: (**a**) The oldest portion of vein fill composed of illite–muscovite (I–M), chlorite (Chl), calcite (Ca), apatite (Ap), and arsenopyrite (Apy). (**b**) Zonal aggregate composed of chlorite (Chl) and dolomite–ankerite (Do) developed around quartz (Q) with minor TiO_2 phase (TiO_2) and illite–muscovite (I–M). (**c**) Cracked arsenopyrite (Apy) with crack-hosted grains of monazite (Mnz), sphalerite (Sp), and galena (Ga). (**d**) Siderite (Sd) and quartz (Q) strongly corroded by calcite (Ca). (**e**) Nest of strongly zoned dolomite–ankerite (Do) hosted by quartz (Q) and containing aggregates of Bi-tellurides (Bi-Te), which are sometimes clearly younger than euhedral crystals of dolomite–ankerite (inset). (**f**) Nest of complexly zoned dolomite–ankerite (Do) hosted by quartz (Q) and illite–muscovite (I–M) gangue. The vug in dolomite–ankerite contains a grain of sphalerite (Sp).

Chlorite is one of major constituents of the arsenopyrite-rich vein. It forms massive irregularly shaped aggregates of up to 0.3 mm in size, which do not exhibit any internal texture in the BSE images (Figure 5a,b). Exceptionally, chlorite veinlets cutting the older aggregates of white mica, quartz (Q-1), or arsenopyrite were observed. Chlorite is usually intimately intergrown with xenomorphic siderite or calcite; the dolomite seems to be always younger.

Monazite is an accessory phase of the arsenopyrite-rich vein, where it occurs either in cracks of arsenopyrite (Figure 5c) in association with chlorite, in younger aggregates of quartz (Q-2, -3) and carbonates, or in the ore-poor quartz-rich portion of the vein. It forms isometric, xenomorphic, sometimes slightly porous grains of up to 10 μm in size.

The **TiO_2 mineral, K-feldspar,** and **apatite** are relatively scarce phases, occurring as irregularly shaped isometric to slightly elongated xenomorphic grains up to 50 μm in size (Figure 5a,b), enclosed in white mica, chlorite, quartz (Q-2, -3), and/or dolomite in the spaces among the crystals of arsenopyrite. No compositional zoning is observed in the BSE images of these phases.

Carbonates are frequent minerals of gangue. **Siderite** is the least frequent carbonate, which occurs either in intergrowths with chlorite or is intimately intergrown with calcite (Figure 5d). The relationship of calcite and siderite in these intergrowths is not unified. Some parts show clear signs of the replacement of the older siderite by younger calcite along the grain boundaries and/or cleavage, but in other cases, based on textural criteria only, one cannot exclude the cogenetic nature of the homogeneously distributed xenomorphic inclusions of siderite enclosed in the calcite. Carbonates of the **dolomite–ankerite** series are a major constituent of the vein. They form both the gangue of the telluride-bearing and barren veinlets as well as nests in the older quartz–phyllosilicate–arsenopyrite vein (Figure 5e,f). The equigranular aggregates of the isometric, xenomorphic grains occur together with the automorphic rhombohedral crystals. In the BSE images, patchy and/or detailed oscillatory zoning is often observed, usually with older brighter parts and younger darker ones (Figure 5e,f). **Calcite** occurs in a quantity comparable to that of dolomite–ankerite. In addition to the above-mentioned intergrowths with siderite (Figure 5d), there also occur aggregates of calcite in the oldest portions of the vein, which are associated with the chlorite (Figure 5a). Finally, tiny calcite veinlets containing base-metal sulfides are also frequent. Corrosion or even replacement of the carbonate of the dolomite–ankerite series by calcite was also observed.

Arsenopyrite is the oldest and most prevalent (>95 vol. %) ore mineral of the vein. Typically, it forms isometric to slightly elongated idiomorphic crystals up to 1 mm in size and crystalline aggregates hosted by quartz, carbonate, and phyllosilicate-rich gangue. The individual crystals often exhibit patchy or oscillatory zoning in the BSE images, usually with brighter cores and darker rims (Figure 6a). Arsenopyrite always shows signs of intense cataclasis manifested by abundant brecciated textures cut by numerous veinlets of other ore and gangue minerals (Figure 6b–d).

Galena is common but minor phase of the studied mineralization. It occurs in the same positions as the Au-Ag alloys, i.e., as individual isometric grains in cracks of arsenopyrite (Figure 6b) that are clearly younger than the grains of the Au-Ag alloy. In addition, it is a younger component of the aggregates of tellurides and base-metal sulfides, where it postdates sphalerite but predates chalcopyrite.

Sphalerite is a rather common mineral of the young sulfide-bearing carbonate veinlets and nests (Figures 5f and 6b). It typically forms isolated isometric grains up to 150 μm in size, enclosed in carbonate, or irregular grains associated with galena and chalcopyrite. No inclusions of other ore minerals (i.e., chalcopyrite "desease") were observed.

Pyrite is minor component of the sulfide paragenesis of the young carbonate veinlets cutting the vein, as well as an accompanying phase in the arsenopyrite-rich portions of the vein. It usually forms irregularly shaped to idiomorphic grains up to 0.3 mm in size, which do not show any zonality in the ore microscope or BSE images.

Chalcopyrite is rare component, whose occurrence is mostly restricted to younger sulfide-bearing carbonate veinlets (Figure 6b,d). It forms irregular aggregates in association with the older galena, sphalerite, and pyrite. No signs of anisotropy were observed in the ore microscope. In the BSE image, it is mostly homogeneous, but in one case, a heterogeneous aggregate with elevated contents of silver was found.

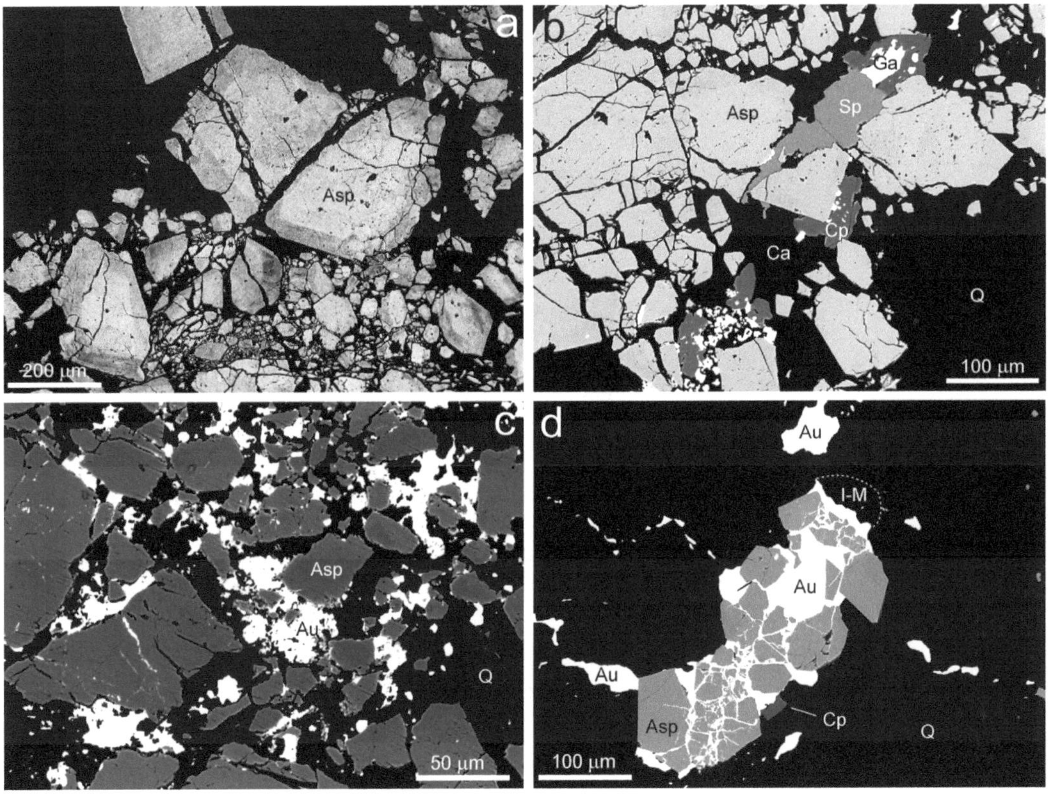

Figure 6. Textures of arsenopyrite-rich portions of the studied Au-bearing vein in BSE images: (**a**) Compositional zoning of fractured arsenopyrite (Asp). (**b**) Arsenopyrite (Asp) hosted by calcite (Ca) and quartz (Q) gangue cut by younger ore veinlet composed of sphalerite (Sp), galena (Ga), and chalcopyrite (Cp). (**c**) Crushed arsenopyrite (Asp) cemented by quartz (Q) with abundant gold (Au) grains. (**d**) Crushed arsenopyrite (Asp) cemented by gold (Au) only. Note the gold veinlet continuing into the host quartz (Q) gangue. I–M—illite–muscovite, Cp—chalcopyrite.

Tetrahedrite and **gudmundite** were observed as single grains only. Isometric tetrahedrite grain, 20 μm in size, occurs as enclosed in Ag-rich chalcopyrite, whereas gudmundite grows in a detachment fissure developed around an arsenopyrite crystal, healed by siderite.

The **Au-Ag alloys** form frequent isometric irregularly shaped grains from 1 to 20 μm in size, which are either present within hair-thin veinlets hosted by the arsenopyrite (Figure 6c,d) or disseminated in the quartz in the vicinity of the arsenopyrite grains (Figures 4A and 6d). The spatial distribution of the Au-Ag grains is irregular. They are usually arranged in clusters of small grains, sprayed in some domains of the vein, whereas the neighboring areas of the gangue, though exhibiting identical mineral composition, are completely free of them. Exceptionally, in the richest parts, the Au-Ag alloy cements the brecciated arsenopyrite (Figure 6d). In reflected light, the Au-Ag alloys show different colors (deep yellow and light yellow to almost white hues) due to the variations in their fineness. In most cases, the Au-Ag alloys with different fineness form individual grains without mutual contact, even in a single cluster (Figure 7a). Only in exceptional cases is it observed that the high-fineness gold is older and overgrown or cut by tiny veinlets of younger low-fineness Au-Ag alloy (Figure 7b,c). These younger silver-rich veinlets are spatially restricted only to the area of the grain itself and never continue into the surrounding gangue. Most of the Au-Ag alloy grains are rimmed and replaced by aurostibite (Figure 7d) and (to a much lesser extent) galena.

Aurostibite overgrows and sometimes completely replaces the grains of Au-Ag alloys (Figure 7d). It forms irregular grains up to 10–40 μm in size, which are spatially associated with grains of non-transformed Au-Ag alloy and are often from rims replaced by the unnamed (Au,Ag)Sb oxide phase (Figure 7e,f). In some parts of the vein, the quantity of grains of aurostibite clearly prevails over that of the Au-Ag alloys. A part of the aurostibite grains is zoned in the BSE images due to the increased content of Bi in their margins.

The unnamed **(Au,Ag)Sb oxide phase** replaces the aurostibite in its contacts with the surrounding gangue. It forms thin rims around the aurostibite (Figures 7f and 8a) and exceptionally complete pseudomorphs after its grains. In reflected light, it is grey to grey-brown with a reflectance similar to that of sphalerite. In BSE images, it is often zoned due to variations in the water contents, Au/Ag ratio, and/or Fe contents (Figure 8a). Typically, darker (in BSE) zones are situated on the outer margin along the contact with the gangue. It is sometimes overgrown by acanthite (Figure 7e).

Dyscrasite forms aggregates up to 50 μm in size, overgrowing, and in places also partly replacing, the grains of the Au-Ag alloys (especially those with a prevalence of Ag over Au; Figure 7c) and/or aurostibite. Isometric grains of dyscrasite are xenomorphic and sometimes show compositional zonality in the BSE images due to variations in the Ag/Au ratios.

Acanthite occurs in relatively large aggregates (up to 100 μm in size), overgrowing the grains of gold, aurostibite, and the unnamed (Au,Ag)Sb oxide phase (Figure 7c). No zoning is observed in the BSE images.

The Bi-tellurides are characteristic accessory constituents of tiny carbonate (zoned dolomite–ankerite) veinlets or small nests in the quartz–arsenopyrite vein (Figure 5e). **Tsumoite** is the prevailing telluride, forming isolated irregularly shaped individual grains up to 80 μm in size (Figure 8b,c). It intensively replaces **tetradymite, Fe-sulphotsumoite, tellurobismutite, pilsenite,** and **hedleyite**, which are sometimes present in the center of the tsumoite aggregates and do not exceed 30 μm in size (Figure 8b–d). Typically, only simple binary telluride assemblages occur in a single grain of tsumoite, although neighboring grains can contain different telluride assemblages. A more complex assemblage, tetradymite + Fe-sulphotsumoite + tsumoite, was found once only (Figure 8c). Isolated monomineral grains of these tellurides, not exceeding 20 μm in size, can also occur sporadically in the dolomite–ankerite gangue. In addition, sporadic, small, and isolated grains of unnamed **Bi_2Te** were found in spatial association but not in direct contact with these tellurides. Unlike the other mentioned tellurides, the latter phase was in one case observed in direct contact with a grain of the Au-Ag alloy (Figure 8e). Bi-tellurides are in places corroded and replaced by hessite.

Hessite forms rare irregularly shaped grains and aggregates up to 60 μm in size, which are always associated with Bi-tellurides. Hessite (in places together with native bismuth) partly replaces or overgrows hedleyite and/or tsumoite (Figure 8d). In the BSE images, hessite appears to be homogeneous.

Native antimony was observed sporadically, sometimes together with small grains or overgrowths of Sb oxide. Both phases overgrow small grains of galena, native gold, or arsenopyrite. No zoning is visible in the BSE images (Figure 8f).

Native bismuth is very rare and was observed in two different positions. The first grain, 5 μm × 15 μm in size, was found to be enclosed in a grain of Au-Ag alloy. Another irregular aggregate of native bismuth was associated with hessite, both partly replacing an aggregate of hedleyite (Figure 8d).

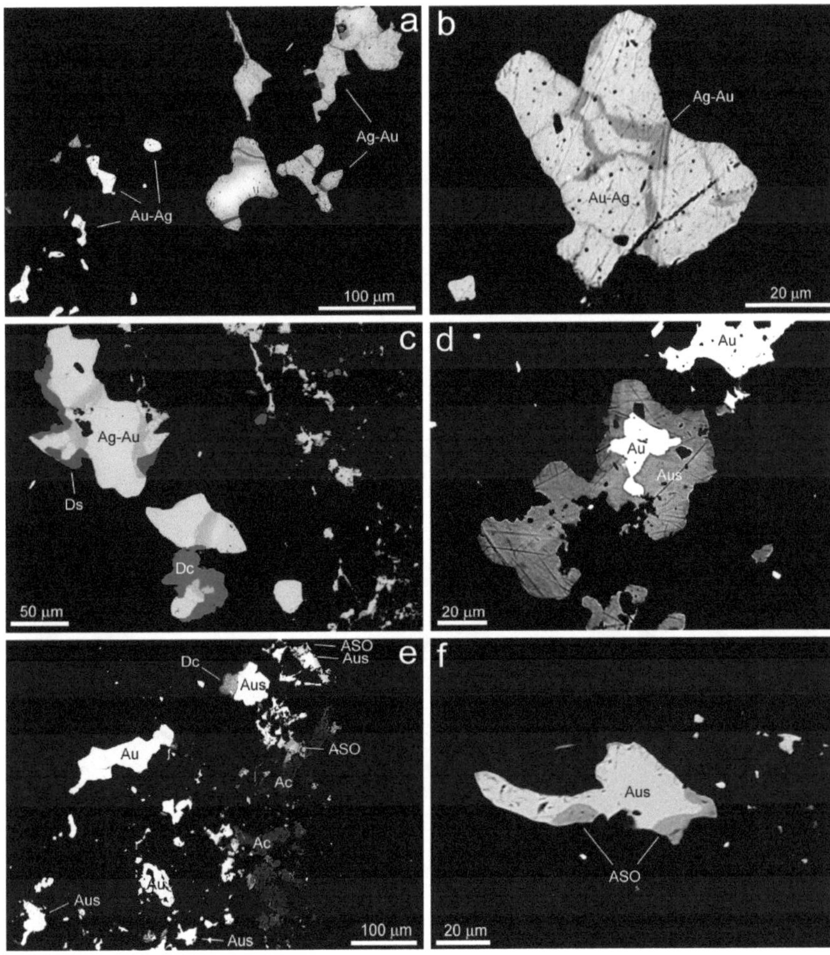

Figure 7. Textures of gold-rich portions of the studied Au-bearing vein in BSE images: (**a**) A group of gold grains with contrasting fineness. The central grain shows diffuse internal zonality with increasing Ag towards the margin and dark veinlets richest in Ag. (**b**) Veinlets of Ag-rich Ag-Au alloy (Ag-Au) cutting the Au-rich Ag-Au alloy (Au-Ag). (**c**) Zoned Ag-Au alloy partly rimmed by dyscrasite (Dc). (**d**) Gold grain (Au) partly replaced by aurostibite (Aus). (**e**) A group of gold grains (Au) partly or completely replaced by aurostibite (Aus), an unnamed (Au,Ag)-Sb oxide (ASO) and acanthite (Ac). (**f**) Grain of aurostibite (Aus) partly replaced by an unnamed (Au,Ag)-Sb oxide (ASO).

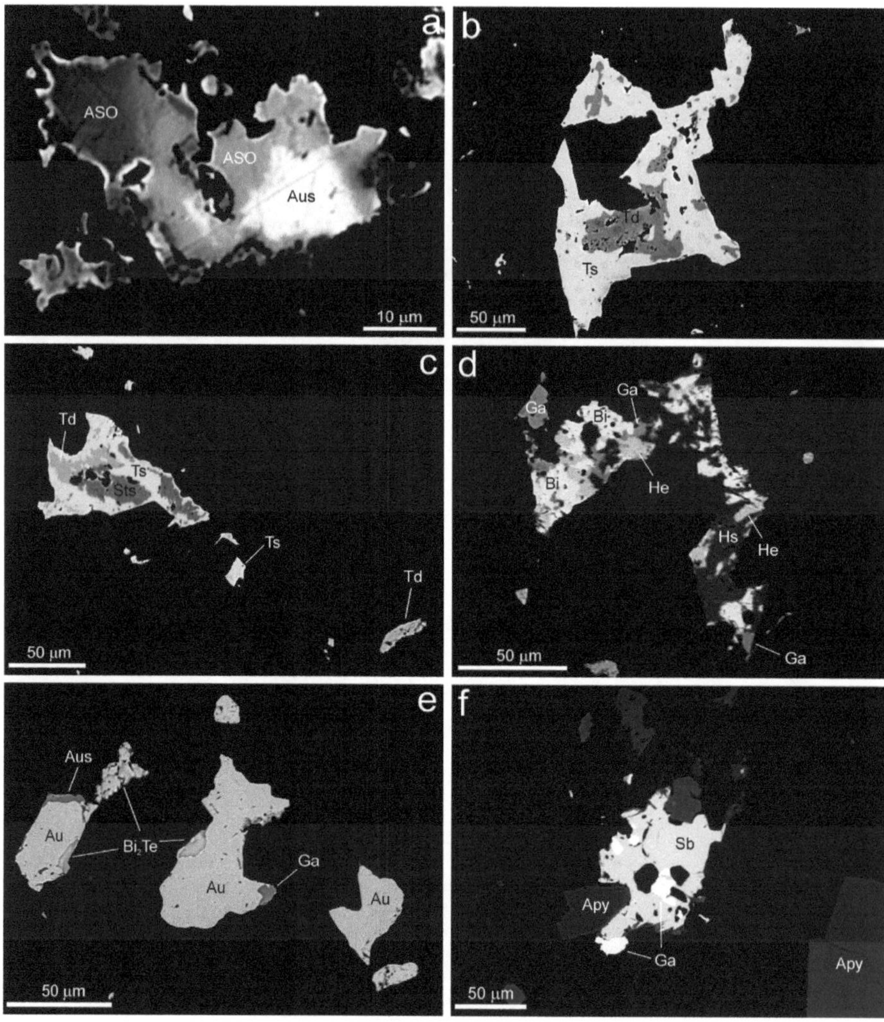

Figure 8. Textures of some ore minerals of the studied Au-bearing vein in BSE images: (**a**) Relic of aurostibite (Aus) replaced by an unnamed strongly zoned (Au,Ag)-Sb oxide (ASO). (**b**) An aggregate of Bi-tellurides composed of tetradymite (Td) rimmed by tsumoite (Ts). (**c**) An aggregate of Bi-tellurides composed of sulphotsumoite (Sts), tetradymite (Td), and tsumoite (Ts). (**d**) Mineralogically rich ore aggregate composed of native bismuth (Bi) rimmed by hedleyite (Hs), galena (Ga), and hessite (He). (**e**) High-fineness gold (Au) grains slightly replaced by aurostibite (Aus), galena (Ga), and an unnamed Bi_2Te (Bi_2Te). (**f**) Intergrowth of galena (Ga) and native antimony (Sb) growing over arsenopyrite (Apy).

4.3. Mineral Composition

4.3.1. Base-Metal Sulfides

During the study of the chemical composition of **arsenopyrite** (See Supplementary Materials, Table S1a,b), we observed, in addition to the main elements (Fe, As, S), only irregular minor traces of Co and Sb, not exceeding 0.004 and 0.001 ***apfu***, respectively. A part of the arsenopyrite grains shows distinct chemical zoning in the BSE images (Figure 6a), caused by AsS_{-1} substitution: in the BSE, the brighter cores are richer in As than the darker rims. All arsenopyrite aggregates are relatively As-poor (Figure 9). The determined range of this substitution (1.02–1.17 *apfu* S) is slightly larger than that published by Kretschmar and Scott [25] or Sharp et al. [26], but it corresponds to that of the arsenopyrite from some other gold deposits (see, e.g., Mikulski [27]; Andráš, Chovan [28]). The cation position of arsenopyrite shows a slight deficiency (0.98–1.00 *apfu* Fe + Co), which does not correlate with the As/S ratio. A similarly slight iron deficiency is usual for both natural and synthetically grown arsenopyrites [25]. The contents of gold range rather uniformly between 0.00041 and 0.00066 *apfu*.

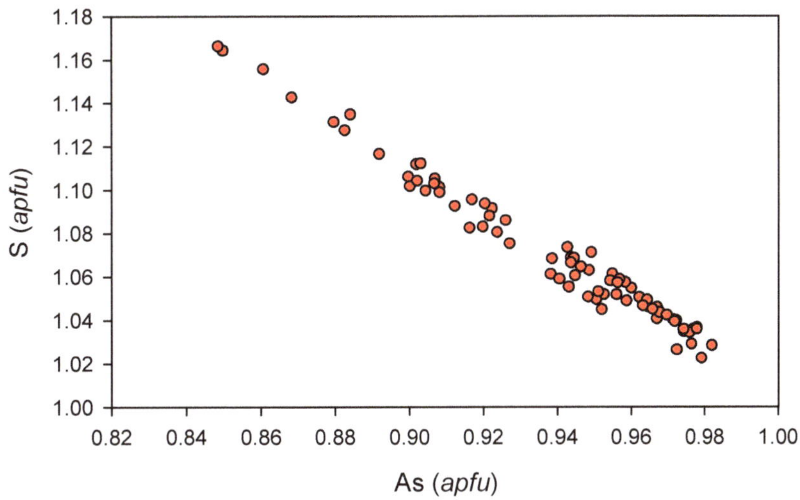

Figure 9. Compositional variation of zonal arsenopyrite in an As versus S (*apfu*) plot.

The chemical composition of **galena** is close to the ideal formula (Table S2); only traces of Ag (up to 0.004 *apfu*), Sb (up to 0.012 *apfu*), and Bi (up to 0.004 *apfu*) were determined; the measured contents of Fe (up to 1.49 wt. %) probably come from the surrounding pyrite. **Pyrite** is stoichiometric without the other elements, including As. **Sphalerite** shows varying chemical composition (Table S3); the most abundant minor elements are Fe, Cd, Mn, and Cu. Iron and cadmium are always present with contents up to 0.06 and 0.003 *apfu*, respectively; their contents positively correlate. The presence of Mn and Cu is irregular and does not reach 0.005 *apfu*. The chemical composition of **chalcopyrite** (Table S4) corresponds to the ideal formula with traces of Mn and Pb; part of its grains contains minor Ag in the range of 0.04–0.09 *apfu*. **Tetrahedrite** is almost As-free (Table S5), with determined Ag contents in the range of 1.11–1.32 *apfu*; Zn (1.11–1.25 *apfu*) is the dominant Me^{2+} element and, according to the recently approved nomenclature of the tetrahedrite group [29], must be named tetrahedrite-(Zn). The chemical composition of **bournonite** agrees well with the ideal stoichiometry (Table S6); Mn and Fe were observed only as traces. The chemical analyses of **gudmundite** (Table S7) show a composition close to the ideal end member, with only a minimal arsenopyrite component (up to 0.004 *apfu* As) and traces of Mn (0.002–0.007 *apfu*).

4.3.2. Tellurides

In the course of the study of the chemical composition of Bi-(sulpho)tellurides, seven mineral phases (Figure 10) were encountered: tellurobismuthite, tetradymite, tsumoite, sulphotsumoite, pilsenite, unnamed Bi_2Te, and hedleyite. The chemical composition of **tellurobismuthite** (Table S8) is near the ideal composition of this mineral (Figure 10). Minor Sb and Pb corresponds to up to 0.12 and 0.01 *apfu*, respectively; S and Se were not detected.

Tetradymite is close to the ideal stoichiometry (Table S8) with Bi partly replaced by Sb up to 0.25 *apfu* (Figure 11) and minor contents of Pb (up to 0.01 *apfu*) and Se (up to 0.09 *apfu*). The range of the Te:(S+Se) ratio of 1.87–2.16 is near the theoretical 2:1.

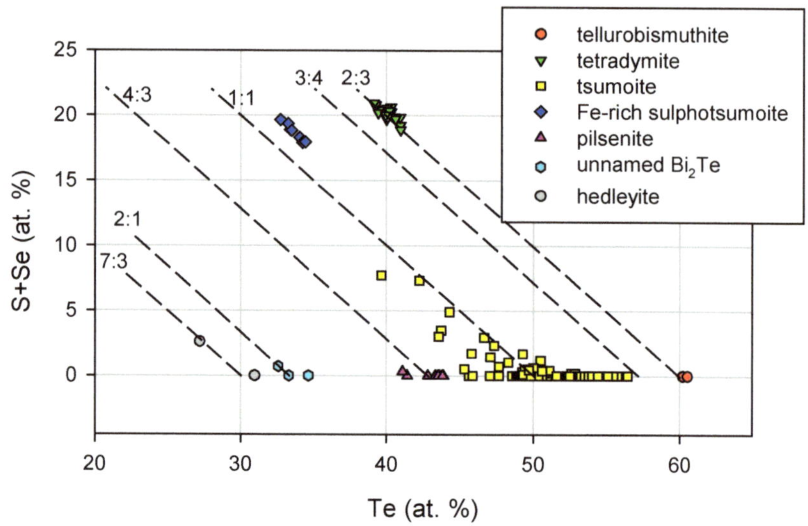

Figure 10. Chemical composition of Bi-(sulpho)tellurides in a Te versus S + Se (at. %) plot; numbers show Me:X ratio.

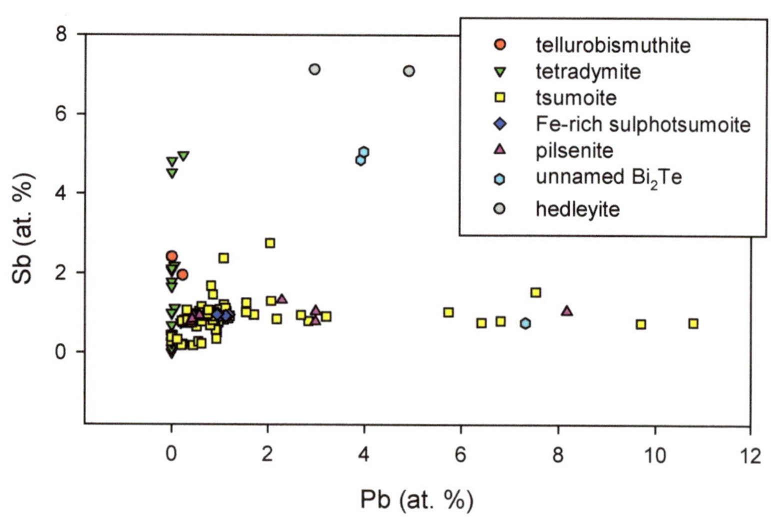

Figure 11. The Pb and Sb contents (at. %) in Bi-(sulpho)tellurides.

The chemical analyses of **tsumoite** show a wide range of the Me:X ratio (Table S9), with contents of Te(+S+Se) from 45.6 to 56.4 at. % (Figures 11 and 12). Abrikosov and Bankina [30] reported in the Bi-Te system a wide solid-solution (γ-phase) from 45 to 55 at. % Te; Shelimova et al. [31] published for this phase the range from 44.7 to 58.1 at. % Te; and Yusa et al. [32] from 43 to 57 at. % Te at 400 °C. These solid-solutions, centered on tsumoite, may help to explain the extensive non-stoichiometry in these parts of the Bi–Te system [33]. A part of the analyzed points (Figure 12) is already close to the composition of

Pb-poor rucklidgeite [33,34], but we do not assume the presence of this phase due to the total extent of the tsumoite solid-solution. Tsumoite usually contains some Sb and Pb (up to 0.06 *apfu*); some grains with increased Pb contents in the range of 0.11–0.22 *apfu* were also observed (Figures 11 and 12). In the anion part of the tsumoite, minor contents of Se (up to 0.02 *apfu*) and S (up to 0.15 *apfu*) were found (Figure 10); the sulfur and lead contents correlate positively.

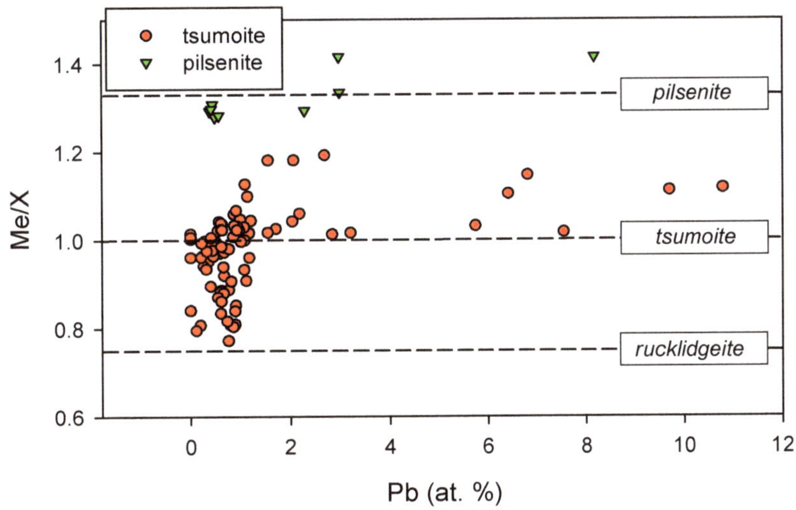

Figure 12. Chemical composition of tsumoite and pilsenite in a Pb (at. %) versus Me/X ratio plot.

One homogeneous grain shows a composition close to **sulfotsumoite**, with increased Fe contents in the range 0.91–0.96 *apfu* (Table S10) and minor contents of Pb and Sb (up to 0.06–0.07 *apfu*). The iron contents in the Bi-(sulpho)tellurides are unusual; due to the grain dimensions and the constant contents of Fe, the influence of the surrounding material is not probable. It is possible that this is a new mineral species with the empirical formula $Fe_{0.93}(Bi_{1.81}Pb_{0.06}Sb_{0.06})_{\Sigma 1.93}Te_{2.02}S_{1.12}$, which is close to the ideal one of $FeBi_2Te_2S$.

Pilsenite is close to the ideal composition (Table S11) with minor contents of Pb up to 0.57 *apfu* (Figure 12) and Sb up to 0.09 *apfu*; traces of S are exceptional. Similarly increased Pb contents were published, e.g., by Ferenc [35] and Mikulski [36]. The chemical composition of the **unnamed Bi_2Te** (Table S12) is near the ideal composition (Figure 10). Minor Pb and Sb correspond to up to 0.22 and 0.15 *apfu*, respectively (Figure 11); Se was not detected, and traces of S were exceptional. An unnamed phase with the formula Bi_2Te has been published by Gamanyin et al. [37,38], Goncharov et al. [39], Huang et al. [40], Luukkonen [41], Gu et al. [42], Cook et al. [33], and Houzar et al. [43]. Phase Bi_2Te was also observed during synthetic studies [30,44], and Shelimova et al. [31] reported a solid-solution (β-phase) in a Bi-Te system with a range from 33.3 to 36.6 at. % Te, which corresponds to ~Bi_2Te. **Hedleyite** is consistent to a 7:3 stoichiometry (Table S13), and the determined contents of Sb up to 0.71 *apfu* and Pb up to 0.49 *apfu* (Figure 11) are higher than those which have been published by Cook et al. [34]. In the anion part of hedleyite, minor contents of S (up to 0.26 *apfu*) were found.

Silver telluride, **hessite**, is close to the ideal composition of this mineral (Table S14); minor contents of Pb, Bi (up to 0.02 *apfu*) and Au, Hg, and Sb (up to 0.01 *apfu*) were sometimes found. In the anion part of hessite, the dominant Te is partly replaced by S (up to 0.08 *apfu*).

4.3.3. Elements and Alloys

The chemical composition of the **minerals of the Au-Ag solid-solution** (Table S15) covers a wide range, from almost pure native gold (4–7 at. % Ag) to Au-rich silver (up to 69 at. % Ag). There can be distinguished two groups of composition (Figures 13 and 14); the first comprises paragenetically early gold with Ag contents in the range of 4–27 at. % (96–73 at. % Au); the second comprises paragenetically late low-fineness Au-Ag alloys (Ag-rich gold, Au-rich silver) with 32–69 at. % Ag (68–30 at. % Au). In addition to the major components (Au and Ag), the minor contents of Hg and Sb were also observed. The determined contents of Hg for both types of Au-Ag are similar (Figure 14), up to 3.7 and 2.6 at. %, respectively. In contrast, increased Sb contents (up to 1.9 at. %) are concentrated only in low-fineness Au-Ag alloys, especially in Au-rich silver (Figure 15).

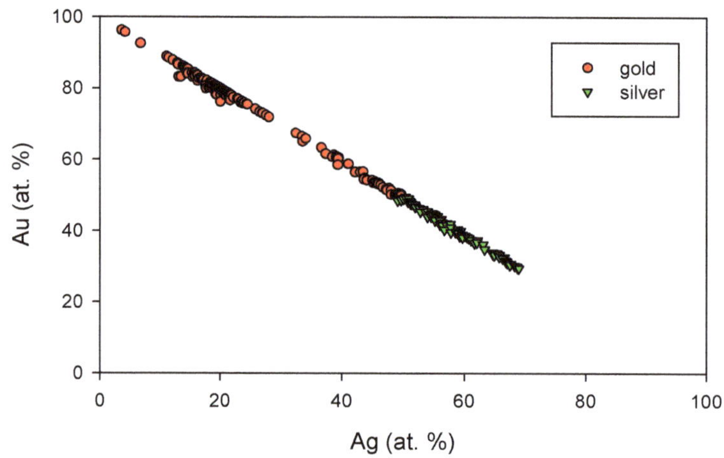

Figure 13. Chemical composition of members of Au-Ag solid-solution in an Ag versus Au (at. %) plot.

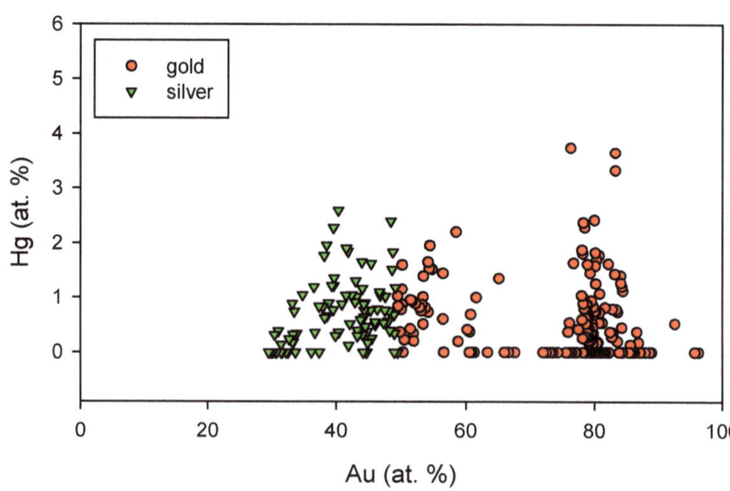

Figure 14. Chemical composition of members of Au-Ag solid-solution in an Au versus Hg (at. %) plot.

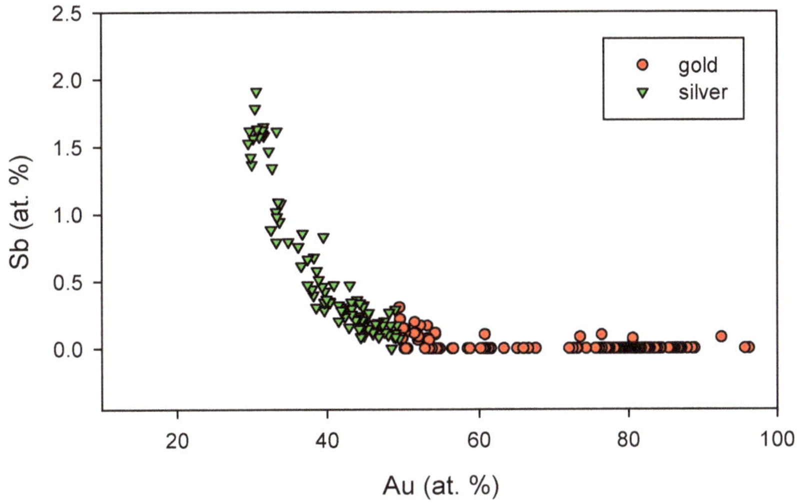

Figure 15. Chemical composition of members of Au-Ag solid-solution in an Au versus Sb (at. %) plot.

The found wide range of the composition of the minerals of the Au-Ag solid-solution is uncommon; a similar range (0–70 at. % Ag) is known only for the Krásná Hora deposit [45,46]. Kalinin et al. [47] described samples from the Oleninskoe deposit comprising two groups of Au-Ag alloys with 5–22 and 53–75 at. % Ag, respectively. From the vein deposits in the Teabaeksan district, Korea, Au-Ag alloys with 36–97 at. % Ag were described [48], but high-fineness Au is not known there.

The chemical analyses of the **native antimony** (Table S16) show, in addition to the dominant Sb, only traces of As and Hg up to 0.006 and 0.004 *apfu*, respectively. **Native bismuth** (Table S16) contains regular contents of Sb in the range of 0.10–0.13 *apfu*.

4.3.4. Au-Ag-Sb Minerals

Aurostibite is usually homogeneous, and its chemical composition corresponds to the ideal $AuSb_2$ formula (Table S17), with traces of Fe (up to 0.05 *apfu*) and locally Hg up to 0.02 *apfu*. Sporadic increased Bi contents (up to 0.05 *apfu*) were found in the marginal parts of some of its aggregates. A complete absence of Ag is characteristic for studied aurostibite, which is similar to that which was stated by Zachariáš and Němec [45] for aurostibite from Krásná Hora (with Ag up to 0.5 wt. %, only).

The chemical analyses of **acanthite** (Table S18) agree with the stoichiometry of this mineral, with minor contents of Fe (up to 0.06 *apfu*) and traces of As, Te, and Au up to 0.009, 0.005, and 0.003 *apfu*, respectively.

Dyscrasite is Sb-deficient with a ratio of (Ag+Au)/(Sb+Hg+As) in the range of 3.83–4.25 (Table S19); a similar situation is well known for this mineral phase, with a proposed ideal formula of $Ag_{3+x}Sb_{1-x}$($x \leq 0.2$, [49,50]). The substantial contents of gold (0.25–0.30 *apfu*) are characteristic for the studied dyscrasite; the found Au contents correlate negatively with Ag (Figure 16). The analogical Au contents are published only for the dyscrasite from the Krásná Hora deposit (0.27–0.32 *apfu* [45] and are comparable with the maximum amount of Au (~0.33 *apfu*) determined in the synthetic phase ε corresponding to dyscrasite [51].

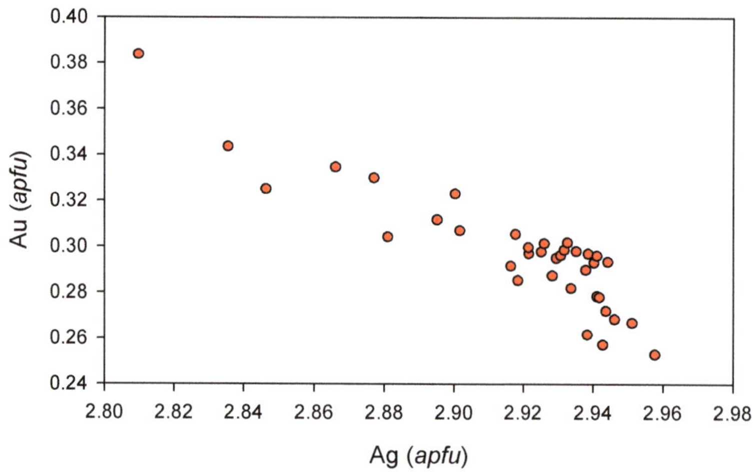

Figure 16. Chemical composition of dyscrasite in an Ag versus Au (*apfu*) plot.

An unnamed **(Au,Ag)Sb oxide** mineral phase with a (Au+Ag)/(Sb+Bi+As+Fe+Hg+Zn) ratio in the range of 0.68–1.50 (mean 1.12), the low microprobe totals of 82.54–95.67 wt. % (Table S20), and a negative Au-Ag correlation (Figure 17) can correspond to the not-approved mineral *"auroantimonate"* [52,53] or $AuSbO_3$ oxide [54]. It is also possible (Figure 18) that it can be close to the so-called *mustard* gold [55,56], but its usual microporous character is not observed in our samples; moreover, the presence of oxygen in this phase was confirmed by the EDS spectra.

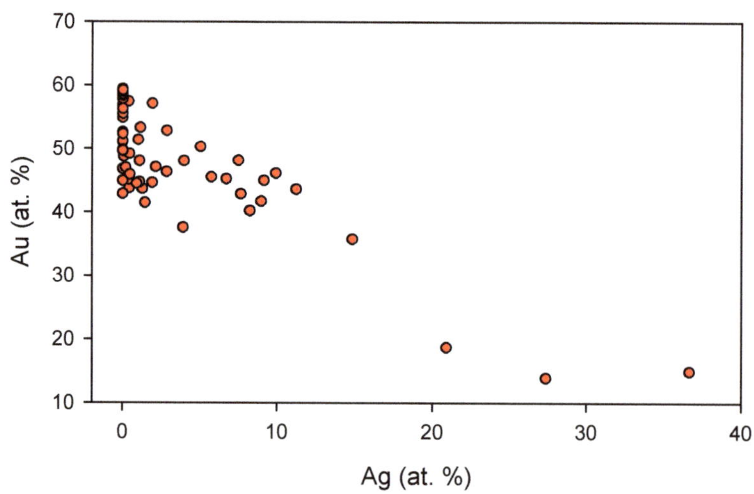

Figure 17. Chemical composition of Au-Sb oxide in an Ag versus Au (*apfu*) plot.

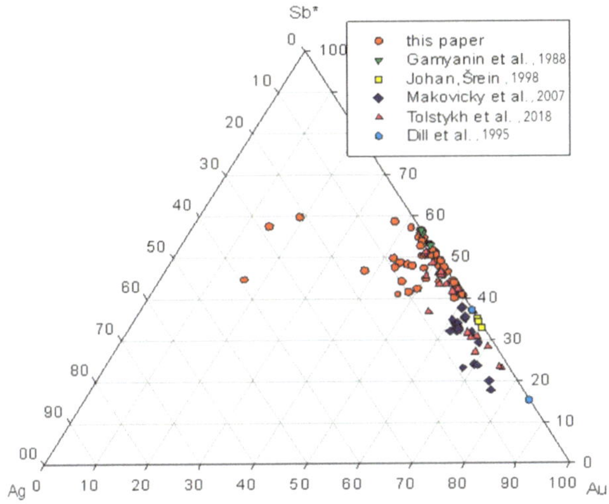

Figure 18. Chemical composition of Au-Sb oxides and mustard gold in a ternary Ag-Sb*-Au (atomic per cent) plot (Sb* = Sb+Bi+As+Te+Fe+Hg+Zn) [52,54–57].

4.3.5. Gangue Minerals

The chemical composition of **white mica** (Table S21) corresponds mostly to **illite** and less frequently to **muscovite**, with respect to the sum of the interlayer cations in the range of 0.58 to 0.92 *apfu* (Figure 19), which are dominated by potassium with small amounts of Na (up to 0.05 *apfu*) and/or Ca (up to 0.04 *apfu*). The typical features include an excess of Si above the theoretical value of 3 (3.19–3.38 *apfu*), coupled with increased amounts of Mg (0.03–0.13 *apfu*) and Fe (0.02–0.17 *apfu*). Al is to a small extent substituted by V and Cr (up to 0.01 *apfu* each). Exceptionally, slightly elevated Zn and Mn contents occurred (up to 0.03 and 0.02 *apfu*, respectively). A pronounced shift of the compositions of white mica towards an illitic composition can be stated in comparison with the other gold deposits, where the presence of regular muscovite is typical (Figure 19).

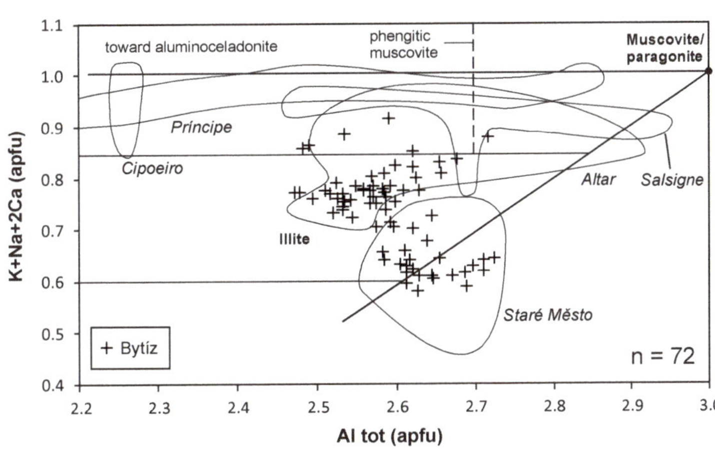

Figure 19. Chemical composition of white mica and its comparison with those of other gold deposits. Diagram is adopted from Maydagán et al. [58], with boundaries for illite modified from Rieder et al. [59]. Note that the low amount of comparative data is caused by the lack of primary analytical data in many publications. Comparative data are from Maydagán et al. [58], de Siqueira Corrêa et al. [60], Klein et al. [61], Demange et al. [62], and Dolníček et al. [63].

Chlorite (Table S22) belongs to the trioctahedral Fe-Mg-Al chlorites. Based on 14 atoms of oxygen, the Si contents range between 2.76 and 3.16 *apfu*, and the Fe/(Fe+Mg) ratio is between 0.72 and 0.90. In the classification scheme of Bayliss [64], they all belong to chamosite; in those of Melka [65], most of them are **chamosites** and two outliers fall to ***delessite*** (Figure 20). A high content of Fe is also manifested in the empirical Liard's diagram, where the projection points of chlorites from Bytíz fall into the fields of chlorites of the iron formations and ooidal ironstone (Figure 20). The contents of Mn, V, and Zn do not exceed 0.05, 0.02, and 0.12 *apfu*, respectively. In ca. 20 analyses, the elevated contents of Na and/or K occurred, pointing to the presence of an admixture of illite/smectite, supported by the positive correlation of the contents of the alkalis and Si ($R^2 = 0.65$). In contrast, a correlation between the contents of Ca (up to 0.27 *apfu*) and Si does not exist ($R^2 = 0.01$), implying that the Ca is bound in chlorite and not in the admixtures of Ca-smectite. When compared with the typical vein of gold deposits (Figure 20), the chlorites from Bytíz belong to the iron- and illite-richest compositions.

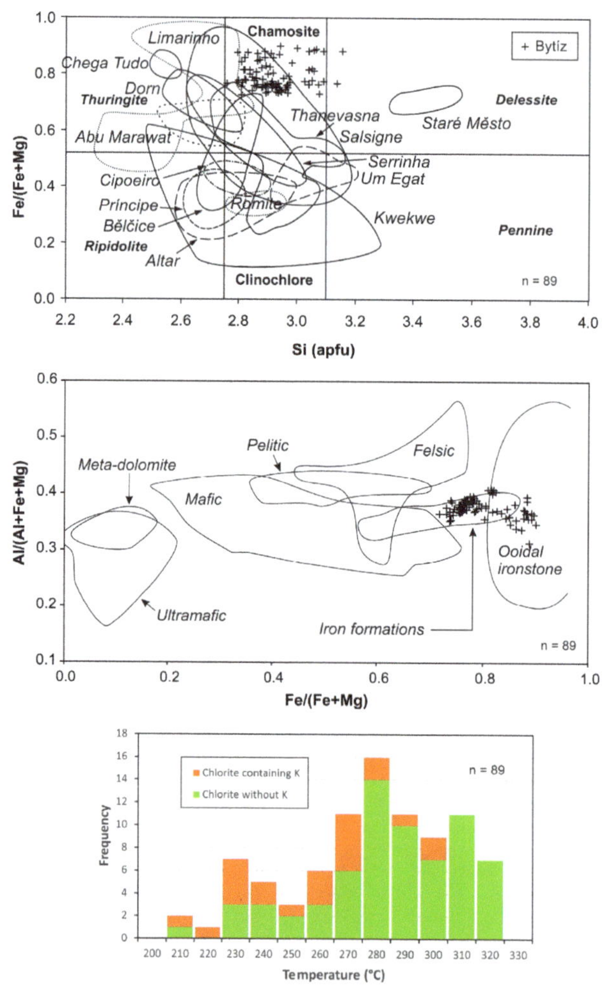

Figure 20. (a) Chemical composition of chlorite in the classification scheme by Melka [65] and its comparison with those of other gold deposits. Comparative data are from Demange et al. [62]; Moura et al. [66]; Buchholz et al. [67]; Klein et al. [61,68]; Zoheir et al. [69,70]; Zachariáš and Novák [71];

Zoheir and Akawy [72]; Crispini et al. [73]; de Siqueira Corrêa et al. [60]; Dora and Randive [74]; Fuertes-Fuente et al. [75], and Dolníček et al. [63]. (**b**) Position of chlorite in the empirical diagram of Liard [76]. (**c**) Histogram of temperatures derived from chlorite compositional thermometry by Cathelineau [77].

The chemical composition of monazite (Table S23) belongs to **monazite-(Ce)**. A part of the phosphorus is substituted by minor Si (up to 0.09 *apfu*) and trace As (up to 0.015 *apfu*) and S (up to 0.007 *apfu*). The REEs are slightly altered, especially by Al (up to 0.10 *apfu*), Ca (up to 0.05 *apfu*), Y (up to 0.03 *apfu*), and Fe (up to 0.025 *apfu*). The contents of Th in monazite are low (up to 0.004 *apfu*), which is in accordance with its hydrothermal origin. Chondrite-normalized REE patterns show a gradual decrease from La to heavier REEs (Figure 21). No tetrad effect is observed in REE patterns in the area of the first tetrad (Figure 21). The Ce anomaly is essentially missing (Ce/Ce* = 0.96–1.11); the Eu anomaly is highly variable (negative, missing, or positive; Eu/Eu = 0.16–1.48). The variable Eu anomaly was observed even within the various domains of a single grain. The La_{CN}/Sm_{CN} ratios are mostly within the range of 3.9–7.5, but three outliers also show significantly higher values up to 23.9 (Figure 21). In comparison with the available literature data on hydrothermal monazites, the samples from Bytíz show a rather wide range of both the REE fractionation and the variability of the Eu anomaly (Figure 21).

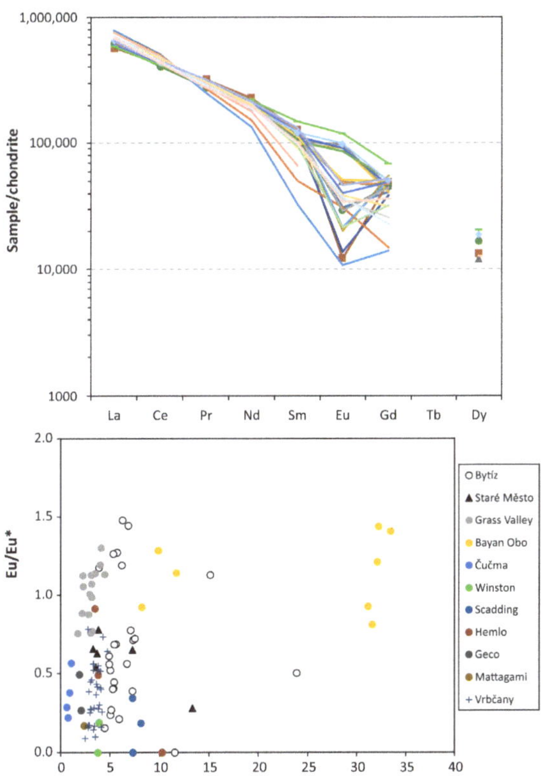

Figure 21. (**a**) Chondrite-normalized REE patterns of monazite. (**b**) The Eu/Eu* vs. La_{CN}/Sm_{CN} plot for monazite from Bytíz in comparison with other hydrothermal monazites. Comparative data are from Zhu and O'Nions [78]; Schandl and Gorton [79]; Števko et al. [80]; Taylor [81]; Dolníček et al. [63] and Dolníček and Ulmanová [82].

The chemical composition of apatite (Table S24) is compatible with **fluorapatite,** containing 0.0–24.7 mol. % hydroxylapatite and 0.0–1.8 mol. % chlorapatite. The excess of fluorine, reported in part of the collected data (the maximum value is 1.27 *apfu*), may be attributed to an overestimation of the fluorine content due to its diffusion during the microprobe analysis of the grains with unsuitable spatial orientation [83]. Phosphorus is substituted by traces of As (up to 0.013 *apfu*) and S (up to 0.007 *apfu*) only; no elevated content of carbon is suggested from the well-balanced stoichiometry of this mineral (P+As+S = 2.99–3.05 *apfu*). The calcium is substituted by minor Fe (up to 0.05 *apfu*) and Mn (up to 0.016 *apfu*).

The WDS analyses of the **TiO_2 mineral** (Table S25) yielded only minor admixtures of Ca (0.006–0.013 *apfu*) and Fe (0.001–0.010 *apfu*) and much lower contents of Cr, V, W, and Zr (up to 0.001 *apfu* each).

The **K-feldspar** (Table S26) is dominated by orthoclase, with minor albite, anorthite, celsian, and slawsonite components ($Or_{91.4–97.1}Ab_{1.3–3.5}An_{0.3–2.5}Cn_{0.9–2.3}Sl_{0.0–0.5}$), including the slightly elevated Fe (up to 0.005 *apfu*).

The **calcite** (Table S27) contains, in addition to the dominating Ca, 0.002–0.142 *apfu* Fe, 0.005–0.125 *apfu* Mn, and up to 0.054 *apfu* Mg. Both the calcite from matrix of the arsenopyrite-rich vein and the calcite from the younger veinlets show largely similar chemical compositions (Figure 22).

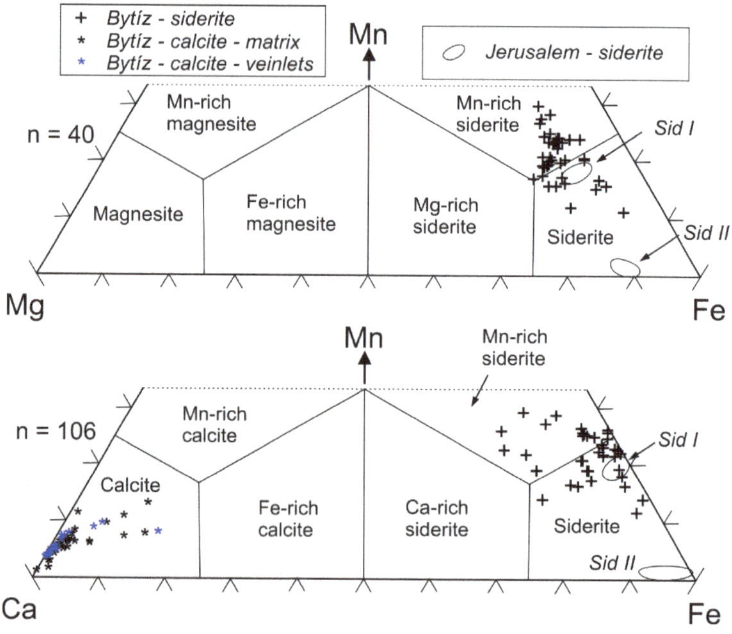

Figure 22. Chemical composition of calcites and siderites from Bytíz in the Mg-Fe-Mn plot and the Ca-Fe-Mn plot (modified from Trdlička, Hoffman [84]. Comparative data are from Sejkora et al. [4].

The chemical composition of **siderite** (Table S28), from the different modes of occurrence, is also characterized by a rather uniform trend. The iron is substituted mainly by Mn (0.111–0.265 *apfu*) and, to a lesser extent, also by Mg (0.054–0.158 *apfu*) and Ca (0.005–0.152 *apfu*). The chemical composition of siderite from the Bytíz deposit corresponds to siderite and Mn-rich siderite (Figure 22) and is comparable to the pre-ore siderite I from the Jerusalem deposit (Příbram uranium and base-metal district) rather than to the post-ore siderite II (cf. Figure 22).

The carbonates of the **dolomite–ankerite** series (Table S29) exhibit a wide range of compositions ($Dol_{7.5–94.0}Ank_{3.3–60.7}Ktn_{0.7–31.8}$), covering the classification fields from the dolomite via Fe-rich dolomite and Mg-rich ankerite up to Mn-rich ankerite in the classification scheme by Trdlička and Hoffman [84]. The almost linear distribution of data in the triangular plot (Figure 23) illustrates the coupled behavior of Fe and Mn. Although significant fluctuations can occur, the early portions of carbonate crystals are in general enriched in Fe and Mn, whereas the younger zones are richer in the dolomite component. This is the opposite trend to that observed by Sejkora et al. [85] in the ore vein from the Brod deposit (Příbram uranium and base-metal district), where, in addition, distinct chemical trends were detected in the two respective generations (Figure 23). The youngest recorded growth zone neighboring with the drusy cavity was found to contain elevated contents of Zn (0.016 *apfu*) in the studied mineralization from Bytíz. It is noteworthy that the carbonates of the dolomite–ankerite series from the matrix of the arsenopyrite–quartz–phyllosilicate vein and those from the younger veinlets cutting the above-mentioned vein do not exhibit any remarkable differences in their chemical compositions (Figure 23).

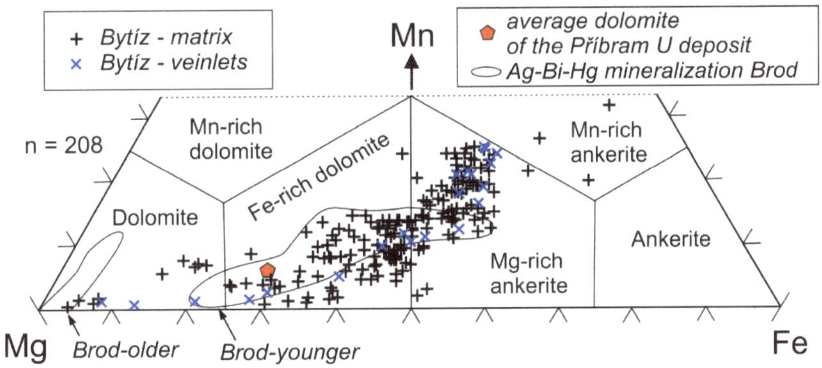

Figure 23. Chemical composition of carbonates of the dolomite–ankerite series from Bytíz in the Mg-Fe-Mn plot (modified from Trdlička and Hoffman [84]). Comparative data are from Cílek et al. [86] and Sejkora et al. [85].

4.4. Fluid Inclusion Study

Fluid inclusions were studied in five specimens. Because of the small size of the fluid inclusions and the poor transparency of most of the studied samples, the microthermometric measurements were complicated. Suitably sized measurable fluid inclusions were found in quartz Q1-Q4, in the calcite from the veinlets, and in the carbonates of the dolomite–ankerite series from the veinlets only.

4.4.1. Fluid Inclusions in Quartz

The fluid inclusions in the quartz Q-1 were difficult to study due to their small size (up to 8 μm) and the poor transparency of the host mineral; therefore, only a few isolated fluid inclusions of uncertain origin (most probably secondary) were measured. These two-phase (F $\approx$ 0.85–0.9) inclusions homogenized to liquid in a wide range between 112 and 226 °C, and the last crystal of ice melted in between −0.8 and −2.2 °C; these values correspond to a salinity of 1.4–3.7 wt. % NaCl eq. We cannot exclude the possibility that these inclusions can contain, at least in part, younger fluids which are common in younger populations of quartz and exhibit the same microthermometric parameters (see below; Table 1).

Table 1. Microthermometric data from fluid inclusions hosted by the studied Au-bearing vein from Bytíz. Temperature parameters in °C, salinity in wt. % NaCl eq. Numbers in parentheses refer to number of measurements.

Sample	Mineral	Genesis	Phase Composition	Th	T_f	T_i	Tm–Ice	Salinity
By-1	Q-2	P/PS	L+V	150–256 (26)	−35		−0.7/−1.6 (18)	1.2–2.7
	Q-2	S	L+V	121–126 (4)	−35		−0.1/−0.2 (2)	0.2–0.4
By-2	Q-2	P/PS	L+V	122–361 (48)	−42/−45		−1.1/−3.2 (41)	1.9–5.3
	Q-3	P/PS	L+V	139–324 (28)	−43/−46		−1.2/−4.4 (26)	2.1–7.0
	Q-4	P/PS	L+V, L	76–104 (4)	−55/−58		−12.1/−15.2 (7)	16.1–18.9
	Dol	P	L+V, L	84–214 (25)	−47/−63	-56 (1)	−1.0/−16.8 (23)	1.7–20.1
By-3	Q-4	P	L+V, L	81–123 (8)	−52/−65		−10.1/−14.9 (9)	14.1–18.6
	Q-4	P	L+V, L	104–161 (7)	−40/−42		−0.1/−0.4 (14)	0.2–0.7
	Dol	P	L+V, L	156–178 (4)	−67		−15.7/−16.0 (4)	19.3–19.5
By-4	Q-2	P/PS	L+V	126–255 (42)	−42/−45		−1.0/−2.4 (29)	1.7–4.0
	Dol	P	L+V, L	122–256 (30)	−46/−72		−0.9/−21.3 (29)	1.6–23.4
By-5	Q-1	?	L+V	112–226 (23)	−40/−46		−0.8/−2.2 (17)	1.4–3.7
	Q-2	P/PS	L+V	120–239 (8)	−40/−45		−0.9/−2.0 (7)	1.6–3.4
	Q-2	PS/S	L+V	66–135 (7)	−40/−41		−0.4/−1.0 (5)	0.7–2.2
	Cal	P/PS	L+V, L	62–178 (19)	−41/−48		−1.5/−3.3 (19)	1.5–3.3

The fluid inclusions in the quartz Q-2 (Figure 24a–c) were studied in thin veinlets and euhedral crystals, where they were situated mostly on short trails and in three-dimensional clusters, suggesting, in the latter case, their primary nature. They were all two-phase and liquid-rich at room temperature (F ≈ 0.85–0.9), with sizes up to 23 μm (mostly between 1 and 6 μm). They homogenized to liquid mostly in between 120 and 248 °C and, exceptionally, up to 361 °C (Table 1). The data below 120 °C very likely represent late (i.e., secondary) inclusions. This idea seems to be supported by subtle differences in their salinities (primary inclusions: 1.2–5.3 wt. % NaCl eq. based on Tm-ice = −0.7 to −3.2 °C; supposed secondary inclusions: 0.2 to 2.2 wt. % NaCl eq. equal to Tm-ice = −0.1 to −1.0 °C). A wide interval of homogenization temperatures (Figure 25a) and a narrow interval of salinities suggests the long-term evolution of one fluid of unchanged composition or the re-filling of part of the fluid inclusions by late fluids with comparably low salinities.

The fluid inclusions in quartz Q-3 (Figure 24d) were measured in a single veinlet only, fortunately crosscutting the Q-1 and Q-2 veins. Their morphology, size, and occurrence are similar to those of Q-2; however, their homogenization temperatures (mostly 245–324 °C) and salinities (2.1–7.0 wt. % NaCl eq.; Tm-ice = −1.2 to −4.4 °C; Table 1) are higher than those of most primary inclusions in the Q-2.

The temperature of initial melting (T_i) was recognized in none of fluid inclusions from the three above-mentioned generations of quartz. The inclusions in Q-1, -2, -3 had frozen on cooling down to −46 °C. This suggests $T_i > -46$ °C and rules out the presence of dissolved $CaCl_2$ in the aqueous fluid. The inclusions contain low-salinity H_2O-NaCl(±$MgCl_2$±$FeCl_2$) fluid.

The fluid inclusions in the late quartz Q-4 (Figure 24e) were mostly monophase. Those which were two-phase and liquid-rich (F ≈ 0.95; up to 20 μm in size) were occasionally identified in the form of discontinuous planes (pseudo-secondary inclusions) or three-dimensional clusters (primary inclusions). Two types of fluids were distinguished: low-salinity (0.2–0.7 wt. % NaCl eq.; Tm-ice = −0.1 to −0.4 °C), with $T_F \approx -40$ °C and a homogenization temperature (Th) in between 104 and 161 °C, and high-salinity (14.1–18.9 wt. % NaCl eq.; Tm-ice = −10.1 to −15.2 °C) with Th = 76–123 °C (Table 1). The freezing temperatures (−52/−65 °C) of the high-salinity inclusions suggest the presence of dissolved $CaCl_2$ in the aqueous fluid. In one case, the low- and high-salinity inclusions occurred as spatially associated in a single relatively large three-dimensional cluster situated around the center of the crystal without clear evidence of the relative chronology of both types of fluids; another quartz crystal contained only high-salinity inclusions. This suggests two separate hydrothermal episodes involving two different types of fluids with contrasting compositions. The monophase fluid inclusions exhibited identical salinities to those of the neighboring two-phase fluid inclusions.

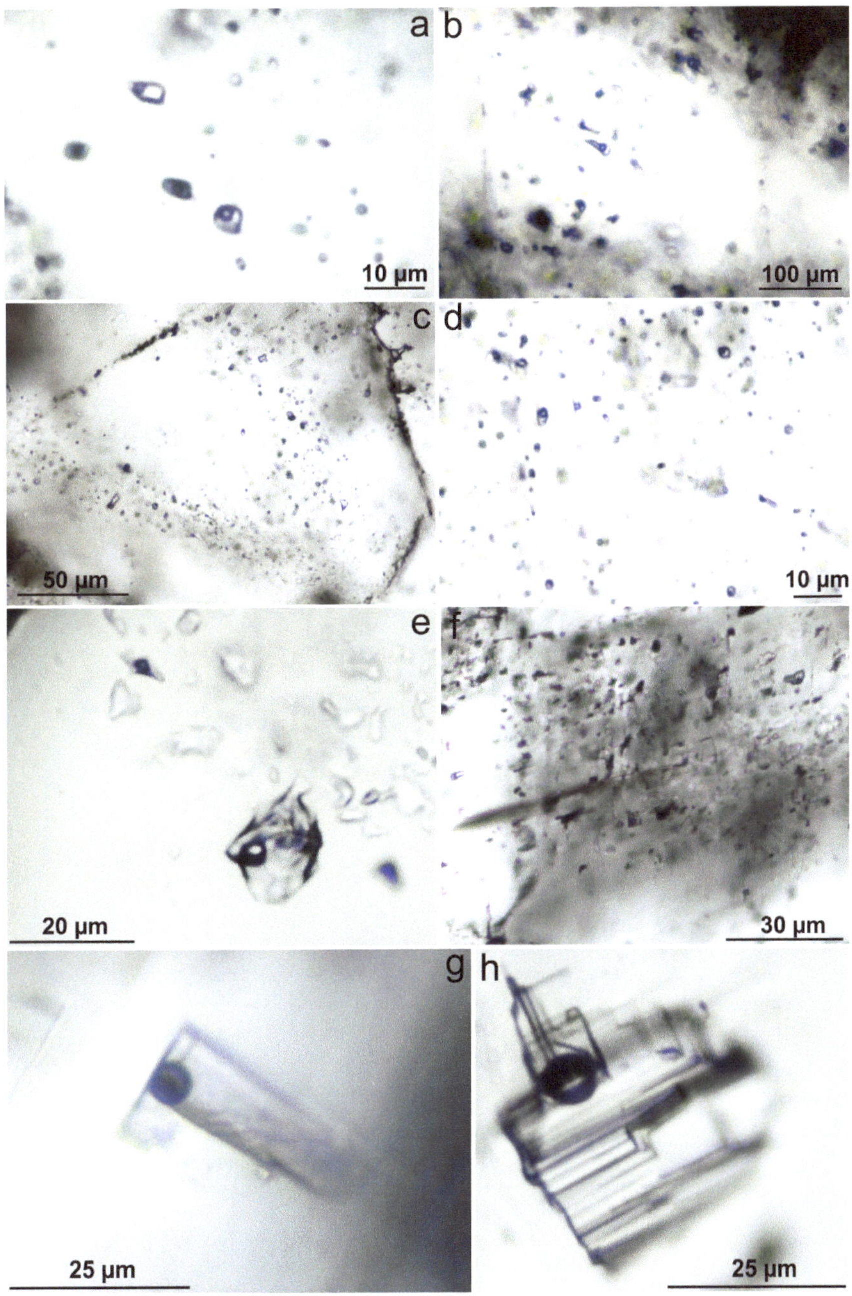

Figure 24. Examples of fluid inclusions from quartz Q-2 (**a**–**c**), quartz Q-3 (**d**), quartz Q-4 (**e**), dolomite (**f**,**g**), and calcite (**h**) from the studied gold-bearing vein.

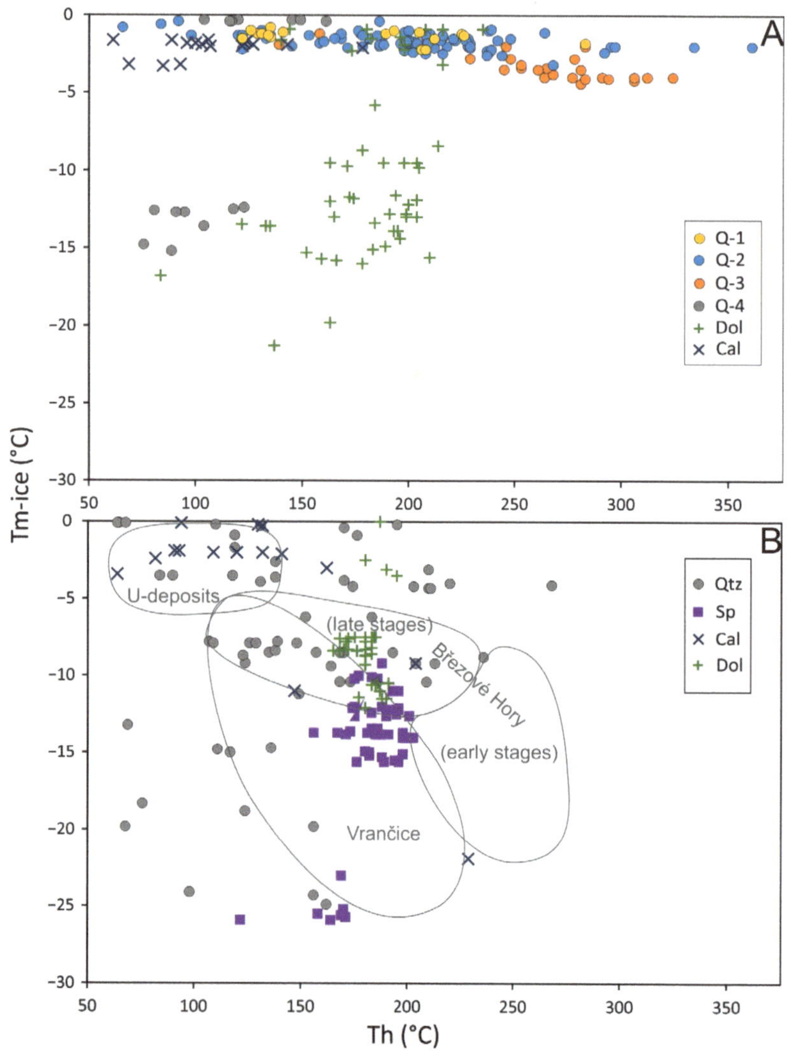

Figure 25. Total homogenization (Th) versus melting temperature of the last crystal of ice (Tm-ice) plot of fluid inclusions. (**A**) Data from the studied Au-bearing vein. (**B**) Comparative data from base-metal and uranium veins of the Příbram ore area. Symbols—data from base-metal vein Bt23C (Ulmanová et al. [5]) cutting the studied Au-bearing vein; fields—published data [87].

4.4.2. Fluid Inclusions in Carbonate Minerals

The primary fluid inclusions in dolomite–ankerite (Figure 24f,g) were situated on the growth zones or uniformly distributed in the crystals as solitary inclusions. They were mostly two-phase and liquid-rich (F ≈ 0.85–0.9) and less often all-liquid, with mostly regular morphology and sizes up to 35 μm. The L+V inclusions homogenized to liquid mostly in the wide range of 84–216 °C; two higher values (235 and 256 °C) probably represent partially leaked inclusions. Their salinities also have a range with a wide interval (1.6–23.4 wt. % NaCl eq. based on Tm-ice = −0.9/−21.3 °C; Table 1). The Ti = −56 °C observed in one high-salinity inclusion indicates the presence of H_2O-NaCl-$CaCl_2$ fluid, which is supported by the low freezing temperatures of most of the inclusions (T_F = −58 to −67 °C). Few of the low-salinity (≤5.3 wt. % NaCl eq.) inclusions froze out at −46 °C, which points to the absence of $CaCl_2$. Within a single growth zone, there occurred inclusions with a limited range of salinities within up to ~5 wt. %. A continuous change of fluid salinity was documented across various growth zones in a single grain of dolomite–ankerite, which is consistent with the evidence on the compositional changes of host mineral phase visible in the BSE images. The available fluid inclusion data suggest the mixing of at least two fluids with contrasting temperatures, salinities, and salt compositions during the crystallization of the dolomite–ankerite (Figure 25a).

The calcite-hosted fluid inclusions (Figure 24h) were mostly two-phase (F $\approx$ 0.85–0.9) and liquid-rich and sometimes monophase all-liquid, with sizes between 8 and 60 μm. They had mostly regular, three-dimensional morphology sometimes close to negative crystal shape and were situated on growth zones (primary inclusions) or within short trails (pseudo-secondary inclusions). They homogenized mostly in the range of 62–127 °C; two higher values (143 and 178 °C) probably represent partially leaked inclusions. The last crystal of ice melted between −1.5 and −3.2 °C, which corresponds to salinities of 2.6–5.4 wt. % NaCl eq. (Table 1). The temperature of the initial melting was not observed; however, freezing temperatures down to −48 °C rule out the presence of dissolved $CaCl_2$ in the aqueous fluid.

5. Discussion

5.1. Paragenetic Sequence

A multistage origin of the studied mineral assemblage is clearly indicated from the textural relationship of the ore and gangue minerals, although the construction of a paragenetic scheme is difficult due to the frequent presence of isolated grains of many phases and thus the absence of mutual contacts among many minerals. The situation is also complicated by frequent replacement of older mineral phases by younger ones, which took place especially in the case of the pairs: Au-Ag alloys→aurostibite, Au-Ag alloys→galena, aurostibite→unnamed (Au,Ag)Sb oxide, tsumoite/hedleyite→hessite, or tellurides→tsumoite. The detailed studies of the chemical compositions of the carbonates present in the various (often complicated) textural positions showed the fully comparable ranges of the chemical compositions of matrix- and veinlet-hosted carbonates, suggesting the presence of only a single generation of each carbonate. This finding stressed the role of the replacement processes involving gangue minerals during the younger stages of the evolution of the studied vein hydrothermal system. Based on the textural relationships, the overall carbonate sequence siderite→dolomite-ankerite→calcite took place. The generalized paragenetic sequence is illustrated in Figure 26.

The available data suggest the presence of two main mineralization events: the older Au(-Bi) stage and the younger Ag-Sb-base metal one. The early Au(-Bi) mineralization comprises two substages; the earlier one is characterized by the crystallization of arsenopyrite and quartz Q-1, which is followed after its cataclasis by bismuth I and Au-rich gold associated with the gangue composed of quartz Q-2 and Q-3, white mica, chlorite, apatite, TiO_2 mineral, monazite, and K-feldspar. The later Ag-Sb-base metal stage was initiated by the invasion of the Ag-Sb-rich fluids, which partly reacted with the native metals deposited during the Au(-Bi) stage, resulting in the formation of Ag-rich Au-Ag alloys, aurostibite, and probably also unnamed (Au,Ag)Sb oxide. Because the formation of these phases is spatially restricted only to the grains of pre-existing Au-rich Au-Ag alloys, the relationship of these ore phases to the oldest carbonate, represented by siderite, remains texturally unconfirmed. However, with respect to the completely absent signs of the alteration of siderite, which is expected to occur during the high-Eh episode associated with the formation of the unnamed (Au,Ag)Sb oxide (see below or Tolstykh et al. [54]), we suggest that siderite postdates the formation of (Au,Ag)Sb oxide. The remobilization of native bismuth, which occurred during the following substage, characterized by the overall formation of quartz (Q-4) and dolomite–ankerite gangue, led to the crystallization of the rare earlier Bi-tellurides (tetradymite, Fe-sulphotsumoite, tellurobismutite, pilsenite, hedleyite), followed by the more abundant tsumoite. The Bi-tellurides were later partly replaced by the assemblage hessite+bismuth II. The last substage is characterized by the formation of calcite gangue containing grains of base-metal sulfides. The position of acanthite and native antimony, which are very scarce in the vein fill, remains unresolved. They could be potentially related to either the activity of Ag-Sb-rich fluids initializing the Ag-Sb-base metal stage or the formation of the late calcite-hosted polymetallic sulfidic assemblage. Similarly, the formation of Sb oxide might be associated with the increase in redox potential associated with the formation of unnamed (Au,Ag)Sb oxide, but the textural evidence for coeval formation of both phases is missing.

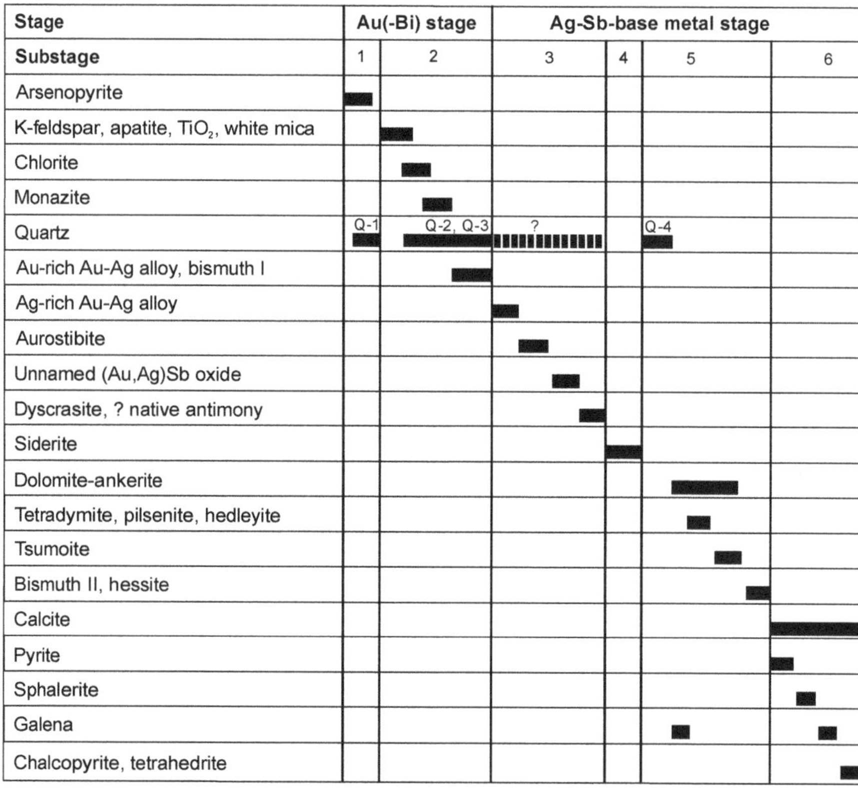

Figure 26. Generalized paragenetic sequence of the studied Au-bearing vein from Bytíz.

5.2. Formation Conditions

All the fluid inclusions present in the studied host minerals show (i) rather stable phase proportions, (ii) limited ranges of homogenization temperatures, and (iii) only a liquid homogenization mode, indicating that the fluids were trapped from homogeneous aqueous solutions. In such a situation, the P–T conditions of the studied mineralization can be interpreted using a combination of the isochores of the fluid inclusions and independent estimates of the temperature and/or pressure (see, e.g., Morishita [88]; Morishita and Nishio [89]).

5.2.1. Au(-Bi) Stage

The earliest portion of the studied vein is composed of arsenopyrite and quartz Q-1. The As content of the arsenopyrite from the gold-bearing quartz veins at Bytíz varies from 28.3 up to 32.7 at. % As (Table S1). Its apparently bimodal distribution with peaks at ~30 and ~32 at. % As may indicate two episodes of arsenopyrite formation. For the application of arsenopyrite thermometry [25,26], knowledge of sulfur fugacity is required. This may represent a difficult task in the absence of coexisting phases such as löllingite, pyrrhotite, or pyrite. We can therefore rely only on the presence of native Bi and on analogy with other gold deposits in the region (e.g., Krásná Hora, Němec, and Zachariáš [46]). This suggests an arsenopyrite formation at temperatures of 400–300 °C (Figure 27). This range is analogous to the formation temperature of early quartz veins with arsenopyrite from the Krásná Hora deposit (28.2–30.7 at. % As [46]). The two deposits thus show significantly lower As contents in the arsenopyrite than in the "intrusion-related gold deposits" of the regions of the Mokrsko (31.0–36.5 at. % As; Zachariáš et al. [53]) and Petráčkova Hora (35.9–38.3 at. % As; Zachariáš et al. [90]) deposits, where the arsenopyrite crystallized at significantly higher temperatures. As the temperatures obtained by arsenopyrite thermometry can be sometimes overestimated (e.g., [25,91,92]), an additional comparison with the available temperature data is necessary. It is evident from the geological situation that the studied Au-bearing vein is a Variscan mineralization, which was formed after the emplacement of the granitic rocks of the Central Bohemian Plutonic Complex (Figure 3). The ***regional*** Variscan thermal overprint of Neoproterozoic sediments is poorly constrained, but the available data suggest that it was comparable or slightly above 100–310 °C, which is suggested by the overlying Paleozoic sediments (Suchý et al. [93]). However, the much more intensive ***local*** thermal overprint of Neoproterozoic rocks is documented in an up to 2 km wide area adjacent to the intrusion of the Central Bohemian Plutonic Complex [94], which also includes the study site (Figure 3). Moreover, there are present hydrothermal veins in the Neoproterozoic sediments of the Teplá-Barrandian unit, which crystallized from fluids, whose temperature was higher than the temperature of the regional Variscan thermal overprint (see, e.g., Žák and Dobeš [87]; Dolníček et al. [95]). Hence, it can be concluded that the temperature estimates based on arsenopyrite thermometry need not be necessarily unrealistic. Because the character of the content of the fluid inclusions hosted by the quartz Q-1 associated with arsenopyrite is questionable, the P–T formation conditions of this mineralizing event cannot be checked by this technique. In any case, it is probable that the pressure during this early stage was comparably high or lower than the pressures calculated for the crystallization of the granitoids of the adjacent part of the Central Bohemian Plutonic Complex (1.7–2.8 kbar; Janoušek et al. [96]).

There are several lines of evidence that show that the younger portion of the Au(-Bi) stage formed at lower temperatures. A temperature range between 210 and 320 °C is indicated from chlorite compositional thermometry, based on the amount of tetrahedral Al [77]. We have excluded from this estimate the temperatures obtained from the alkali-enriched chlorite analyses, in which the amount of Si is shifted by the admixture of illite and/or smectite, giving rise to underestimated temperature values (Figure 20). In accordance with this temperature range is also the presence of K-feldspar in the vein paragenesis, which is stable at temperatures of at least ca. 250 °C under hydrothermal conditions [97]. A temperature variability during this younger portion of the Au(-Bi) stage may be also indicated by the highly variable Eu anomaly [98] in monazite. The negative Eu anomaly is typical for a high-temperature environment and/or reducing conditions, in which Eu is present in a divalent state (Eu^{2+}), which has, in comparison with Eu^{3+}, too large size to be incorporated into the crystal structure of many hydrothermal minerals. The evolution towards the positive Eu anomaly may thus suggest a decrease in the temperature of the hydrothermal fluid below the value, under which the thermochemical oxidation of Eu^{2+} to Eu^{3+} occurs. The exact temperature depends on the chemical composition of the fluid, but for common (non-extreme) fluid compositions, it is close to ca. 200 °C [99–101], which is in line with the lower temperature limit suggested by chlorite thermometry. Alternatively, a change in the character of the Eu anomaly can be explained by an increase in the redox potential of the initially low-Eh fluids (see, e.g., Lee et al. [102]), which can also be compatible with the elevated contents of the redox-sensitive elements present in the high oxidation state (i.e., S^{6+} and As^{5+}) in some of the studied monazites. Nevertheless, the available data suggest a long-lasting evolution of the hydrothermal system at this stage, associated with a progressive cooling from ca. 320 °C to ca. 180 °C (Figure 28). This interpretation is also supported by the fluid inclusion data from the contemporaneous Q-2 and Q-3 quartz: the distinct sub-horizontal data distribution in the Th-Tm plot (Figure 25a) can be associated with the changes in temperature, pressure, or both variables during the long-lasting evolution of a single fluid. The pressure conditions were estimated using the above-mentioned independent thermometers and isochors of the fluid inclusions present in the associated quartz (Figure 28). At this substage, we again presuppose that the formation pressure did not exceed those attained during the crystallization of the granitoids, which seems reasonable in the context of the late-Variscan evolution of the area commonly characterized by rapid uplift and erosion (Dörr and Zulauf [103]).

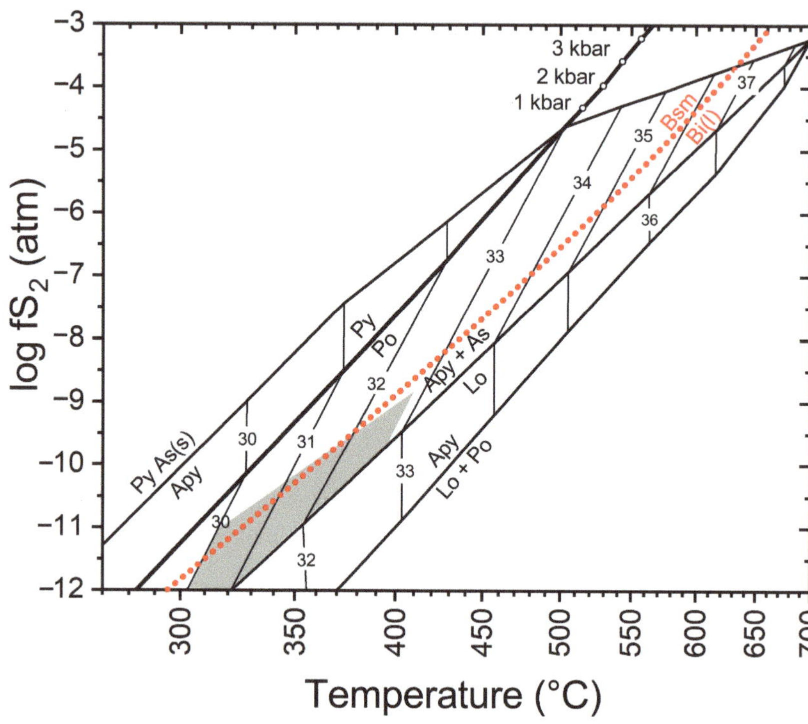

Figure 27. Phase diagram of the Fe-As-S system showing the correlation between the As content of arsenopyrite (at. %) and temperature and sulfur fugacity. The red dotted line shows the stability of bismuthinite (Bsm) and liquid bismuth (Bi (l)). The shaded field indicates estimated conditions of formation of arsenopyrite from the studied vein, based on the most typical range of As (at. %) content.

The unusually large spread of the La_N/Sm_N ratios of monazite, exceeding those of most occurrences of hydrothermal monazites (Figure 21), suggests a wide range of the fractionation of REE. This is possible to achieve in hydrothermal fluids enriched in strong REE-fractionating ligands, especially fluoride, hydroxide, or carbonate anions, which form, according to the experimental works, the strongest complexes with REE, having thus the best potential for fractionation of REE (see, e.g., Wood [104]; Bau and Möller [101]; Migdisov and Williams-Jones [105]). The mineral composition of the studied mineralization together with the chemical composition of the fluid inclusions indicate that hydroxide and carbonate anions were probably not the case as they are stable in strongly alkaline environment only. By contrast, the fluoride anions may be suitable candidates since elevated contents of F were present in the parent fluids, as is suggested by the chemical composition of apatite.

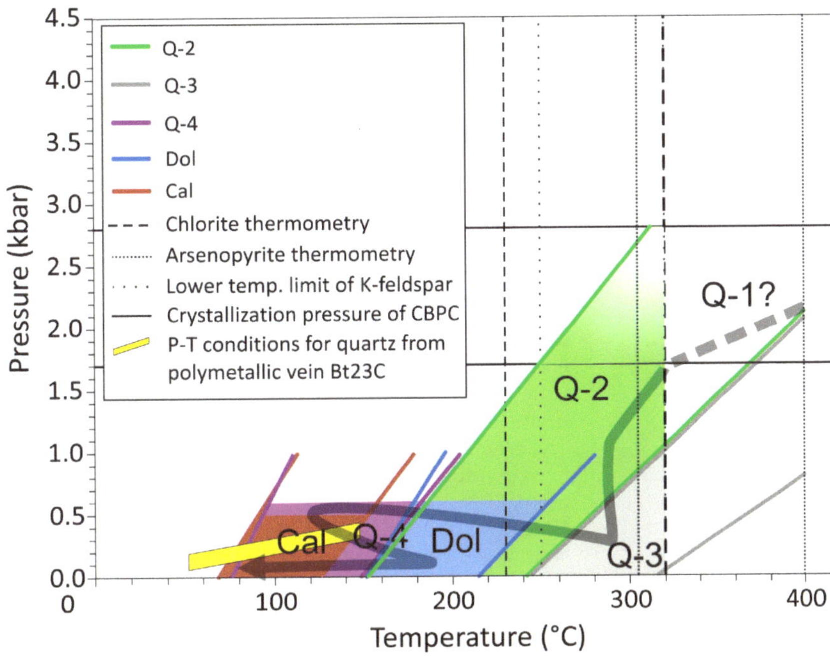

Figure 28. Estimated P–T conditions for the studied Au-bearing vein based on a combination of isochors of fluid inclusions and independent temperature/pressure estimates. Thick grey curve illustrates possible evolution trend. The P–T field for late quartz from base-metal vein Bt23C is based on fluid inclusions containing immiscible aqueous-carbonic fluids (Ulmanová et al. [5]).

The ultimate primary source of gold for the discussed early mineralization remains unclear. Arsenopyrite contains an elevated content of gold in the range between 0.00041 and 0.00066 *apfu*. Such rather homogeneous values suggest that Au atoms are bound in the structure of arsenopyrite (i.e., "refractory gold") and not as (sub)microscopic inclusions of the native metal. The Au-bearing arsenopyrite may thus be potentially suggested as a source of Au for the subsequent deposition of the relatively coarse-grained, high-fineness native metal, assuming the remobilization of Au from arsenopyrite by younger fluids. Although the arsenopyrite was the subject of intense fracturing and cataclasis, there are observed no marks of a pronounced dissolution of this mineral phase by the younger fluids. Therefore, it seems probable that the Au for the precipitation of high-fineness Au-Ag alloys was brought by hydrothermal fluid. The clear spatial association of the arsenopyrite and the grains of high-fineness Au-Ag alloys might thus be explained by arsenopyrite as a suitable geochemical (or electrochemical) substrate for Au precipitation from Au-bearing fluid.

5.2.2. Ag-Sb-Base Metal Stage

Although the variations in the chemical composition of Au-Ag alloys can be caused by changes in some parameters (Au/Ag ratio, temperature, sulfur fugacity, pH, Eh, chloride concentration) of a single fluid (Shikazono and Shimizu [106]; Morrison et al. [107]; Gammons and Williams-Jones [108]), we suggest that our mineralogical and chemical evidence suggests the participation of another fluid with a contrasting composition. The formation of the early portion of the younger Ag-Sb-base metal stage (Figure 26) was initiated by the infiltration of Sb-rich fluids, which altered the grains of the early Au-Ag alloy, giving rise to replacement rims or total pseudomorphs formed by aurostibite. The increase in Ag content in fluids due to leaching of silver from dissolved high-fineness Au-Ag grains and the precipitation of virtually Ag-absent aurostibite, possibly combined with the primary high content of Ag in the Sb-rich fluids, led to the crystallization of the younger population of Au-rich dyscrasite and Au-Ag alloys with lower fineness and sometimes also increased the content of Sb. The end of this episode was characterized by an increase in the oxygen fugacity of the fluid, allowing the formation of unnamed (Au,Ag)Sb oxide (Tolstykh et al. [54]) and possibly also native antimony and an Sb oxide. The direct evidence of temperature conditions during this early portion of the Ag-Sb-base metal stage is missing.

The mineral composition of the younger portion of the Ag-Sb-base metal stage can be largely parallelized with the typical vein base-metal mineralization of the Příbram ore district. The base-metal mineralization is usually characterized by the predominance of carbonates over quartz in the vein fill, with the typical carbonate sequence siderite→dolomite-ankerite→calcite and the presence of common Pb-Zn(-Cu-Ag) sulfides. The only difference is in the presence of tellurides in the dolomite gangue from the studied Au-bearing vein. Nevertheless, accessory tellurides locally appeared in the earlier gold-bearing vein mineralization of this district [8–10], from which they could be potentially remobilized. The chemical composition of the carbonates and sulfides also follows identical trends in both the studied Au mineralization and the typical base-metal ore veins of the Příbram ore district. Genetic similarities confirmed the data from the fluid inclusion study by Ulmanová et al. [5], which included samples from base-metal vein Bt23C crosscutting the studied Au-bearing vein, which yielded broadly overlapping data clusters (Figure 25). There is also the coincidence with the data published by Žák and Dobeš [87] for base-metal ore mineralization from other localities in the Příbram ore district (Figure 25b). Taken together, the paragenetic, chemical, and fluid inclusion data clearly indicate that the younger portion of the mineral assemblage of the studied Au-bearing vein originated from fluids giving rise to the base-metal mineralization of the Příbram ore district. The temperature conditions decreased from ca. 250 to 50 °C during the paragenetic evolution of the base-metal mineralization (cf. Žák and Dobeš [87]; Dolníček et al. [6]), which is consistent with our fluid inclusion data (Figure 28). The uppermost temperature limit at 250 °C is in accordance with the nature of assemblage of the native bismuth and hedleyite (Figure 8d), whose textural arrangement is far from those indicating their crystallization from the melt, which is stable above 266 °C [75]. The pressure conditions were below ca. 0.5 kbar during this stage (Žák and Dobeš [87]; Ulmanová et al. [5]).

5.2.3. Origin of Fluids

The composition of the fluids involved in the formation of the early Au(-Bi) stage at Bytíz is different from those characterizing the gold-bearing veins in the wider area, in which the circulation of aqueous solutions postdates the activity of the aqueous-carbonic fluids (see, e.g., Zachariáš et al. [90,109,110]). The source of low-salinity, gas-absent aqueous fluids was likely not in the magmatogenic fluids exsolved during the crystallization of the granitoids of the Central Bohemian Plutonic Complex (see, e.g., Zachariáš et al. [90]; Žák et al. [110]), but such fluids can be potentially generated by the thermal alteration of the host Neoproterozoic sediments caused by the intrusion of granitoid rocks. Such a scenario would also be consistent with the relatively high temperatures inferred from arsenopyrite thermometry. Nevertheless, future specialized (i.e., isotopic) investigations are needed to test this hypothesis.

The origin of base-metal fluids is better constrained by the previous fluid inclusion and stable isotope studies. In this case, the participation of three fluid endmembers is suggested in the Příbram ore area. The early base-metal mineralization of the siderite–sulfidic stage crystallized from the aqueous fluids characterized by high temperatures (200–300 °C), high salinities (14–23 wt. % NaCl eq.), strongly positive $\delta^{18}O$ values (+6 to +10 ‰ V-SMOW), and a well-mixed $\delta^{13}C$ signature (−16 to −15 ‰ V-PDB), which are interpreted to be sourced from the deep circulation in the hot crustal rocks [5,6,87]. The superimposed uranium mineralization was formed from low-temperature (<150 °C), low-salinity (<5 wt. % NaCl eq.), near-zero $\delta^{18}O$ (−3 to +3 ‰ V-SMOW) fluids, which were dominated by shallowly circulating surficial waters [5,6,87]. Finally, the participation of low-temperature (<150 °C), high-salinity (up to 30 wt. % NaCl eq.) Ca-Na-Cl solutions was found during the late stages of hydrothermal activity, representing either external brines of marine provenance or locally produced "shield brines" [5,6,87].

6. Conclusions

(1) The studied gold-bearing quartz–sulfide vein encountered in the deep mine works of the Bytíz deposit shows several anomalous features when compared with the typical gold-bearing veins in the wider area: (i) an extremely wide range of composition of gold (4–69 at. % Ag); (ii) the presence of Bi-tellurides bound to younger dolomite–ankerite veinlets; (iii) anomalously high amounts of base metals, Ag, and Sb bound especially to the younger portion of the vein fill; and (iv) the contrasting characteristics of the fluid inclusions from early and late portions of the given vein mineralization.

(2) The commonly observed disequilibrium textures, manifesting various stages of replacement of older ore and gangue minerals by younger ones, are consistent with an idea of the intense re-working of old gold-bearing veins by younger Ag,Sb-rich fluids in distinct physico-chemical conditions. The most common is the replacement of gold by aurostibite and an unnamed (Au,Ag)-Sb oxide, Bi-tellurides by hessite, and siderite by calcite. Such textures recorded at paragenetically distinct phases are indicative of the repeated dissolution and/or re-mobilization of previously deposited ores.

(3) The fluid inclusion study confirmed the differences in the composition and temperatures of fluids from the older and younger portion of mineralization. Whereas the early mineralization originated from progressively cooled high-temperature (ca. 400 to ca. 180 °C), low-salinity (1.2–7.0 wt. % NaCl eq.) aqueous fluids, the younger portion of mineralization formed from aqueous fluids with highly variable salinity (0.2–23.4 wt. % NaCl eq.) and homogenization temperatures decreasing from ca. 250 °C to <50 °C.

(4) The fluid characteristics of the late mineralization of the studied Au-bearing vein compare well to those of the base-metal mineralization of the vein Bt23C, which crosscut the Au-bearing vein at the sampling site. The data also fit those from other base-metal veins of the Příbram ore area. Hence, it is evident that the observed mineralogical changes were associated with activity of the fluids, giving rise to the common late-Variscan base-metal vein mineralization of the Příbram area.

Supplementary Materials: The following supporting information can be downloaded at: https://www.mdpi.com/article/10.3390/min12121539/s1, Tables S1–S29: chemical composition of individual mineral phases.

Author Contributions: Conceptualization, J.S. and Z.D.; methodology, J.S., Z.D. and P.Š.; investigation, J.S., Z.D., J.Z., J.U. and V.Š.; resources, V.Š. and P.Š.; writing—original draft preparation, J.S. and Z.D.; writing—review and editing, J.S. All authors have read and agreed to the published version of the manuscript.

Funding: This research was financially supported by grant GAČR 19-16218S (Czech Science Foundation) and by the Ministry of Culture of the Czech Republic (long-term project DKRVO 2019–2023/1.I.d; National Museum, 00023272).

Data Availability Statement: All data are contained in this work and Supplementary Materials.

Acknowledgments: The authors thank Jaroslav Jindra (Příbram) for providing research samples. We must also thank Jiří Litochleb (deceased, National Museum, Praha) for his cooperation in the ore microscopy and succession studies. The referees as well as the academic editor Yuichi Morishita are acknowledged for comments and suggestions that helped to improve the early version of the manuscript.

Conflicts of Interest: The authors declare no conflict of interest. The funders had no role in the design of the study; in the collection, analyses, or interpretation of data; in the writing of the manuscript; or in the decision to publish the results.

References

1. Ettler, V.; Sejkora, J.; Drahota, P.; Litochleb, J.; Pauliš, P.; Zeman, J.; Novák, M.; Pašava, J. Příbram and Kutná Hora mining districts—From historical mining to recent environmental impact. *Acta Miner.–Petr. Field Guide Ser.* **2010**, *7*, 1–23.
2. Litochleb, J.; Černý, P.; Litochlebová, E.; Sejkora, J.; Šreinová, B. The deposits and occurrences of mineral raw materials in the Střední Brdy Mts. and the Brdy piedmont area (Central Bohemia). *Bull. Mineral.–Petrol. Oddělení Národního Muz. V Praze* **2003**, *12*, 57–86. (In Czech)
3. Kutina, J. Genetische Diskussion der Makrotexturen bei der geochemischen Untersuchung des Adalbert-Hauptganges in Příbram. *Chem. Erde* **1955**, *17*, 241–323.
4. Sejkora, J.; Škácha, P.; Plášil, J.; Dolníček, Z.; Ulmanová, J. Hrabákite, Ni_9PbSbS_8, a new member of the hauchecornite group from Příbram, Czech Republic. *Mineral. Mag.* **2021**, *85*, 189–196. [CrossRef]
5. Ulmanová, J.; Dolníček, Z.; Škácha, P.; Sejkora, J. Fluid inclusions in Zn-Pb mineralization of the vein Bt23C, Bytíz deposit, Příbram uranium and base-metal ore district, Czech Republic: The role of immiscible aqueous-carbonic fluids. *Mineral. Depos. submitted* 2022.
6. Dolníček, Z.; Ulmanová, J.; Sejkora, J.; Knížek, F.; Škácha, P. Mineralogy and genesis of the Pb-Zn-Sb-Ag vein H32A in the Příbram uranium and base-metal district, Bohemian Massif, Czech Republic. *Ore Geol. Rev.* **2022**. *submitted.*
7. Komínek, J. *Geology of the Wide Surroundings and of the Deposit, Part I and II*; Final Report on the uranium district Příbram; Unpublished Report DIAMO; DIAMO: Příbram, The Czech Republic, 1995; 418p. (In Czech)
8. Litochleb, J.; Knížek, F.; Šrein, V. Gold-bearing mineralization in the Příbram uranium deposit. *Sbor. konf. Hornická Příbram ve vědě a Tech. Sekce Ložisková Geol.* **1986**, 318–325. (In Czech)
9. Litochleb, J.; Černý, P.; Šrein, V.; Lantora, M.; Sejkora, J. Mineralogy of the gold-bearing mineralization in the underground gas reservoir at Háje near Příbram (central Bohemia). *Bull. Mineral.-Petrolog. Odd. Nár. Muz.* **2000**, *8*, 195–201. (In Czech)
10. Litochleb, J.; Šrein, V. The chemical composition of bismuth and tellurium minerals from the gold-bearing veins of the Příbram uranium deposit. *Acta Univ. Carol. Geol.* **1989**, *4*, 511–519. (In Czech)
11. Škácha, P.; Sejkora, J.; Plášil, J. Selenide mineralization in the Příbram uranium and base-metal district (Czech Republic). *Minerals* **2017**, *7*, 91. [CrossRef]
12. Knížek, F.; Litochleb, J.; Šrein, V. Dyscrasite and allargentum from the Háje vein bundle of the Příbram uranium deposit. *Věst. Ústř. Úst. Geol.* **1990**, *65*, 321–328.
13. Sejkora, J.; Litochleb, J.; Knížek, F.; Škácha, P. The characteristic of the Ag-occurrences on the H61Z vein under the 6th level, shaft No. 21, Příbram uranium and base-metal district. *Bull. Mineral.-Petrolog. Odd. Nár. Muz.* **2007**, *14–15*, 107–113. (In Czech)
14. Škácha, P.; Sejkora, J.; Knížek, F.; Slepička, V.; Litochleb, J.; Jebavá, I. The occurrences of the unique monometallic silver ores on the vein H14F3 between 7th and 9th level of the shaft No. 21 Háje, Příbram uranium and base-metal district (Czech Republic). *Bull. Mineral.-Petrolog. Odd. Nár. Muz.* **2012**, *20*, 230–254. (In Czech)
15. Anderson, E.B. *Isotopic-Geochronological Investigation of the Uranium Mineralization of Czechoslovakia*; Unpublished Czechoslovak Uranium Industry Report; Archiv of the State Enterprise Diamo: Příbram-Bytíz, Czech Republic, 1987; pp. 1962–1987.
16. Řídkošil, T.; Knížek, F. Zeolite mineralization and occurrence of whewellite at the Příbram uranium deposit. *Sbor. Semin. "Miner. uránovych a s nimi súvisiacich nerastných surovín" Čingov* **1987**, 126–130. (In Czech)
17. Žák, K.; Skála, R. Carbon isotopic composition of whewellite ($CaC_2O_4 \cdot H_2O$) from different geological environments and its significance. *Chem. Geol.* **1993**, *106*, 123–131. [CrossRef]
18. Knížek, F.; Litochleb, J. Calcite drusy cavities with whewellite on the Bytíz uranium deposit (the Příbram uranium-polymetallic district, central Bohemia). *Bull. Mineral.-Petrolog. Odd. Nár. Muz.* **2005**, *13*, 73–77. (In Czech)
19. Pouchou, J.L.; Pichoir, F. "PAP" (φρZ) procedure for improved quantitative microanalysis. In *Microbeam Analysis*; Armstrong, J.T., Ed.; San Francisco Press: San Francisco, CA, USA, 1985; pp. 104–106.
20. Anders, E.; Grevesse, N. Abundances of the elements: Meteoritic and solar. *Geochim. Cosmochim. Acta* **1989**, *53*, 197–214. [CrossRef]
21. McLennan, S.M. Rare earth elements in sedimentary rocks: Influence of provenance and sedimentary processes. *Rev. Mineral.* **1989**, *21*, 169–200.
22. Bodnar, R.J. Revised equation and table for determining the freezing point depression of H_2O-NaCl solutions. *Geochim. Cosmochim. Acta* **1993**, *57*, 683–684. [CrossRef]
23. Bakker, R.J. Package FLUIDS 1. Computer programs for analysis of fluid inclusion data and for modelling bulk fluid properties. *Chem. Geol.* **2003**, *194*, 3–23. [CrossRef]
24. Knight, C.L.; Bodnar, R.J. Synthetic fluid inclusions: IX. Critical PVTX properties of NaCl-H_2O solutions. *Geochim. Cosmochim. Acta* **1989**, *53*, 3–8. [CrossRef]
25. Kretschmar, U.; Scott, S.D. Phase relations involving arsenopyrite in the system Fe-As-S and their application. *Can. Mineral.* **1976**, *14*, 364–386.
26. Sharp, Z.D.; Essene, E.J.; Kelly, W.C. A re-examination of the arsenopyrite geothermometer: Pressure considerations and applications to natural assemblages. *Can. Mineral.* **1985**, *23*, 517–534.

27. Mikulski, S.Z. The arsenopyrite geothermometer—A difficult application to arsenopyrite from the Radzimowice abandoned Au-As-Cu deposit (Kaczawa Mountains). *Mineral. Polonica Spec. Papers* **2005**, *26*, 209–212.
28. Andráš, P.; Chovan, M. Gold incorporation into sulphide minerals from the Tatric Unit, the Western Carpathians, with respect to their chemical composition. *J. Czech Geol. Soc.* **2005**, *50*, 143–155. [CrossRef]
29. Biagioni, C.; George, L.G.; Cook, N.J.; Makovicky, E.; Moëlo, Y.; Pasero, M.; Sejkora, J.; Stanley, C.J.; Welch, M.D.; Bosi, F. The tetrahedrite group: Nomenclature and classification. *Am. Mineral.* **2020**, *105*, 109–122. [CrossRef]
30. Abrikosov, N.C.; Bankina, V.F. Study of phase diagram of the system Bi–Te. *Zh. Neorg. Chim.* **1958**, *3*, 659–667. (In Russian)
31. Shelimova, L.E.; Karpinsky, O.G.; Kosyakov, V.I.; Shestakov, V.A.; Zemskov, V.S.; Kuznetsov, F.A. Homologous series of layered tetradymite-like compounds in Bi–Te and GeTe–Bi_2Te_3 systems. *J. Struct. Chem.* **2000**, *41*, 81–87. [CrossRef]
32. Yusa, K.; Kitakaze, A.; Sugaki, A. Synthesized bismuth—Tellurium—Sulfur system minerals—Synthetic sulfide minerals (IX). *Sci. Rep. Tohoku Univ. 3rd Ser.* **1979**, *14*, 121–133.
33. Cook, N.J.; Ciobanu, C.L.; Wagner, T.; Stanley, C.J. Minerals of the system Bi-Te-Se-S related to the tetradymite archetype: Review of classification and compositional variation. *Can. Mineral.* **2007**, *45*, 665–708. [CrossRef]
34. Cook, N.J.; Ciobanu, C.L. Bismuth tellurides and sulphosalts from the Larga hydrothermal system, Metaliferi Mts., Romania: Paragenesis and genetic significance. *Mineral. Mag.* **2004**, *68*, 303–321. [CrossRef]
35. Ferenc, Š. New occurrences of Te minerals in the western part of the Slovenské rudohorie Mts. *Mineral. Slov.* **2004**, *36*, 317–322. (in Slovak).
36. Mikulski, S.Z. Te-Bi-Au-Ag-Pb-S mineral assemblages within the late Hercynian polymetallic deposit in the western Sudetes (Poland). In *Guidebook of the International Field Workshop of IGCP-486*; Alba Iulia, Romania, 2004; pp. 242–244.
37. Gamyanin, G.N.; Leskova, N.V.; Vyalsov, L.N.; Laputina, I.P. Bismuth tellurides—Bi_2Te and BiTe—in deposits of northeastern USSR. *Zap. Vses. Mineral. Obshchest.* **1980**, *109*, 230–235. (In Russian)
38. Gamyanin, G.N.; Leskova, N.V.; Vyalsov, L.N.; Laputina, I.P. Bismuth tellurides—Bi_2Te and BiTe—in the deposits of the northeast USSR. *Int. Geol. Rev.* **1982**, *24*, 451–456. [CrossRef]
39. Goncharov, V.I.; Alshevskiy, A.V.; Vortsepenev, V.V.; Sergeyeva, N.Y.; Savva, N.Y. New discovery of the bismuth telluride Bi_2Te in northeastern USSR. *Dokl. Akad. Nauk SSSR* **1984**, *275*, 717–720. (In Russian)
40. Huang, D.; Ding, X.; Wu, C.; Zhang, C. Mineral characteristics and occurrence of gold, silver and bismuth of the Caijaying lead—zinc—silver deposit, Hebei Province. *Acta Geol. Sin.* **1991**, *65*, 127–140. (In Chinese)
41. Luukkonen, A. Main geological features, metallogeny and hydrothermal alteration phenomena of certain gold and gold—tin—tungsten prospects in southern Finland. *Geol. Surv. Finl. Bull.* **1994**, *377*, 1–153.
42. Gu, X.P.; Watanabe, M.; Hoshino, K.; Shibata, Y. Mineral chemistry and associations of Bi–Te(S,Se) minerals from China. *Neues Jahrbuch für Mineralogie–Monatshefte.* **2001**, 289–309.
43. Houzar, S.; Litochleb, J.; Sejkora, J.; Cempirek, J.; Cicha, J. Unusual mineralization with niobian titanite and Bi-tellurides in scheelite skarn from Kamenné doly quarry near Písek, Moldanubian Zone, Bohemian Massif. *J. Geosci.* **2008**, *53*, 1–16. [CrossRef]
44. Okamoto, K.; Tanner, L.E. Bi–Te (bismuth–tellurium). In *Binary Alloy Phase Diagrams*; Massalski, T.B., Ohamoto, K., Eds.; ASM International: Materials Park, OH, USA, 1990; pp. 800–801.
45. Zachariáš, J.; Němec, M. Gold to aurostibite transformation and formation of Au-Ag-Sb phases: The Krásná Hora deposit, Czech Republic. *Mineral. Mag.* **2017**, *81*, 987–999. [CrossRef]
46. Němec, M.; Zachariáš, J. The Krásná Hora, Milešov, and Příčovy Sb-Au ore deposits, Bohemian Massif: Mineralogy, fluid inclusions, and stable isotope constraints on the deposit formation. *Mineral. Depos.* **2018**, *53*, 225–244. [CrossRef]
47. Kalinin, A.A.; Savchenko, Y.E.; Selivanova, E.A. Mustard Gold in the Oleninskoe Gold Deposit, Kolmozero–Voronya Greenstone Belt, Kola Peninsula, Russia. *Minerals* **2019**, *9*, 786. [CrossRef]
48. Pak, S.J.; Choi, S.G.; Choi, S.H. Systematic mineralogy and chemistry of gold-silver vein deposits in the Taebaeksan district, Korea: Distal relatives of a porphyry system. *Mineral. Mag.* **2004**, *68*, 467–487. [CrossRef]
49. Scott, J.D. Refinement of the crystal structure of dyscrasite, and its implications for the structure of allargentum. *Can. Mineral.* **1976**, *14*, 139–142.
50. Cipriani, C.; Corazza, M.; Mazzetti, G. Reinvestigation of natural silver antimonides. *Eur. J. Mineral.* **1997**, *8*, 1347–1350. [CrossRef]
51. Zoro, E.; Servant, C.; Legendre, B. Thermodynamic modeling of the Ag-Au-Sb ternary system. *J. Phase Equilibr. Diffus.* **2007**, *28*, 250–257. [CrossRef]
52. Gamyanin, G.N.; Nekrasov, I.J.; Zhdanov, J.J.; Leskova, N.V. Auroantimonate, a new natural compound of gold. *Dokl. Akad. Nauk SSSR* **1988**, *301*, 947–950. (In Russian)
53. Zachariáš, J.; Morávek, P.; Gadas, P.; Pertoldová, J. The Mokrsko-West gold deposit, Bohemian Massif, Czech Republic: Mineralogy, deposit setting and classification. *Ore Geol. Rev.* **2014**, *58*, 238–263. [CrossRef]
54. Tolstykh, F.N.; Vymazalová, A.; Tuhý, M.; Shapovalova, M. Conditions of formation of Au–Se–Te mineralization in the Gaching ore occurrence (Maletoyvayam ore field), Kamchatka, Russia. *Mineral. Mag.* **2018**, *82*, 649–674. [CrossRef]
55. Dill, H.G.; Weiser, T.; Bernhardt, I.R.; Kilibarda, C.R. The composite gold-antimony vein deposit at Kharma (Bolivia). *Econ. Geol.* **1995**, *90*, 51–66. [CrossRef]
56. Makovicky, E.; Chovan, M.; Bakos, F. The stibian mustard gold from the Kriváň Au deposit, Tatry Mts., Slovak Republic. *Neues Jahrbuch für Mineralogie–Abhandlungen* **2007**, *184*, 207–215. [CrossRef]

57. Johan, Z.; Šrein, V. Un nouvel oxyde naturel de Au et Sb. *Comptes Rendus De L'académie Des Sci. -Ser. IIA-Earth Planet. Sci.* **1998**, *326*, 533–538. [CrossRef]
58. Maydagán, L.; Franchini, M.; Impiccini, A.; Lentz, D. Phyllosilicates geochemistry and distribution in the Altar porphyry Cu-(Au) deposit, Andes Cordillera of San Juan, Argentina: Applications in exploration, geothermometry, and geometallurgy. *J. Geoch. Explor.* **2016**, *167*, 83–109. [CrossRef]
59. Rieder, M.; Cavazzini, G.; D´yakonov, Y.S.; Kamenetskii, V.A.F.; Gottardi, G.; Guggenheim, S.; Koval´, P.V.; Mueller, G.; Neiva, A.M.R.; Radoslovich, E.W.; et al. Nomenclature of micas. *Can. Mineral.* **1998**, *36*, 905–912. [CrossRef]
60. de Siqueira Corrêa, R.; Oliveira, C.G.; Vidotti, R.M.; da Silva Souza, V. Regional-scale pressure shadow-controlled mineralization in the Príncipe Orogenic Gold Deposit, Central Brazil. *Ore Geol. Rev.* **2015**, *71*, 273–304. [CrossRef]
61. Klein, E.L.; Harris, C.; Giret, A.; Moura, C.A.V. The Cipoeiro gold deposit, Gurupi Belt, Brazil: Geology, chlorite geochemistry, and stable isotope study. *J. S. Am. Earth Sci.* **2007**, *23*, 242–255. [CrossRef]
62. Demange, M.; Pascal, M.-L.; Raimbault, L.; Armand, J.; Forette, M.C.; Serment, R.; Touil, A. The Salsigne Au-As-Bi-Ag-Cu deposit, France. *Econ. Geol.* **2006**, *101*, 199–234. [CrossRef]
63. Dolníček, Z.; Nepejchal, M.; Kapusta, J.; Ulmanová, J.; Fojt, B. Primary gold mineralization in the Andělské údolí Valley near Staré Město pod Sněžníkem (Czech Republic). *Bull. Mineral. Petrolog.* **2018**, *26*, 12–27. (In Czech)
64. Bayliss, P. Nomenclature of the trioctahedral chlorites. *Can. Mineral.* **1975**, *13*, 178–180.
65. Melka, K. A proposal to classification of chlorite minerals. *Věst. Ústř. Úst. Geol.* **1965**, *40*, 23–27. (In Czech)
66. Moura, M.A.; Botelho, N.F.; Olivo, G.M.; Kyser, T.K. Granite-related Paleoproterozoic, Serrinha gold deposit, southern Amazonia, Brazil: Hydrothermal alteration, fluid inclusion and stable isotope constraints on genesis and evolution. *Econ. Geol.* **2006**, *101*, 585–605. [CrossRef]
67. Buchholz, P.; Oberthür, T.; Lüders, V.; Wilkinson, J. Multistage Au-As-Sb mineralization and crustal-scale fluid evolution in the Kwekwe District, Midlands Greenstone Belt, Zimbabwe: A combined geochemical, mineralogical, stable isotope, and fluid inclusion study. *Econ. Geol.* **2007**, *102*, 347–378. [CrossRef]
68. Klein, E.L.; Ribeiro, J.W.A.; Harris, C.; Moura, C.A.V.; Giret, A. Geology and fluid characteristics of the Mina Velha and Mandiocal orebodies and implications for the genesis of the orogenic Chega Tudo gold deposit, Gurupi Belt, Brazil. *Econ. Geol.* **2008**, *103*, 957–980. [CrossRef]
69. Zoheir, B.A.; El-Shazly, A.K.; Helba, H.; Khalil, K.I.; Bodnar, R.J. Origin and evolution of the Um Egat and Dungash orogenic gold deposits, Egyptian Eastern Desert: Evidence from fluid inclusions in quartz. *Econ. Geol.* **2008**, *103*, 405–424. [CrossRef]
70. Zoheir, B.A. Controls on lode gold mineralization, Romite deposit, South Eastern Desert, Egypt. *Geosci. Front.* **2012**, *3*, 571–585. [CrossRef]
71. Zachariáš, J.; Novák, T. Gold-bearing quartz veins of the Bělčice ore district, Bohemian Massif: Evidence for incursion of metamorphic fluids into a granodiorite body and for isothermal mixing between two types of metamorphic fluids. *J. Geosci.* **2009**, *54*, 57–72. [CrossRef]
72. Zoheir, B.A.; Akawy, A. Genesis of the Abu Marawat gold deposit, central Eastern Desert of Egypt. *J. Afr. Earth. Sci.* **2010**, *57*, 306–320. [CrossRef]
73. Crispini, L.; Federico, L.; Capponi, G.; Talarico, F. The Dorn gold deposit in northern Victoria Land, Antarctica: Structure, hydrothermal alteration, and implications for the Gondwana Pacific margin. *Gondw. Res.* **2011**, *19*, 128–140. [CrossRef]
74. Dora, M.L.; Randive, K.R. Chloritisation along the Thanewasna shear zone, Western Bastar Craton, Central India: Its genetic linkage to Cu–Au mineralisation. *Ore Geol. Rev.* **2015**, *70*, 151–172. [CrossRef]
75. Fuertes-Fuente, M.; Cepedal, A.; Lima, A.; Dória, A.; dos Anjos Ribeiro, M.; Guedes, A. The Au-bearing vein system of the Limarinho deposit (northern Portugal): Genetic constraints from Bi-chalcogenides and Bi–Pb–Ag sulfosalts, fluid inclusions and stable isotopes. *Ore Geol. Rev.* **2016**, *72*, 213–231. [CrossRef]
76. Liard, J. Chlorites: Metamorphic petrology. *Rev. Mineral.* **1988**, *19*, 405–447.
77. Cathelineau, M. Cation site occupancy in chlorites and illites as a function of temperature. *Clay Miner.* **1988**, *23*, 471–485. [CrossRef]
78. Zhu, X.K.; O'Nions, R.K. Monazite chemical composition: Some implications for monazite geochronology. *Contrib. Mineral. Petrol.* **1999**, *137*, 351–363. [CrossRef]
79. Schandl, E.S.; Gorton, M.P. A textural and geochemical guide to the identification of hydrothermal monazite: Criteria for selection of samples for dating epigenetic hydrothermal ore deposits. *Econ. Geol.* **2004**, *99*, 1027–1035. [CrossRef]
80. Števko, M.; Uher, P.; Ondrejka, M.; Ozdín, D.; Bačík, P. Quartz–apatite–REE phosphates–uraninite vein mineralization near Čučma (eastern Slovakia): A product of early Alpine hydrothermal activity in the Gemeric Superunit, Western Carpathians. *J. Geosci.* **2014**, *59*, 209–222. [CrossRef]
81. Taylor, R.D. Orogenic gold formation and tectonic evolution of the Grass Valley gold district and temporal correlations of gold deposits in California. Ph.D. Thesis, Colorado School of Mines, Golden, CO, USA, 2015; p. 149, *unpublished*.
82. Dolníček, Z.; Ulmanová, J. Mineralogy of two types of hydrothermal veins containing REE minerals from the quarry near Vrbčany (Kutná Hora Crystalline Complex). *Bull. Mineral. Petrolog.* **2019**, *27*, 331–345. (In Czech)
83. Stormer, J.C., Jr.; Pierson, M.J.; Tacker, R.C. Variation of F and Cl X-ray intensity due to anisotropic diffusion of apatite during electron microprobe analysis. *Am. Mineral.* **1993**, *78*, 641–648.

84. Trdlička, Z.; Hoffman, V. Untersuchungen der chemischen Zusammensetzung der Gangkarbonate von Kutná Hora (ČSSR). *Freib. Forsch.* **1975**, *6*, 29–81.
85. Cílek, V.; Prokeš, S.; Škubal, M.; Hladíková, J.; Šmejkal, V.; Žák, K. Geochemistry of hydrothermal carbonates of the Příbram uranium deposit. *Vlast. Sbor. Podbrdska* **1984**, *26*, 79–102. (In Czech)
86. Žák, K.; Dobeš, P. Stable isotopes and fluid inclusions in hydrothermal deposits: The Příbram ore region. *Rozpr. ČSAV* **1991**, *101*, 1–109.
87. Morishita, Y. Fluid evolution and geobarometry on the Ohtani and Kaneuchi tungsten-quartz vein deposits, Japan: Oxygen and carbon isotopic evidence. *Mineral. Depos.* **1991**, *26*, 40–50. [CrossRef]
88. Morishita, Y.; Nishio, Y. Ore genesis of the Takatori tungsten–quartz vein deposit, Japan: Chemical and isotopic evidence. *Minerals* **2021**, *11*, 765. [CrossRef]
89. Zachariáš, J.; Pertold, Z.; Stein, H.; Markey, R.; Pudilová, M.; Pertoldová, J.; Stein, H.; Markey, R. Geology and genesis of Variscan porphyry-style gold mineralization, Petráčkova hora deposit, Bohemian Massif, Czech Republic. *Mineral. Depos.* **2001**, *36*, 517–541. [CrossRef]
90. Bortnikov, N.S. On the reliability of arsenopyrite and arsenopyrite-sphalerite geothermometers. *Geol. Ore Depos.* **1993**, *35*, 159–172.
91. Kerestedjian, T. Chemical and morphological features of arsenopyrite, concerning its use as a geothermometer. *Miner. Petrol.* **1997**, *60*, 231–243. [CrossRef]
92. Suchý, V.; Sandler, A.; Slobodník, M.; Sýkorová, I.; Filip, J.; Melka, K.; Zeman, A. Diagenesis to very low-grade metamorphism in lower Palaeozoic sediments: A case study from deep borehole Tobolka 1, the Barrandian Basin, Czech Republic. *Int. J. Coal Geol.* **2015**, *140*, 41–62. [CrossRef]
93. Žák, J.; Holub, F.V.; Verner, K. Tectonic evolution of continental magmatic arc from transpression in the upper crust to exhumation of mid-crustal orogenic root recorded by episodically emplaced plutons: The Central Bohemian Plutonic Complex, Bohemian Massif. *Int. J. Earth Sci.* **2005**, *94*, 385–400. [CrossRef]
94. Dolníček, Z.; Stöhr, P.; Ulmanová, J.; Vrtiška, L.; Malíková, R. Dickite-bearing hydrothermal mineralization in Ordovician claystones of the Bohdalec Formation from the tunnel of subway Line D in Prague-Pankrác. *Bull. Mineral. Petrolog.* **2020**, *28*, 116–125. (In Czech) [CrossRef]
95. Janoušek, V.; Braithwaite, C.J.R.; Bowes, D.R.; Gerdes, A. Magma-mixing in the genesis of Hercynian calc-alkaline granitoids: An integrated petrographic and geochemical study of the Sázava intrusion, Central Bohemian Pluton, Czech Republic. *Lithos* **2004**, *78*, 67–99. [CrossRef]
96. Deer, W.A.; Howie, R.A.; Zussman, J. *Rock-Forming Minerals: Feldspars;* Volume 4A, Geological Society of London: London, UK, 2001.
97. Liu, W.; Etschmann, B.; Migdisov, A.; Boukhalfa, H.; Testemale, D.; Müller, H.; Hazemann, J.-L.; Brugger, J. Revisiting the hydrothermal geochemistry of europium (II/III) in light of new in-situ XAS spectroscopy results. *Chem. Geol.* **2017**, *459*, 61–74. [CrossRef]
98. Sverjensky, D.A. Europium redox equilibria in aqueous solution. *Earth Planet. Sci. Lett.* **1984**, *67*, 70–78. [CrossRef]
99. Bau, M. Rare-earth element mobility during hydrothermal and metamorphic fluid-rock interaction and the significance on the oxidation state of europium. *Chem. Geol.* **1991**, *93*, 219–230. [CrossRef]
100. Bau, M.; Möller, P. Rare earth element fractionation in metamorphogenic hydrothermal calcite, magnesite and siderite. *Mineral. Petrol.* **1992**, *45*, 231–246. [CrossRef]
101. Lee, S.G.; Lee, D.H.; Kim, Y.; Chae, B.G.; Kim, W.Y.; Woo, N.C. Rare earth elements as indicators of groundwater environment changes in a fractured rock system: Evidence from fracture—Filling calcite. *Appl. Geoch.* **2003**, *18*, 135–143. [CrossRef]
102. Dörr, W.; Zulauf, G. Elevator tectonics and orogenic collapse of a Tibetan-style plateau in the European Variscides: The role of the Bohemian shear zone. *Int. J. Earth Sci. (Geol. Rundsch.)* **2010**, *99*, 299–325. [CrossRef]
103. Wood, S.A. The aqueous geochemistry of rare earth elements and yttrium. 2. Theoretical predictions of speciation in hydrothermal solutions to 350 °C at saturated water pressure. *Chem. Geol.* **1990**, *88*, 99–125. [CrossRef]
104. Migdisov, A.; Williams-Jones, A.E. Hydrothermal transport and deposition of the Rare Earth Elements by fluorine-bearing aqueous liquids. *Mineral. Depos.* **2014**, *49*, 987–997. [CrossRef]
105. Shikazono, N.; Shimizu, M. The Ag/Au ratio of native gold and electrum and the geochemical environment of gold vein deposits in Japan. *Miner. Depos.* **1987**, *22*, 309–314. [CrossRef]
106. Morrison, G.W.; Rose, W.J.; Jaireth, S. Geological and geochemical controls on the silver content (fineness) of gold in gold-silver deposits. *Ore Geol. Rev.* **1991**, *6*, 333–364. [CrossRef]
107. Gammons, C.H.; Williams-Jones, A.E. Hydrothermal geochemistry of electrum: Thermodynamic constraints. *Econ. Geol.* **1995**, *90*, 420–432. [CrossRef]
108. Zachariáš, J.; Paterová, B.; Pudilová, M. Mineralogy, fluid inclusion and stable isotope constraints on the genesis of the Roudný Au-Ag deposit, Bohemian Massif. *Econ. Geol.* **2009**, *104*, 53–72. [CrossRef]
109. Zachariáš, J.; Žák, K.; Pudilová, M.; Snee, L.W. Multiple fluid sources/pathways and severe thermal gradients during formation of the Jílové orogenic gold deposit, Bohemian Massif, Czech Republic. *Ore Geol. Rev.* **2013**, *54*, 81–109. [CrossRef]
110. Žák, K.; Svojtka, M.; Breiter, K.; Ackerman, L.; Zachariáš, J.; Pašava, J.; Veselovský, F.; Litochleb, J.; Ďurišová, J.; Haluzová, E. Padrť Stock (Teplá-Barrandian Unit, Bohemian Massif): Petrology, geochemistry, U-Pb zircon dating of granodiorite, and Re-Os age and origin of related molybdenite mineralization. *J. Geosci.* **2014**, *59*, 351–366. [CrossRef]

minerals MDPI

Article

Genesis and Evolution of Hydrothermal Fluids in the Formation of the High-Grade Hishikari Gold Deposit: Carbon, Oxygen, and Sulfur Isotopic Evidence

Yuichi Morishita [1,2,3,4,*] and Yoriko Yabe [4]

1 Center for Integrated Research and Education of Natural Hazards, Shizuoka University, Shizuoka 422-8529, Japan
2 Geological Survey of Japan, National Institute of Advanced Industrial Science and Technology (AIST), Tsukuba 305-8567, Japan
3 Museum of Natural and Environmental History, Shizuoka 422-8017, Japan
4 Faculty of Science, Shizuoka University, Shizuoka 422-8529, Japan
* Correspondence: morishita.yuichi@shizuoka.ac.jp

Abstract: The Hishikari low-sulfidation epithermal gold (Au) deposit in Kyushu, Japan, is world-famous for its premium ore. It has been hypothesized that magmatic contributions to the hydrothermal fluid during early stages of mineralization is possible, even if the hydrothermal fluids for many Au occurrences near the Hishikari deposit are of meteoric origin and are influenced by basement sedimentary rocks. The purpose of this study is to obtain constraints on the genesis and evolution of hydrothermal fluids in the formation of the high-grade Hishikari Au deposit by carbon and oxygen isotope ratios of calcite-bearing samples. Since the microanalysis of carbon and oxygen isotope ratios in every 12 μm of the calcite-bearing sample along the growth direction (corresponding to 10 years of the Hishikari mineralization) scatter in a particular range, the fluid evolution might not be a gradual change from a magmatic to a meteoric origin. Alternatively, a rapid turnover of two fluids might be happening locally. The average sulfur isotope ratio of hydrothermal pyrite is similar to that of the adjacent magma. However, according to the secondary ion mass spectrometry (SIMS) microanalysis, local pyrite with extremely low sulfur isotope ratios may interact with basement sedimentary rocks. Unlike other epithermal Au deposits in the vicinity, rapid local mixing of the magmatic-origin deep fluid and meteoric-origin fluid reacted with organic matter containing basement sedimentary rocks might cause gold precipitation at the Hishikari deposit.

Keywords: Hishikari deposit; genesis and evolution; hydrothermal fluid; oxygen; carbon and sulfur isotope ratios; calcite; pyrite; basement sedimentary rock; isotope ratio mass spectrometry (IRMS); secondary ion mass spectrometry (SIMS)

Citation: Morishita, Y.; Yabe, Y. Genesis and Evolution of Hydrothermal Fluids in the Formation of the High-Grade Hishikari Gold Deposit: Carbon, Oxygen, and Sulfur Isotopic Evidence. *Minerals* **2022**, *12*, 1595. https://doi.org/10.3390/min12121595

Academic Editor: Pei Ni

Received: 10 November 2022
Accepted: 8 December 2022
Published: 12 December 2022

1. Introduction

It is crucial to research the Hishikari gold (Au) deposit in Kyushu, Japan, as an example of vein-type high-grade Au deposits because of its highly regarded high-grade ore. The Sumitomo Metal Mining Co., Ltd. has been mining the Hishikari deposit since 1985. The Hishikari mine has produced 260 metric tons of Au, and the ore reserve amounts to 160 metric tons. The Au production in 2021 is 6 metric tons [1]. The average grade of ore is 34 g/t Au, which is significantly higher than those of neighboring and average mines worldwide. An important question to ask is why the Hishikari deposit is of a higher grade among Au deposits in the same district. The early-stage hydrothermal fluid in the Hishikari deposit was enriched in ^{18}O over local meteoric water, despite its low-sulfidation epithermal deposit [2]. Several studies [2–6] have pointed out the possibility of magmatic contribution to the hydrothermal fluid in the earlier stage of mineralization. These previous studies focus on the genesis and evolution of hydrothermal fluids in forming the high-grade Hishikari Au deposit.

Fluid-rock reactions play a vital role in the carbon and oxygen isotope budgets of geologic settings of ore-forming hydrothermal systems in the Earth's crust. Although many studies on isotopes in calcites have been undertaken, few researchers have expressed an interest in utilizing the isotope data of hydrothermal calcite. This is partly because calcite often occurs in the last stage of hydrothermal mineralization; therefore, a calcite vein is considered a barren vein in mining districts. However, fertile calcite veins may also exist in the exact location of barren calcite veins. Therefore, this study proposes using calcite's isotope data. Although oxygen isotope ratios of quartz from the Hishikari veins have been reported [2–6], the veins contain silicate minerals such as adularia other than quartz. Therefore, merely analyzing pure quartz is not an easy task. However, calcite isotope data can be obtained using acid from quartz veins containing other silicate minerals.

Furthermore, the calcite isotope measurements still include specific details about the fluid that coexisted with it. Therefore, a set of calcite carbon and oxygen isotope ratios has better information than the oxygen isotope ratio of vein quartz. Consequently, isotope ratio mass spectrometry (IRMS) was used to examine the carbon and oxygen isotope ratios of samples that contained calcite.

On the other hand, pyrite is ubiquitous in the Hishikari veins. Therefore, Morishita et al. [7,8] conducted a microanalysis of Au and As concentrations in pyrite grains to indicate that the Au concentration positively correlates with the As concentration in a small area. Furthermore, sulfur isotope ratios of pyrite help constrain the source of sulfur. Therefore, sulfur isotope ratios of pyrite were measured using secondary ion mass spectrometry (SIMS).

The purpose of this study is to obtain constraints on the genesis and evolution of the hydrothermal fluids in the formation of the high-grade Hishikari Au deposit. Additionally, the isotopic characteristics of the fluid in the Hishikari deposit will be discussed.

2. Geology and Mineralogy in the Hishikari Mining Area

2.1. Geologic Setting and Au Deposits

Kyushu is located at the eastern margin of the Eurasian plate facing the Philippine Sea plate. Numerous epithermal Au deposits from the Pliocene to Pleistocene are found in Late Cenozoic subaerial andesitic to dacitic volcanic rocks that resulted from the subduction of the Philippine Sea plate beneath the Eurasian plate (Figure 1). The Hokusatsu district of southwestern Kyushu (Figure 1) has the highest Au production in Japan. Most of the ore deposits in this district, such as the Hishikari and Kushikino deposits, sit above a typical basement of the Shimanto Group.

The Hishikari low-sulfidation epithermal Au deposit (Figure 1) is the largest in Japan. Hosono et al. [9] determined the Sr–Nd–Pb isotope compositions of Late Pliocene to Pleistocene volcanic rocks around the Hishikari deposit to elucidate their source characteristics of magma related to gold mineralization. The basement rocks in the Hishikari mining area are sedimentary rocks from the Lower Shimanto Group (Cretaceous accretionary wedge), which are visible underground but do not outcrop close to the deposit. The Hishikari Lower Andesites consist mainly of andesitic pyroclastic rocks. They cover the basement sedimentary rocks unconformably and have been dated by the K–Ar method to be from 1.62 ± 0.09 to 0.98 ± 0.10 Ma for 15 whole-rock samples from within 2 km of the Hishikari mine [10,11]. Several studies have previously described the Hishikari deposit's geology (e.g., [9–16]).

2.2. Gold-Bearing Veins and Mineral Paragenesis

The Hishikari deposit consists of the Sanjin, Honko, and Yamada ore zones (Figure 2). Each ore zone has multiple vein swarms that extend roughly 1 km from NE to SW. The Hishikari veins strike N50° E and dip steeply north (70–90°). The Sanjin and Honko ore zones comprise high-grade Au-bearing quartz veins in the basement sedimentary rocks of the Cretaceous Shimanto Group and in the andesitic rocks of the Hishikari Lower Andesites that overlie the basement sedimentary rocks. However, relatively low-grade Au-bearing quartz veins in the Yamada ore zone occur in the andesitic rocks. K–Ar ages show that the Hishikari deposit's mineralization started at 1.15 Ma and ended at 0.60 Ma [11]. Sanematsu et al. [18] determined the $^{40}Ar/^{39}Ar$ ages of the Hishikari adularia–quartz veins. The mineralization ages in the Honko and Sanjin zones range from 1.04 to 0.75 Ma, whereas the majority of results are focused in the range of 1.01–0.88 Ma and those in the Yamada zone range from 1.21 to 0.64 Ma [18].

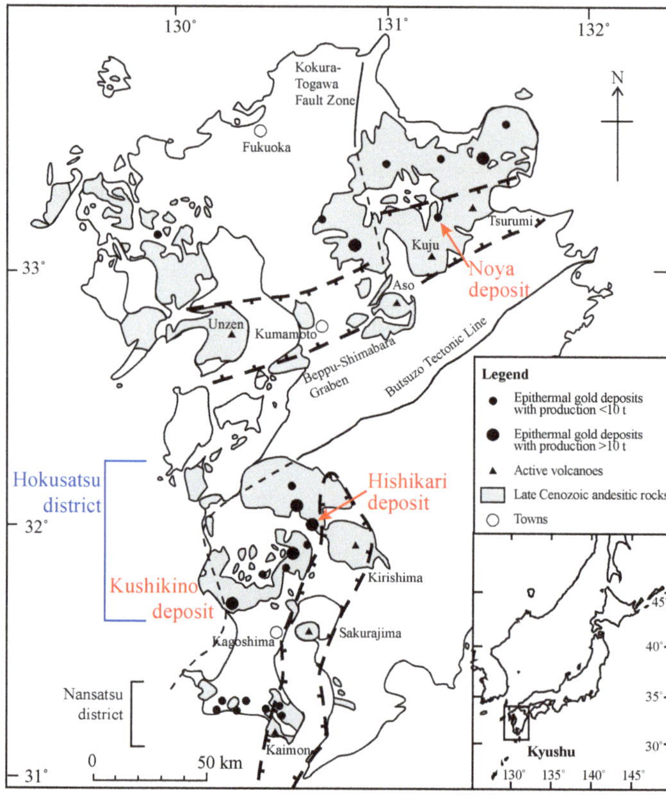

Figure 1. Epithermal gold deposits in Kyushu, southwestern Japan (produced from [12,17]). In southern Kyushu, epithermal gold deposits occur in Late Cenozoic subaerial andesitic to dacitic volcanic rocks in the Hokusatsu district. The basement rocks in the district are sedimentary rocks of the Lower Shimanto Group (Cretaceous accretionary wedge), which do not outcrop near the deposits. The location of the Hishikari, Kushikino, and Noya deposits are shown.

The Hishikari veins consist mainly of quartz, adularia, and smectite. The metallic minerals in the Hishikari veins are pyrite, marcasite, electrum, naumannite, aguilarite, pyrargyrite, and chalcopyrite [10], of which pyrite is the most abundant. The host rocks also contain hydrothermal pyrite [7,8]. Additionally, the veins contain minor amounts of calcite and truscottite [10]. The homogenization temperatures for quartz in veins within the basement sedimentary rocks average 213 °C, with most values in the range of 195–230 °C [10]. The strontium and carbon isotopic ratios of calcite reveal that the Hishikari ore fluid was affected by the sedimentary basement rocks [19]. Bladed quartz (lamellar quartz) is found everywhere in the Hishikari deposit [10,20]. When initially produced, calcite reacts with later low-temperature fluids; and it dissolves because calcite's solubility rises as the hydrothermal fluid's temperature falls, leaving cavities for later quartz to precipitate as bladed quartz [10,20]. Bladed quartz signifies that bladed calcite was once deposited by boiling fluid and then dissolved during cooling. Boiling occurred during the Hishikari mineralization based on the observations of fluid inclusions [10,14,20]. However, other studies [4,21] do not recognize the evidence of boiling in the fluid inclusion observations, suggesting that not all areas of boiling were subject to it. Calcite precipitation from the hydrothermal fluid occurs either with increasing temperature or pH or with decreasing fugacity of CO_2 [22]. Among them, loss of CO_2 because of boiling might happen in the Hishikari hydrothermal fluid, and cause calcite precipitation.

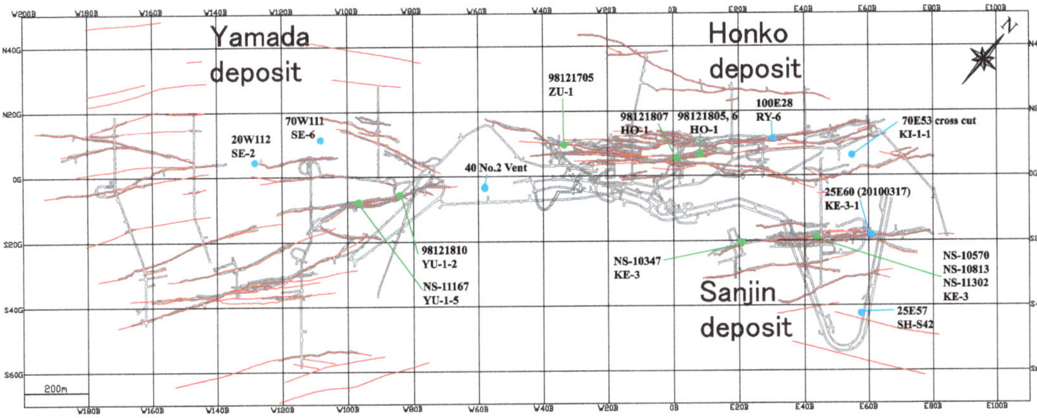

Figure 2. Sampling sites are shown on the vein system (red) of the Hishikari deposit. The number and vein name of the sample is attached next to the symbol. Blue symbols are for carbon and oxygen isotope ratios and green symbols are for sulfur isotope ratios.

Additionally, mixing fluids might increase temperature or pH, resulting in calcite precipitation. Therefore, calcite might initially form by boiling or mixing fluids. Ore fluids that formed the banding structure of veins, have passed through the veins and they might dissolve most of the calcite precipitated from earlier fluids, because the solubility of calcite increases as the temperature drops. Because of this, calcite is rarely observed in the veins of the Hishikari deposit, and there haven't been many investigations on the carbon and oxygen isotopes of calcite. In this study, we attempted to find as much calcite as possible in the Hishikari deposit and we analyzed the calcite-bearing samples' carbon and oxygen isotope ratios. The presence of ubiquitous pyrite shows that the Hishikari hydrothermal system was under reducing conditions. The mixing of deep fluids with shallow groundwater is thought to have considerably altered the temperature and oxidation conditions of the hydrothermal fluids at the boundary between the basement rocks and the overlying andesites [10].

The Hishikari veins frequently have symmetrical formations, which suggests that vein minerals near the wall rock precipitate earlier than the minerals inside the vein. The outer band close to the wall rock consists mainly of adularia and quartz, and the innermost band in the center consists mainly of quartz [14]. However, in the Hishikari veins, asymmetrical formations produced by cutting with later-stage veins are not uncommon. According to the observed crosscutting connections of the veins, Sekine et al. [16] divided the vein system of the Hishikari deposit into early and late veins according to the observed crosscutting relationships of veins. The late veins commonly display symmetrical banding, whereas banding in the early veins is often obscured.

3. Materials and Methods

3.1. Materials

3.1.1. Calcite-Bearing Hand Specimens

Calcite-bearing vein samples obtained were investigated for their mineral compositions using an X-ray diffractometer (XRD) SmartLab (Rigaku, Tokyo, Japan) at Shizuoka University. Figure 2 and Table 1 show calcite-bearing samples' locality and mineral composition. The approximate abundance of calcite was obtained from the XRD calibration curves. Electrum is very fine-grained for microscopic observations. Photos of representative vein samples are shown in Figure 3.

Table 1. Descriptions and isotope data of calcite-bearing vein samples from the Hishikari deposit.

Sample No.	Locality *1	Vein	Level *2	Mineral Composition (%) *3	$\delta^{13}C$ (‰)	$\delta^{18}O$ (‰)
31001a	70W111SE-6E	Seisen-6	70 mL	Calcite (100)	−10.3	6.1
31001b	70W111SE-6E	Seisen-6	70 mL	Quartz (62), Adularia (12), Calcite (26)	−4.3	10.6
31001c	70W111SE-6E	Seisen-6	70 mL	Quartz (59), Adularia (6), Calcite (35)	−7.9	8.1
31002c	25E57S42HW	Shosen S42	25 mL	Quartz (5), Calcite (95)	−9.2	3.8
31002d	25E57S42HW	Shosen S42	25 mL	Quartz (32), Calcite (68)	−8.6	5.2
31003-1e	20W112SE-2W	Seisen-6	20 mL	Quartz (93), Calcite (7)	−3.6	12.4
31003-2a	20W112SE-2W	Seisen-6	20 mL	Quartz (90), Calcite (10)	−4.7	10.5
31003-2b	20W112SE-2W	Seisen-6	20 mL	Quartz (6), Calcite (94)	−2.6	11.8
31003-2c	20W112 SE-2W	Seisen-6	20 mL	Quartz (45), Adularia (21), Calcite (34)	−3.3	11.8
31506b	20W112SE-2W	Seisen-6	20 mL	Quartz (66), Calcite (34)	−6.5	11.1
31506c	20W112SE-2W	Seisen-6	20 mL	Quartz (57), Calcite (43)	−23.4	17.9
31004a	40 No.2 Vent		40 mL	Quartz (75), Calcite (25)	−4.8	7.9
31004b	40 No.2 Vent		40 mL	Quartz (71), Calcite (29)	−5.5	9.8
31004c	40 No.2 Vent		40 mL	Quartz (35), Calcite (65)	−5.6	8.2
31005c	70E53 cross cut	Kinsen-1-1	70 mL	Quartz (36), Calcite (64)	−4.5	10.6
31615c	70E53 cross cut	Kinsen-1-1	70 mL	Quartz (69), Calcite (31)	−25.1	17.1
31610b	100E28RY-6E	Ryosen-6	100 mL	Calcite (100)	−6.2	5.7
31611b	100E28RY-6E	Ryosen-6	100 mL	Quartz (70), Adularia (13), Calcite (17)	−5.2	14.6
20100317	25E60KE-3-1E	Keisen-3-1	25 mL	Quartz, Calcite: See Table 2	See Tables 2–4.	

The vein width is not shown because each calcite-containing sample is from a branch veinlet of the main vein. Several calcite-bearing samples are obtained by the mine geologists. Branch signs like a, b, etc. show that they are from the same hand specimen. Different sample numbers with the same locality were collected at different times. *1: E or W indicates the direction of the drift. *2: "mL" corresponds to the altitude above sea level. *3: Approximate mineral composition was obtained from the XRD calibration curve.

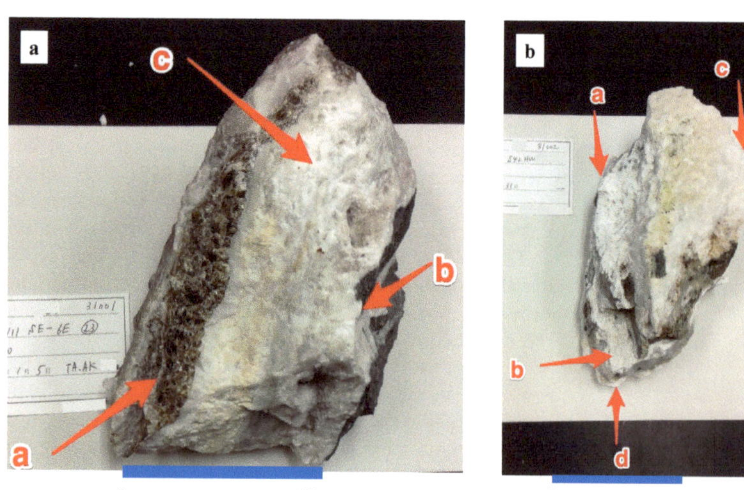

Figure 3. Photos of representative vein samples. (**a**) Sample 31001 with the analysis localities a, b, and c (Table 1) are shown. (**b**) Sample 31002 with the analysis localities a and b (Table 1) are shown.

3.1.2. A Calcite-Dominant Vein Sample for Microanalysis

The Hishikari deposit's high-grade vein sample 20100317 (Figure 4a), taken from a 25 mL E60 Keisen 3-1E, contains continuous calcite in the growth direction. The vein width is about 50 cm. The Keisen No. 3 vein swarms are representative fertile veins in the Sanjin ore zone [8,16]. Therefore, vein sample 20100317 is suitable for investigating the fluid evolution in the Hishikari ore-forming hydrothermal system, and no other appropriate sample has been found so far. The vein was expanding from right to left in Figure 4a because its proper end is close to the wall rock, and its left end corresponds to the vein's center. This sample was cut into 10 mm widths and named for each from 1743 to 1749 (10 mm-width sample set), as shown in Figure 4b and Table 2. Then 1761, 1762, and 1763 samples were further cut to a width of 0.9 mm (Figure 4c) with a cutting machine, Minitom (Struers, Denmark) at Shizuoka University and obtained 176-1 to 176-21 sections (0.9 mm-width sample set; Table 3). The 1763 sample corresponds to the same stage as 1743 (Figure 4b). Therefore, compared to the 10 mm-width sample set, the 0.9 mm-width sample set (1761–1763) indicates earlier stages than the 10 mm-width sample set. The blue lines in Figure 4c represent space, which corresponds to the thickness of the blade, 0.3 mm, so the sample interval in the growth direction is 1.2 mm. The quartz-adularia vein's approximate calcite abundance is determined via the XRD calibration curve (Table 3). The 176-1 section consists of 69% of quartz and 31% of adularia (determined by XRD calibration curve) with a minor amount of fine-grained pyrite and clay minerals, which causes the gray color in the early-stage mineral precipitations.

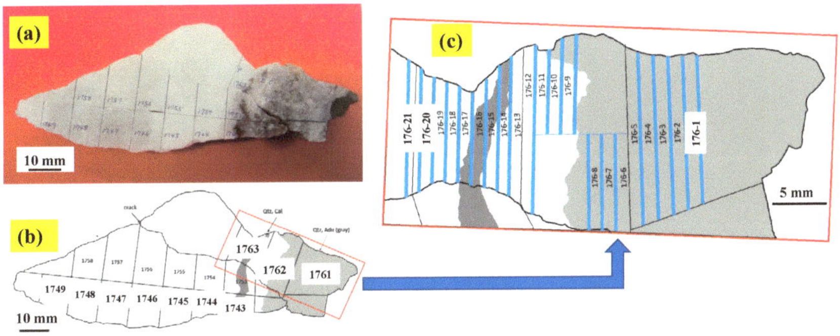

Figure 4. (**a**) Photo for the 20100317 high-grade vein sample taken from the 25 mL KE-3-1 vein. The vein width is about 50 cm. The right end of the vein is adjacent to the wall rock, and the left is around the center of the vein. In the early stage of mineralization, the gray part contains fine-grained pyrite and clay minerals. The calcite-bearing samples covering early to late mineralization are rare in the Hishikari deposit. (**b**) The vein sample 20100317 is cut every 10 mm in the growing direction, and every sample is numbered. The rectangle shows the area for Figure 3c. (**c**) Sections measuring 0.9 mm-width are made from 1761, 1762, and 1763 samples, numbered from 176-1 to 176-21. Blue lines designate areas cut off with a 0.3 mm thick blade. The 12 μm-width powder samples for microanalysis are scraped from 176-20 and 176-21 sections.

Table 2. Mineral composition and isotope data of calcite-bearing 20100317 vein sample (10 mm-width sample set; Figure 4b) from 25 mL E60 KE-3-1E.

Sample No.	Mineralization Order	Mineral Composition (%) *	$\delta^{13}C$ (‰)	$\delta^{18}O$ (‰)
1743	(Early)	Quartz (83), Calcite (17)	−7.7	4.3
1744	↓	Quartz (19), Calcite (81)	−6.5	3.5
1745		Quartz (29), Calcite (71)	−6.2	3.4
1746		Quartz (86), Calcite (14)	−6.3	3.9
1747		Quartz (94), Calcite (6)	−4.8	8.2
1748	(Late)	Quartz (100)	-	-
1749	Center of the vein	Quartz (100)	-	-

*: Approximate mineral composition was obtained from the XRD calibration curve.

Finer samples in the subsequent work were obtained to increase the time resolution. Although cutting the vein sample on a micron scale is impractical, Shizuoka University's biological microtome (FX-801N, Yamato Kohki Industrial Co., Ltd., Saitama, Japan) can be used to scrape off a small portion of the calcite-rich soft sample. The 176-20 and 176-21 sections (Figure 4c) contain more than 70% of calcite, and they are soft enough to scrape off calcite-dominant powder from the samples with a stainless steel blade of the microtome. The average scraped sample size is roughly 0.4 mg, corresponding to a sample width in the original sample's growth direction of about 12 μm. The continuous samples (12 μm-width sample set) in the growth direction obtained from 176-20 and 176-21 sections are listed in Table 4. The scraped thickness is calculated from the weight of the powder scraped, and Table 4's distance from the microtome cutting's starting point (the 'origin' that is the line separating 176-19 and 176-20) is displayed. The number attached to the original sample number (176-19 and 176-20) indicates the order of crystallization; namely, 176-20-1 is the earliest end of the 176-20 section, and 176-20-12 is the latest grown sample, about 0.14 mm from the origin. 176-20 section with a thickness of about 0.76 mm remained unscraped since the sample was broken during the subsequent cutting. 176-21 section was divided into 17 samples, from 176-21-1 to 176-21-17. Each scraped size was similar to that for 176–20, and the distance from the same 'origin' is shown in Table 4. The 176-21 section was broken during the 18th cutting after 17 sample powders were obtained. Notably, 29 mineral samples were collected, despite the difficulty of cutting vein samples with a biological microtome. Be aware that the 176-20-12 and 176-21-1 sample differ by approximately 1.06 mm (the blade thickness and the remainder of the 176-20 sample).

Table 3. Calcite abundance and isotope data of calcite-bearing 20100317 vein sample (0.9 mm-width sample set; Figure 4b,c).

Sample No.	Mineralization Order	Calcite Abundance *	$\delta^{13}C$ (‰)	$\delta^{18}O$ (‰)
176-1	Wall rock side	0%		
176-2	(Early)	0%		
176-3	↓	0%		
176-4		0%		
176-5		4%	−5.0	6.0
176-6		0%		
176-7		0%		
176-8		16%	−6.9	4.7
176-9		5%	−6.3	3.7
176-10		3%	−7.2	3.8
176-11		9%	−7.7	4.4
176-12		17%	−7.5	3.0
176-13		31%	−5.4	4.2
176-14		25%	−6.3	4.5
176-15		17%	−6.5	5.6
176-16		11%	−5.6	5.8
176-17		6%	−4.7	7.0
176-18		33%	−8.4	5.5
176-19		73%	−7.4	5.2
176-20		70%	−7.2	4.8
176-21	(Late)	72%	−6.1	4.5

*: Approximate calcite abundance in quartz-adularia vein was obtained from the XRD calibration curve.

Table 4. Microanalyses of 176-20 and 176-21 cutoff (12 μm-width sample set; Figure 4c) from 20100317 sample.

Sample No.	Distance (μm) from the 'Origin' *	Mineralization Order	$\delta^{13}C$ (‰)	$\delta^{18}O$ (‰)
176-20-1	7	Early	−6.3	8.4
176-20-2	15	↓	−8.4	5.9
176-20-3	26		−9.1	5.0
176-20-4	37		−7.7	4.8
176-20-5	42		−4.5	9.9
176-20-6	57		−5.5	7.4
176-20-7	76		−8.2	5.2
176-20-8	87		−7.7	6.5
176-20-9	107		−9.5	5.1
176-20-10	121		−8.0	5.7
176-20-11	137		−6.9	6.5
176-20-12	144		−2.5	10.9
176-21-1	1207		−9.1	5.0
176-21-2	1212		−7.9	3.6
176-21-3	1222		−7.8	5.3
176-21-4	1234		−8.5	4.5
176-21-5	1241		−8.5	4.5
176-21-6	1247		−8.4	4.8
176-21-7	1255		−9.1	5.4
176-21-8	1274		−9.1	5.0
176-21-9	1285		−8.2	4.3
176-21-10	1301		−7.2	5.7
176-21-11	1315		−8.0	4.9
176-21-12	1319		−6.9	4.9
176-21-13	1324		−6.9	4.7
176-21-14	1343		−7.4	6.0
176-21-15	1361		−7.6	5.0
176-21-16	1381		−5.6	7.5
176-21-17	1388	Late	−7.4	5.2

*: The origin is the boundary between 176-19 and 176-20.

3.1.3. Pyrite Samples

Pyrite samples were collected from the main veins of the three ore zones of the Hishikari deposit. The grain size of pyrite is fine, it is mostly between 10 and 30 μm. Figure 2 and Table 5 show the locality of analyzed veins, which are Keisen veins (Sanjin ore zone), Zuisen and Hosen veins (Honko ore zone), and Yusen veins (Yamada ore zone). Additionally, Table 5 includes two chalcopyrite grains for analysis. Figure 5 shows backscattered electron (BSE) images of representative pyrite and chalcopyrite grains. Additionally, pyrite grains in the host rock (shale) adjacent (within 2 cm from the vein boundary) to the Hosen No. 1 vein were also used for analysis. Au and As concentrations were analyzed using SIMS [8,23] on the same pyrite samples used in this study (Table 5).

Table 5. Locality and the sulfur isotope ratio of pyrite from the Hishikari deposit.

Ore Zone	Sample Name	Vein *1	Level *2	Sample Locality	Host Rock	Analysis Spot	Analysis No.	$\delta^{34}S$ (‰)
Sanjin	NS-10347	Keisen No.3	40 mL	E21	andesite	13-52	SMB3*76.ais	−0.4
Sanjin	NS-10570	Keisen No.3	25 mL	E45	shale	6-26	SMB3*73.ais	2.1
Sanjin	NS-10813	Keisen No.3	10 mL	E45	shale	1-2	SMB3*69.ais	−10.2
						1-3	SMB3*70.ais	−15.1
						1-1	SMB3*71.ais	−16.4
						2-5	SMB3*72.ais	−10.1
Sanjin	NS-11302	Keisen No.3	−5 mL	E45	shale	1-1	SMB2*47.ais	−2.1
						2-1	SMB2*48.ais	−16.5
						2-2	SMB2*49.ais	−12.1
						3-1	SMB2*50.ais	1.7
						4-1	SMB2*51.ais	3.5

Table 5. *Cont.*

Ore Zone	Sample Name	Vein *1	Level *2	Sample Locality	Host Rock	Analysis Spot	Analysis No.	$\delta^{34}S$ (‰)
Honko	98121705	Zuisen No.1	25 mL + 2sl *3	W33	andesite	1-18 *4	SMB3*80.ais	−1.4
						1-1	SMB3*81.ais	0.5
						2-4	SMB3*82.ais	−3.4
						2-6	SMB3*83.ais	0.2
						2-19 *4	SMB3*84.ais	1.0
						4-13	SMB3*86.ais	−0.4
						4-15	SMB3*87.ais	−2.0
Honko	98121807	Hosen No.1	55 mL + 2sl	W1 (E1) *5	shale	1-1	SMB1*30.ais	0.4
						1-2	SMB1*31.ais	0.1
						1-3	SMB1*32.ais	−1.4
						2-2	SMB1*33.ais	−1.6
						2-3	SMB1*34.ais	0.6
						2-4	SMB1*35.ais	−0.7
						2-5	SMB1*36.ais	−1.2
Honko	98121806	Hosen No.1	55 mL + 2sl	W1 (E8) *5	shale	1-1	SMB3*92.ais	0.3
						1-2	SMB3*93.ais	0.6
						1-4	SMB3*94.ais	−0.1
						1-6	SMB3*95.ais	1.9
						1-9	SMB3*96.ais	1.5
						1-10	SMB3*97.ais	2.1
Honko	98121805	Shale Host rock for the Hosen No.1 vein	55 mL + 2sl	W1 (E8) *5		3-2	SMB1*21.ais	3.2
						3-3	SMB1*22.ais	2.2
						3-5	SMB1*23.ais	2.7
						1-1	SMB1*24.ais	2.0
						4-1	SMB1*25.ais	0.7
						5-1	SMB1*26.ais	1.3
Yamada	98121810	Yusen No.1-2	30 mL	W85	andesite	2-23	SMB3*101.ais	1.2
						2-22	SMB3*102.ais	0.7
						2-18	SMB3*103.ais	2.3
						2-17	SMB3*104.ais	1.7
						2-13	SMB3*105.ais	−1.0
						3-30	SMB3*106.ais	4.5
						3-36	SMB3*107.ais	3.1
						3-41	SMB3*108.ais	4.3
						3-47	SMB3*109.ais	1.5
						3-48	SMB3*110.ais	4.4
Yamada	NS-11167	Yusen No.1-5	40 mL	W95	andesite	2-1	SMB2*58.ais	4.0
						2-2	SMB2*59.ais	2.9
						2-3	SMB2*60.ais	3.4
						6-2	SMB2*64.ais	0.8
						7-1	SMB2*65.ais	−2.0

*1: Samples are from quartz veins except for one sample (98121805) from the basement shale. *2: "mL" corresponds to the altitude above sea level. *3: "sl", which corresponds to +3.75 m, stands for the slice level for mining. *4: The two grains are chalcopyrite that were confirmed by electron probe microanalysis (EPMA). *5: W1 is a working name. E1 or E8 is the actual location.

3.2. Analytical Methods

3.2.1. Carbon and Oxygen Isotope Ratios of Calcite

IRMS determined calcite-bearing samples' carbon and oxygen isotope ratios using the method described in [24]. Several aliquots of the powder sample of 0.15 mg each are kept in steel thimbles and dropped down one-by-one into phosphoric acid at 60 °C in an online reaction chamber under vacuum conditions [24]. Evolved gas is cryogenically purified to retain CO_2. The reaction chamber is connected to the inlet system of a Finnigan MAT-250 mass spectrometer (Thermo Finnigan, Bremen, Germany) at Shizuoka University. Isotope ratios are reported in standard δ notation in per mil (‰) relative to the Vienna standard mean ocean water (SMOW) for $\delta^{18}O$ and relative to the Vienna Pee Dee Belemnite (PDB) for $\delta^{13}C$. Reproducibility was approximately ±0.1‰ (2σ) for both the $\delta^{13}C_{PDB}$ and $\delta^{18}O_{SMOW}$ values of calcite. Using measurements on a laboratory working standard and the NBS 19, measured isotope ratios were normalized to limestone reference material [25].

3.2.2. Sulfur Isotope Ratios of Pyrite

Sulfur isotope ratios of pyrite were determined using an ims-1270 SIMS (Cameca, France) with a multi-collector at the Geological Survey of Japan (GSJ), National Institute of Advanced Industrial Science and Technology (AIST). Analysis was conducted using positive primary ions and negative secondary ions. A 10 kV defocused Cs+ primary beam (Köhler illumination) was restricted to 20 μm in diameter by a circular aperture to obtain a homogeneous primary beam of about 1 nA. Additionally, an electron flood gun was deployed on the sample surface for charge compensation to achieve better stability of the secondary ions. A 20 μm-diameter circle that served as the primary beam bombardment's target area is the area that was examined on the sample surface. Sputtered negative secondary ions with an accelerating voltage of −10 kV were extracted from the carbon-coated sample surface. To guarantee that the same energy band of secondary ions was consistently chosen for experiments, an energy window of 50 eV was adjusted to the energy distribution curve. The secondary $^{32}S^-$ and $^{34}S^-$ ions were detected simultaneously without energy filtering using Faraday cups of the multi-collection system. The primary beam intensity was adjusted to around 1 nA to generate a high intensity of secondary ions ($^{32}S = 2 \times 10^8$ cps) for high-accuracy sulfur isotope studies. The measured data were normalized using a working standard sample (Ak pyrite, $\delta^{34}S$ = +5.1‰) and are reported utilizing the standard δ notation in per mil (‰) relative to Canyon Diablo Troilite for $\delta^{34}S$. The accuracy is less than ±0.2‰ (2σ).

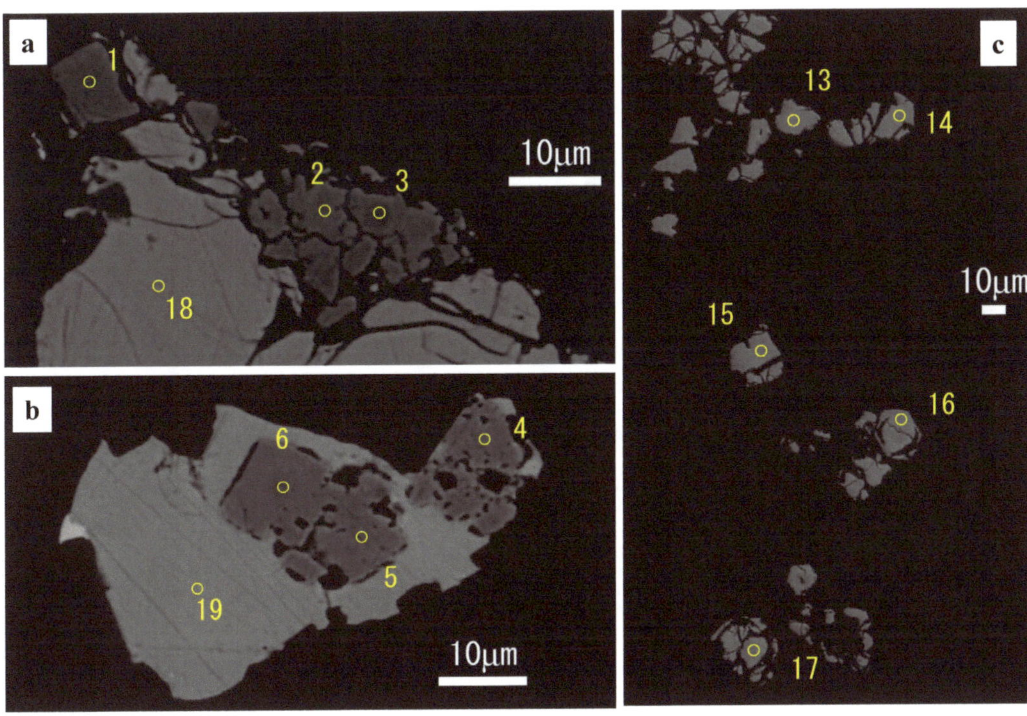

Figure 5. BSE images of pyrite and chalcopyrite grains from the Zuisen No. 1 vein (sample name 98121705 in Table 5) of the Hishikari ore deposit. (**a**) The analysis spot 1-1 (pyrite grain 1) and analysis spot 1-18 (chalcopyrite grain 18) are shown. (**b**) The analysis spots 2-4 and 2-6 (pyrite grains 4 and 6) and analysis spot 2-19 (chalcopyrite grain 19) are shown. (**c**) The analysis spots 4-13 and 4-15 (pyrite grains 13 and 15) are shown. Most pyrite grains are fine.

4. Results

4.1. Carbon and Oxygen Isotope Ratios of Calcite-Bearing Hand Specimens

Table 1 lists calcite's $\delta^{13}C$ and $\delta^{18}O$ values from the Hishikari deposit. The $\delta^{13}C$ and $\delta^{18}O$ values of calcite range from −25.1‰ to −2.6‰ and from +3.8‰ to +17.9‰, respectively. The ranges of the $\delta^{13}C$ and $\delta^{18}O$ values of calcite, when two samples with extraordinarily low $\delta^{13}C$ values are excluded, are −10.3‰ to −2.6‰ and +3.8‰ to +14.6‰, respectively.

4.2. Microanalysis of Calcite from a Calcite-Dominant Vein

Continuous carbon and oxygen isotope analysis of the calcite-dominant vein (20100317 sample) was determined. Samples from 1743 to 1749 (10 mm-width sample set; Figure 4b) were first assessed to understand the isotopic trend through the growing direction. The most central portion of the vein, the 1748 and 1749 samples, have just 100% quartz and no calcite. The $\delta^{13}C$ and $\delta^{18}O$ values of calcite range from −7.7‰ to −4.8‰ and from +3.4‰ to +8.2‰, respectively (Table 2). The value of $\delta^{13}C$ tends to increase in the growth direction, although the value of $\delta^{18}O$ does not significantly change except for the latest 1747 sample having a higher value.

Sections from 176-1 to 176-21 (0.9 mm-width sample set; Figure 4c) provide every 1.2 mm interval isotope data along the growth direction. The earlier stages of samples contain no calcite (Table 3). The $\delta^{13}C$ and $\delta^{18}O$ values of the sections range from −8.4‰ to −4.7‰ and from +3.0‰ to +7.0‰, respectively (Table 3).

Every 12 μm of the sample made using a microtome (12 μm-width samples) is microanalyzed for $\delta^{13}C$ and $\delta^{18}O$ values in the direction of growth. The $\delta^{13}C$ and $\delta^{18}O$ values range from −9.5‰ to −2.5‰ and from +3.6‰ to +10.9‰, respectively (Table 4). Additionally, the distance in μm from the boundary between 176-19 and 176-20 is shown to understand each sample's positional relationship (Table 4).

4.3. Sulfur Isotope Ratios of Pyrite

Sulfur isotope ratios of pyrite in quartz veins from three ore zones were determined using SIMS. The $\delta^{34}S$ values of pyrite from the Sanjin ore zone range from −16.5‰ to +3.5‰ (Table 5). The $\delta^{34}S$ values of pyrite from veins and sedimentary rocks in the Honko ore zone range from −3.4‰ to +2.1‰ and from +0.7‰ to +3.2‰, respectively (Table 5). The $\delta^{34}S$ values from the Yamada ore zone range from −2.0‰ to +4.5‰ (Table 5). The $\delta^{34}S$ values of two chalcopyrite grains from the Zuisen No. 1 vein are −1.4‰ and +1.0‰. The average $\delta^{34}S$ value of the whole data is −0.7‰. The average $\delta^{34}S$ values of pyrites in veins and sedimentary rocks are −1.1‰ and +2.0‰, respectively. As the grain size of pyrite is small (mostly between 10 and 30 μm) relative to the analysis area (20 μm), separate analytical values for the rim and core of pyrite are unavailable.

5. Discussion

5.1. Role of Basement in the Epithermal Au Deposits in Kyushu

Fluid-rock dynamics significantly impact the carbon and oxygen isotope budgets of the geologic contexts of ore-forming hydrothermal systems. First, let's consider the geologic settings of the Kushikino deposit adjacent to the Hishikari deposit (Figure 1). The Kushikino deposit occurs in Pliocene andesitic rocks, and the basement rocks of the deposit are sedimentary rocks of the Lower Shimanto Group (a Cretaceous accretionary wedge). On the other hand, the Noya deposit is found in Pleistocene andesitic rocks in northern Kyushu (Figure 1). The deposit's basement rocks are high-temperature metamorphic and granitic rocks of the Cretaceous period. The influence of basement rocks on the ore-forming fluid is examined in the Kushikino and Noya deposits. The $\delta^{13}C$ values in the fluid for the Noya and Kushikino gold deposits in Kyushu were estimated to be −6.5‰ [26] and −10.8‰ [19,27], respectively. The calculated $\delta^{13}C$ value of about −7‰ for the Noya deposit is regarded as average crustal carbon (e.g., [28,29]) and as deep-seated sources (e.g., [30,31]). Both deposits have andesitic host rocks, although the basement rocks are mostly granitic for the Noya deposit and sedimentary accretionary rocks for the Kushikino deposit. The difference in basement rocks might cause the difference in $\delta^{13}C$ value between the two deposits. The $\delta^{13}C$ value in the Kushikino fluid might be lowered by about 4‰ by reactions with organic carbon in the basement sedimentary rocks [19].

The isotopic equilibrium fractionation lines for calcite from the Kushikino and Noya deposits are shown in Figure 6. The $\delta^{13}C$ and $\delta^{18}O$ values of most calcite from the Kushikino deposit lie on the isotopic equilibrium fractionation lines, the Kushikino trend (Figure 6), which suggests that the reaction with the basement sedimentary rocks occurred not locally but in a vast reservoir. There is a discrepancy between the two trends in Figure 6 because the basement rocks for the two deposits are different. Calcites from the Lower Shimanto Group sedimentary basement rocks that are part of the epithermal Au deposits in the Hokusatsu district (Figure 1) are located on the Kushikino trend [19,32]. The widely distributed basement rocks of the Shimanto Group may have exerted regional control over the district's relatively low and homogeneous $\delta^{13}C$ values of ascending ore-forming fluids [32]. The Sr isotope ratios ($^{87}Sr/^{86}Sr$) of the Naya's veins are the same as adjacent andesite (0.704). However, the ratios of the Kushikino and Hishikari veins are both high (higher than 0.706), which is attributed to reactions with basement Cretaceous sedimentary rocks with high Sr isotope ratios [26]. So, the $\delta^{13}C$ depletion and $^{87}Sr/^{86}Sr$ enrichment show the basement influence on Kyushu's epithermal ore-forming fluid.

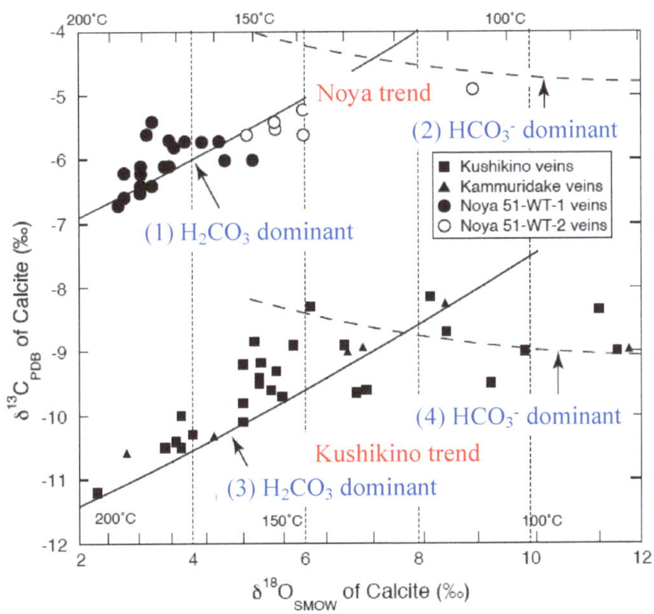

Figure 6. The $\delta^{13}C$ and $\delta^{18}O$ values of vein calcite for the Noya and Kushikino deposits (Figure 1). Calculated as a function of temperature, curves (1) and (2) represent equilibrium fractionation trends for calcite under the assumption that H_2CO_3 [33] and HCO_3^- [34] are the predominant carbon species in the fluid, respectively. The $\delta^{13}C$ and $\delta^{18}O$ values of the Noya fluid for both curves are obtained to be −6.5‰ and −7.5‰, respectively [26]. The temperatures marked on the upper *x*-axis are isotopic equilibrium temperatures [35]. Curves (3) and (4) are equilibrium fractionation trends for calcite from the Kushikino deposit. The $\delta^{13}C$ and $\delta^{18}O$ values of the Kushikino fluid for both curves are obtained to be −10.8‰ and −7.0‰, respectively [19]. At about 140 °C with decreasing temperature, the predominant carbon species in the Kushikino fluid changed from H_2CO_3 to HCO_3^- [27]. Samples of the Kushikino veins are from the Kushikino deposit, and those of the Kammuridake veins are from a peripheral area of the deposit. The oxygen isotopic equilibrium temperatures for the Kushikino hydrothermal fluid are marked on the lower *x*-axis.

5.2. Hydrothermal Fluids for the Hishikari Deposit

Figure 7 shows calcite's $\delta^{13}C$ and $\delta^{18}O$ values from the Hishikari deposit. The figure excludes two calcite samples with very low $\delta^{13}C$ values of −23.4‰ and −25.1‰ (Table 1). The average global $\delta^{13}C$ value of C3 plants is approximately −28.5‰ [36]. Local pyrolysis or microbial decomposition of these organic materials in the basement sedimentary rocks can result in the production of CO_2. The fluid temperature at the precipitation of the very low $\delta^{13}C$ calcite is calculated to be around 45 °C using the $\delta^{18}O$ values of the calcites and calibration curve of O'Neil et al. [35], assuming that the $\delta^{18}O$ value of −7‰ for the fluid, which is similar to that of the Kushikino ore fluid ($\delta^{18}O$ = −7‰) and regional meteoric water ($\delta^{18}O$ = −7.1‰) [19,37]. Therefore, the calcites with very low $\delta^{13}C$ values might precipitate after the ore mineralization.

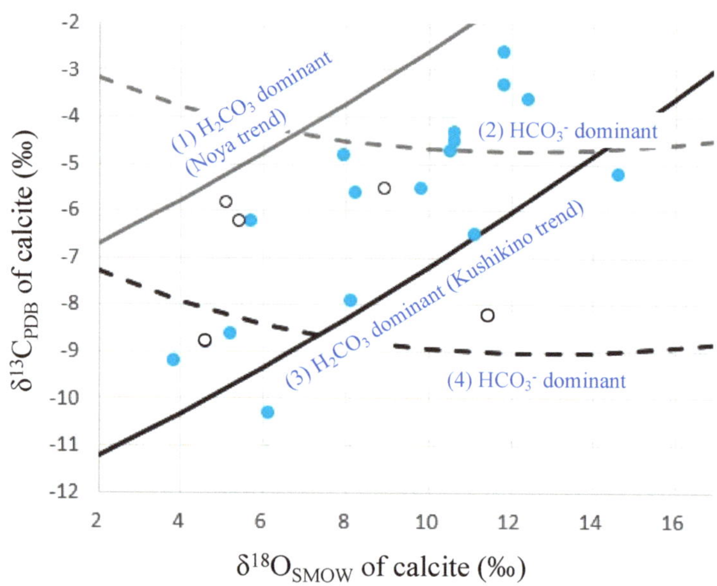

Figure 7. The $\delta^{13}C$ and $\delta^{18}O$ values of calcite from the Hishikari deposit are shown. Solid circles are from hand specimens taken in this study (Table 1), and open circles are from an author's previous work [19]. The equilibrium fractionation trends of calcite for the Noya and Kushikino deposits explained in Figure 6 are also shown.

Since the Hishikari and Kushikino deposits both occur in the same Hokusatsu district and share the same basement sedimentary rocks, it is expected that the calcite's $\delta^{13}C$ and $\delta^{18}O$ values will follow the Kushikino pattern. Although several isotope data are along the Kushikino trend, the others lie between the Noya and Kushikino trends (Figure 7). These particular characteristics of the Hishikari fluid will be investigated next.

First, the $\delta^{13}C$ and $\delta^{18}O$ values of the 10 mm-width sample set of the calcite-dominant vein (Section 3.1.2) are obtained to get a rough idea of the isotope evolution from the early to late stages of vein formation. In Figure 8, the $\delta^{13}C$–$\delta^{18}O$ combination moves slightly from 1743 through 1746 samples. The last sample, 1747, has a high $\delta^{18}O$ value, probably because the temperature drops at the later stage of calcite mineralization, since the oxygen isotope fractionation between calcite and water increases with decreasing temperature. Next, we take finer samples (0.9 mm-width sample set: from 176-1 to 176-21) from the early mineralization stage than the 10 mm-width sample set (Figure 4b,c). A more precise and definite evolutional isotopic trend of the fluid is expected to appear. However, no such trend is found. The $\delta^{13}C$ and $\delta^{18}O$ values of the 0.9 mm-width samples move irregularly over time, but the range is limited (Figure 8).

Finally, maximum efforts to increase the time resolution were made. The $\delta^{13}C$ and $\delta^{18}O$ values of calcite from 176-20 and 176-21 sections (12 μm-width sample set; Table 4) are shown in Figure 9a,b, respectively. The range of $\delta^{13}C$ and $\delta^{18}O$ values for the 176-20 section is slightly bigger than that for the 176-21 section. However, they are similar to the range of 10 mm-width and 0.9 mm-width samples (Figure 9a,b). The change in $\delta^{18}O$ values might reflect temperature changes. However, isotope ratios and growth order do not correlate with one another. The $\delta^{13}C$ and $\delta^{18}O$ values of calcite jump around in a particular range

(Figure 9a,b), and it is difficult to explain the fluid evolution by a simple change of the fluid in temperature or pH.

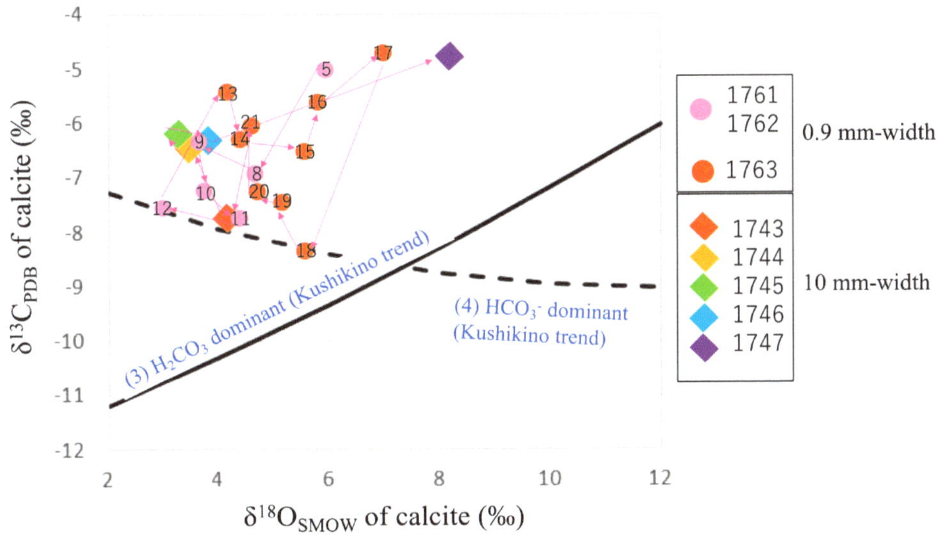

Figure 8. The $\delta^{13}C$ and $\delta^{18}O$ values of calcite (10 mm-width sample set and 0.9 mm-width sample set) from the 20100317 sample (Table 3). The number with the symbol represents the growth order in the 0.9 mm-width sample set. The equilibrium fractionation trends of calcite for the Kushikino deposits are shown.

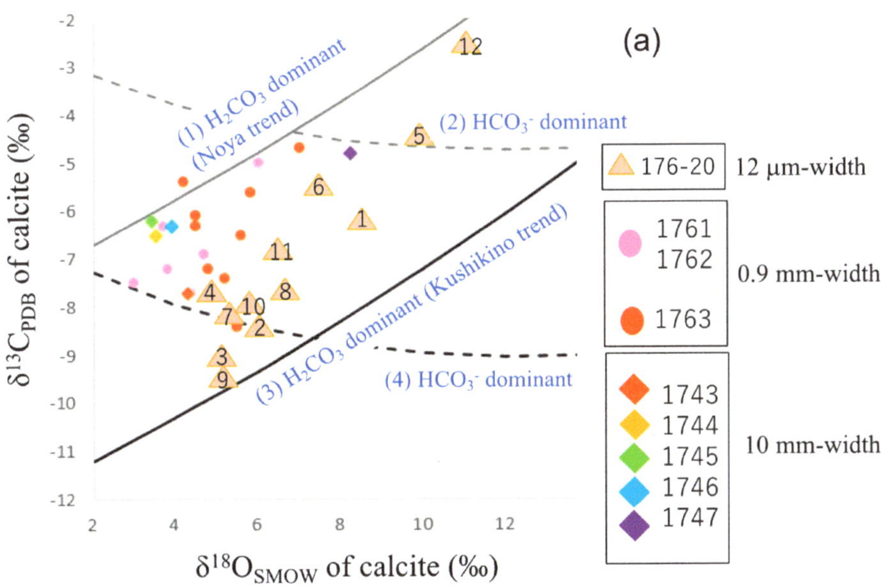

Figure 9. *Cont.*

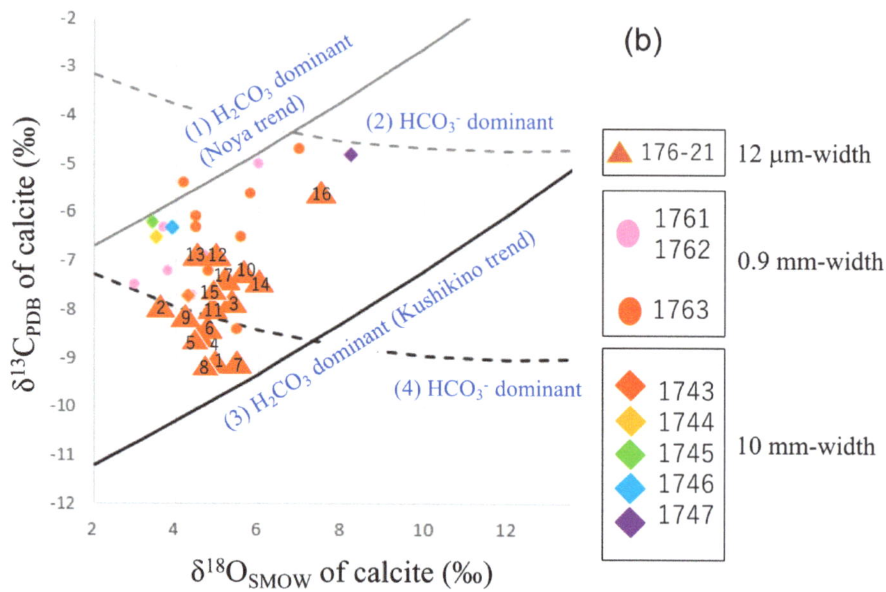

Figure 9. The results of microanalysis. The $\delta^{13}C$ and $\delta^{18}O$ values of calcite (micro-scale samples obtained by a microtome; Table 4) are superimposed in Figure 8. The number with the symbol represents the growth order. The equilibrium fractionation trends of calcite for the Noya and Kushikino deposits are also shown. (**a**) The 12 μm-width samples (176-20-1 to 176-20-12) were taken from the 176-20 section. (**b**) The 12 μm-width samples (176-21-1 to 176-21-17) were taken from the 176-21 section.

5.3. Sulfur Isotopes

A histogram of sulfur isotope ratios of pyrite from the Hishikari deposit is shown in Figure 10. The analysis area using the SIMS microanalysis is 20 μm. The whole data's average $\delta^{34}S$ value is −0.7‰. The $\delta^{34}S$ values of pyrites in the host rock are similar to those in the adjacent ore vein at the same location of the host rock (E8) and Hosen-1 (E8) in Figure 10, which indicates that the same fluid forms both hydrothermal pyrites. The pyrite grains in the host rock (shale), within 2 cm from the vein, occur near the vein by the ore fluid that flows into the host rock. Late Cenozoic volcanic rocks in the Hokusatsu district are of the magnetite series (similar to I-type granitoids; [38]) that are defined by Ishihara [39]. Ishihara et al. [40] measured $\delta^{34}S$ values of three vein ores and six basement sedimentary rocks from the Hishikari deposit by the conventional method, and the average $\delta^{34}S$ value is around +0.3‰. The $\delta^{34}S$ value of stibnite from the Hishikari deposit is −0.2‰ [41], and the $\delta^{34}S$ values of seven stibnite samples, which are from the latest stage of the Hishikari mineralization, are narrow-ranged (from −0.2‰ to +0.7‰) among several different veins [42]. The near-zero values of $\delta^{34}S$ from the Hishikari hydrothermal system are consistent with that from magnetite series magma [43].

Very low $\delta^{34}S$ values from −16.5‰ to −10.1‰ of pyrites are found only in the high-grade Keisen–3 vein of the Sanjin ore zone (Figure 10; Table 5). The very low $\delta^{34}S$ values, first obtained by the SIMS microanalysis, are apart from the pile of other values (Figure 10). Sanematsu et al. [18] proposed that the fractures corresponding to the Keisen veins were a significant channel for the ascending hot hydrothermal fluids. The $\delta^{34}S$ values of unaltered Shimanto sedimentary rocks are from −21.5‰ to −1.4‰ [43]. Therefore, the pyrite with a very low $\delta^{34}S$ value might form locally that reacted with the Shimanto sedimentary rocks. The reaction with the sedimentary rocks might occur in the microscopic region because relatively high $\delta^{34}S$ values are also obtained from the same hand specimen of the Keisen–3 vein (Figure 10).

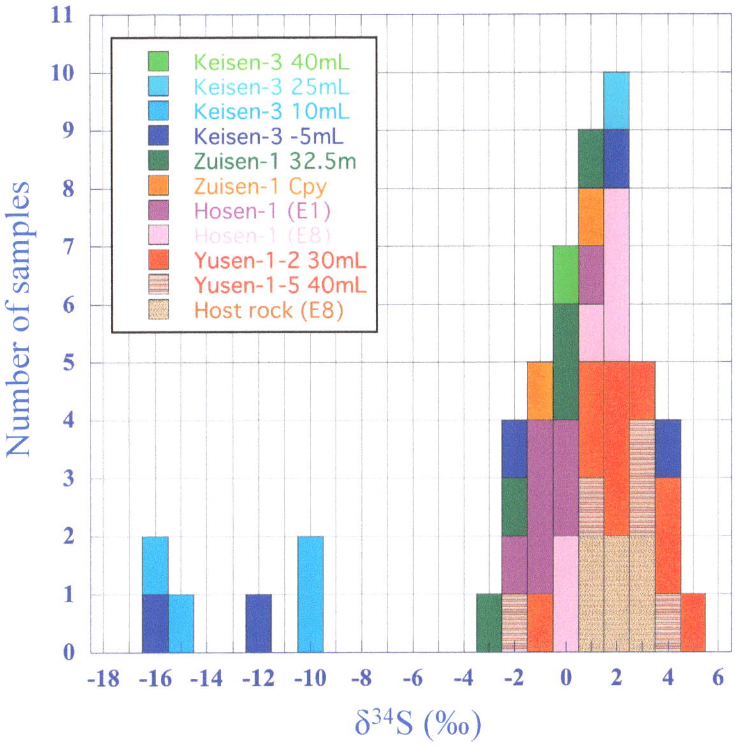

Figure 10. A histogram of $\delta^{34}S$ values of pyrite and chalcopyrite from the Hishikari deposit using the SIMS microanalysis is shown based on Table 5. The vertical axis is the number of samples. The Keisen vein is from the Sanjin ore zone, the Zuisen and Hosen veins with the host rocks are from the Honko ore zone, and the Yusen veins are from the Yamada ore zone. The $\delta^{34}S$ value for each vein is color-coded. The average $\delta^{34}S$ value of whole data is −0.7‰, and the average $\delta^{34}S$ value of pyrites in veins is −1.1‰. Pyrite grains in the sedimentary rocks (Host rock in the legend) have slightly high $\delta^{34}S$ values than those in quartz veins (Hosen-1) at the same location of E8. Cpy: chalcopyrite.

5.4. Genesis and Evolution of Hydrothermal Fluids in the Formation of the Hishikari Deposit

The Hishikari deposit's ore-forming fluids may mix meteoric and magmatic fluids. Particularly along the unconformity between the basement sedimentary rocks and overlying andesites [44]. The fluids are formed from meteoric water undergoing isotopic exchange with basement rocks [6]. The isotope results in this study also suggest that the ore-forming fluid for the Hishikari deposit is a mixture of two fluids. We refer to the two fluids as Fluid 1 (along the Noya trend), which the basement sedimentary rocks may not affect, and Fluid 2 (along the Kushikino trend), which they may constrain. Fluid 2 is common to the deposits in the Hokusatsu district (Figure 1); that is, the $\delta^{13}C$ and $\delta^{18}O$ values of calcite from most Au deposits in the Hokusatsu district are along Fluid 2 [19,32]. According to past research [2–6], the Hishikari fluid may have received some magmatic fluid input, particularly during the earliest stages of mineralization. The genesis of the Hishikari hydrothermal fluid is the most of meteoric origin. However, some sporadically input of magmatic origin to the fluid has occurred. The Hishikari deposit also has Fluid 2 [19,21,32], like any other deposit in the district, but it should be emphasized that Fluid 1 is peculiar to the Hishikari deposit. Calcite samples from Fluids 1 and 2 are everywhere in the deposit.

Figure 11 shows the $\delta^{13}C$ and $\delta^{18}O$ values of vein calcite with/without electrum [21]. Although the $\delta^{13}C$ values of samples with electrum are relatively high compared to those without electrum, the $\delta^{13}C$ and $\delta^{18}O$ values of calcites with electrum lie either near the Noya trend or Kushikino trend. Additionally, since vein calcite samples without electrum in Figure 11 are from Au-rich veins, they are also recognized as fertile. Therefore, at the Hishikari deposit, Fluids 1 and 2 are both created from fluids capable of forming ore (Figure 11). The 20100317 sample is from the Au-rich Keisen 3-1 vein, and the earliest formed calcite in the 20100317 sample (176-5; Table 3) lies on the Noya trend (Fluid 1), which is apart from the Kushikino trend (Fluid 2 in Figure 11). Fluids 1 and 2 are both fertile ore-forming fluids. Fluid 2 is thought to have reacted well with basement sedimentary rocks that contain organic matter, and Fluid 1, on the other hand, is not affected on a large scale by sedimentary rocks.

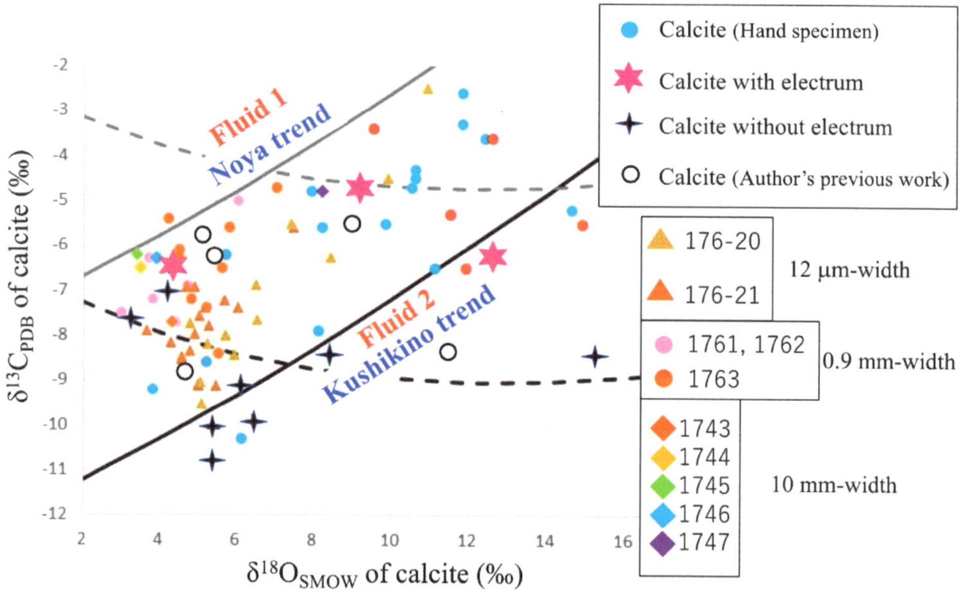

Figure 11. The $\delta^{13}C$ and $\delta^{18}O$ values of calcite from the Hishikari deposit. Blue solid circles indicate isotope data of calcite from hand specimens. The data from the 10 mm-width, 0.9 mm-width, and 12 μm-width samples are represented by each symbol. Isotope data of calcite with/without electrum (purple star and black cross, respectively) are from [21], and data (open circles) from [19] are shown. The equilibrium fractionation trends of calcite for the Noya and Kushikino deposits are also shown.

The previous studies [2–6] have suggested that some input from magmatic fluid was found in the Hishikari hydrothermal fluid in the earlier stages of mineralization. In other words, the genesis of the Hishikari fluid is mostly of meteoric origin, but some sporadical input of magmatic fluid to the hydrothermal system has occurred. Additionally, the unusually low $\delta^{34}S$ readings in the Keisen veins suggest that magmatic fluid only locally and in a small volume interacted with sedimentary rocks. The microanalysis results of $\delta^{13}C$ and $\delta^{18}O$ values in every 12 μm of the calcite-bearing sample along the growth direction scatter in a particular range (Figure 9a,b). The duration of 12 μm-thickness mineralization is about 10 years of the Hishikari mineralization according to the $^{40}Ar/^{39}Ar$ ages of adularia [45]. Since the very high temporal resolution cannot find the isotope evolution of the hydrothermal fluid, the evolution is not a gradual transition from magmatic to meteoric origin. Still, a rapid turnover of two fluids might be locally happening.

The average pyrite $\delta^{34}S$ value of −0.7‰ from veins and sedimentary rocks is similar to that of the adjacent magma, consistent with previous studies. However, the SIMS microanalysis reveals that very low $\delta^{34}S$ values from −16.5‰ to −10.1‰ and positive $\delta^{34}S$ values of pyrites are found in the same hand specimen from the high-grade Keisen-3 vein of the Sanjin ore zone (Figure 10). Therefore, the pyrite having a very low $\delta^{34}S$ value might form locally that was reacted with the Shimanto sedimentary rocks. The evidence from sulfur isotope data proves that the reaction of the hydrothermal fluid with basement sedimentary rocks occurs locally in the microscopic region.

Fluid 2, the mainstream in the Hokusatsu district, is supposed to flow continuously in the Hishikari deposit during the mineralization. However, Fluid 1, a magmatic-origin deep fluid with little reaction with the basement sedimentary rocks, might inject intermittently into the Fluid 2-filled vein system. Intermittent opening of the top of a vein lowers the pressure in the vein, resulting in the introduction of deep Fluid 1 to the epithermal system and subsequent mixing with Fluid 2 and boiling in some cases. Au was probably transported as $Au(HS)_2^-$ [3] and might be deposited in response to changes in the temperature, pH, log fO_2, and total sulfur concentrations [46]. Boiling would fractionate H_2S strongly into the vapor, lower HS^- activity, and increase pH by losing H_2S, which causes the deposition of calcite and Au [47].

The characteristics of the high-grade Hishikari deposit, compared to the other Au deposits in the Hokusatsu district, might be caused by Fluid 1. Unlike other epithermal Au deposits in the district, the rapid local mixing of Fluid 1 and 2 might cause the Au-rich ore deposition at the Hishikari deposit.

6. Concluding Remarks

The Hishikari gold deposit in Kyushu, Japan, is world-famous for its high-grade ore. The Hishikari deposit is one of the numerous epithermal gold occurrences in the Hokusatsu district. It is located in the basement rocks of the Lower Shimanto Group (Cretaceous accretionary wedge). The average pyrite $\delta^{34}S$ value of −0.7‰ from veins and sedimentary rocks is similar to that of the adjacent magma, consistent with previous studies. However, the SIMS microanalysis reveals that very low $\delta^{34}S$ values and positive $\delta^{34}S$ values of pyrites are found in the same hand specimen from the high-grade Keisen-3 vein. Therefore, sulfur isotope measurements demonstrate that the hydrothermal fluid's reaction with the basement sedimentary rocks occurs locally because the pyrite with a very low $\delta^{34}S$ value might form locally.

Although the basement rocks regulate the feature of the hydrothermal fluids for the deposits, it has been inferred that the incorporation of magmatic-origin deep fluid to the meteoric-origin hydrothermal fluid might cause the gold and calcite precipitation at the Hishikari deposit. Although this scenario applies to the Hishikari deposit generally, the microanalysis of $\delta^{13}C$ and $\delta^{18}O$ values of vein calcite reveals that the fluid evolution was not gradually occurring, and a rapid turnover of two fluids might be locally happening. Thus, the $\delta^{13}C$ and $\delta^{18}O$ values of hydrothermal calcite can provide insights into the genesis and evolution of the epithermal deposits.

Author Contributions: Conceptualization, Y.M.; methodology, Y.M.; validation, Y.M. and Y.Y.; formal analysis, Y.M.; investigation, Y.M. and Y.Y.; resources, Y.M.; writing—original draft preparation, Y.M. and Y.Y.; writing—review and editing, Y.M.; visualization, Y.M. and Y.Y.; supervision, Y.M.; project administration, Y.M.; funding acquisition, Y.M. All authors have read and agreed to the published version of the manuscript.

Funding: This research was funded by JSPS Kakenhi, grant numbers JP18K03758 and JP22K03736.

Data Availability Statement: Not applicable.

Acknowledgments: The authors are grateful to the Sumitomo Metal Mining Co. Ltd. for facilitating underground surveys and for permission to publish this paper, specifically to Yoshinori Okaue, Takayuki Seto, Koji Morimoto, Ryota Sekine and Yu Yamato. Several calcite-bearing vein samples are collected by the mine geologists. We thank for their efforts. The authors are also indebted to Nobutaka

Shimada and Kazuhiko Shimada for providing several pyrite-bearing samples. Two anonymous reviewers are thanked for their careful and constructive comments, which improved the manuscript.

Conflicts of Interest: The authors declare no conflict of interest.

References

1. Sumitomo Metal Mining Co., Ltd. Facts at a glance of the Hishikari Mine. 2022; *Unpublished Company Brochure*. 1p. (In Japanese)
2. Matsuhisa, Y.; Aoki, M. Temperature and oxygen-isotope variations during formation of the Hishikari epithermal gold-silver veins, Southern Kyushu, Japan. *Econ. Geol.* **1994**, *89*, 1608–1613. [CrossRef]
3. Shikazono, N.; Nagayama, T. Origin and depositional mechanism of the Hishikari gold-quartz-adularia mineralization. *Resour. Geol. Sp. Issue* **1993**, *14*, 47–56.
4. Hayashi, K.; Maruyama, T.; Satoh, H. Submillimeter scale variation of oxygen isotope of vein quartz at the Hishikari deposit, Japan. *Resour. Geol.* **2000**, *50*, 141–150. [CrossRef]
5. Hayashi, K.; Maruyama, T.; Satoh, H. Precipitation of gold in a low-sulfidation epithermal gold deposit: Insights from a submillimeter scale oxygen isotope analysis of vein quartz. *Econ. Geol.* **2001**, *96*, 211–216. [CrossRef]
6. Faure, K.; Matsuhisa, Y.; Metsugi, H.; Miozota, C.; Hayashi, S. The Hishikari Au–Ag epithermal deposit, Japan: Oxygen and hydrogen isotopic evidence in determining the source of paleohydrothermal fluids. *Econ. Geol.* **2002**, *97*, 481–498. [CrossRef]
7. Morishita, Y.; Shimada, N.; Shimada, K. Invisible gold and arsenic in pyrite from the high-grade Hishikari gold deposit, Japan. *Appl. Surf. Sci.* **2008**, *255*, 1451–1454. [CrossRef]
8. Morishita, Y.; Shimada, N.; Shimada, K. Invisible gold in arsenian pyrite from the high-grade Hishikari gold deposit, Japan: Significance of variation and distribution of Au/As ratios in pyrite. *Ore Geol. Rev.* **2018**, *95*, 79–93. [CrossRef]
9. Hosono, T.; Nakano, T.; Murakami, H. Sr-Nd-Pb isotopic compositions of volcanic rocks around the Hishikari gold deposit, southwest Japan: Implications for the contribution of a felsic lower crust. *Chem. Geol.* **2003**, *201*, 19–36. [CrossRef]
10. Izawa, E.; Urashima, Y.; Ibaraki, K.; Suzuki, R.; Yokoyama, T.; Kawasaki, K.; Koga, A.; Taguchi, S. The Hishikari gold deposit—High-grade epithermal veins in Quaternary volcanics of Southern Kyushu, Japan. *J. Geochem. Explor.* **1990**, *36*, 1–56. [CrossRef]
11. Izawa, E.; Etoh, J.; Honda, M.; Motomura, Y.; Sekine, R. Hishikari gold mineralization: A case study of the Hosen No. 1 vein hosted by basement Shimanto sedimentary rocks, southern Kyushu, Japan. *SEG Guideb. Ser. (CD-ROM)* **2001**, *34*, 21–30.
12. Izawa, E.; Urashima, Y. Quaternary gold mineralization and its geologic environments in Kyushu, Japan. *SEG Guideb. Ser. (CD-ROM)* **2001**, *34*, 1–9.
13. Ibaraki, K.; Suzuki, R. Gold-silver-quartz-adularia veins of the Main, Yamada and Sanjin deposits, Hishikari gold mine; A comparative study of their geology and ore deposits. *Resour. Geol. Sp. Issue* **1993**, *14*, 1–11.
14. Nagayama, T. Precipitation sequence of veins at the Hishikari deposit, Kyushu, Japan. *Resour. Geol. Sp. Issue* **1993**, *14*, 13–27.
15. Yamato, Y.; Ohno, M.; Ushirone, N. Exploration and development of the Yusen No.1 vein group based on the vein structure, Yamada deposit, Hishikari Mine. *Shigen Chishitsu* **2002**, *52*, 11–17, (In Japanese with English Abstract).
16. Sekine, R.; Izawa, E.; Watanabe, K. Timing of fracture formation and duration of mineralization at the Hishikari deposit, southern Kyushu, Japan. *Resour. Geol.* **2002**, *52*, 395–404. [CrossRef]
17. Geological Survey of Japan, AIST. Geological map of Japan 1:2,000,000. In *1:2,000,000 Map*, Series No. 4; 5th ed.; Geological Survey of Japan: Tsukuba, Japan, 2002.
18. Sanematsu, K.; Duncan, R.; Imai, A.; Watanabe, K. Geochronological constraints using $^{40}Ar/^{39}Ar$ dating on the mineralization of the Hishikari epithermal gold deposit, Japan. *Resour. Geol.* **2005**, *55*, 249–266. [CrossRef]
19. Morishita, Y.; Nakano, T. Role of basement in epithermal deposits: The Kushikino and Hishikari gold deposits, southwestern Japan. *Ore Geol. Rev.* **2008**, *34*, 597–609. [CrossRef]
20. Etoh, J.; Izawa, E.; Taguchi, S. A fluid inclusion study on columnar adularia from the Hishikari low-sulfidation epithermal gold deposit, Japan. *Resour. Geol.* **2002**, *52*, 73–78. [CrossRef]
21. Imai, A.; Uto, T. Association of electrum and calcite and its significance to the genesis of the Hishikari gold deposits, southern Kyusyu, Japan. *Resour. Geol.* **2002**, *52*, 381–394. [CrossRef]
22. Fournier, R.O. The behavior of silica in hydrothermal solution. *Rev. Econ. Geol.* **1985**, *2*, 45–61.
23. Morishita, Y.; Hammond, N.Q.; Momii, K.; Konagaya, R.; Sano, Y.; Takahata, N.; Ueno, H. Invisible gold in pyrite from epithermal, banded-iron-formation-hosted, and sedimentary gold deposits: Evidence of hydrothermal influence. *Minerals* **2019**, *9*, 447. [CrossRef]
24. Wada, H.; Niitsuma, N.; Saito, T. Carbon and oxygen isotopic measurements of ultra-small samples. *Geosci. Rep. Shizuoka Univ.* **1982**, *7*, 35–50, (In Japanese with English Abstract).
25. Morishita, Y.; Matsuhisa, Y. Measurement of carbon and oxygen isotope ratios of carbonate reference samples. *Bull. Geol. Surv. Jpn.* **1984**, *35*, 69–79, (In Japanese with English Abstract).
26. Morishita, Y.; Takeno, N. Nature of the ore-forming fluid at the Quaternary Noya gold deposit in Kyushu, Japan. *Resour. Geol.* **2010**, *60*, 359–376. [CrossRef]
27. Matsuhisa, Y.; Morishita, Y.; Sato, T. Oxygen and carbon isotope variations in gold bearing hydrothermal veins in the Kushikino mining area, southern Kyushu, Japan. *Econ. Geol.* **1985**, *80*, 283–293. [CrossRef]
28. Craig, H. The geochemistry of the stable carbon isotopes. *Geochim. Cosmochim. Acta* **1953**, *3*, 53–92. [CrossRef]

29. Fuex, A.N.; Baker, D.R. Stable carbon isotopes in selected granitic, mafic, and ultramafic igneous rocks. *Geochim. Cosmochim. Acta* **1973**, *37*, 2509–2521. [CrossRef]
30. Taylor, H.P., Jr.; Frechen, J.; Degens, E.T. Oxygen and carbon isotope studies of carbonatites from the Laacher See district, Sweden. *Geochim. Cosmochim. Acta* **1967**, *31*, 407–430. [CrossRef]
31. Pineau, F.; Javoy, M.; Bottinga, Y. 13C/12C ratios of rocks and inclusions in popping rocks of the problem of isotopic composition of deep-seated carbon. *Earth Planet. Sci. Lett.* **1976**, *29*, 413–421. [CrossRef]
32. Morishita, Y. Carbon and oxygen isotopic characteristics of epithermal veins in the Hokusatsu gold district, southern Kyushu, Japan. *Resour. Geol. Sp. Issue* **1993**, *14*, 103–114.
33. Bottinga, Y. Calculation of fractionation factors for carbon and oxygen isotopic exchange in the system calcite–carbon dioxide–water. *J. Phys. Chem.* **1968**, *72*, 800–808. [CrossRef]
34. Mook, W.G.; Bommerson, J.C.; Staverman, W.H. Carbon isotope fractionation between dissolved bicarbonate and gaseous carbon dioxide. *Earth Planet. Sci. Lett.* **1974**, *22*, 169–176. [CrossRef]
35. O'Neil, J.R.; Clayton, R.N.; Mayeda, T.K. Oxygen isotope fractionation in divalent metal carbonates. *J. Chem. Phys.* **1969**, *51*, 5547–5558. [CrossRef]
36. Kohn, M.J. Carbon isotope compositions of terrestrial C3 plants as indicators of (paleo)ecology and (paleo)climate. *Proc. Natl. Acad. Sci. USA* **2010**, *107*, 19691–19695. [CrossRef] [PubMed]
37. Matsubaya, O.; Ueda, A.; Kusakabe, M.; Matsuhisa, Y.; Sakai, H.; Sasaki, A. An isotopic study of the volcanoes and the hot springs in Satsuma Iwo-jima and some areas in Kyushu. *Bull. Geol. Surv. Jpn.* **1975**, *26*, 375–392.
38. Chappell, B.W.; White, A.J.R. Two contrasting granite types. *Pacific Geol.* **1974**, *8*, 173–174.
39. Ishihara, S. The magnetite-series and ilmenite-series granitic rocks. *Mining Geol.* **1977**, *27*, 293–305.
40. Ishihara, S.; Sakamaki, Y.; Sasaki, A.; Teraoka, Y.; Terashima, S. Role of the basement in the genesis of the Hishikari gold-quartz vein deposit, southern Kyushu, Japan. *Mining Geol.* **1986**, *36*, 495–509.
41. Imai, A.; Shikazono, N.; Shimizu, M.; Shimazaki, H. Sulfur isotope study on Hg and Sb deposits in Japan. *Resour. Geol.* **2006**, *56*, 37–48. [CrossRef]
42. Shimizu, T. Sulfur isotopic ratios and mode of occurrence of stibnite at the Hishikari epithermal Au-Ah deposit, Japan. *Bull. Geol. Surv. Japan* **2017**, *68*, 111–117. [CrossRef]
43. Sasaki, A.; Ishihara, S. Sulfur isotopic composition of the magnetite-series and ilmenite-series granitoids in Japan. *Contrib. Mineral. Petrol.* **1979**, *68*, 107–115. [CrossRef]
44. Takahashi, R.; Tagiri, R.; Blamey, N.J.F.; Imai, A.; Watanabe, Y.; Takeuchi, A. Characteristics and behavior of hydrothermal fluids for gold mineralization at the Hishikari deposits, Kyushu, Japan. *Resour. Geol.* **2017**, *67*, 279–299. [CrossRef]
45. Sanematsu, K.; Watanabe, K.; Duncan, R.A.; Izawa, E. The history of vein formation determined by $^{40}Ar/^{39}Ar$ dating of adularia in the Hosen-1 vein at the Hishikari epithermal gold deposit, Japan. *Econ. Geol.* **2006**, *101*, 685–698. [CrossRef]
46. Seward, T.M. Thio complexes of gold and the transport of gold in hydrothermal ore solutions. *Geochim. Cosmochim. Acta* **1973**, *37*, 379–399. [CrossRef]
47. Drummond, S.E.; Ohmoto, H. Chemical evolution and mineral deposition in boiling hydrothermal systems. *Econ. Geol.* **1985**, *80*, 126–147. [CrossRef]

minerals

MDPI

Article

Isotope Geochemistry of the Heihaibei Gold Deposit within the Kunlun River Area in the Eastern Kunlun Orogen in Northwest China and Its Metallogenic Implications

Hai-Feng Lu [1], Tong Pan [2], He Jiao [1], Qing-Feng Ding [3,*], Xuan Zhou [3] and Rui-Zhe Wu [3]

1 Qinghai Provincial Key Laboratory of Salt Lake Resources Exploration and Research in Qaidam Basin, Qaidam Integrated Geological Exploration Institute of Qinghai Province, Golmud 816099, China
2 Bureau of Geological Exploration and Development of Qinghai Province, Xining 810008, China
3 College of Earth Science, Jilin University, Changchun 130061, China
* Correspondence: dingqf@jlu.edu.cn

Abstract: The Heihaibei gold deposit is located in the Eastern Kunlun Orogen in Northwest China. The gold mineralization here occurs predominantly in quartz veins within faulted granite zones. The sulfide mineral assemblage is dominated by pyrite and arsenopyrite, with minor chalcopyrite, galena, sphalerite, tetrahedrite, and micro-native gold. Weak alterations in Heihaibei granites include silicification and sericitization, with minor chloritization and carbonatization. The measured δD_{H2O} and $\delta^{18}O_{quartz}$ values of quartz in auriferous quartz veins range from −104.2‰ to −81.1‰ and +9.2‰ to +13.9‰, respectively. The $\delta^{34}S$ values of sulfides in auriferous quartz veins range from +7.60‰ to +8.65‰, and the lead isotope compositions of sulfides in ores range from 18.7219 to 19.0007 for $^{206}Pb/^{204}Pb$, 15.6959 to 15.7062 for $^{207}Pb/^{204}Pb$, and 37.7359 to 38.8055 for $^{208}Pb/^{204}Pb$. The Pb isotope compositions of potassic feldspars from Heihaibei granites vary from 18.3532 to 19.4864 for $^{206}Pb/^{204}Pb$, 15.6475 to 15.6812 for $^{207}Pb/^{204}Pb$, and 37.1750 to 38.4598 for $^{208}Pb/^{204}Pb$. Collectively, the isotope (H, O, S, and Pb) geochemistry suggests that the ore-forming fluid was a special metamorphic water evolved from the deep slab-derived fluids, and the sulfur and lead were predominantly sourced from such metamorphic fluids, and from the deep parts of the Heihaibei granites. Therefore, the Heihaibei gold deposit can be classified as an orogenic gold deposit, which is closely associated with the subduction of the Paleo-Tethys oceanic plate, and even the final closure of this ocean by the Later Triassic.

Keywords: slab-derived metamorphic water; orogenic gold deposit; Kunlun River gold field

Citation: Lu, H.-F.; Pan, T.; Jiao, H.; Ding, Q.-F.; Zhou, X.; Wu, R.-Z. Isotope Geochemistry of the Heihaibei Gold Deposit within the Kunlun River Area in the Eastern Kunlun Orogen in Northwest China and Its Metallogenic Implications. *Minerals* **2023**, *13*, 274. https://doi.org/10.3390/min13020274

Academic Editor: Dominique Gasquet

Received: 23 December 2022
Revised: 9 February 2023
Accepted: 13 February 2023
Published: 15 February 2023

1. Introduction

The Eastern Kunlun Orogen (EKO) extends from east to west for more than 1200 km, with a width of 50–200 km along the northern margin of the Qinghai–Tibetan Plateau, and records the earliest amalgamation history of the Qinghai–Tibetan Plateau (Figure 1) [1,2]. The EKO, which is bounded by the Qaidam block (QDM) on the north and the Bayan Har–Songpanganzi Terrane (BH–SG) on the south, is usually considered to be composed of three sub-parallel tectonic belts from north to south, the North Kunlun belt (NKL), the Middle Kunlun belt (MKL), and the South Kunlun belt (SKL), separated by the North Kunlun fault (N. KLF), the Middle Kunlun fault (M. KLF), and the South Kunlun–Aryan Maqin fault (S. KLF–AMF), respectively [3]. Since 1999, numerous ore prospecting projects within the EKO have been deployed by the China Geology Survey [4]. So far, the EKO has become an important ore belt in which ca. 140 mineral occurrences regarding Au, Fe, Cu, Pb, Zn, W, Mo, Ni, Co, and Sb were discovered, and ca. 26 of them have become different sizes of mineral deposits in the EKO and the adjacent BH–SG (Figure 1b) [5,6]. Especially, the EKO hosts numerous gold deposits, as best represented by those in the Wulonggou gold field [4,7–13], Gouli gold field [14–19], and Kaihuangbei gold field [14,20].

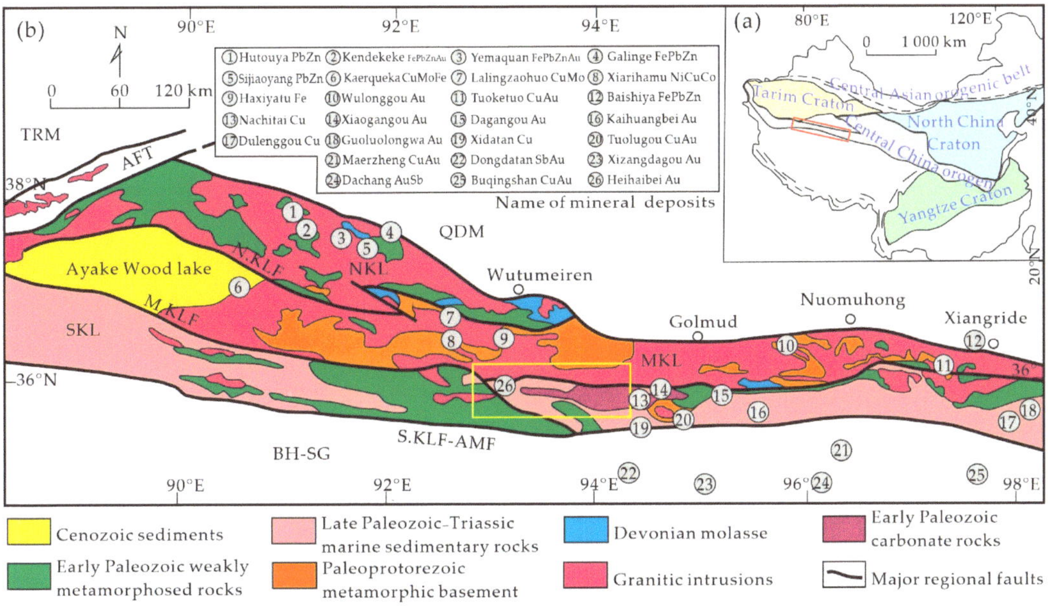

Figure 1. (**a**) Sketch map of China showing the tectonic location of the EKO, revised after [21]; (**b**) schematic geological map of the EKO, simplified and modified after [2]. EKO: Eastern Kunlun Orogen; QDM: Qaidam Block; TRM: Tarim Basin; NKL: North Kunlun Belt; MKL: Middle Kunlun Belt; SKL: South Kunlun Belt; BH–SG: Bayan Har–Songpanganzi Terrane; ATF: Altyn Tagh fault; N.KLF: North Kunlun Fault (South Qiman Tagh Fault); M.KLF: Middle Kunlun Fault; S.KLF–AMF: South Kunlun–Aryan Maqin Fault. Major mineral deposits after [5,6].

In recent years, the Kunlun River goldfield (KLRA) in the SKL has gradually become an important gold mineralization belt because more than 10 gold deposits or occurrences were discovered throughout this area (Figure 2), among which the Heihaibei gold deposit is the largest [22–25]. Thus far, few geologists have carried out geological research on the Heihaibei gold deposit or on other deposits in the KLRA. Most of the published papers or reports emphasize the geological characteristics and gold prospecting potentials for the Heihaibei gold deposit [22,24–27]. Because the dominant ore types are auriferous quartz veins, it was thought that the Heihaibei gold deposit belongs to a hydrothermal-vein-type gold deposit [22], a magmatic hydrothermal gold deposit [24], or an orogenic gold deposit [25]. Based on the geological characteristics and a few stable isotope compositions, some other gold deposits in the KLRA such as the Heihainan, Heicigou, and Dazaohuo gold deposits were also considered to belong to orogenic gold deposits [28,29] or altered-rock-type gold deposits [30], and the ore-forming fluids and materials were derived predominantly from metamorphic water and wall rocks, respectively [28,29]. However, Yu et al. (2022) argued that the Heihaibei gold deposit was a granitic intrusion-related gold deposit rather than an orogenic gold deposit, and the ore-forming fluids and materials were derived from a magmatic hydrothermal source that was closely related to the Heihaibei granites, according to their three sets of H, O, S, and Pb isotope compositions [23]. In a word, the origins of the ore-forming fluids, sources of metals, and genetic types responsible for the Heihaibei gold deposit or other gold mineralization in the Kunlunhe River area remain controversial. Systematic isotope studies on the Heihaibei gold deposit have not been carried out, and especially, a controversy between the orogenic gold deposit and the granitic intrusion-related gold deposit has persisted for a long time.

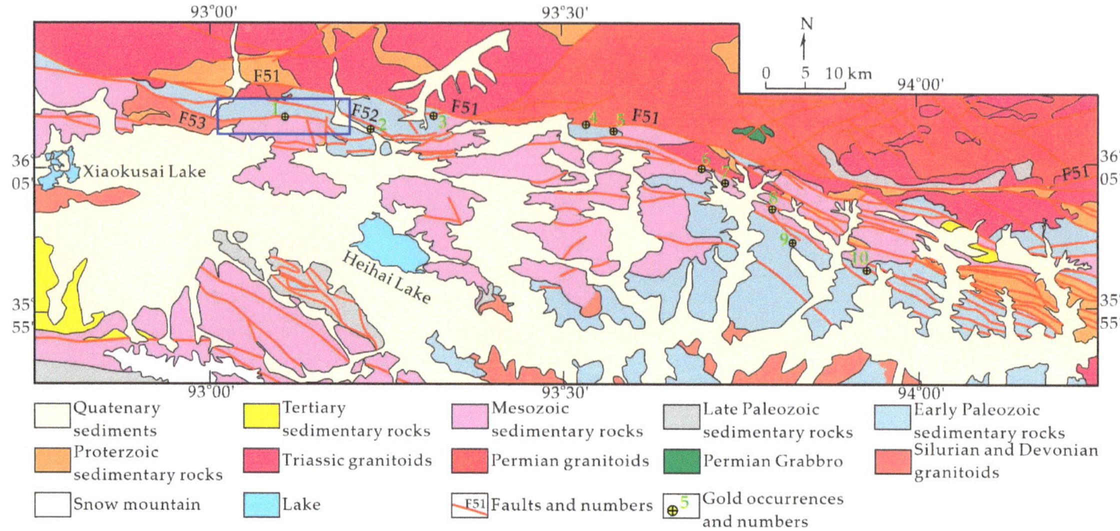

Figure 2. Geological map of the Kunlun River area in EKO (modified after [22]). Numbers and names of gold occurrences: 1—Heiaibei, 2—Lalingzaohuo, 3—Suhaitu, 4—Jiaxi, 5—Xiangyanggou, 6—Jiaodong, 7—Dazaohuo–Heicigou, 8—Dazaohuodong, 9—Yaodongshan, 10—Heicigou.

A systematic study on the H, O, S, and Pb isotope geochemistry for auriferous quartz vein ores in the Heihaibei gold deposit was carried out. Comparing our new data and a few previously published data on the Heihaibei gold deposit, we can better interpret the sources of H, O, S, and Pb in the Heihaibei gold deposit, determine the genetic type of it, and even further constrain the sources of ore-forming fluids and materials for gold deposits within the whole KLRA.

2. Geological Background and Ore Deposit Geology

2.1. Geological Background

Usually, it is thought that the EKO has stages of orogenies that are closely related to the subduction and closure of the Proto-Tethys and Paleo-Tethys oceans [1,3,31–37]. Two ophiolite belts, which lay along the M.KLF and S.KLF–AMF, represent the suture zones of the Neo-Proterozoic to Early Devonian Proto-Tethys, and the Carboniferous to Late Triassic Paleo-Tethys oceans, respectively [3,31,37]. The EKO is considered to be divided into three fault-bounded tectonic belts, i.e., NKL, MKL, and SKL (Figure 1b) [3]. The NKL and MKL have dual tectonic layers of a common Precambrian craton, which are characterized by the Paleo-Proterozoic basement, with intermediate to high-grade metamorphism, and the Meso- to Neo-Proterozoic cover strata, with low-grade metamorphism [2]. The NKL is characterized by thick Ordovician clastic and carbonate sedimentary rocks interlayered with mafic volcanic rocks, which were unconformably overlain by the Devonian molasse sediments [38,39]. Widespread granitoids intruded into these Ordovician and Devonian strata, primarily in the late Paleozoic, in this belt [40,41]. The MKL, in which the Precambrian metamorphic rocks vary from gneiss to schist to phyllite [3,39], was predominantly emplaced by the widespread Neo-Proterozoic, early Cambrian–Early Devonian, and Late Permian–Early Triassic granitic intrusions [13,42–46]. The SKL is dominated by Precambrian to Jurassic siliciclastic and carbonate rocks [3,43], and it was locally intruded upon by the Paleozoic and Triassic granites [23,47,48]. The KLRA is tectonically located in the central segment of the SKL within the EKO (Figure 1b).

2.2. Ore Deposit Geology

The KLRA, as the voluminous area to the north bank of the Kunlun River is usually referred to, is a recently discovered gold mineralization belt in the SKL (Figure 2). The northwest-striking Fault F51, which is a segment of the M.KLF, is a regional fault that bounds the MKL and SKL in the KLRA. To the north of Fault F51, the part of the MKL is characterized by widespread Permian to Triassic granitoids [22]. To the south of Fault F51, part of the SKL is composed of sedimentary rocks which were deposited predominantly in the Early Paleozoic and Quaternary, with minor parts in the Proterozoic and Tertiary. A few Silurian and Triassic granitoids were intruded in the Early Paleozoic and Mesozoic sedimentary rocks (Figure 2). Subsidiary faults of Fault F51 that were usually considered to be the products of the closure of the Proto-Tethys Ocean [49] mostly affect the SKL. Moreover, Fault F51 and its subsidiary faults in the KLRA would have been reactivated from the Early Triassic onwards because of the subduction of the Paleo-Tethys oceanic plate [2,49]. Ore-forming hydrothermal fluids can migrate upward through such reactivated faults. More than 10 gold deposits are controlled by the WNW-striking subsidiary faults to the south of Fault F51 in the KLRA (Figure 2) [22,25].

The Heihaibei gold deposit, which is located ca. 20 km NW of the Heihai Lake, is the westernmost of 10 gold deposits in the KLRA (Figure 3). The dominant sedimentary rocks in the Heihaibei area are mainly composed of Silurian–Ordovician Nachitai Group metamorphic feldspathic quartz sandstone and sandy slate in the north, and Triassic Hongshuichuan Formation feldspathic quartz sandstone in the south, with minor Middle-Neo-Proterozoic Wanbaogou Group sandy limestone and basalt, and Triassic Naocangjiangou Formation feldspathic quartz sandstone in the southeast (Figure 3a) [22,23]. An EW- to WNW-striking >15 km long and >100 m wide monzogranite–granodiorite dyke (i.e., the Heihaibei granites) was intruded along the south of the Silurian–Ordovician Nachitai Group. We argued that this monzogranite–granodiorite dyke was classified as A-type granite [50] and was emplaced in the Early Devonian (416 Ma) in a post-collisional setting [50]), rather than in the Ordovician–Silurian (442 Ma [27]), 443 Ma [25], and 453 Ma [23]). The Triassic Hongshuichuan Formation feldspathic quartz sandstone and other minor Triassic sediments were deposited over older sedimentary rocks and the Heihaibei granites [24]. Subsidiary faults in the Heihaibei area predominantly include the NW-striking Fault F52 and Fault F53 (Figure 3a). Fault F52, ca. 18 km long and 10–100 m wide, generally strikes WNW across the Nachitai Group sandstone in the northeast part of the Heihaibei mining area and dips 25°–52° to the north; meanwhile, Fault F53, ca. 15 km long and 10–100 m wide, generally strikes EW across the central part of the study area and dips 30°–63° to the north [23,24]. Cataclasites, fault breccias, and fault gouges are developed in the faulted zones of Faults F52 and F53. Fault F53 with secondary branch faults, i.e., Fault F53-1, is commonly developed along contacts of different geological bodies in the Heihaibei area, i.e., between the granites and the Triassic Hongshuichuan Formation, and the granites and the Nachitai Group, as well as the slate and sandstone in the Nachitai Group. Some auriferous quartz veins, pyritizations, and weak hydrothermal alterations are developed along the faulted monzogranite zone of Fault F53, but little mineralization occurs in Fault F52, which is limited in the Nachitai Group sandstone (Figure 3b).

Weak hydrothermal alterations within monzogranite wall rocks (Figure 4a) are mostly silicification and sericitization, with minor chloritization and carbonatization. In monzogranites, biotite is basically altered into chlorite, plagioclase is partially altered into muscovite, and hydrothermal quartz and/or calcite fill into fine fractures (Figure 4b,c). More than 20 gold ore bodies (with Au abundances larger than the cut-off grade) and 30 auriferous bodies (with Au abundances below the cut-off grade) were delineated in the faulted monzogranite zone within the fault F53, and they yielded a total reserve of 5.7 t Au, with an average grade of 4.8 g/t, in which the 660 m long and average 5 m wide ore body M3 contributes to ca. 90% of the Au reserves [22]. Most of the gold ore bodies strike EW to ENE (east-northeast) and occur as lenses along the faulted monzogranite zone within Fault F53 (Figures 3b and 5). Ore bodies consist of predominant smoky grey quartz veins with minor stockwork quartz veinlets or breccias in the altered monzogranite, and smoky grey quartz veins are usually bordered by fault breccia belts in the faulted monzogranite belt [22,24].

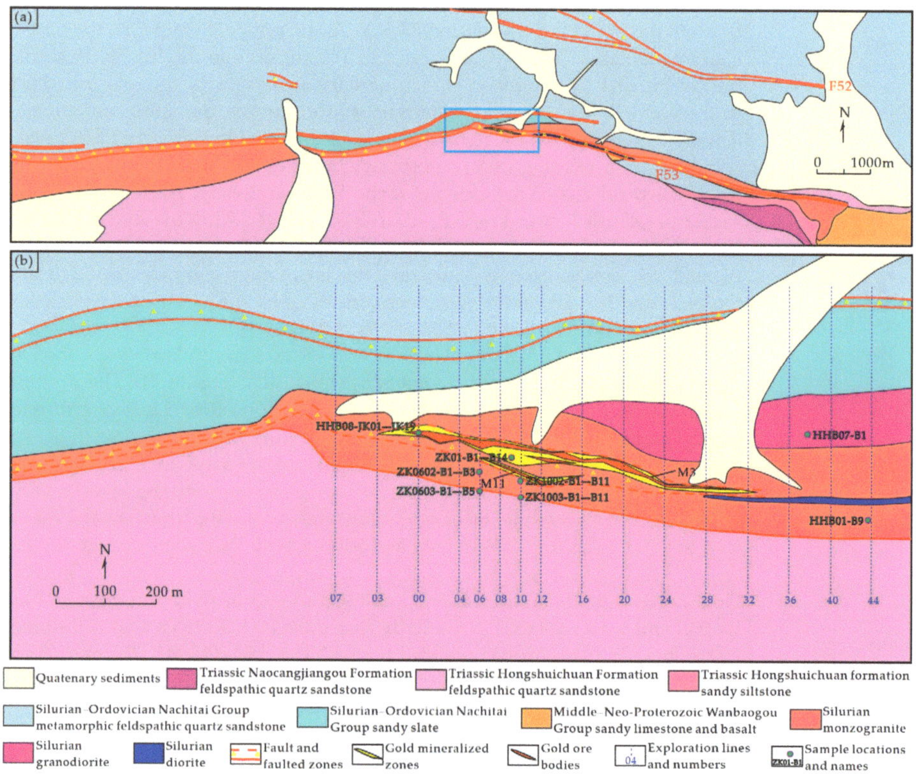

Figure 3. (**a**) Geological map of the Heihaibei and part of Lalingzaohuo area in the KLRA (simplified and revised after [22]). (**b**) Geological map of the Heihaibei gold deposit in the KLRA (simplified and revised after [22]).

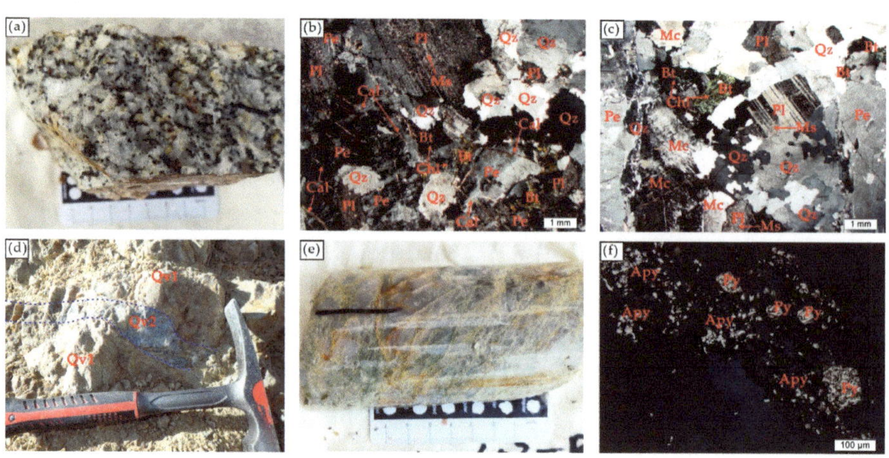

Figure 4. *Cont.*

Figure 4. Photographs and photomicrographs showing the morphologies and textural features of the ores and monzogranite wall rocks in the Heihaibei gold deposit. Photo (**a**) and photomicrographs (**b**,**c**) of monzogranite wall rock sample with weak hydrothermal alterations; (**d**) photograph of quartz veins, white quartz veins (Qv1) were cut by a smoky grey quartz vein (Qv2); (**e**) photograph and (**f**) photomicrograph of Qv1 type sample HHBZK0602-B2; (**g**) photograph and (**h**) photomicrograph of Qv2 type sample HHB08-JK11; (**i**,**j**) photomicrographs of typical Qv2 type samples; (**k**) photograph and (**l**) photomicrograph of samples HHBZK1003-B11 from breccia ores; (**m**) photograph and photomicrograph (**n**,**o**) of Qv2 type sample HHB08-JK13. Apy—arsenopyrite, Py—pyrite, Gn—galena, Sph—sphalerite, Cpy—chalcopyrite, Au—native gold, Co—covellite, Mc—microcline, Pe—perthite, Pl—plagioclase, Bt—biotite, Ms—muscovite, Chl—chlorite, Qz—quartz, Cal—calcite.

Three mineralized stages were proposed in the Heihaibei gold deposit, i.e., the K-feldspar-quartz-pyrite-arsenopyrite-sericite-epidote (I), the quartz-pyrite-native gold-chlorite (II), and the quartz-carbonate (III) by [23]. Based on the field survey and the optical microscopic study, we argued that the so-called K-feldspar-quartz-pyrite-arsenopyrite-sericite-epidote stage (I) represents silicified granites with few pyrite veinlets, and that the K-feldspar in such ores is a primary rock-forming mineral rather than a hydrothermal mineral in the Heihaibei granites. Additionally, we cannot find any typical products (e.g., quartz calcite veins) of the so-called quartz-carbonate stage (III) in the ore deposit. Therefore, we considered that the paragenetic sequences in the Heihaibei gold deposit might be divided into three major stages, i.e., the hydrothermal stage, the deformation stage, and the supergene stage (Figure 6). The hydrothermal stage can be subdivided into the sulfide–poor–quartz epoch (I) and the polymetallic–sulfide–quartz epoch (II). A few white quartz veins (here called Qv1), which are composed of predominant quartz with minor sericite, pyrite, and arsenopyrite, represent the sulfide–poor–quartz ore epoch (I) (Figure 4e,f), and some white quartz veinlets or silicification in weakly altered monzogranite wall rocks (Figure 4b,c) are the products of this epoch. Predominant smoky grey quartz veins (here called Qv2), which represent the ores from the polymetallic–sulfide–quartz epoch (II), usually contain predominant quartz with a few sulfide veinlets, disseminations, or massive aggregations, which consist of pyrite and arsenopyrite, with minor native gold, chalcopyrite, galena, sphalerite, and tetrahedrite (Figure 4g–j,m–o). Micro-native gold is mostly trapped in pyrite, and arsenopyrite in the Qv2 (Figure 4n). Usually, the contents of sulfides in the Qv2 are less than 5%. Some pyrite–quartz-bearing veinlets and stockworks, which can be found in the weakly altered monzogranite wall rock or Qv1, constitute gold mineralized zones, and some of them may be delineated to ore bodies. Comparatively, the sulfide–poor–quartz epoch (I) and polymetallic–sulfide–quartz epoch (II) in this study are roughly referred to as the stage I and stage II, respectively, as classified by [23]. Partially, both Qv1 and Qv2 can be deformed within the faulted monzogranite zone of Fault F53 and become breccia ores (Figure 4k,l). The oxidized ores from the supergene stage mostly consist of goethite, lepidocrocite, and covellite, which result from the replacement of the hydrothermal stage of primary Qv1 and Qv2 (Figures 4n and 6). In a word, the geological characteristics of the Heihaibei gold deposit show similar features to those of the orogenic gold deposits defined by [51].

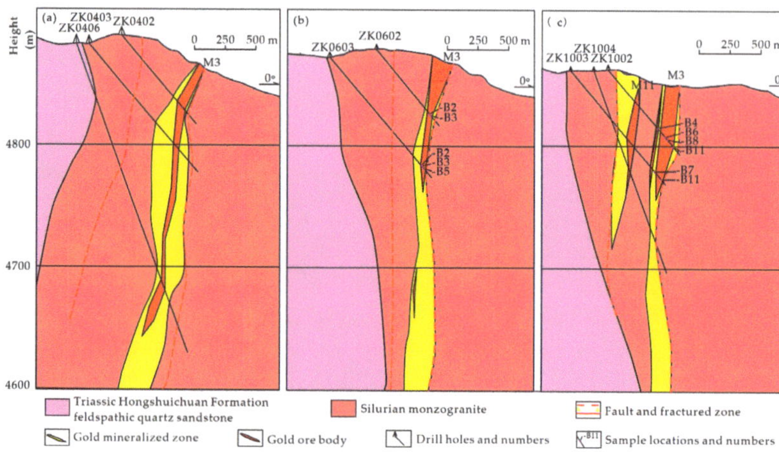

Figure 5. Typical sections of the Heihaibei gold deposit (Simplified and revised from [22]). (**a**) Prospecting line 4, (**b**) Prospecting line 6, (**c**) Prospecting line 8.

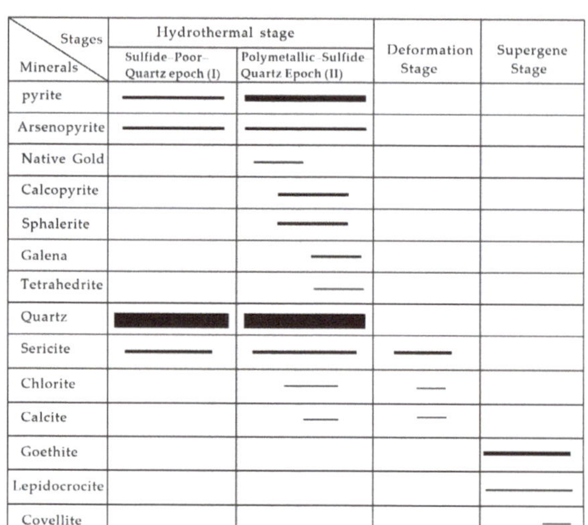

Figure 6. Paragenetic sequences of minerals in different ore-forming stages in the Heihaibei gold deposit. The length and width of the square represent the time and amount of mineral growth, respectively.

3. Sampling and Analytical Methods

3.1. Sampling

The samples for isotope analyses were collected from drill cores, i.e., ZK0602, ZK0603, ZK01, ZK1002, and ZK1003, or surface outcrop sites, i.e., HHB01, HHB07, and HHB08. Detailed sample locations and numbers can be found in Figures 3b and 5, and Table 1. Fifteen auriferous quartz samples (Qv1 and Qv2) were collected for oxygen and hydrogen isotope analysis (Table 2). Seventeen sulfide-bearing ore samples (Qv1 and Qv2) were collected for sulfur isotope analysis (Table 3). Four sulfide-bearing ore samples (Qv1 and Qv2) and five potassic feldspars in the Heihaibei granites were collected for lead isotope analysis (Table 4). Previously published data for other gold deposits in the EKO, compared with data in this study, can be found in the Supplementary Materials (Tables S1–S3).

Table 1. Sample list in the Heihaibei gold deposit.

No.	Sample Name	Rock Type	Ore Type	Mineral Assembly	Location
1	HHBZK0602-B2	brecciated white quartz vein	Qv1	quartz with minor sericite, pyrite, and arsenopyrite	69.7 m of the drillhole core of the ZK1002
2	HHBZK1002-B4	brecciated white quartz vein	Qv1		62.9 m of the drillhole core of the ZK1004
3	HHBZK1002-B6	brecciated white quartz vein	Qv1		72.3 m of the drillhole core of the ZK1004
4	HHBZK1002-B8	brecciated white quartz vein	Qv1		75.5 m of the drillhole core of the ZK1004
5	HHB08-JK01	white quartz vein	Qv1		93°06′39″ E, 36°08′35″ N
6	HHB08-JK10	white quartz vein	Qv1		93°06′39″ E, 36°08′35″ N
7	HHB08-JK17	white quartz vein	Qv1		93°06′39″ E, 36°08′35″ N
8	HHBZK01-B3	brecciated smoky grey quartz vein	Qv2	quartz with some pyrite and arsenopyrite, and minor native gold, chalcopyrite, galena, and sphalerite, as well as tetrahedrite	26.4 m of the drillhole core of the ZK01
9	HHBZK01-B5	brecciated smoky grey quartz vein	Qv2		29.6 m of the drillhole core of the ZK01
10	HHBZK01-B6	brecciated smoky grey quartz vein	Qv2		30.3 m of the drillhole core of the ZK01
11	HHBZK0602-B3	smoky grey quartz vein	Qv2		70.8 m of the drillhole core of the ZK0602
12	HHBZK0603-B2	smoky grey quartz vein	Qv2		114.7 m of the drillhole core of the ZK0603
13	HHBZK0603-B3	brecciated smoky grey quartz vein	Qv2		115.4 m of the drillhole core of the ZK0603
14	HHBZK0603-B5	brecciated smoky grey quartz vein	Qv2		117.0 m of the drillhole core of the ZK0603
15	HHBZK1003-B11	brecciated smoky grey quartz vein	Qv2		115.9 m of the drillhole core of the ZK1003
16	HHBZK1002-B11	brecciated smoky grey quartz vein	Qv2		82.8 m of the drillhole core of the ZK1002
17	HHB08-JK02	smoky grey quartz vein	Qv2		93°06′39″ E, 36°08′35″ N
18	HHB08-JK03	smoky grey quartz vein	Qv2		93°06′39″ E, 36°08′35″ N
19	HHB08-JK05	smoky grey quartz vein	Qv2		93°06′39″ E, 36°08′35″ N
20	HHB08-JK08	smoky grey quartz vein	Qv2		93°06′39″ E, 36°08′35″ N
21	HHB08-JK13	smoky grey quartz vein	Qv2		93°06′39″ E, 36°08′35″ N
22	HHB08-JK15	smoky grey quartz vein	Qv2		93°06′39″ E, 36°08′35″ N
23	HHB08-JK18	smoky grey quartz vein	Qv2		93°06′39″ E, 36°08′35″ N
24	HHB08-JK19	smoky grey quartz vein	Qv2		93°06′39″ E, 36°08′35″ N
25	HHBZK01-B11	monzogranite	wall rock	plagioclase, 25–40% alkali feldspar, 25–45% quartz, <5% biotite	175.0 m of the drillhole core of the ZK01
26	HHB01-B9	monzogranite	wall rock		93°07′14″ E, 36°08′30″ N
27	HHBZK01-B14	granodiorite	wall rock	35–50% plagioclase, 35–45% quartz, 10–15% alkali feldspar, <5% biotite	209.4 m of the drillhole core of the ZK01
28	HHBZK1003-B7	granodiorite	wall rock		107.2 m of the drillhole core of the ZK1003
29	HHB07-B1	granodiorite	wall rock		93°07′09″ E, 36°08′36″ N

Table 2. H and O isotope data for quartz in the Heihaibei gold deposit. δD was measured in fluid inclusions; $\delta^{18}O$ was measured in quartz; $\delta^{18}O$ of water in equilibrium with quartz was calculated at the temperature indicated by Th [23] using the equation of $1000\ln\alpha = 3.38 \times 10^6/T^2 - 3.4$ [52].

No.	Sample Name	Rock Type	Ore Type	Mineral	δDVSMOW (‰)	δ^{18}OVSMOW (‰)	δ^{18}OH2O (‰)	Average Th (°C)	References
1	HHBZK0602-B2	brecciated white quartz vein	Qv1	Quartz	−97.2	13.4	7.8	340.0	this study
2	HHBZK0602-B3	smoky grey quartz vein	Qv2	Quartz	−91.0	12.1	2.7	241.0	this study
3	HHBZK0603-B2	smoky grey quartz vein	Qv2	Quartz	−101.8	12.6	3.2	241.0	this study
4	HHBZK0603-B3	brecciated smoky grey quartz vein	Qv2	Quartz	−97.7	12.8	3.4	241.0	this study
5	HHBZK0603-B5	brecciated smoky grey quartz vein	Qv2	Quartz	−102.3	13.3	3.9	241.0	this study
6	HHB08-JK01	white quartz vein	Qv1	Quartz	−81.1	9.2	3.6	340.0	this study
7	HHB08-JK02-1	smoky grey quartz vein	Qv2	Quartz	−94.1	13.7	4.3	241.0	this study
8	HHB08-JK02-2	smoky grey quartz vein	Qv2	Quartz	−94.6	13.7	4.3	241.0	this study
9	HHB08-JK03	smoky grey quartz vein	Qv2	Quartz	−89.9	11.3	1.9	241.0	this study
10	HHB08-JK05	smoky grey quartz vein	Qv2	Quartz	−90.4	13.6	4.2	241.0	this study
11	HHB08-JK08	smoky grey quartz vein	Qv2	Quartz	−97.4	11.6	2.2	241.0	this study
12	HHB08-JK10	white quartz vein	Qv1	Quartz	−92.8	13.9	8.3	340.0	this study
13	HHB08-JK15	smoky grey quartz vein	Qv2	Quartz	−97.6	11.6	2.2	241.0	this study
14	HHB08-JK17	white quartz vein	Qv1	Quartz	−96.8	11.0	5.4	340.0	this study
15	HHB08-JK18	smoky grey quartz vein	Qv2	Quartz	−104.2	13.0	3.6	241.0	this study
16	HHB08-JK19	smoky grey quartz vein	Qv2	Quartz	−101.5	11.0	1.6	241.0	this study
17	K1	silicified monzogranite	Qv1+wall rocks	Quartz	−84.1	15.1	9.5	340.0	[23]
18	K5	smoky grey quartz vein	Qv2	Quartz	−81.1	14.1	4.7	241.0	[23]
19	K7	smoky grey quartz vein	Qv2	Quartz	−84.0	14.0	4.6	241.0	[23]

Table 3. Sulfur isotope data for sulfides in the Heihaibei deposit. $\delta^{34}S_{H2S}$ means $\delta^{34}S_{VCDT}$ values for H_2S in equilibrium with sulfides. The assumed temperatures of equilibrium for different ore types are the same as Th [23] listed in Table 1.

No.	Sample Name	Rock Type	Ore Type	Mineral	$\delta^{34}S_{CDT}$(‰)	$\delta^{34}S_{H2S}$(‰)	References
1	HHBZK01-B3	brecciated smoky grey quartz vein	Qv2	Pyrite	8.5	7.0	this study
2	HHBZK01-B5	brecciated smoky grey quartz vein	Qv2	Pyrite	8.0	6.5	this study
3	HHBZK01-B6	brecciated smoky grey quartz vein	Qv2	Pyrite	7.7	6.2	this study
4	HHBZK0603-B2	smoky grey quartz vein	Qv2	Pyrite	8.0	6.5	this study
5	HHBZK0603-B3	brecciated smoky grey quartz vein	Qv2	Pyrite	7.6	6.1	this study
6	HHBZK0603-B5	brecciated smoky grey quartz vein	Qv2	Pyrite	8.0	6.5	this study
7	HHBZK1003-B11	brecciated smoky grey quartz vein	Qv2	Pyrite	7.6	6.1	this study

Table 3. *Cont.*

No.	Sample Name	Rock Type	Ore Type	Mineral	$\delta^{34}S_{CDT}$(‰)	$\delta^{34}S_{H2S}$(‰)	References
8	HHBZK1002-B4-1	brecciated white quartz vein	Qv1	Pyrite	8.1	7.0	this study
9	HHBZK1002-B4-2	brecciated white quartz vein	Qv1	Pyrite	8.0	6.9	this study
10	HHBZK1002-B6	brecciated white quartz vein	Qv1	Pyrite	8.0	6.9	this study
11	HHBZK1002-B8	brecciated white quartz vein	Qv1	Pyrite	8.1	7.0	this study
12	HHBZK1002-B11	brecciated smoky grey quartz vein	Qv2	Pyrite	7.8	6.3	this study
13	HHB08-JK02	smoky grey quartz vein	Qv2	Pyrite	8.5	7.0	this study
14	HHB08-JK08	smoky grey quartz vein	Qv2	Pyrite	8.0	6.5	this study
15	HHB08-JK13	smoky grey quartz vein	Qv2	Pyrite	8.3	6.8	this study
16	HHB08-JK15	smoky grey quartz vein	Qv2	Pyrite	8.2	6.7	this study
17	HHB08-JK17	white quartz vein	Qv1	Pyrite	8.7	7.6	this study
18	K1	silicified monzogranite	Qv1 + wall rocks	Arsenopyrite	8.5		[23]
19	K5	smoky grey quartz vein	Qv2	Pyrite	7.7		[23]
20	K8	smoky grey quartz vein	Qv2	Pyrite	7.9		[23]

Table 4. Pb isotope compositions for sulfides in the Heihaibei gold deposit.

No.	Sample	Rock Type	Ore Type	Mineral	$^{206}Pb/^{204}Pb$	$^{207}Pb/^{204}Pb$	$^{208}Pb/^{204}Pb$	References
1	HHBZK01-B3	brecciated smoky grey quartz vein	Qv2	pyrite	18.8122	15.7002	38.8055	this study
2	HHBZK1002-B4	brecciated white quartz vein	Qv1	pyrite	19.0007	15.7062	38.7655	this study
3	HHB08-JK08	smoky grey quartz vein	Qv2	pyrite	18.7219	15.6959	38.7356	this study
4	HHB08-JK19	smoky grey quartz vein	Qv2	pyrite	18.8262	15.6998	38.7836	this study
5	HHBZK01-B11	monzogranite	wall rock	K-feldspar	18.4283	15.6789	38.4342	this study
6	HHBZK01-B14	granodiorite	wall rock	K-feldspar	18.4363	15.6808	38.4532	this study
7	HHBZK1003-B7	granodiorite	wall rock	K-feldspar	18.4864	15.6475	38.1750	this study
8	HHB01-B9	monzogranite	wall rock	K-feldspar	18.4368	15.6812	38.4598	this study
9	HHB07-B1-1	granodiorite	wall rock	K-feldspar	18.3532	15.6750	38.3713	this study
10	HHB07-B1-2	granodiorite	wall rock	K-feldspar	18.3537	15.6753	38.3724	this study
11	K1	silicified monzogranite	Qv1 + wall rocks	pyrite	19.188	15.713	39.235	[23]
12	K5	smoky grey quartz vein	Qv2	pyrite	18.722	15.695	38.728	[23]
13	K12	smoky grey quartz vein	Qv2	pyrite	18.724	15.699	38.746	[23]

3.2. Oxygen and Hydrogen Isotope Analysis

Oxygen isotope analyses were carried out with the bromine pentafluoride method of [53] at the Key Laboratory of Metallogeny and Mineral Assessment, Institute of Mineral Resources, Chinese Academy of Geological Sciences (CAGS), China. Oxygen was liberated from 6 mg of quartz via reaction with BrF_5 and converted to CO_2 on a platinum-coated carbon rod. The $\delta^{18}O$ determinations were made on a Thermo Scientific 253 plus mass spectrometer, with analytical precisions of $\pm 0.2‰$, and the detailed analytical procedures are described by [54]. Hydrogen isotope analyses were carried out at the Beijing Createch Testing Technology Co., Ltd, Beijing, China (BCTT). Hydrogen isotopic ratios were measured on water released via decrepitation (at 1420 °C), of crushed and sieved (to the size range of 60 mesh) 10–20 mg quartz, according to the method of [55]. Samples were first degassed by heating under a vacuum at 90 °C for 12 h; then, the water was released from fluid inclusions by heating the samples to 1420 °C in a Thermo Scientific Flash 2000 HT elemental analyzer. The subsequently released water was trapped, reduced to H_2 by glassy carbon, and then analyzed for δD values using a Thermo Scientific 253 plus mass spectrometer, with a precision of $\pm 1‰$. The isotope results are expressed in the delta (δ) notation as the per mil (‰) deviation relative to the Vienna Standard Mean Ocean Water (V-SMOW) standard. The isotopic fractionation of oxygen between quartz and water was calculated using the equation of $1000\ln\alpha = 3.38 \times 10^6/T^2 - 3.4$ [52].

3.3. Sulfur Isotope Analysis

Sulfur isotope analyses of sulfide separates were carried out at BCTT using the conventional combustion method [56]. The detailed analytical procedures are described by [57]. Approximately 15 mg of sulfides were combusted at 960 °C for quantitative conversion to SO_2, which were then analyzed for sulfur isotope with a Thermo Scientific 253 plus mass spectrometer. The results are reported in the δ notation as the ‰ deviation of the isotope ratio ($^{34}S/^{32}S$) relative to the Vienna Canyon Diablo Troilite (V-CDT) standard, with an analytical precision of $\pm$ 0.2‰.

3.4. Lead Isotope Analysis

Lead isotope analyses were conducted on four pyrite separates and five potassic feldspar separates at BCTT, using the analytical procedure as described in [58]. About 150 mg of the pyrite or feldspar sample were completely dissolved in an ultrapure acid mixture of 1 mL HNO_3 + 2 mL HF. After drying, the residue was redissolved in concentrated HNO_3 and dried again. Finally, the sample residue was dissolved in 1mL 3.5 M HNO_3 and then loaded into a column with 100–150 μm of Triskem Sr-Spec resins to separate the Pb fraction by washing with 3.5 M HNO_3, 8 M HCl, and Milli-Q water. Finally, the extracted sample residue was dissolved in 1 mL 2% HNO_3. Isotopic ratios of Pb analyses were undertaken using a Thermo Scientific Neptune Plus Multi Collector-Inductively Coupled Plasma Mass Spectrometer (MC-ICP-MS), and the results were corrected using $^{203}Tl/^{205}Tl = 0.418922$ [58]. The data quality was monitored via repeated analyses of the CAGS Pb reference material, and the error margin (2σ) was <0.002 for all three ratios.

4. Results

4.1. Oxygen and Hydrogen Isotope Results

Quartz samples from quartz veins show a $\delta^{18}O_{quartz}$ range from +9.2 to +13.9‰, with an average of +12.3‰ (Table 2, Figure 7). It is difficult to calculate the precise $\delta^{18}O_{H2O}$ values of fluids that were in equilibrium with quartz, due to a large variation of temperatures [23] obtained from fluid inclusions (e.g., [59]). According to the microthermometry of fluid inclusions in auriferous quartz from the Heihaibei gold deposit finished by [23], the homogenization temperatures of Qv1 type of quartz from epoch I range from 268 °C to 412 °C, with an average of 340 °C; meanwhile, the homogenization temperatures of the Qv2 type of quartz from epoch II range from 183 °C to 288 °C, with an average of 241 °C. Usually, the trapping temperature could be defined by the average homogenization temperature to calculate the isotopic fractionation values. Given 340 °C as the trapping temperature for epoch I, the $\delta^{18}O_{H2O}$ values of the ore-forming fluids in the fluid inclusions from the quartz in the Qv1 samples range from +3.6‰ to +8.3‰ (Table 2, Figure 7b). Meanwhile, given 241 °C as the trapping temperature for epoch II, the $\delta^{18}O_{H2O}$ values of the ore-forming fluids for fluid inclusions from quartz in the Qv2 samples range from +1.6‰ to +4.3‰ (Table 2, Figure 7b). Generally, the trapping temperatures are most likely higher than the homogenization temperatures for the orogenic fluids.

The hydrogen isotope results for the fluid inclusions from the quartz samples are also shown in Table 2, in which the δD_{VSMOW} values range from −104.2‰ to −81.1‰, with an average of −95.3‰ (Figure 7). The quartz in the Qv1 samples has δD_{VSMOW} values ranging from −97.2 ‰ to −81.1‰, and the quartz in the Qv2 samples has δD_{VSMOW} values ranging from −104.2‰ to −89.9‰ (Figure 7b).

Consequently, the analytical hydrogen isotopes and the calculated $\delta^{18}O_{H2O}$ values for the fluid inclusions from the quartz in Qv1 and Qv2 samples are different from each other (Figure 7b). Three sets of hydrogen–oxygen isotope compositions for the Heihaibei gold deposits were previously reported by [23], in which two samples from the smoky grey quartz veins (Qv2) have similar hydrogen–oxygen isotope compositions as our data from Qv1. However, the sample K1 is slightly different from our data from Qv1, likely because it might have been collected in the mineralized Heihaibei granitic wall rock, thus representing a mixed sample of Qv1 plus the granitic wall rock.

4.2. Sulfur Isotope Results

Only pyrite separates had been selected from 16 ore samples and analyzed for sulfur isotope compositions. Other sulfide separates were insufficient for analysis because the ore samples were collected from drill cores and had limited volumes. The $\delta^{34}S_{VCDT}$ values of 17 pyrite separates from 16 ore samples in the Heihaibei gold deposit range from +7.6‰ to +8.7‰, with a narrow variation and an average of 8.1‰ (Table 3, Figure 8a). Pyrite separates from the Qv1 type of ores have almost invariable $\delta^{34}S$ values ranging from +8.0‰ to +8.7‰, with an average of + 8.2‰. Pyrite separates from the Qv2 type of ores have similar invariable $\delta^{34}S$ values ranging from +7.6‰ to +8.5‰, with an average of +8.0‰. The $\delta^{34}S$ values of all pyrites have very narrow variations, and the $\delta^{34}S$ values of different ores have a comparably small range, usually indicating a relatively unique source.

Furthermore, our sulfur isotope data, ranging from +7.6‰ to +8.7‰, essentially coincide with the three previously published $\delta^{34}S_{VCDT}$ values, i.e., +7.7‰, +7.9‰, and +8.5‰ [23].

4.3. Lead Isotope Results

Only four pyrites separated from four ore samples in the Heihaibei gold deposit were large enough and yielded relatively invariable Pb isotope compositions, with $^{206}Pb/^{204}Pb$ ratios of 18.7219 to 19.0007, $^{207}Pb/^{204}Pb$ ratios of 15.6959 to 15.7062, and $^{208}Pb/^{204}Pb$ ratios of 37.7359 to 38.8055 (Table 4, and Figure 9a,b). Different types of quartz veins have relatively similar Pb isotopic compositions (Figure 9c,d). One pyrite separate from the Qv1 type of ores has $^{206}Pb/^{204}Pb$ ratios of 19.0007, $^{207}Pb/^{204}Pb$ ratios of 15.7062, and $^{208}Pb/^{204}Pb$ ratios of 38.7655. Three pyrite separates from the Qv2 type of ores have invariable Pb isotopic ratios, with $^{206}Pb/^{204}Pb$ ratios from 18.7219 to 18.8262, $^{207}Pb/^{204}Pb$ ratios from 15.6959 to 15.7002, and $^{208}Pb/^{204}Pb$ ratios from 38.7356 to 38.8055. Three sets of Pb isotope compositions for the Heihaibei gold deposits have previously been reported by [23], in which two samples from smoky grey quartz veins (Qv2) have similar Pb compositions as our data from Qv2, but the sample K1 has more radiogenic Pb isotope than our data from Qv1.

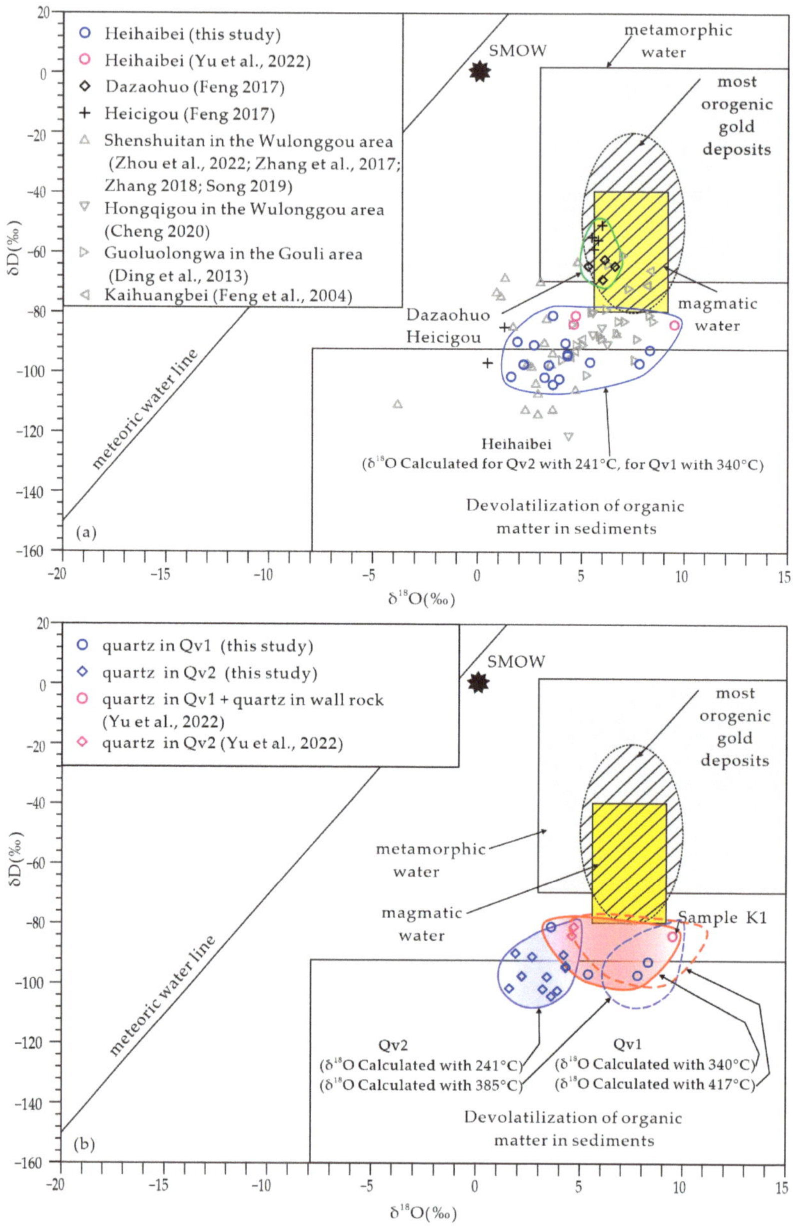

Figure 7. (**a**) Plot of analytical δD $_{VSMOW}$ vs. calculated $\delta^{18}O_{H2O}$ for ore-forming fluids from auriferous quartz veins (for equilibrium at 241 °C for Qv2, 340 °C for Qv1) in the Heihaibei gold deposit, compared with quartz in other gold deposits (data after [4,8–10,14,19,23,29,60], Table S1) in the EKO; (**b**) Plot of analytical δD vs. calculated $\delta^{18}O_{H2O}$ (with homogenization temperatures of 241 °C for Qv2 and 340 °C for Qv1, comparing with arsenopyrite geothermometer temperatures of 385 °C for Qv2 and 417 °C for Qv1) for ore-forming fluids from auriferous quartz veins in the Heihaibei gold deposit. Magmatic, metamorphic, and organic (e.g., devolatilization of organic matter in sediments) water fields after [61]; field for most orogenic gold deposits revised after [59].

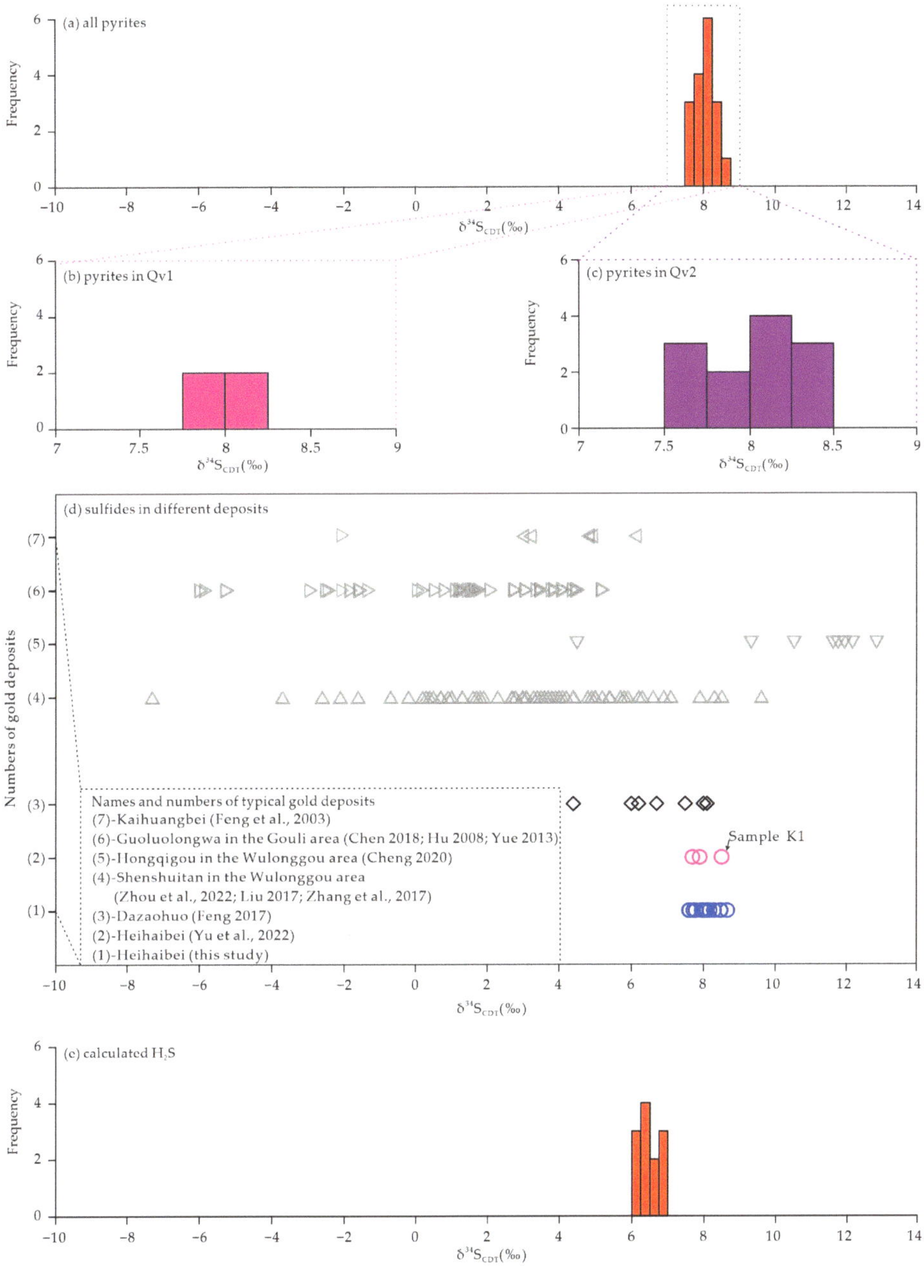

Figure 8. Sulfide isotope for sulfide separates from auriferous quartz veins in the Heihaibei gold deposit. (**a**) Histogram of sulfur isotopes for pyrites from all types of ores; (**b**) Histogram of sulfur isotopes for pyrites in Qv1 type of ores; (**c**) Histogram of sulfur isotopes for pyrites in Qv2 type of ores; (**d**) Plot of sulfur isotopes for sulfides in the Heihaibei gold deposit compared with those in other gold deposits (data after [4,7,9,15–17,20,23,29,60], Table S2) in the EKO; (**e**) Histogram of calculated H_2S isotopes ($\delta^{34}S_{H2S}$) in equilibrium with sulfides in the Heihaibei gold deposit. The assumed temperatures of equilibrium for different ore types are the same as Th [23] listed in Table 1.

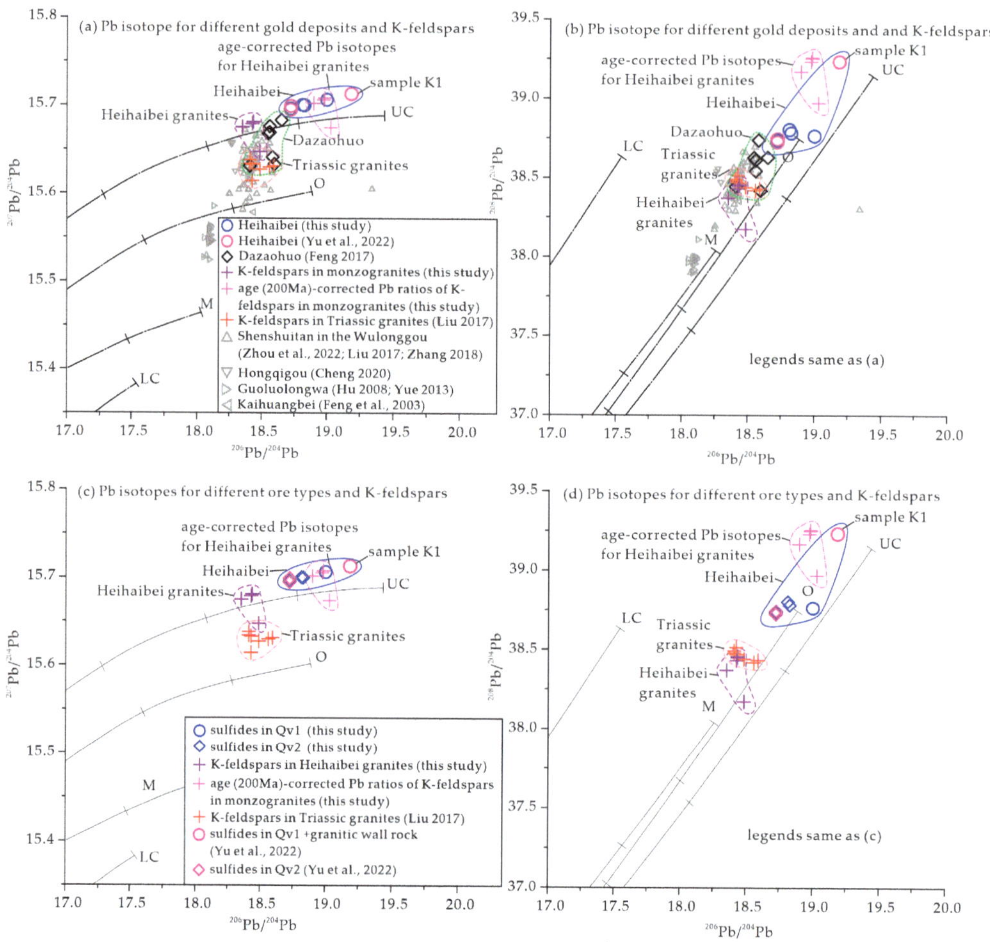

Figure 9. Lead isotopic compositions of sulfide separates in auriferous quartz veins (Qv1 and Qv2), K-feldspars in granite wall rocks, and age (200 Ma)-corrected Pb ratios of K-feldspars in monzogranites in the Heihaibei gold deposit. (**a**) Diagram of $^{207}Pb/^{204}Pb$ vs. $^{206}Pb/^{204}Pb$; (**b**) Diagram of $^{208}Pb/^{204}Pb$ vs. $^{206}Pb/^{204}Pb$ for pyrites and K-feldspars in the Heihaibei gold deposit, compared with sulfides in other gold deposits (data after [4,7,9,16,17,20,23,29,60], Table S3) in the EKO; (**c**) Diagram of $^{207}Pb/^{204}Pb$ vs. $^{206}Pb/^{204}Pb$; and (**d**) Diagram of $^{208}Pb/^{204}Pb$ vs. $^{206}Pb/^{204}Pb$ for different ores compared with K-feldspars in granite wall rocks in the Heihaibei gold deposit. UC, Upper crust; O, Orogen; M, Mantle; LC, Lower crust. The average growth curve is from [62], and tick marks along each curve indicate progressively older times in 0.4 b.y. increments.

Six potassic feldspar separates from five Heihaibei granite wall rocks have invariable Pb isotope compositions, with $^{206}Pb/^{204}Pb$ ratios of 18.3532 to 19.4864, $^{207}Pb/^{204}Pb$ ratios of 15.6475 to 15.6812, and $^{208}Pb/^{204}Pb$ ratios of 37.1750 to 38.4598, indicating the initial lead isotope for the granite wall rocks when they were emplaced at ca. 416 Ma [50].

5. Discussion

5.1. Origins of the Ore-Forming Fluid

The $\delta^{18}O_{VSMOW}$ values (+9.2–+13.9‰) of quartz separates from quartz veins at the Heihaibei gold deposit are basically similar to those of the Phanerozoic orogenic gold deposits elsewhere ($\delta^{18}O$ = +7–+13‰, [63,64]). Given the average homogenization temperatures of 340 °C and 241 °C as the trapping temperature for the Qv1 and Qv2 samples, the calculated $\delta^{18}O_{H2O}$ values of the fluids in epoch I and epoch II show relatively large variations, from +3.6 to +8.3‰ and +1.6 to +4.3‰, respectively (Table 2, the solid contour lines in Figure 7b). Generally, the trapping temperatures are most likely higher than the homogenization temperatures for the ore-forming fluids. Therefore, the calculated $\delta^{18}O_{H2O}$ values of the fluids would be slightly higher than the values of +1.6–+8.3‰ in the Heihaibei gold deposit. In this regard, application of the arsenopyrite geothermometer, calculated using our unpublished electron microprobe analysis (EMPA) data, even higher average temperatures of 417 °C are obtained for epoch I and 385 °C for epoch II, which could also roughly represent the trapping temperature for the pyrite + arsenopyrite assembly in the Qv1 and Qv2, respectively. Given 417 °C and 385 °C as the trapping temperatures for Qv1 and Qv2, the calculated $\delta^{18}O_{H2O}$ values of the fluids in epoch I and epoch II would be obviously higher than the previously calculated ones, and they would be up to +5.5–+10.2‰ and +6.6–+9.3‰, respectively (the dashed contour lines in Figure 7b). Sheppard (1986) indicates that the $\delta^{18}O_{H2O}$ values for fluids of magmatic origin are mostly less than +9‰, whereas the metamorphic fluids are usually higher than +3‰ [61]. Therefore, the relatively high and variable $\delta^{18}O_{H2O}$ values of the ore-forming fluids in the quartz separates in the Heihaibei gold deposit are consistent with a metamorphic origin or alternatively a magmatic fluid that was related to the Heihaibei granites hosting the ore, as proposed by [23]. However, we argue that the ore-bearing Heihaibei granites are not the potential source granites because of their initial Pb isotope ratios, which are totally different from the Pb isotopes of ores (see below). Therefore, the ore-forming water is not possibly derived from the magmatic water related to the Heihaibei granites. A deep magmatic water is still possible, but there is no evidence of another granitic intrusion deeper than the Heihaibei granites, and usually, magmatic water that potentially contributes to hydrothermal alteration and corresponding mineralization should be spatio-temporally related to a nearby-granitic intrusion, i.e., the Heihaibei granites. Moreover, the ore-forming fluids in the Heihaibei gold deposit are characterized by low to intermediate salinity, and intermediate to high homogenization temperatures, and no daughter mineral was trapped in the fluid inclusions [23], also lowering the possibilities of magmatic water. Apparently, higher calculated $\delta^{18}O_{H2O}$ values of the fluids are closer to those of most of the orogenic gold deposits suggested by [59]. In a word, we argue a likely metamorphic origin rather than a magmatic water one, although we cannot totally preclude a magmatic origin. The calculated $\delta^{18}O_{H2O}$ values of the Qv1 type of samples (+3.6–+8.3‰, calculated with 340 °C) are commonly higher than those of the Qv2 type of samples (+1.6–+4.3‰, calculated with 240 °C) (Figure 7b), perhaps indicating the mixing of minor meteoric water in the epoch II, because the mixing of metamorphic and meteoric fluids will make the $\delta^{18}O_{H2O}$ values lower. If equilibrium calculations are made at the temperatures indicated by the arsenopyrite geothermometer, both the calculated $\delta^{18}O_{H2O}$ values of the Qv1 type of samples (+5.5–+10.2‰, calculated with 417 °C) and the comparably similar ones of the Qv2 type of samples (+6.6–+9.3‰, calculated with 385 °C) (Figure 7b) are all closer to the field for most orogenic gold deposits suggested by [59], further suggesting a metamorphic origin with relatively high but variable $\delta^{18}O_{H2O}$ values.

Fluid inclusions in 16 quartz separates within the quartz veins at the Heihaibei gold deposit have variable δD values varying from −104.2‰ to −81.1‰ (Table 2), which are apparently lower than those of most orogenic gold deposits suggested by [59], and they are partially plotted in the field of the devolatilization of organic matter in sediments (Figure 7a). However, no report shows organic matter in the sedimentary wall rock in the Heihaibei gold deposit. In some cases, bulk fluids extracted from quartz may be a mixture of primary, pseudosecondary, and secondary inclusions, then low δD values from the bulk extraction of fluid inclusions are often considered to reflect the secondary inclusions formed in the presence of meteoric water during the uplift of the deposits [65–68]. Therefore, we believe that the secondary inclusions formed in the presence of meteoric water during the uplift of the deposits contribute to the depletion of H isotope compositions from the bulk extraction of fluid inclusions within the quartz in the Heihaibei gold deposit. In this way, the potentially higher δD values may be closer to the δD values of the primary metamorphic ore-forming fluids (Figure 7b). Moreover, low δD values due to the presence of H_2 and/or CH_4 in the aqueous fluid [69] cannot be precluded either; unfortunately, laser Raman spectroscopy has not been carried out. Finally, meteoric water mixing is possible, in order to lower the δD values. Actually, the δD_{VSMOW} values of quartz in the Qv1 type samples are slightly higher (−97.2‰ to −81.1‰) than those of the Qv2 type samples (−104.2‰ to −89.9‰), suggesting mixing with minor meteoric water in epoch II. Overall, both the δD and $\delta^{18}O$ values of the ore-forming fluids showed a slight decrease from epoch I (Qv1 ores) to epoch II (Qv2) (Figure 7b), further indicating mixing with minor meteoric water, especially in epoch II.

The Dazaohuo and Heicigou gold deposits in KLRA have relatively different H and O isotope compositions compared to each other [29], and both are totally different from those of the Heihaibei gold deposit (Figure 7a), suggesting a different source of ore-forming fluids for these gold deposits in the KLHA. However, other gold deposits, i.e., Shenshuitan [4,8–10], Guoluolongwa [19] in the MKL, and Kaihuangbei [14] in the SKL, in which there are both quartz veins and phyllic rock-type ores, have relatively variable H and O isotope compositions relative to gold deposits which have predominant quartz vein ores, such as the Heihaibei gold deposit (this study), as well as the Hongqigou gold deposit [60], perhaps due to higher degrees of water–rock reaction/mixing, and more meteoric water mixing during their mineralization. By the way, the H and O isotope compositions (Figure 7b) for sample K1 are slightly different from other ore samples, perhaps because sample K1 (Qv1 plus Heihaibei granitic wall rock) had experienced a greater extent of water–rock reaction than ore samples in the QV1 or QV2 veins.

5.2. Sulfur Sources

The pyrites from the gold-bearing quartz veins of the Heihaibei gold deposit have almost invariable $\delta^{34}S_{VCDT}$ values ranging from +7.6‰ to +8.7‰, with an average of +8.1‰ (Figure 8a), which span three previously published values of 7.7‰, 7.9‰, and 8.5‰ reported by [23]. Moreover, our sulfur isotopic compositions are within the published $\delta^{34}S$ values for most orogenic gold deposits elsewhere (typically $\delta^{34}S_{VCDT}$ = 0–9‰, [51,67,70,71]). Pyrites from the Qv1 type of samples in the study area have almost invariable $\delta^{34}S$ values ranging from +8.0‰ to +8.7‰ (Figure 8b), which are totally similar to those of pyrites from the Qv2 type samples ranging from +7.6‰ to +8.5 (Figure 8c). As for the large similarities among the sulfides in the Qv1 type samples and those in the Qv2 samples (Figure 8f), as well as the filling mineralization of quartz veins in the Heihaibei gold deposit, we argue that the sulfur in the ores may be derived directly from an ore-forming fluid that had only experienced a temperature decrease and mixing with minor meteoritic water, rather than the apparent sulfur isotope exchange with the wall rock from epoch I to epoch II. The presence of pyrite + arsenopyrite, and the weak alteration assemblage of quartz + sericite in the Heihaibei gold deposit, together with the lack of oxidized phases (hematite) and sulfate minerals, indicate that sulfur was present in the hydrothermal fluids, mainly as reduced sulfur (H_2S) [72,73]. The $\delta^{34}S_{VCDT}$ values for H_2S ($\delta^{34}S_{H2S}$) in equilibrium with sulfides were also estimated by evaluating the $\delta^{34}S_{VCDT}$ values of pyrite using the equilibrium isotopic fractionation factor of sulfide for H_2S suggested by [73], and the average temperature of the hydrothermal fluid during the gold–sulfide mineralization event is inferred as 340 °C for Qv1 (epoch I), and 241 °C for Qv2 (epoch II) (based on the homogenization temperatures reported by [23]). The calculated $\delta^{34}S_{H2S}$ values for H_2S in the hydrothermal fluids ranging from +6.9‰ to +7.6‰, with an average value of +7.08‰ in epoch I, are also very similar to those in the hydrothermal fluids, ranging from +6.1‰ to +7.0‰ in epoch II (Table 3, Figure 8g). Although we did not analyze all types of sulfides in the Heihaibei gold deposit, pyrites dominate the sulfur mass balance, and such an evaluation for the $\delta^{34}S_{H2S}$ values of H_2S in the hydrothermal fluids using the sulfur isotopes of pyrites might be roughly reliable.

Previous researchers have proposed that the relatively high $\delta^{34}S_{VCDT}$ values of the ores were probably derived from the Heihaibei granites [23]. However, if the ore-forming fluids collected and exchanged sulfur from the neighboring Heihaibei granites, the ores in the Heihaibei gold deposit would possibly have more various $\delta^{34}S$ values rather than a narrow range, and different ores would have different $\delta^{34}S$ values because the sulfur isotope ranges of typical granite reservoirs have apparently various $\delta^{34}S$ values and could even range from −13.4‰ to +26.7‰ [74]. Unfortunately, we did not analyze $\delta^{34}S$ values of primary sulfides in the Heihaibei granites, so we are only making a rough guess here. We argue that the relatively high and invariable $\delta^{34}S$ values of sulfides for the Heihaibei gold deposit likely indicate a unique sulfur source directly from deep ore-forming fluids, rather than the neighboring granitic wall rock, because both Qv1 and Qv2 formed directly from primary ore-forming fluids, rather than from the apparent reaction between the ore-forming fluid and the granitic host rock during mineralization. Apparently, prior to upward migration, the deep ore-forming fluids leached sulfur, from the deep source rock in the upper crust, that was homogenized before mineralization.

In the KLHA, compared with the high and invariable $\delta^{34}S$ values of sulfides for the Heihaibei gold deposit, the Dazaohuo gold deposit has more variable $\delta^{34}S$ values of sulfides (Figure 8d, [29]), which are consistent with its phyllic rock-type ores, rather than the quartz veins. As we know, an apparent water–rock reaction which formed phyllic rock-type ores and related hydrothermal alterations possibly resulted in widespread sulfur mixing and exchange, as well as sulfur isotope fractionations in different ores. Therefore, other gold deposits which have predominant phyllic rock-type-ores, i.e., Shenshuitan [4,7,8,15,16], Guoluolongwa [15–17] in the MKL, and Kaihuangbei [20] in the SKL, likely have more variable $\delta^{34}S$ values of sulfides than those in the Heihaibei gold deposit (Figure 8d). Additionally, the Hongqigou gold deposit, which has predominant quartz vein ores, has relatively invariable heavy $\delta^{34}S$ values of sulfides that are similar to this study, indicating an original metamorphic S-bearing fluid [4]. So, once more, the relatively high and invariable $\delta^{34}S$ values of sulfides suggest that the sulfur source is derived directly from a deep metamorphic origin itself, which is consistent with the filling mineralization by the quartz vein, rather than an apparent water–rock reaction.

5.3. Lead Sources

Four pyrite minerals from quartz veins in the Heihaibei gold deposit have relatively invariable Pb isotope compositions (Figure 9a,b). Combined with three previously published sets of Pb isotope ratios ([23] in Figure 9), most of the Pb isotope compositions for sulfides in the Heihaibei gold deposit roughly cluster as a linear group (the blue contour lines in Figure 9a,b), close to the upper crust lead reservoir in the uranogenic plot (Figure 9), which might suggest two-component mixing between a high radiogenic end-member, and another low radiogenic end-member. Apparently, the fields for the Pb isotope compositions of sulfides in the Heihaibei gold deposit are totally different from the fields of the initial Pb isotopes (i.e., the Pb isotopes of K-feldspars) of the Heihaibei granites (the pink dashed lines in Figure 9a,b), and the ores (sulfides) show more radiogenic Pb isotopes and plot roughly along the upper crust growth line (Figure 9), indicating that Au-mineralization was not caused by the Heihaibei granites at ca. 416 Ma. In that way, a granitic intrusion-related gold deposit related to the Heihaibei granites proposed by [23] can be precluded for the Heihaibei gold deposit. Based on the below discussions, we argue that the gold mineralization might have occurred in the Triassic, which was likely around 200 Ma after the formation of the Heihaibei granites that host the veins. The opinion that Pb sources were derived from the neighbor Triassic granites in the north of Fault F51 within the MKL, as shown in Figure 2, might not be a good option, because the Pb isotope compositions of sulfides in the Heihaibei are also apparently different from the Pb isotope compositions of K-feldspars from some Triassic granites in the MKL (e.g., Triassic granites in the Wulonggou area within the MKL reported by [7]). Alternatively, apart from the K-feldspars, the rest of the minerals in the bulk Heihaibei granites will continue to evolve to higher $^{206}Pb/^{204}Pb$, $^{207}Pb/^{204}Pb$, and $^{208}Pb/^{204}Pb$ due to the U and Th decay for ca. 200 Ma, since they were intruded in the Early Devonian. Using the average U, Th, and Pb abundances of the Heihaibei granite [50] and the possible mineralization age (Triassic, ca. 200 Ma younger than the Early Devonian emplacement), the age (200 Ma)-corrected Pb isotopic ratios, roughly representing the Triassic Pb isotope compositions for the Early Devonian Heihaibei granitic wall rocks, can be estimated (initial Pb isotopes of K-feldspars, plus radiogenic Pb isotopes after ca. 200 Ma decay), and the results show that the age-corrected Triassic Pb isotope groups for the Devonian Heihaibei granites are partially localized within the ore Pb isotope group (Figure 9), indicating that the high radiogenic Pb isotope end-member might be the Triassic Pb isotope group of the Devonian Heihaibei granites. Furthermore, the sulfides within the Sample K1 from [23] have apparently higher radiogenic Pb isotope compositions than other sulfides (Figure 9c,d), perhaps indicating more Pb sources from granitic wall rocks, because Sample K1 was collected in the mineralized granitic wall rock, which would experience more water–rock reactions than the filling ores in QV1 or QV2. Finally, the low radiogenic Pb isotope end-member should be the deep ore-forming fluid itself.

Meanwhile, the Pb isotope compositions of sulfides in the Heihaibei gold deposit are also different from the Dazaohuo gold deposit, which has predominant phyllic rock ores [29] in the KLRA, or other gold deposits in the EKO (Figure 9a,b), also showing they have different Pb sources among them. The variable Pb isotope compositions in the Heihaibei gold deposit have more uranogenic and thorogenic Pb isotope compositions than other gold deposits, which have predominant phyllic rock ores, i.e., Shenshuitan [4,7,9], Guoluolongwa [16,17] in the MKL, and Kaihuangbei [20] in the SKL, also suggesting that Pb sources are distinctive for the Heihaibei gold deposit. Possibly in the Heihaibei gold deposit, the primary ore-forming fluids would have washed through deep parts of the Heihaibei granites with high levels of U/Pb and Th/Pb in the upper crust before final gold mineralization, because the apparent exchanging of the Pb isotopes between the ore-forming fluid and the Heihaibei granitic wall rock is relatively difficult during the gold filling mineralization accompanying the weak water–rock reaction in the Triassic. Pb isotope compositions of sulfides are more variable than sulfur isotopes, likely because the Heihaibei granites commonly contain more variable radiogenic Pb isotopes, which were continuously increased from U and Th decay than from sulfur stable isotopes. Moreover, the Qv1 type of ore has slightly more uranogenic and thorogenic Pb isotopes than the Qv2 type (Figure 9c,d), likely indicating less of a mixing of the Triassic Pb source of the Heihaibei granites in epoch II than epoch I, because the low-temperature ore-forming fluid of the epoch II might have experienced less Pb isotope exchanging with the deep parts of the Heihaibei granites or the upper crust before gold mineralization.

5.4. Genetic Type and Possible Metallogenic Model

Genetic classification of the Heihaibei gold deposit remains controversial: a hydrothermal-vein-type gold deposit [22], a magmatic hydrothermal gold deposit [24], an orogenic gold deposit [25], or a granitic intrusion-related gold deposit related to the Heihaibei granites [23] have all been proposed. However, as mentioned above, the ore bodies in the Heihaibei gold deposit are usually delineated in quartz veins along subsidiary faults within the Heihaibei granites, and the Heihaibei granites (monzogranites and granodiorites) are wall rocks for gold mineralization. Furthermore, many of the geological and geochemical features of the Heihaibei gold deposit are similar to those of the orogenic gold deposits, as described by [51]. These features include: (1) the geodynamic setting of the Heihaibei gold deposit related with the Triassic evolution of the EKO; (2) the gold mineralization controlled and hosted by the subsidiary faults of Fault F51 (a segment of the M. KLF) within the Heihaibei area, as well as in the KLRA; (3) the major sulfide minerals (contents < 5%) represented by pyrite and arsenopyrite, with minor chalcopyrite, galena, sphalerite, tetrahedrite, and micro-native gold; (4) the weak alteration assemblage, including silicification, and sericitization, with minor chloritization and carbonatization, similar to some typical orogenic gold deposits; (5) the major ore-forming fluids showing H_2O-NaCl fluids with low salinity (3.69~16.63 equiv. wt% NaCl), and intermediate to high homogenization temperatures from 183 to 412 °C [23]; (7) the δD values (−104.2 to −81.1‰) and estimated $\delta^{18}O$ values (+1.6 to +8.3‰, or even from +5.5 to +10.2‰) for ore-forming fluids, indicating a metamorphic water rather than a magmatic water; (8) the $\delta^{34}S$ values of sulfides from +7.6‰ to +8.7‰, suggesting a sulfur source that is directly from the deep ore-forming fluids themselves, with sulfur isotope homogenization; and (9) the Pb isotopic compositions of sulfides, with $^{206}Pb/^{204}Pb$ ratios of 18.7219–19.0007, $^{207}Pb/^{204}Pb$ ratios of 15.6959–15.7062, and $^{208}Pb/^{204}Pb$ ratios of 37.7359–38.8055, showing partial similarities to the age-corrected Triassic Pb isotope compositions of the Devonian Heihaibei granitic wall rocks, rather than to the initial Pb isotope compositions of the granite at its time of emplacement 416 Ma ago, indicating a mixed Pb source between the deep ore-forming fluids and the Triassic Pb isotope compositions of the deep Devonian Heihaibei granitic wall rock, and some affinities to upper crust lead reservoirs [62]. Based on the results described above, and the comparison with typical orogenic gold deposits, we propose that the Heihaibei gold deposit belongs to the orogenic gold category, as described by [51], rather than a granitic intrusion-related gold deposit proposed by [23].

To date, mineralization in the Heihaibei gold deposit, whose timing has not been determined directly by radiometric dating but must be later than the Early Devonian (416 Ma, [50]) of the monzogranite wall rocks. As mentioned above, the EKO has undergone at least two orogenies that are closely related to the consumption of the Neo-Proterozoic to Early Devonian Proto-Tethys Ocean, which lays along the M.KLF, and the Carboniferous to Late Triassic Paleo-Tethys Ocean, which lays along the S.KLF–AMF, respectively [3,31,37]. We have previously considered that the EKO during the Late Early Silurian–Middle Devonian (430–389 Ma) was in the post-collisional stage related to the closure of the Proto-Tethys Ocean [75], and that the A-type Heihaibei granites were exact products of post-collisional magmatic emplacement during the post-collisional stage [50]. The northward subduction of the Paleo-Tethys oceanic plate was not initiated until the Carboniferous [3,31,37]. From the Middle Devonian (389 Ma) to the Carboniferous, the tectonism and magmatism in the EKO would have been relatively rare, as well as the gold mineralization here. Actually, numerous magmatic emplacements which are related to the evolution (the subduction and subsequent collision) of the Paleo-Tethys Ocean between the SKL and BH–SG happened predominantly from the Late Permian to the Triassic [42,44,46,75,76]. Additionally, it was in the Early to Middle Triassic that the EKO was strongly affected by the northward subduction of the Paleo-Tethys oceanic plate, and it was in the Late Triassic that the EKO was dominated by the closure of the Paleo-Tethys oceans between the EKO and BH–SG [46]. Accompanying the Early to Middle Triassic subduction and the Late Triassic collision, most typical gold deposits within the EKO have predominant Triassic mineralization ages varying from 237 Ma to <215 Ma, e.g., Shenshuitan (zircon U-Pb dating, <215 Ma, [13]; sericite Ar-Ar dating, 237.0 Ma and 230.8 Ma, [8]), Shihuigou (sericite Ar-Ar dating, 236.5 Ma, [77]; zircon fission track thermochronology, 235–216 Ma [12]), Hongqigou (zircon fission track thermochronology, 223 Ma, [12]), Guoluolongwa (sericite Ar-Ar dating, 202.7 Ma for the late stage of mineralization [78]; and Re-Os dating, 375 Ma and 354 Ma for the early stage of mineralization [18]). With this in mind, we consider that a Triassic age of gold mineralization in the Heihaibei gold deposit may be reasonable.

We previously argued that the ancient metamorphic fluid, perhaps produced during the Precambrian metamorphism in the EKO, cannot explain the predominant Triassic orogenic gold mineralization found in the Wulonggou area in the EKO [4]. Even in the Late Triassic, the soft collision between EKO and BH–SG [33] cannot produce metamorphism of the deep rocks and subsequent metamorphic fluids at depth in the lower crust [4]. If this is the case, the slab devolatilization model proposed for the Jiaodong gold province [79] may be also applicable to the Heihaibei gold deposit, as with the Shenshuitan gold deposit (e.g., [4]). In their model, the late-orogenic metamorphic devolatilization of stalled subduction slabs and oceanic sediments may directly produce metamorphic fluids, and the overpressured slab-derived metamorphic fluids might also be channeled directly into crustal-scale faults with subsequent upward flow [63,79]. Based on the above discussion, the Heihaibei gold deposit may be associated with the subduction of the Paleo-Tethys oceanic plate in the Early to Middle Triassic, or even the collision between BH–SG and EKO in the Later Triassic. Therefore, during the Triassic in the EKO, the slab-derived metamorphic fluids may be initiated following the EKO Triassic geodynamic evolution, and they might have migrated through the first-order crustal-scale faults (i.e., M.KLF) upward into the subsidiary faults at higher levels (i.e., Fault F53) (Figure 10). During the migration of these deep slab-derived metamorphic fluids, sulfur, lead, and gold may be leached from various types of surrounding source rocks in the upper crust. According to the highly radiogenic Pb isotope ratios in this study, we likely infer that such slab-derived fluids may wash up high U/Pb and Th/Pb source rock, i.e., the deep part of the Heihaibei granites, in the upper crust, before upward gold mineralization; then, such slab-derived fluid actually becomes a special ore-forming fluid, since it migrates across such a "strongly leached zone" in the upper crust (Figure 10). Eventually, in the Heihaibei gold deposit, such evolved slab-derived metamorphic water with leached ore-forming materials (such as gold, sulfur, arsenic, iron, lead, copper, zinc, etc.) may continue to migrate upward, become a special ore-forming metamorphic fluid, and periodically deposit quartz veins within shallow subsidiary faults, likely because of a sudden change of physico-chemical conditions rather than strong water–rock reactions between ore-forming fluids and wall rocks. A similar filling mineralization of quartz veins in subsidiary faults could be found in other gold deposits, such as the Heicigou gold deposit [29] in the KLRA, and the Hongqigou gold deposit [18] in the MKL, etc. During the predominant filling of quartz veins in subsidiary faults in the Heihaibei gold deposit, minor water–rock interactions may also occur in the granitic wall rocks and contribute to weak hydrothermal alterations within brecciated granitic wall rocks, minor ores of stockwork quartz veinlets, or breccias in such altered granites. The late stage of tectonic movements due to the soft collision between EKO and BH–SG may deform previously formed quartz veins (Qv1 and Qv2) along fault belts in the Heihaibei gold deposit and form breccia ores, indicating that the deformation stage occurred. Given that there is a strong water–rock interaction between the ore-forming metamorphic water and wall rocks, important phyllic-type ores and strong hydrothermal alterations would have occurred, such as in other gold deposits, e.g., Dazaohuo [29], and the Shenshuitan gold deposit in the Wulonggou gold field [4,7–10]. Certainly, both the filling of quartz veins and phyllic rocks could be found simultaneously in certain other gold deposits in the EKO, such as the Guoluolongwa gold deposit [14–19] and the Kaihuangbei gold deposit [14,20], mostly because of various characteristics of the ore-forming faults.

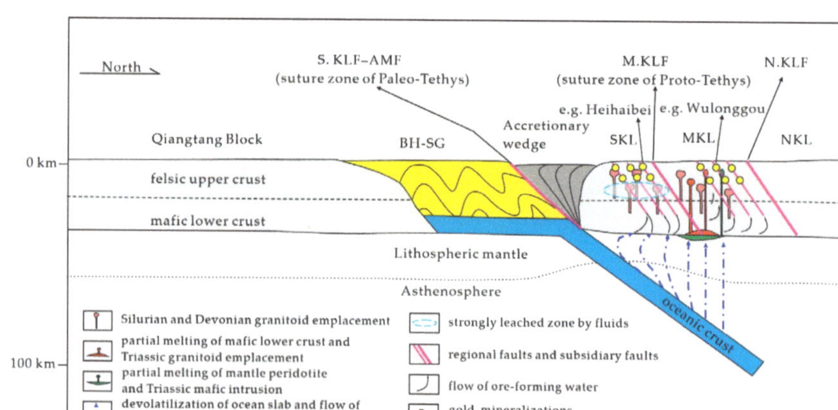

Figure 10. Metallogenic model for the Heihaibei gold deposit, as well as gold deposits in the EKO. Abbreviations are the same as those in Figure 1.

6. Conclusions

From the H, O, S, and Pb isotope compositions of ores from the Heihaibei gold deposit in the KLRA reported in this study the following conclusions can be derived.

The measured $\delta^{18}O_{quartz}$ values for the quartz from the auriferous quartz veins in the Heihaibei gold deposit range from +9.2‰ to +13.9‰, and the estimated $\delta^{18}O$ values for ore-forming fluids range from +1.6‰ to +8.3 ‰, or even up to +5.5 to +10.2‰. The δD_{H2O} values for the fluid inclusions of the quartz range from -104.2‰ to -81.1‰. The results suggest that the ore-forming fluid is a special metamorphic water, perhaps somewhat modified by interaction with meteoric waters.

The $\delta^{34}S$ values of sulfide separates from auriferous quartz veins in the Heihaibei gold deposit range from +7.6‰ to +8.7‰, and the calculated $\delta^{34}S_{H2S}$ values for H_2S in the ore-forming fluids range from +6.1‰ to +7.6‰, suggesting that the sulfur sources may be directly derived from the deep ore-forming fluids themselves.

Six potassic feldspar separates from the Heihaibei granites have apparently invariable Pb isotope compositions, with $^{206}Pb/^{204}Pb$ ratios of 18.3532 to 19.4864, $^{207}Pb/^{204}Pb$ ratios of 15.6475 to 15.6812, and $^{208}Pb/^{204}Pb$ ratios of 37.1750 to 38.4598. Additionally, the Pb isotope compositions of sulfide separates from the auriferous quartz veins in the Heihaibei gold deposit range, from 118.7219 to 19.0007 for $^{206}Pb/^{204}Pb$, 15.6959 to 15.7062 for $^{207}Pb/^{204}Pb$, and 37.7359 to 38.8055 for $^{208}Pb/^{204}Pb$, showing partial similarities to the age (200 Ma)-corrected Triassic Pb isotope compositions of the Devonian Heihaibei granitic wall rocks, rather than the initial Pb isotope compositions of the granites, indicating a mixed Pb source between the deep ore-forming fluids and the Triassic Pb isotope compositions of the deep parts of the Devonian Heihaibei, and showing some affinities to the upper crust lead reservoir.

The Heihaibei gold deposit may be classified as an orogenic gold deposit, which is associated with the subduction of the Paleo-Tethys oceanic plate in the Early to Middle Triassic, or even the collision between BH–SG and EKO in the Later Triassic. The slab-derived metamorphic fluids may be generated during the above Triassic geodynamic evolution and subsequently leached ore-forming materials from high U/Pb and Th/Pb source rock, i.e., the deep parts of the Heihaibei granites and their comagmatic granitoids in the upper crust. Then, they continuously migrated upward to the subsidiary faults in the KLRA or even the whole EKO, mixed with minor meteoric water, especially during the later epoch of gold mineralization, and periodically deposited quartz veins within shallower subsidiary faults in the Heihaibei gold deposit. Late stages of tectonic movements may deform previously formed quartz veins along fault belts and form breccia ores.

Supplementary Materials: The following are available online at https://www.mdpi.com/article/10.3390/min13020274/s1, Table S1: H and O isotope data for quartz in the Heihaibei gold deposit; Table S2: Sulfur isotope data for sulfides in the Heihaibei deposit; Table S3: Pb isotope compositions for sulfides in the Heihaibei gold deposit.

Author Contributions: Author Contributions: Conceptualization, Q.-F.D.; methodology, H.J., X.Z. and R.-Z.W.; software, X.Z. and R.-Z.W.; validation, H.-F.L., Q.-F.D., R.-Z.W. and T.P.; formal analysis, H.-F.L., X.Z., R.-Z.W. and Q.-F.D.; investigation, H.-F.L., H.J., R.-Z.W., X.Z. and Q.-F.D.; resources, H.-F.L., T.P., H.J. and Q.-F.D.; data curation, H.-F.L. and Q.-F.D.; writing—original draft preparation, H.-F.L. and Q.-F.D.; writing—review and editing, H.-F.L. and Q.-F.D.; visualization, H.-F.L., X.Z. and Q.-F.D.; supervision, T.P. and Q.-F.D.; project administration, H.-F.L. and Q.-F.D.; funding acquisition, H.-F.L. and Q.-F.D. All authors have read and agreed to the published version of the manuscript.

Funding: This research was funded by the Action Plan of Kunlun Talents, Qinghai Talent Office file (2020) No. 18.

Institutional Review Board Statement: Not applicable.

Informed Consent Statement: Not applicable.

Data Availability Statement: The datasets generated during and/or analyzed during the current study are available in full within the published article and its supplementary materials.

Acknowledgments: We are grateful to the analytical Laboratories at CAGS and BCTT for their help with the stable isotope analyses.

Conflicts of Interest: The authors declare no conflict of interest.

References

1. Bian, Q.T.; Li, D.H.; Pospelov, I.; Yin, L.M.; Li, H.S.; Zhao, D.S.; Chang, C.F.; Luo, X.Q.; Gao, S.L.; Astrakhantsev, O.; et al. Age, geochemistry and tectonic setting of Buqingshan ophiolites, North Qinghai–Tibet Plateau, China. *J. Asian Earth Sci.* **2004**, *23*, 577–596. [CrossRef]
2. Xu, Z.Q.; Yang, J.S.; Li, H.B.; Zhang, J.X.; Wu, C.L. *Terrane Amalgamation, Collision and Uplift in the Qinghai–Tibet Plateau*; Geological Publishing House: Beijing, China, 2007; p. 458.
3. Jiang, C.F.; Yang, J.S.; Feng, B.G.; Zhu, Z.X.; Zhao, M.; Chai, Y.C.; Shi, X.D.; Wang, H.D. *Opening Closing Tectonics of Kunlun Shan*; Geological Publishing House: Beijing, China, 1992; p. 217.
4. Zhou, X.; Pan, T.; Ding, Q.-F.; Cheng, L.; Song, K.; Liu, F.; Gao, Y. Isotope Geochemistry of the Shenshuitan Gold Deposit within the Wulonggou Gold Field in the Eastern Kunlun Orogen, Northwest China: Implications for Metallogeny. *Minerals* **2022**, *12*, 339. [CrossRef]
5. Li, Z.M.; Xue, C.J.; Wang, X.H.; Tang, H.; Tu, Q.J.; Teng, J.X.; Li, R.S. Features of regional mineralization and analysis of the exploration development in the Eastern Kunlun Mountains. *Geol. Rev.* **2007**, *53*, 708–718, (In Chinese with English abstract).
6. Du, Y.L.; Jia, Q.Z.; Han, S.F. Mesozoic tectono-magmatic-mineralization and copper-gold polymetallic ore prospecting research in East Kunlun Metallogenic Belt in Qinghai. *Northwestern Geol.* **2012**, *45*, 69–75, (In Chinese with English abstract). [CrossRef]
7. Liu, F. Research on the Geological, Geochemical Characteristics and Genesis of Shenshuitan Gold Deposit in the Wulonggou Ore Concentration District in the Eastern Kunlun Orogenic Belt, Qinghai Province. Master's Thesis, Jilin University, Changchun, China, 2017.
8. Zhang, J.Y.; Ma, C.Q.; Li, J.W.; Pan, Y.M. A possible genetic relationship between orogenic gold mineralization and post-collisional magmatism in the eastern Kunlun Orogen, western China. *Ore Geol. Rev.* **2017**, *81*, 342–357. [CrossRef]
9. Zhang, Y.T. Research on Metallogenesis of Gold Deposits in the Wulongou Ore Concentration Area, Central Segment of the East Kunlun Mountains, Qinghai Province. Ph.D. Thesis, Jilin University, Changchun, China, 2018.
10. Song, K. Research on the Metallogenic Geological Background and Genesis of Huanglonggou Ore Block in Shenshuitan Gold Deposit in the Wulonggou Ore Concentration Area in the Eastern Kunlun Orogenic Belt, Qinghai Province. Master's Thesis, Jilin University, Changchun, China, 2019.
11. Tian, C.S. Research on Gold Mineralization and Metallogenic Prognosis of Wulonggou Goldfield in the Eastern Kunlun Orogen. Ph.D. Thesis, China University of Geosciences, Beijing, China, 2012.
12. Yuan, W.M.; Mo, X.X.; Zhang, A.K.; Chen, X.N.; Duan, H.W.; Li, X.; Hao, N.N.; Wang, X.M. Fission Track thermochronology evidence for multiple periods of mineralization in the Wulonggou Gold Deposits, Eastern Kunlun Mountains, Qinghai province. *J. Earth Sci.* **2013**, *24*, 471–478. [CrossRef]
13. Ding, Q.F.; Jiang, S.Y.; Sun, F.Y. Zircon U–Pb geochronology, geochemical and Sr–Nd–Hf isotopic compositions of the Triassic granite and diorite dikes from the Wulonggou mining area in the Eastern Kunlun Orogen, NW China: Petrogenesis and tectonic implications. *Lithos* **2014**, *205*, 266–283. [CrossRef]
14. Feng, C.Y.; Zhang, D.Q.; Wang, F.C.; Li, D.X.; She, H.Q. Geochemical characteristics of ore-forming fluids from the orogenic Au (and Sb) deposits in the eastern Kunlun area, Qinghai province. *Acta Petro Sin.* **2004**, *20*, 949–960, (In Chinese with English abstract). [CrossRef]
15. Chen, J.J. Paleozoic-Mesozoic Tectono-Magmatic Evolution and Gold Mineralization in Gouli Area, East End of East Kunlun Orogen. Ph.D. Thesis, China University of Geosciences, Wuhan, China, 2018.
16. Hu, R.G. Research on Geological-Geochemial Characteristics and Genesis of the Guoluolongwa Gold Deposit in Qinghai Province. Master's Thesis, Central South University, Changsha, China, 2008.
17. Yue, W.H. Study on Metallogenic Mechanism and Geological-Geochemical Characteristics of Typical Deposits in Gouli Gold Ore Concentration Area in the Eastern Segment of the Eastern Kunlun Orogen. Ph.D. Thesis, Kunming University of Science and Technology, Kunming, China, 2013.
18. Chen, J.; Fu, L.; Selby, D.; Wei, J.; Zhao, X.; Zhou, H. Multiple episodes of gold mineralization in the East Kunlun Orogen, western Central Orogenic Belt, China: Constraints from Re-Os sulfide geochronology. *Ore Geol. Rev.* **2020**, *123*, 103587. [CrossRef]
19. Ding, Q.F.; Jin, S.K.; Wang, G.; Zhang, B.L. Ore-forming fluid of the Guoluolongwa gold deposit in Dulan County, Qinghai province. *J. Jilin Univ. Earth Sci. Ed.* **2013**, *42*, 415–426, (In Chinese with English abstract).
20. Feng, C.Y.; Zhang, D.Q.; Li, D.X.; She, H.Q. Sulfur and lead isotope geochemistry of the orogenic gold deposits in East Kunlun area, Qinghai province. *Acta Geosci. Sin.* **2003**, *24*, 593–598, (In Chinese with English abstract). [CrossRef]
21. Wang, Q.Y.; Dong, Y.J.; Pan, Y.M.; Liao, F.X.; Guo, X.W. Early Paleozoic Granulite-Facies Metamorphism and Magmatism in the Northern Wulan Terrane of the Quanji Massif: Implications for the Evolution of the Proto-Tethys Ocean in Northwestern China. *J. Earth Sci.* **2018**, *29*, 1081–1101. [CrossRef]
22. QIGEIQP. *Report of Detailed Exploration for Heihaibei Gold Deposit in Golmud City, Qinghai Province*; Qaidam Integrated Geological Exploration Institute of Qinghai Province (QIGEIQP): Golmud, China, 2021; pp. 1–328, (In Chinese with English abstract).

23. Yu, L.; Sun, F.; Beier, C.; Wu, D.; Li, L.; Wang, L.; Huang, G.; Fan, X.; Xu, C. Geology, U-Pb geochronology and stable isotope geochemistry of the Heihaibei gold deposit in the southern part of the Eastern Kunlun Orogenic Belt, China: A granitic intrusion-related gold deposit? *Ore Geol. Rev.* **2022**, *144*, 104859. [CrossRef]
24. Huang, G.; Ma, W.; Li, C.; Jiao, H. Geological characteristics and prospecting prospects of Heihaibei gold deposit in Kunlunhe area, Qinghai province. *Gold* **2021**, *42*, 26–30, (In Chinese with English abstract).
25. Jiao, H.; Kang, J.Z.; Huang, G.B.; Jia, J.T.; Peng, J. Magmatism, metallogenic characteristics, and prospecting prediction for gold deposits in the north of Kunlun River area, Qinghai, China. *J. Geomech.* **2022**, *28*, 383–405, (In Chinese with English abstract). [CrossRef]
26. Han, G. A Study on the Geological Characteristics of Gold Mineralization in Heihaibei Region, East Kunlun, Qinghai Province. Master's Thesis, China University of Geosciences, Beijing, China, 2013.
27. Li, J. Metallogenic Regularity and Metallogenic Prognosis of Gold Deposit in the East Kunlun Orogen, Qinghai Province. Ph.D. Thesis, Chang'an University, Xi'an, China, 2017.
28. Shen, H. Geochemical Characteristics and Ore Genesis of the Heihainan Gold Deposit in the Kunlun River District, Qinghai Province. Master's Thesis, China University of Geosciences, Beijing, China, 2016.
29. Feng, L. Geological and Geochemical Characteristics and Ore Genesis of the Gold Deposits in the Kunlun River District, Qinghai Province. Master's Thesis, China University of Geosciences, Beijing, China, 2017.
30. An, H.; Han, G.; Wu, P.; Wu, M.; Han, W. Deposit geological characteristics and prospecting potential of Dazaohuogou-Heicigou gold mine in Qinghai province. *Metal Mine* **2018**, *9*, 127–136, (In Chinese with English abstract).
31. Yang, J.S.; Robinson, P.T.; Jiang, C.F.; Xu, Z.Q. Ophiolites of the Kunlun Mountains, China and their tectonic implications. *Tectonophysics* **1996**, *258*, 215–231. [CrossRef]
32. Pan, Y.S.; Zhou, W.M.; Xu, R.H.; Wang, D.A.; Zhang, Y.Q.; Xie, Y.W.; Chen, T.E.; Luo, H. Geological characteristics and evolution of the Kunlun Mountains region during the early Paleozoic. *Sci. China Ser. D* **1996**, *39*, 337–347.
33. Yin, A.; Harrison, T.M. Geologic evolution of the Himalayan-Tibetan orogen. *Annu. Rev. Earth Planet. Sci.* **2000**, *28*, 211–280. [CrossRef]
34. Roger, F.; Arnaud, N.; Gilder, S.; Tapponnier, P.; Jolivet, M.; Brunel, M.; Malavieille, J.; Xu, Z.Q.; Yang, J.S. Geochronological and geochemical constraints on Mesozoic suturing in east central Tibet. *Tectonics* **2003**, *22*, 1037. [CrossRef]
35. Harrowfield, M.J.; Wilson, C.J.L. Indosinian deformation of the Songpan Garze Fold Belt, northeast Tibetan Plateau. *J. Struct. Geol.* **2005**, *27*, 101–117. [CrossRef]
36. Reid, A.J.; Wilson, C.J.L.; Liu, S. Structural evidence for the Permo-Triassic tectonic evolution of the Yidun Arc, eastern Tibetan plateau. *J. Struct. Geol.* **2005**, *27*, 119–137. [CrossRef]
37. Yang, J.S.; Shi, R.D.; Wu, C.L.; Wang, X.B.; Robinson, P.T. Dur'ngoi Ophiolite in East Kunlun, Northeast Tibetan Plateau: Evidence for Paleo-Tethyan Suture in Northwest China. *J. Earth Sci.* **2009**, *20*, 303–331. [CrossRef]
38. Pan, G.T.; Ding, J.; Wang, L.; Zhuang, Y.; Wang, Q.; Zhai, G.; Wang, D. Important new progress in regional geological survey of Qinghai Tibetan Plateau. *Geol. Bull. China* **2002**, *21*, 787–793, (In Chinese with English abstract).
39. Liu, Y.J.; Genser, J.; Neubauer, F.; Jin, W.; Ge, X.H.; Handler, R.; Takasu, A. Ar-40/Ar-39 mineral ages from basement rocks in the Eastern Kunlun Mountains, NW China, and their tectonic implications. *Tectonophysics* **2005**, *398*, 199–224. [CrossRef]
40. Gao, Y.; Li, W.; Li, Z.; Wang, J.; Hattori, K.; Zhang, Z.; Geng, J. Geology, geochemistry, and genesis of Tungsten-Tin deposits in the Baiganhu district, Northern Kunlun Belt, Northwestern China. *Econ. Geol.* **2014**, *109*, 1787–1799. [CrossRef]
41. Meng, F.C.; Cui, M.H.; Wu, X.K.; Ren, Y.F. Heishan mafic-ultramafic rocks in the Qimantag area of Eastern Kunlun, NW China: Remnants of an early Paleozoic incipient island arc. *Gondwana Res.* **2015**, *27*, 745–759. [CrossRef]
42. Liu, C.D.; Mo, X.X.; Luo, Z.H.; Yu, X.H.; Chen, H.W.; Li, S.W.; Zhao, X. Mixing events between the crust- and mantle-derived magmas in Eastern Kunlun: Evidence from zircon SHRIMP II chronology. *Chin. Sci. Bull.* **2004**, *49*, 828–834. [CrossRef]
43. Yuan, C.; Sun, M.; Xiao, W.; Wilde, S.; Li, X.; Liu, X.; Long, X.; Xia, X.; Ye, K.; Li, J. Garnet-bearing tonalitic porphyry from East Kunlun, Northeast Tibetan Plateau: Implications for adakite and magmas from the MASH Zone. *Int. J. Earth Sci.* **2009**, *98*, 1489–1510. [CrossRef]
44. Zhang, J.Y.; Ma, C.Q.; Xiong, F.H.; Liu, B. Petrogenesis and tectonic significance of the Late Permian-Middle Triassic calc-alkaline granites in the Balong region, eastern Kunlun Orogen, China. *Geol. Mag.* **2012**, *149*, 892–908. [CrossRef]
45. Huang, H.; Niu, Y.L.; Nowell, G.; Zhao, Z.D.; Yu, X.H.; Zhu, D.C.; Mo, X.X.; Ding, S. Geochemical constraints on the petrogenesis of granitoids in the East Kunlun Orogenic belt, northern Tibetan Plateau: Implications for continental crust growth through syn-collisional felsic magmatism. *Chem. Geol.* **2014**, *370*, 1–18. [CrossRef]
46. Ding, Q.F.; Liu, F.; Yan, W. Zircon U–Pb geochronology and Hf isotopic constraints on the petrogenesis of Early Triassic granites in the Wulonggou area of the Eastern Kunlun Orogen, Northwest China. *Int. Geol. Rev.* **2015**, *57*, 1735–1754. [CrossRef]
47. Chen, N.; Sun, M.; Wang, O.; Zhao, G.; Chen, Q.; Shu, G. EMP chemical ages of monazites from Central Zone of the eastern Kunlun Orogen: Records of multi-tectonometamorphic events. *Chin. Sci. Bull.* **2007**, *52*, 2252–2263. [CrossRef]
48. Xin, W.; Sun, F.Y.; Li, L.; Yan, J.M.; Zhang, Y.T.; Wang, Y.C.; Shen, T.S.; Yang, Y.J. The Wulonggou metaluminous A2-type granites in the Eastern Kunlun Orogenic Belt, NW China: Rejuvenation of subduction-related felsic crust and implications for post-collision extension. *Lithos* **2018**, *312–313*, 108–127. [CrossRef]

49. Gu, X.X.; Zhang, Y.M.; Feng, L.Q.; He, G.; Wang, J.L.; Kang, J.Z.; Wang, B.Z. Recognition, age determination and tectonic significance of the Kunlun River ductile shear zone in the East Kunlun Mountains. *Geol. Bull. China* **2018**, *37*, 345–355, (In Chinese with English abstract).
50. Lu, H.L.; Pan, T.; Jiao, H.; Ding, Q.F.; Zheng, T.; Zhou, X.; Wu, R. Geochronology, geochemistry, and Hf isotopic compositions of the Early Devonian Heihaibei granite in the East Kunlun Orogen, Northwest China. *Can. J. Earth Sci.* **2022**, *59*, 566–579. [CrossRef]
51. Groves, D.I.; Goldfarb, R.J.; Gebre-Mariam, M.; Hagemann, S.G.; Robert, F. Orogenic gold deposits: A proposed classification in the context of their crustal distribution and relationship to other gold deposit types. *Ore Geol. Rev.* **1998**, *13*, 7–27. [CrossRef]
52. Clayton, R.N.; O'Neil, J.R.; Mayeda, T.K. Oxygen isotope exchange between quartz and water. *J. Geophys. Res.* **1972**, *77*, 3057–3067. [CrossRef]
53. Clayton, R.N.; Mayeda, T.K. The use of bromine pentafluoride in the extraction of oxygen from oxides and silicates for isotopic analysis. *Geochim. Cosmochim. Acta* **1963**, *27*, 43–52. [CrossRef]
54. Ding, Q.F.; Jiang, S.Y.; Sun, F.Y.; Qian, Y.; Wang, G. Origin of the Dachang gold deposit, NW China: Constraints from H, O, S, and Pb isotope data. *Int. Geol. Rev.* **2013**, *55*, 1885–1901. [CrossRef]
55. Gong, B.; Zheng, Y.F.; Chen, R.X. An online method combining a thermal conversion elemental analyzer with isotope ratio mass spectrometry for the determination of hydrogen isotope composition and water concentration in geological samples. *Rapid Commun. Mass Spectrom.* **2007**, *21*, 1386–1392. [CrossRef] [PubMed]
56. Robinson, B.W.; Kusakabe, M. Quantitative preparation of sulfur dioxide, for sulfur-34/sulfur-32 analyses, from sulfides by combustion with cuprous oxide. *Anal. Chem.* **1975**, *47*, 1179–1181. [CrossRef]
57. He, Z.; Zhang, X.; Deng, X.; Hu, H.; Li, Y.; Yu, H.; Archer, C.; Li, J.; Huang, F. The behavior of Fe and S isotopes in porphyry copper systems: Constraints from the Tongshankou Cu-Mo deposit, Eastern China. *Geochim. Cosmochim. Acta* **2020**, *270*, 61–83. [CrossRef]
58. Chen, D.; Han, Y.; Wang, Z.; Chen, K.; Cai, B.; Liu, S. The prestigious tin–lead horse fittings in the Warring States Period: Evidence from funerary artefacts of a Warring States Tomb at Shouxian, Anhui. *Archaeometry* **2021**, *63*, 1290–1305. [CrossRef]
59. Goldfarb, R.J.; Ayuso, R.; Miller, M.L.; Ebert, S.W.; Marsh, E.E.; Petsel, S.A.; Miller, L.D.; Bradley, D.; Johnson, C.; McClelland, W. The Late Cretaceous Donlin Creek gold deposit, southwestern Alaska: Controls on epizonal ore formation. *Econ. Geol.* **2004**, *99*, 643–671. [CrossRef]
60. Cheng, L. Study on the Geological and Geochemical Characteristics and Genesis of Hongqigou Gold Deposit, Wulonggou Ore Concentration Area, East Kunlun, Qinghai Province. Master's Thesis, Jilin University, Changchun, China, 2020.
61. Sheppard, S. Characterization and isotopic variations in natural waters. *Rev. Mineral. Geochem.* **1986**, *16*, 165–183.
62. Zartman, R.E.; Doe, B.R. Plumbotectonics—The model. *Tectonophysics* **1981**, *75*, 135–162. [CrossRef]
63. Goldfarb, R.J.; Groves, D.I. Orogenic gold: Common or evolving fluid and metal sources through time. *Lithos* **2015**, *233*, 2–26. [CrossRef]
64. Bierlein, F.P.; Crowe, D.E. Phanerozoic orogenic lode gold deposits. In *Gold in 2000. Reviews in Economic Geology*; Hagemann, S.G., Brown, P.E., Eds.; Society of Economic Geologists, Inc.: Littleton, CO, USA, 2000; Volume 13, pp. 103–139.
65. Pickthorn, W.J.; Goldfarb, R.J.; Leach, D.L. Comment and Reply on "Dual origins of lode gold deposits in the Canadian Cordillera": COMMENT. *Geology* **1987**, *15*, 471–472. [CrossRef]
66. Goldfarb, R.J.; Snee, L.W.; Miller, L.D.; Newberry, R.J. Rapid dewatering of the crust deduced from ages of mesothermal gold deposits. *Nature* **1991**, *354*, 296–298. [CrossRef]
67. McCuaig, T.C.; Kerrich, R. P–T–t-deformation-fluid characteristics of lode gold deposits: Evidence from alteration systematics. *Ore Geol. Rev.* **1998**, *12*, 381–453. [CrossRef]
68. Jiang, S.H.; Nie, F.J.; Hu, P.; Lai, X.R.; Liu, Y.F. Mayum: An orogenic gold deposit in Tibet, China. *Ore Geol. Rev.* **2009**, *36*, 160–173. [CrossRef]
69. Fyon, J.A.; Crocket, J.H.; Schwartz, H.P. The Carshaw and Malga Iron-Formation Hosted Gold Deposits of the Timmins Area. In *The Geology of Gold in Ontario*; Colvine, A.C., Ed.; Ontario Geological Survey: Sudbury, ON, Canada, 1984; pp. 98–110.
70. Kerrich, R.; Goldfarb, R.; Groves, D.; Garwin, S.; Jia, Y. The characteristics, origins, and geodynamic settings of supergiant gold metallogenic provinces. *Sci. China Ser. D* **2000**, *43*, 1–68. [CrossRef]
71. Goldfarb, R.J.; Groves, D.I.; Gardoll, S. Orogenic gold and geologic time: A global synthesis. *Ore Geol. Rev.* **2001**, *18*, 1–75. [CrossRef]
72. Ohmoto, H.; Rye, R.O. Isotopes of Sulfur and Carbon. In *Geochemistry of Hydrothermal Ore Deposits*, 2nd ed.; Barnes, H.L., Ed.; John Wiley: New York, NY, USA, 1979; pp. 509–567.
73. Ohmoto, H.; Goldhaber, M.B. Sulfur and carbon Isotopes. In *Geochemistry of Hydrothermal Ore Deposits*, 3rd ed.; Barnes, H.L., Ed.; John Wiley: New York, NY, USA, 1997; pp. 517–611.
74. Zheng, Y.F.; Chen, J.F. *Stable Isotope Geochemistry*; Science Press: Beijing, China, 2000; p. 316.
75. Ding, Q.F.; Song, K.; Zhang, Q.; Yan, W.; Liu, F. Zircon U–Pb geochronology and Hf isotopic constraints on the petrogenesis of the Late Silurian Shidonggou granite from the Wulonggou area in the Eastern Kunlun Orogen, Northwest China. *Int. Geol. Rev.* **2019**, *61*, 1666–1689. [CrossRef]
76. Mo, X.X.; Luo, Z.H.; Deng, J.F.; Yu, X.H.; Liu, C.D.; Chen, H.W.; Yuan, W.M.; Liu, Y.H. Granitic intrusions and crustal growth in the East-Kunlun orogenic belt. *Geol. J. China Univ.* **2007**, *13*, 403–414, (In Chinese with English abstract). [CrossRef]

77. Zhang, D.Q.; Dang, X.Y.; She, H.Q.; Li, D.X.; Feng, C.Y.; Li, J.W. Ar–Ar dating of orogenic gold deposits in the northern margin of Qaidam and East Kunlun Mountains and its geological significance. *Miner. Depos.* **2005**, *24*, 87–98, (In Chinese with English abstract). [CrossRef]
78. Xiao, Y.; Feng, C.Y.; Li, D.X.; Liu, J.N. Chronology and fluid inclusions of the Guoluolongwa gold deposit in Qinghai province. *Acta Geol. Sin.* **2014**, *88*, 895–902, (In Chinese with English abstract).
79. Groves, D.I.; Santosh, M. The giant Jiaodong gold province: The key to a unified model for orogenic gold deposits? *Geosci. Front.* **2016**, *7*, 409–417. [CrossRef]

minerals

MDPI

Article

Sulfur and Carbon–Oxygen Isotopic Geochemistry and Fluid Inclusion Characteristics of the Yolindi Cu-Fe Skarn Mineralization, Biga Peninsula, NW Turkey: Implications for the Source and Evolution of Hydrothermal Fluids

Mustafa Kaya [1,*], Mustafa Kumral [1], Cihan Yalçın [2] and Amr Abdelnasser [1,3]

[1] Department of Geological Engineering, Faculty of Mines, Istanbul Technical University, Istanbul 34469, Turkey; kumral@itu.edu.tr (M.K.); amr.khalil@fsc.bu.edu.eg (A.A.)
[2] General Directorate of Industrial Zones, Ministry of Industry and Technology, Ankara 06510, Turkey; cihan.yalcin@sanayi.gov.tr
[3] Department of Geology, Faculty of Science, Benha University, Benha 13518, Egypt
* Correspondence: kayamusta@itu.edu.tr

Abstract: The current study sought to investigate the physiochemical conditions and fluid evolution within the Yolindi Cu-Fe skarn mineralization located in the Biga Peninsula, NW Turkey. This was accomplished through a comprehensive investigation of geological and mineralogical data, along with isotopic analyses of sulfur (δ^{34}S), carbon (δ^{13}C), and oxygen (δ^{18}O) of sulfide and calcite minerals, respectively, as well as fluid inclusion data pertaining to various minerals (e.g., andradite, quartz, and calcite). The Yolindi area features a complex geological framework, including the Paleozoic Kalabak Group (which includes the Torasan, Yolindi, and Sazak formations) and the Triassic Karakaya Complex. These formations were subsequently intruded via Early Miocene Şaroluk granitoids and Hallaçlar volcanics. Skarn formation is zoned into endoskarn and exoskarn types (being categorized into proximal, intermediate, and distal zones), with distinct mineral assemblages indicating concentric and contact metamorphic alteration patterns around the western part of Şaroluk granitoid intrusion in contact with the Torasan formation. The ore mineralogy and paragenesis suggest three distinct stages of evolution: an initial phase of prograde metasomatism characterized by the formation of magnetite and pyrite alongside anhydrous calc-silicate minerals; a subsequent phase of retrograde alteration marked by the formation of epidote, actinolite, and scapolite, accompanied by the occurrence of chalcopyrite and specular hematite; and finally, a post-metasomatic stage involving oxidation processes that led to the development of secondary mineral assemblages containing cerussite, covellite, and malachite. Sulfur isotopes (δ^{34}S) of sulfides from endoskarn (from +0.27 to +0.57‰$_{VCDT}$) to intermediate exoskarn (from −9.44 to −5.46‰$_{VCDT}$) zones indicate a diverse sulfur source, including magmatic, sedimentary, and possibly organic matter. δ^{34}S values in hydrothermal fluids suggest a magmatic–hydrothermal origin, with endoskarn and proximal zone fluids showing a slight negative signature and intermediate zone fluids indicating a strong influence from organic-rich or metamorphic sulfur reservoirs. Carbon and oxygen isotopic compositions (δ^{13}C and δ^{18}O) of calcite revealed a progression from marine carbonate signatures in marble samples (from +1.89 to +2.23‰$_{VPDB}$; from +21.61 to +21.73‰$_{VSMOW}$) to depleted values in prograde (from −6.0 to +0.09‰$_{VPDB}$; from +6.22 to +18.14‰$_{VSMOW}$) and retrograde skarns (from −3.8 to −2.25‰$_{VPDB}$; from +0.94 to +3.62‰$_{VSMOW}$), reflecting interactions with high-temperature magmatic fluids and meteoric water mixing. The fluid inclusions in prograde minerals generated under the conditions of fluid boiling exhibited high temperatures, reaching up to 412 °C, and salinities up to 26 wt.% NaCl equivalent. Conversely, the fluid inclusions in retrograde minerals, which were generated due to fluid mixing, exhibited lower temperatures (with an average of 318 °C) and salinities with an average of 4.9 wt.% NaCl equivalent. This indicated that the cooler and more diluted fluids mix with meteoric waters and interact with organic materials in the host rocks. This suggests a multifaceted origin involving various sources and processes. Therefore, this study concluded that the skarn mineralization in the Yolindi area resulted from complex interactions between magmatic, metamorphic, and meteoric

Citation: Kaya, M.; Kumral, M.; Yalçın, C.; Abdelnasser, A. Sulfur and Carbon–Oxygen Isotopic Geochemistry and Fluid Inclusion Characteristics of the Yolindi Cu-Fe Skarn Mineralization, Biga Peninsula, NW Turkey: Implications for the Source and Evolution of Hydrothermal Fluids. *Minerals* **2023**, *13*, 1542. https://doi.org/10.3390/min13121542

Academic Editor: Yuichi Morishita

Received: 6 November 2023
Revised: 5 December 2023
Accepted: 11 December 2023
Published: 14 December 2023

fluids, reflecting a dynamic ore-forming environment with implications for the regional metallogeny of Cu-Fe skarn deposits.

Keywords: mineralogy; skarn evolution and zonation; sulfur isotopes ($\delta^{34}S$); carbon isotopes ($\delta^{13}C$); oxygen isotopes ($\delta^{18}O$); fluid inclusions; Yolindi Cu-Fe Skarn; Biga Peninsula; NW Turkey

1. Introduction

Skarn deposits are a noteworthy category of mineral resources that result from the complex geological interaction of magmatic, hydrothermal, and metasomatic processes [1]. The Biga Peninsula in northwestern Turkey is particularly rich in such deposits, with Fe skarns and Cu skarns being prominent [2–5]. Fe skarns are often associated with the regional Oligocene to Lower Miocene granitoids and often occur with porphyry copper or epithermal gold deposits [3,6–9]. These skarns have a high iron content and may also display copper enrichment, especially in areas where magnetite undergoes alteration to hematite. In addition, Cu skarns have often been identified by the presence of garnet–pyroxene and wollastonite–garnet skarns in the endoskarn zones and wollastonite–garnet skarns in the exoskarn zones [3]. The formation of these deposits has been closely associated with intrusions, such as granodiorite, monzodiorite, and quartz diorite, that interact with their surrounding sedimentary and volcanosedimentary rocks [3]. The observed variation in mineral composition, which ranges from endoskarns within the intrusive bodies to exoskarns in the surrounding country rocks, is a direct consequence of the differences in fluid composition, pressure, and temperature that occurred during their formation [1,3,10].

The Yolindi area is situated in the northeastern part of the Biga Peninsula, which is located in the western sector of the Sakarya Zone in northwest Turkey (Figure 1). The tectonic activity of the Biga Peninsula is mainly influenced by the movement taking place along the North Anatolian Fault Zone [11]. This fault zone splits into two main branches, known as the northern and southern branches, which are situated in the eastern region of Marmara. The main tectono-stratigraphic formations in the area of the Biga Peninsula consist of the Denizgoren ophiolite and the Çetmi ophiolitic mélange, located in the southwestern and northern parts of the Biga Peninsula, respectively [2,12]. During the Mesozoic period, the peninsula saw significant tectonic activities, such as sedimentation, the emplacement of ophiolites, and the formation of volcanic island arcs. The process of sedimentation played a crucial role in shaping the geological features of the region throughout the Triassic period [12]. It included the accumulation of carbonate rocks, namely limestones and dolomites, in shallow marine environments. During the transition from the Mesozoic era to the Jurassic and Cretaceous periods, there was a spreading of ophiolite complexes, flysch, and mélange deposits due to tectonic activity. In addition to the metamorphic basement rocks, the Biga Peninsula encompasses a comprehensive geological history of magmatic and volcanic activity that extends from the pre-Cenozoic to the Cenozoic periods [12]. The emplacement of granitic intrusions and volcanic rocks occurred in the Biga Peninsula during the Late Paleozoic period as a result of the closing of the Paleo-Tethys Ocean and the evolution of the supercontinent Pangea [13]. The volcanic rocks, together with their associated mineral deposits, provide crucial insights into the tectonic and magmatic processes that occurred in the area throughout the Late Paleozoic period. In the Cenozoic era, there were several intrusions that occurred from the Eocene to the Oligo-Miocene periods. These intrusions include Dikmen, Evciler, Eybek, Kestanbol, Kuşçayir, Şaroluk, and Yenice [7,8,14–19]. They were mostly found in the middle and southern areas of the peninsula. Meanwhile, the Ayvacik volcanic rocks, Ezine basalt, and Taştepe basalt represent the Cenozoic volcanic rocks. These rocks exhibit calc-alkaline to alkaline affinities, suggesting a varied history of magmatic activity [15,20,21]. The Cenozoic plutons and volcanic successions of the Biga Peninsula have both geological and economic importance.

They are responsible for the bulk of the metallic mineral resources and industrial materials that have contributed to this region's economy [2,22].

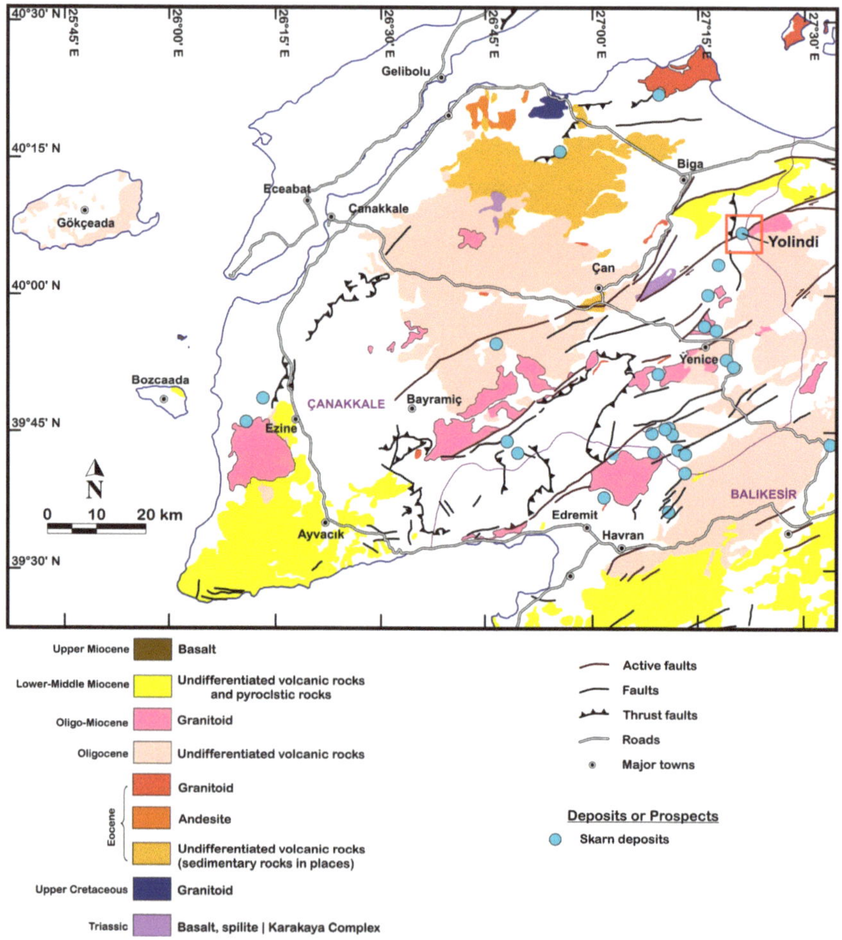

Figure 1. The spatial distribution of skarn deposits on the Biga Peninsula, located in northwestern Turkey, modified according to Yigit [2], Kuşcu [3], and Kaya et al. [4]. The red box refers to Figure 2.

This study utilized a multidisciplinary method, integrating microthermometry of fluid inclusions and several isotopic data types, to investigate the physiochemical conditions and fluid evolution within the Yolindi Cu-Fe skarn mineralization. In sulfide minerals, sulfur isotopes ($\delta^{34}S$) can be used to identify sulfur sources, whereas carbon ($\delta^{13}C$) and oxygen isotopes ($\delta^{18}O$) in calcite minerals provide information regarding the composition and temperature of fluid–rock interactions. Moreover, the fluid inclusions within various minerals formed at distinct phases of skarn mineralization provide a direct record of the physical and chemical conditions, allowing to figure out how pressure, temperature, and compositional evolution have changed throughout the skarn formation.

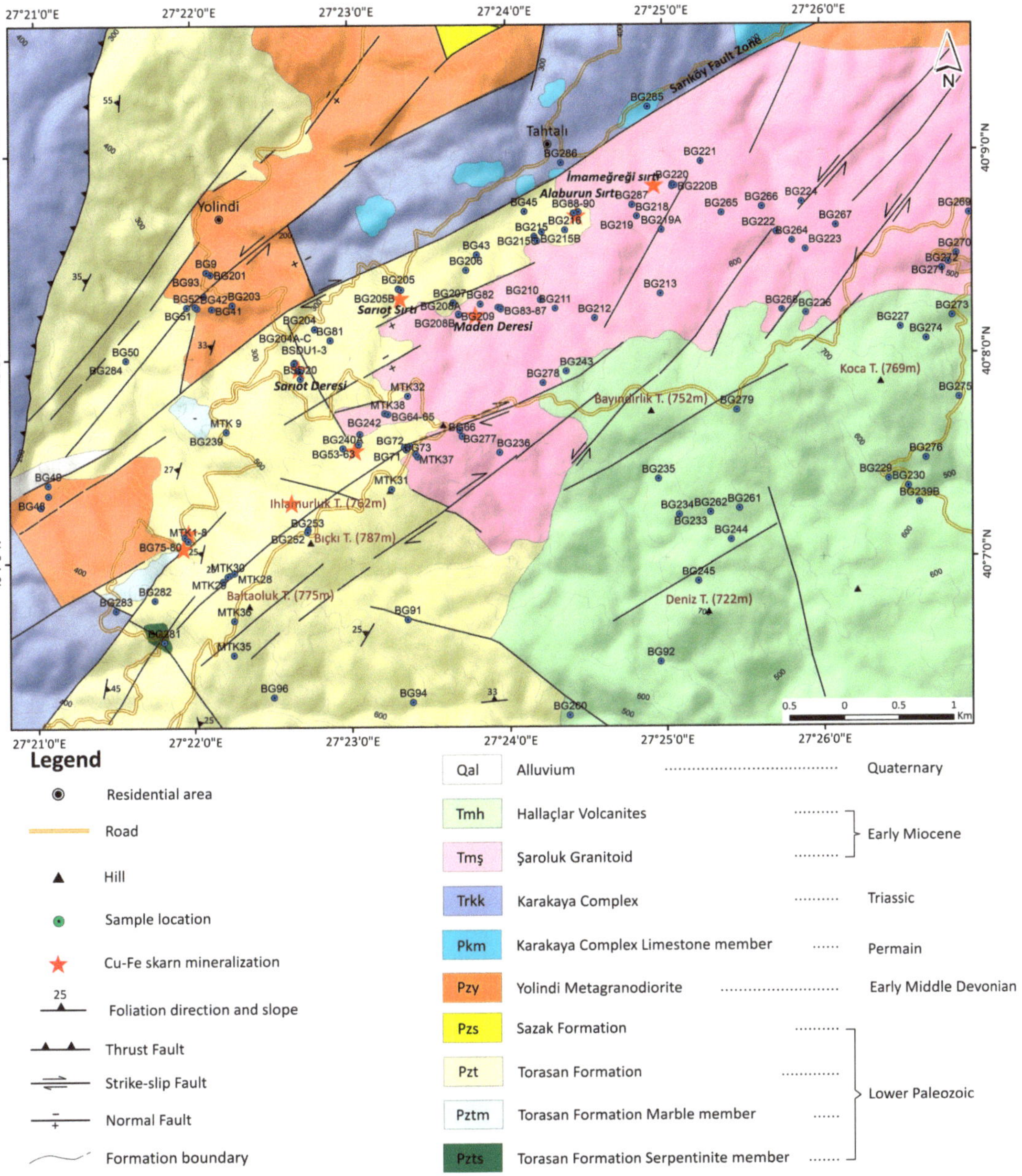

Figure 2. Geologic map of the Yolindi area (modified according to Duru et al. [23], Aysal [24], Aysal et al. [25], and Kaya et al. [4]).

2. Geology of the Yolindi Area

The Yolindi area is located NE of the Biga Peninsula in Çanakkale province, NW Turkey (Figure 1). It comprises the Paleozoic Kalabak Group and Triassic Karakaya Complex, which were subsequently intruded by the Early Miocene Şaroluk granitoids and Hallaçlar volcanics (Figure 2).

The Kalabak Group forms the basement of the Sakarya Zone, which is a distinct tectonic unit that includes the Torasan formation at the base and the Sazak formation at the top, later intruded by Yolinidi metagranodiorite (Figure 2). The Lower Paleozoic Torasan formation (Pzt) is located in the western study area, in close proximity to Yolindi village, Bıçkı, and Baltaoluk hill (Figure 2). Phyllites, muscovite, biotite, talc, tremolite, and chlorite schists, along with their associated tuffs, constitute the majority of this formation, along with actinolite hornfels, metasandstone, and metasiltstone. Phyllites contain serpentinized peridotites (Pzts) (Figure 3a) and marble members (Pztm) as their blocks and lenses, respectively. Phyllites have a distinct foliation consisting of muscovite, biotite, and quartz, together with opaque minerals (Figure 3b). Actinolite hornfels are composed of actinolite, quartz, clinozoisite, clinopyroxene, and opaque minerals (Figure 3c). The composition of serpentinite lenses primarily consists of antigorite, chlorite, and clinopyroxene, along with a small amount of opaque minerals (Figure 3d). Marble and recrystallized limestone are characterized by a fine-grained texture and are primarily composed of calcite (Figure 3e). The Yolindi metagranitoids (Pzy) have been determined to be of early Middle Devonian ages using U-Pb zircon SHRIMP and LA-ICP-MS dating techniques, as reported by Aysal et al. [25]. The study area in Yolindi village has prominent outcrops to the north (Figures 2 and 3f). These rocks include biotite gneiss and metagranodiorite. Biotite gneiss consists of biotite, quartz, sericitized plagioclase, and opaque minerals (Figure 3g). Metagranodiorite is composed of plagioclase that has undergone sericitization, together with quartz, biotite, and opaque minerals (Figure 3h). The Triassic Karakaya Complex (Trkk) is located in the middle and western parts of the study area, close to the villages of Inova and Tahtali (Figure 2). It is made up of marble, metasandstone, phyllite, metavolcanics, and serpentinites [4].

The Şaroluk granitoid (TmŞ) is a 20 km^2 elliptical pluton that is oriented in an east–west direction (Figure 2). The U-Pb zircon dating conducted by Aslan et al. [26] revealed that the Şaroluk pluton originated from magmatic activity during the Early Miocene period, between 22.18 $\pm$ 0.40 and 21.51 $\pm$ 0.37 Ma. Aysal [27] reported an accurate date of 23.97 $\pm$ 0.53 Ma for the same event. A hornfels zone evolved in the Torasan formation along the western boundary of the Şaroluk intrusion (Figure 3i). The Şaroluk pluton mostly consists of quartz monzonite, with some granodiorite. Quartz monzonite consists of biotite, tremolite, actinolite, K-feldspar, quartz, and biotite with tremolite (Figure 3j). Granodiorite is composed of plagioclase (50%–70%), quartz (20%–30%), microperthite (5%–15%), hornblende, biotite, small quantities of pyroxene, and opaque minerals (Figure 3k).

The Hallaçlar formation (Tmh) is found in the southwest of the study area, around Beyoluk village, Bayındırlık Hill, Koca Hill, and Deniz Hill (Figure 2). It was dated at 23.6 Ma by Krushensky [28], 26.5 $\pm$ 1.1 Ma by Dönmez et al. [29], and 22.6 $\pm$ 0.8 Ma by Karacik et al. [30], indicating volcanic activity from the Late Oligocene period to the Early Miocene period. Rhyodacite, felsic tuffs, tuffaceous siltstone, and pyroclastics are included. Porphyritic andesite mainly comprises hornblende, biotite, and plagioclase phenocrysts embedded in fine-grained microcrystalline groundmass of plagioclase laths, quartz, and opaque minerals (Figure 3l).

Figure 3. (**a**) Serpentinized peridotites within the Torasan formation. (**b**) Phyllite exhibits well-developed muscovite, biotite, and quartz foliation with opaque minerals. (**c**) Actinolite and quartz in actinolite hornfels with clinozoisite, clinopyroxene, and opaque minerals. (**d**) Serpentinites are made up of antigorite, chlorite, clinopyroxene remnants, and other less frequent opaque minerals. (**e**) The major component of marble is calcite. (**f**) General view of the Yolindi metagranitoids. (**g**) Biotite gneiss contains biotite in addition to quartz, sericitized plagioclase, and other opaque minerals. (**h**) Sericitized plagioclase, quartz, biotite, and opaque minerals are found in metagranodiorite. (**i**) At the border of the Şaroluk granitoids, a hornfels zone formed inside the Torasan formation. (**j**) The majority of quartz monzonite is composed of plagioclase, K-feldspar, quartz, biotite, actinolite, and tremolite. (**k**) Granodiorite contains plagioclase, quartz, and microperthite, as well as hornblende, biotite, and other opaque minerals. (**l**) Porphyritic andesite with hornblende, actinolite, biotite, and plagioclase phenocrysts embedded in fine-grain microcrystalline plagioclase laths, biotite, hornblende, quartz, and opaque minerals. Abbreviations: actinolite (act), antigorite (atg), biotite (bt), calcite (cal), chlorite (chl), clinopyroxene (cpx), clinozoisite (czo), hornblende (hb), microperthite (mp), muscovite (ms), orthoclase (or), opaque mineral (opq), plagioclase (pl), quartz (qz), and sericite (ser).

3. Deposit Geology of the Yolindi Cu-Fe Skarn Mineralization

The Yolindi Cu-Fe (±Zn ± Pb) skarn mineralization exhibits zonal alteration patterns that can occur either in a concentric manner or perpendicular to the contact between the Şaroluk granitoid intrusion and the Torasan formation, as well as carbonate-rich rock or limestone. The zones are categorized as endoskarns and exoskarns, which can be further subdivided into three zones based on their distance from the intrusive body: proximal, intermediate, and distal zones (Figure 4). These zones exhibit distinct mineral assemblages (Figure 4).

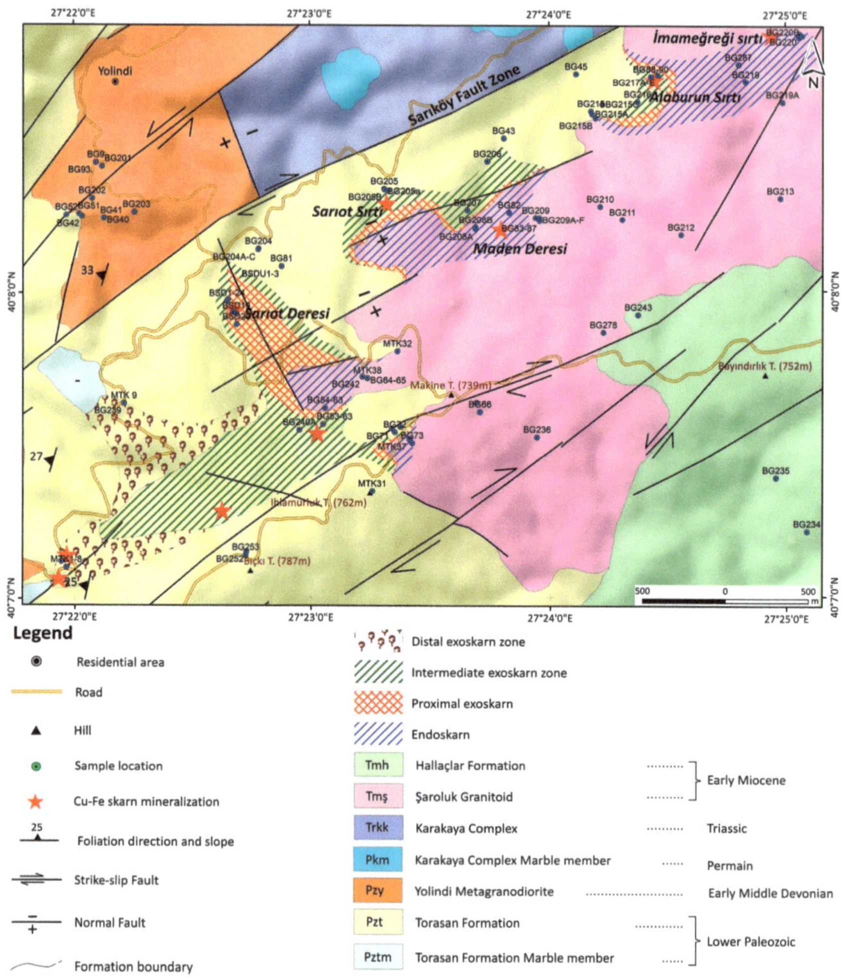

Figure 4. The zonation map of the Yolindi Cu-Fe Skarn mineralization illustrates several skarn zones, including the endoskarn and the proximal, intermediate, and distal exoskarn zones [4].

The Şaroluk quartz monzonite has undergone bleaching along the boundaries between the Torasan formation and the Şaroluk intrusion. This is evident in the field through the presence of the endoskarn zone (Figures 4 and 5a). The formation of endoskarn occurs within the Şaroluk intrusive body itself, usually in its outer parts. It has been specifically noticed near the Maden Deresi and İmameğreği Sırtı areas (Figure 4). On a mesoscopic scale, the occurrences of skarnized granitoids (which include andradite) can refer to the garnet formation, followed by epidote and sericite, during metasomatism. In the Maden Deresi area, there was a transition from granodiorite containing magnetite that has undergone a few alterations to endoskarn containing magnetite, with the degree of alterations rising towards the contact region (Figure 5a). The magnetite-bearing granodiorites comprise quartz, plagioclase, microperthite, biotite, hornblende, pyroxene, and sphene, along with magnetite (Figure 5b). The magnetite-bearing endoskarn is characterized by the presence of andradite and epidote minerals, together with magnetite and goethite (Figure 5c). The İmameğreği Sırtı area is recognized by the presence of magnetite- and pyrite-bearing granodiorite with quartz, plagioclase, microperthite, biotite, diopside, actinolite, and sphene, along with magnetite, pyrite, and hematite (Figure 5d).

Figure 5. (**a**) General characteristics of the endoskarn zone within the Şaroluk intrusion in Maden Deresi. (**b**) Magnetite-bearing granodiorites. (**c**) Magnetite-bearing endoskarns with andradite and

epidote. (**d**) Magnetite- and pyrite-bearing granodiorite containing quartz, plagioclase, microperthite, biotite, and actinolite as opaques, as well as magnetite and pyrite. (**e**,**f**) Garnet- and magnetite-bearing prograde proximal zones within the Torasan formation. (**g**) In the proximal zone, andradite, calcite, and quartz occur. (**h**) The retrograde proximal zone's greenish tint, which has been cut with a quartz carbonate vein. (**i**) In the retrograde proximal zone, andradite–grossular intergrowth, calcite, and quartz occur. (**j**) Green and brown color alteration of the intermediate exoskarn zone. (**k**,**l**) In the intermediate exoskarn zone, garnet (andradite–grossular), pyroxenes, scapolite, and calcite coexist with opaque minerals. (**m**–**o**) Sericite and carbonate alteration of the distal exoskarn zone. Abbreviations: actinolite (act), andradite (adr), biotite (bt), calcite (cal), chlorite (chl), clinopyroxene (cpx), dolomite (dol), epidote (ep), feldspar (fls), grossular (grs), hematite (hem), hornblende (hb), magnetite (mag), microperthite (mp), orthopyroxene (opx), plagioclase (pl), pyroxene (px), pyrite (py), quartz (qz), scapolite (scp), and sericite (ser).

The exoskarns located around the Şaroluk granitoid body in the Torasan formation usually have magnetite, pyrite, epidote, scapolite, garnet, pyroxene, chalcopyrite, bornite, hematite, and sometimes sphalerite and galena. Meinert [31] classified these zonal patterns into three subzones: proximal, intermediate, and distant zones.

The earliest phase of the skarn system is the proximal prograde exoskarn zone, characterized by a high concentration of garnet and magnetite in the Sarıot Deresi area (Figure 4). It has been distinguished by zones of dark garnet (zoned andradite) and magnetite zones with subordinate hematite (martitization) throughout the hornfels rocks of the Torasan formation (Figure 5e–g). The proximal retrograde exoskarn zone located in the northwest of the Sarıot Deresi area is distinguished by its greenish color and having retrograde minerals like epidote and actinolite, along with clinopyroxene relics and andradite–grossular intergrowth that are intersected by quartz–carbonate veins (Figure 5h,i).

The intermediate exoskarn zone underwent development in the Sarıot Sırtı, Ihlamurluk Tepe, and Alaburun Sırtı areas within the Torasan formation (Figure 5j). This area is rich in garnet, pyroxene, calcite, and scapolite. This zone exhibits distinct characteristics that indicate a gradual transition from the hot, magma-influenced area located closer to the source to the cooler, more distant area with retrograde processes. Minerals such as epidote, scapolite, and chlorite are frequently formed (Figure 5k,l), suggesting a shift from high-temperature minerals to lower-temperature and retrograde minerals, which is caused by changes in temperature and fluid composition during the evolution of skarn, including chalcopyrite, pyrite, and specular hematite.

In addition, the distal exoskarn zone is located farthest from the Şaroluk granitoid pluton inside the Torasan formation in the Bıçkı Tepe area (Figure 4). It encompasses distinct mineral compositions, including epidote, chlorite, calcite, dolomite, and sericite (Figure 5m–o), along with galena, sphalerite, and a lesser amount of chalcopyrite with cerussite, covellite, and malachite.

4. Sampling and Analytical Methods

A comprehensive fieldwork project was conducted in the Yolindi area to gain a thorough comprehension of the geological setting. In order to ensure that the samples were representative, they were systematically collected from both the main host rocks and the altered rocks that exhibited mineralization. A total of 121 thin sections and 88 polished sections were meticulously prepared. These sections underwent a petrographic examination using a polarizing microscope to ascertain the mineralogical and textural relationships within the rocks.

Thirty-four mineralized samples were carefully selected from various skarn zones with the specific aim of evaluating sulfur isotope data on sulfide minerals. A total of six pyrite samples were obtained from the endoskarn zone, consisting of three pyrite and three chalcopyrite samples from the prograde proximal exoskarn zone, ten pyrite and three chalcopyrite samples from the retrograde proximal exoskarn zone, and seven pyrite and two chalcopyrite specimens from the intermediate exoskarn zone. The initial preparation of these selected samples comprised the procedure of grinding and sieving to obtain a consistent mesh size of 2 mm, ensuring uniformity in the granularity of these samples. Individual mineral grains were carefully handpicked under a binocular microscope to guarantee the purity of the sample set for isotopic analysis. The total sulfur content and sulfur isotopic composition ($\delta^{34}S/\delta^{32}S$ ratios) of the samples were determined using elemental analysis coupled with isotope ratio mass spectrometry (EA/IRMS) at the Iso-Analytical Laboratory in Sandbach, UK. The isotope data were expressed as $\delta^{34}S$ values that were standardized to the Vienna Cañon Diablo Troilite (VCDT) standard. The addition of vanadium pentoxide (V_2O_5) to the samples was performed in order to enhance the combustion characteristics. The International Atomic Energy Agency (IAEA) distributes NBS-127, IAEA-S-1, and IAEA-SO-5 as inter-laboratory comparison standards, which have globally recognized $\delta^{34}S$ values. For cross-validation, the NBS-127, IAEA-S-1, and IAEA-SO-5 were used to accomplish analytical calibration. The measurement precision of the δ34S results was verified to be within ±0.3‰ for all standards, with six duplicates each, thereby confirming correctness.

Isotopic compositions of carbon ($\delta^{13}C$) and oxygen ($\delta^{18}O$) in calcite minerals were determined from a diverse suite of samples, including marble (n = 5), retrograde skarn (n = 9), and prograde skarn (n = 11). Each carbonate sample, precisely weighed between 0.2 and 0.6 mg using a sensitive balance, underwent a reaction with approximately 0.1 mL of 99% orthophosphoric acid (MERCK) at a constant temperature of 25 °C on an autosampler table. The resultant CO_2 gas, after being liberated from the carbonate samples, was processed using the gas bench continuous flow isotope ratio mass spectrometry technique. The isotopic ratios were analyzed utilizing a Thermo Finnigan Deltaplus XP isotope ratio mass spectrometer (IRMS) (Thermo Fisher Scientific, Waltham, MA, USA), with the isotopic ion ratios converted into raw isotope data using ISODAT 2.0 software. The international standard NBS19 Limestone (NIST), with known $\delta^{13}C$ of 1.95‰ and $\delta^{18}O$ of −2.20‰, served as the primary reference for calibrating the crude isotope measurements to true values, anchored to the VPDB (Vienna Pee Dee Belemnite) standard. Measurements were conducted with a precision of ±0.2‰ for both isotopes, reflecting high reproducibility and accuracy. These analyses were carried out at the METU Central Laboratory within the Stable Isotope Ratio Mass Spectroscopy Laboratory (DIL), ensuring robustness in the methodological approach and data integrity.

The fluid inclusion investigations were conducted in the Istanbul Technical University (ITU) fluid inclusions laboratory on 80-μm-thick, double-polished, thin sections of andradite, grossular, quartz, and calcite minerals. The measurements were acquired using a Linkam THMG-600 stage, which had the ability to both heat and freeze and was mounted on a Leica DM2500P optical microscope (Leica, Wetzlar, Germany). The microscope was outfitted with a video camera and a display. The temperature range for the measurements was from +600 to −200 °C. The calibration of this step was conducted with potassium dichromate, pure H_2O, CO_2, and H_2O–NaCl standards. The microthermometric measurements for heating and freezing were performed with an accuracy of 0.4 °C for the freezing stage and 5.0 °C for the heating stage, using the standard approaches outlined by Roedder [32] and Shepherd et al. [33]. This study evaluated and recorded the homogenization temperature (T_h), eutectic temperature (T_e), and final melting temperature of ice ($T_{m\text{-}ice}$).

5. Results

5.1. Ore Mineralogy and Paragenesis

The ore mineralogy includes magnetite, pyrite, chalcopyrite, and specular hematite, with a subordinate amount of galena, sphalerite, cerussite, malachite, and goethite (Figure 6). Magnetite mostly occurred within the endoskarn in the magnetite-bearing granodiorite (Figure 6a) and magnetite-bearing endoskarn (Figure 6b), as well as throughout the proximal prograde exoskarn zone (Figure 6c), in association with pyrite, andradite, and augite minerals. It is distinguished by brownish crystals, which vary in shape from subhedral to anhedral and may reach diameters of up to 2 mm. Occasionally, it undergoes martitization, a process in which magnetite is substituted with hematite as a result of oxidation (Figure 6c). In the study area, chalcopyrite was the predominant copper–iron sulfide mineral. It occurs in the proximal retrograde exoskarn zone in the northwest of Sarıot Deresi, as well as in the intermediate zone. Subhedral crystals of a yellowish color are formed together with pyrite, hematite, and sphalerite (Figure 6d–f). Pyrite is made up of subhedral to euhedral light-yellow crystals that are 10 to 200 microns across in length. It frequently forms with magnetite in the endoskarn (Figure 6b), as well as in the proximal and intermediate exoskarn zones associated with chalcopyrite, hematite, and bornite (Figure 6d,e,g). Specular hematite is a kind of hematite that can be identified via its clusters of shiny, metallic, flat flakes or tabular crystals (Figure 6h). It is formed within the intermediate zone of the exoskarn, associated with chalcopyrite and pyrite (Figure 6e,f). Galena is distinguished by the occurrence of sizable, separate subhedral light gray crystals in the distal zone, which is encountered with sphalerite and undergoes an alteration to form cerussite (Figure 6j,k). In addition, the distal zone undergoes the alterations of galena, sphalerite, and chalcopyrite, leading to the formation of cerussite, covellite, and malachite, respectively (Figure 6j–l).

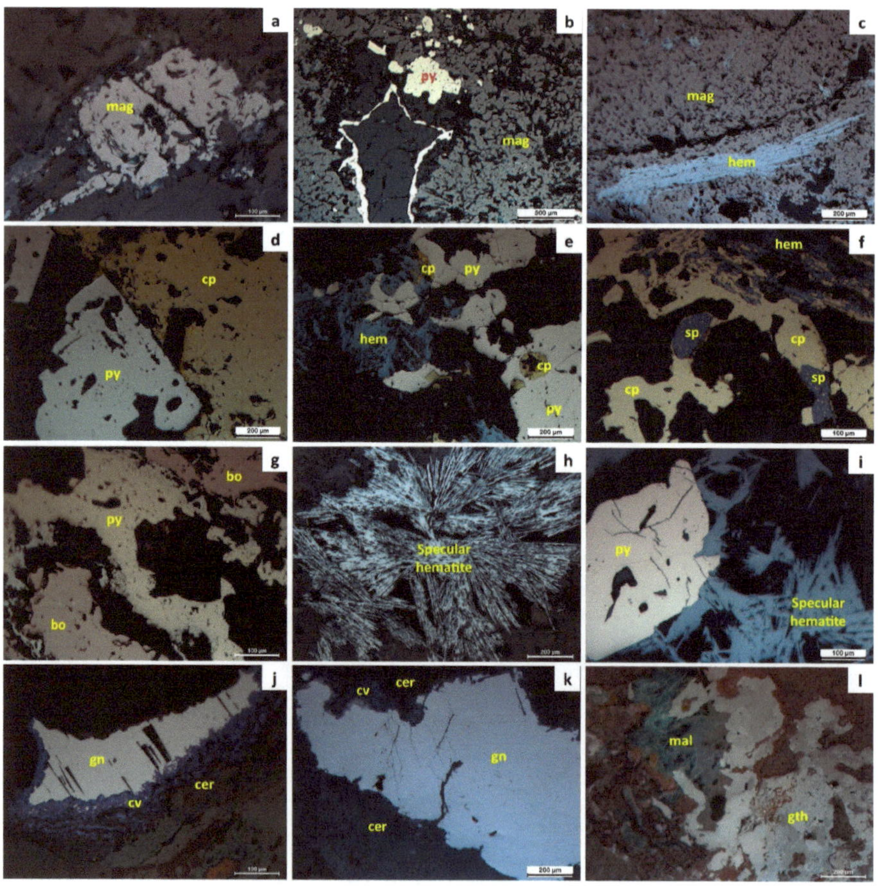

Figure 6. (**a**) Magnetite occurred within the granodiorite in Maden Deresi. (**b**) Magnetite and pyrite in magnetite-containing endoskarn. (**c**) Magnetite's martitization to hematite in the proximal exoskarn zone. (**d**) Chalcopyrite and pyrite in the proximal retrograde zone. (**e**) In the retrograde proximal

zone, there is chalcopyrite, pyrite, and specular hematite. (**f**) Chalcopyrite occurs, along with sphalerite and specular hematite. (**g**) Pyrite with bornite in the proximal retrograde zone. (**h**) Clusters of specular hematite flakes that are shiny, metallic, and flat. (**i**) Specular hematite is seen in association with pyrite. (**j**,**k**) Cerussite and covellite are the alteration products of galena in the distal zone. (**l**) Malachite and goethite are the alteration products of chalcopyrite and pyrite, respectively, in the distal zone. Abbreviations: bornite (bo), cerussite (cer), chalcopyrite (cp), covellite (cv), galena (gn), goethite (gth), hematite (hem), magnetite (mag), malachite (mal), pyrite (py), and sphalerite (sp).

Paragenesis studies on the Yolindi Cu-Fe skarn mineralization process revealed that there were three distinct mineralization stages that were identified based on the contact relationships between the metamorphic, alteration, and ore minerals, as shown in Figure 7. The prograde stage was observed in the Yolindi area, which was distinguished by the formation of magnetite, pyrite, and medium-to-coarse-grained Fe-rich anhydrous calc-silicate minerals (e.g., andradite and augite). These minerals occurred in the endoskarn and proximal exoskarn zones. Magnetite was generated concurrently with pyrite during this phase (Figure 7). The occurrence of minerals, such as actinolite, epidote, scapolite, calcite, and chlorite pseudomorph, after garnet and pyroxene suggested that a retrograde stage occurred at lower temperatures and pressures. In addition, at this particular stage, the andradite–grossular intergrowth is associated with the hedenbergite. The aforementioned minerals are often formed in association with ore minerals, such as chalcopyrite, pyrite, specular hematite, galena, and sphalerite (Figure 7). Cerussite and malachite are characteristics of the supergene stage (Figure 7). This showed the process of metals moving and leaching from nearby areas, where primary metal sulfides break down and then turn into carbonates when the pH and redox conditions change.

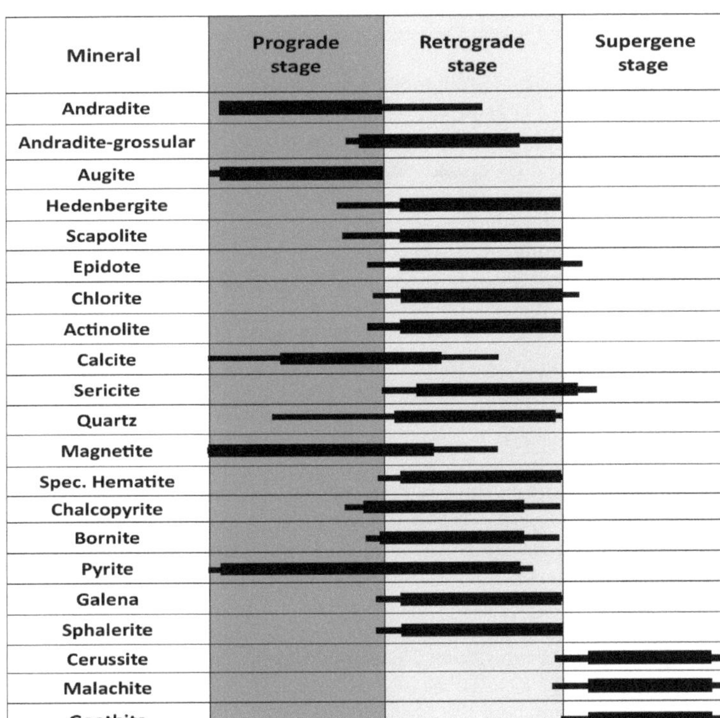

Figure 7. A paragenetic sequence of metamorphic, altered, and ore minerals in the Yolindi Fe-Cu skarn mineralization.

5.2. Fluid Inclusion Studies

5.2.1. Fluid Inclusion Petrography

The Yolindi Cu-Fe skarn mineralization exhibits a substantial prevalence of fluid inclusions, mostly found inside andradite, grossular, quartz, and calcite crystals. The FIs may be categorized into two phases of inclusion, namely liquid H_2O (L) and vapor H_2O (V), with one type being rich in liquid and the other type being rich in vapor, in accordance with the classifications of Roedder [32] and Van den Kerkhof and Hein [34]. Multiple instances have led to the identification of the primary and secondary fluid inclusions in these samples. On the other hand, the primary fluid inclusions exhibit a sub-rounded, elliptical, elongated, and irregular shape, and have a maximum size of 350 μm (Figure 8). The andradite minerals that occurred in the prograde proximal zone have FIs exhibiting sub-rounded, elliptical, and irregular forms with diameters ranging from 40 to 108 μm (Figure 8a,b). At room temperature, the vapor bubble typically fills around 15%–20% of the inclusion volume, as shown in Figure 8a,b. This mineral has vapor-rich inclusions that are presently sub-rounded and irregular in shape. These inclusions have diameters ranging from 65 to 102 μm and constitute 55%–65% of the volume of the vapor bubble at room temperature (Figure 8c). In contrast, the grossular, quartz, and calcite minerals generated during the retrograde proximal stage include liquid-rich L+V FIs. In grossular, quartz, and calcite minerals, these FIs have prismatic and irregular forms, with their diameters ranging from 16 to 154 μm, from 31 to 357 μm, and from 16 to 116 μm, respectively (Figure 8d–i). Furthermore, the vapor bubble accounts for 20%–25%, 9%–22%, and 10%–20% of the volume of FIs in grossular, quartz, and calcite FIs, respectively (Figure 8d–i). The quartz crystals in the intermediate retrograde zone include FIs with diameters ranging from 3 to 110 μm in diameter. These FIs are of the liquid-rich L+V type, as shown in Figure 8j. However, their morphologies exhibit irregularities, varying in diameter from 23 to 290 μm in the quartz minerals of the distal zone (Figure 8k,l).

5.2.2. Microthermometry Results

Extensive analysis was performed on a total of 119 fluid inclusions that occurred in andradite, grossular, quartz, and calcite minerals formed throughout the prograde and retrograde phases of skarn mineralization in the proximal, intermediate, and distal exoskarn zones. The main objective of this analysis was to determine the temperatures at which homogenization and freezing took place for each inclusion (Table 1). The eutectic temperatures (T_e) of liquid-rich and vapor-rich (L+V) fluid inclusions varied across various minerals (Table 1). The primary fluid inclusions of andradite, grossular, quartz, and calcite minerals from the proximal zone experienced T_e values ranging from −37.80 to −12.10 °C, from −8.80 to −1.60 °C, from −26.70 to −0.80 °C, and from −17.80 to −1.10 °C, respectively (Table 1). The T_e of the fluid inclusions (FIs) that were detected in quartz minerals from the intermediate and distal zones varied between −26.60 and −1.70 °C and between −8.60 and −1.90 °C, respectively (Table 1). These findings indicated the presence of a H_2O–NaCl system. In contrast, the final melting temperature (T_m) of the primary fluid inclusions in andradite, grossular, quartz, and calcite minerals from the proximal zone varied between −25.80 and −8.50 °C, between −5.30 and −0.40 °C, between −13.50 and −0.50 °C, and between −12.30 and −0.20 °C, respectively (Table 1). The T_m values of the FIs found in quartz minerals from the intermediate and distal zones ranged from −17.60 to −0.40 °C and from −3.40 to −0.90 °C, respectively (Table 1). The homogenization temperature (T_h) values of the primary FIs ranged from 359.00 to 412.20 °C (an average of 390.15 °C) for andradite minerals, from 280.10 to 372.60 °C (an average of 348.13 °C) for grossular minerals, from 256.00 to 394.30 °C (an average of 304.89 °C) for quartz minerals, and from 197.70 to 317.80 °C (an average of 280.45 °C) for calcite minerals (Table 1). The temperature range for the T_h of the FIs observed in quartz minerals from the intermediate and distal zones was between 299.80 and 370.00 °C, with an average temperature of 337.95 °C, and between 296.10 and 336.50 °C, with an average temperature of 319.47 °C, respectively (Table 1).

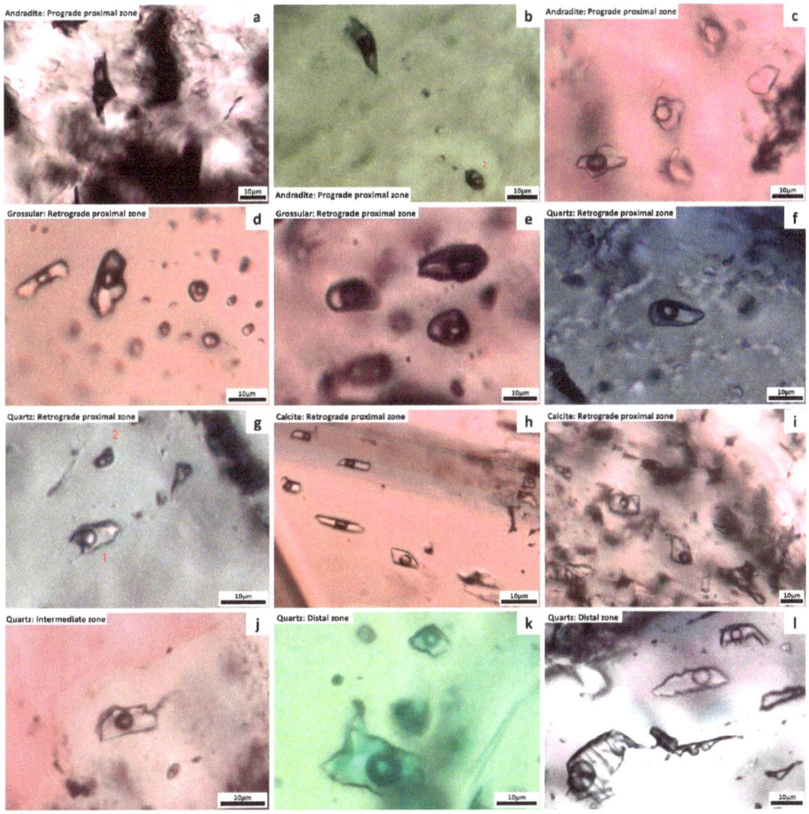

Figure 8. (**a**,**b**) Liquid-rich primary (liquid–vapor) fluid inclusions in andradite minerals from the prograde proximal exoskarn zone. (**c**) Vapor-rich primary (L+V) fluid inclusions with liquid-rich ones in andradite minerals from the prograde proximal exoskarn zone. (**d**,**e**) Liquid-rich primary (L+V) fluid inclusions in grossular minerals from the retrograde proximal exoskarn zone. (**f**,**g**) Liquid-rich primary (L+V) fluid inclusions in quartz minerals from the retrograde proximal exoskarn zone. (**h**,**i**) Liquid-rich primary (L+V) fluid inclusions in calcite minerals from the retrograde proximal exoskarn zone. (**j**) Liquid-rich primary (L+V) fluid inclusions in quartz minerals from the intermediate exoskarn zone. (**k**,**l**) Liquid-rich primary (L+V) fluid inclusions in quartz minerals from the distal exoskarn zone.

The salinities were calculated using the final ice melting temperatures ($T_{m\text{-}ice}$), as suggested by Darling [35], Archer [36], Bodnar [37], and Steele-MacInnis et al. [38], along with the utilization of the BULK computer software developed by Bakker [39] (Table 1). The salinities of the primary L+V FIs in the andradite minerals from the prograde proximal exoskarn zone varied from 12.28% to 26.07 wt.% NaCl equivalent, with an average of 20.27 wt.% NaCl equivalent (Table 1). In contrast, the grossular, quartz, and calcite minerals occurring in the retrograde proximal exoskarn zone had lower salinities in their fluid inclusions. The salinity of FIs in grossular minerals varied from 0.70 to 8.28 wt.% NaCl equivalent; in quartz, it ranged from 0.88 to 17.34 wt.% NaCl equivalent, and in calcite, it ranged from 0.35 to 16.24 wt.% NaCl equivalent (Table 1). The salinity range for FIs in quartz minerals was lower in the intermediate and distal zones compared to the proximal zones. It varied between 0.70 and 20.67 wt.% NaCl equivalent, with an average of 9.35 wt.% NaCl equivalent in the intermediate zone, and between 0.35 and 16.24 wt.% NaCl equivalent, with an average of 3.63 wt.% NaCl equivalent in the distal zone (Table 1). The total densities of FIs were computed using the equation of state proposed by Zhang and Frantz [40]. The primary FIs in the andradite crystals from the prograde proximal zone have a density ranging from 0.69 to 0.91 g/cm^3, with an average of 0.81 g/cm^3. The densities of FIs in the grossular, quartz, and calcite minerals from the retrograde proximal zone range from 0.58 to 0.77 g/cm^3 (with an average of 0.64 g/cm^3), from 0.56 to 0.88 g/cm^3 (with an average of 0.75 g/cm^3), and from 0.71 to 0.89 g/cm^3 (with an average of 0.78 g/cm^3), respectively (Table 1). The overall densities of FIs in the quartz minerals from the intermediate and distal zones range from 0.65 to 0.86 g/cm^3 and from 0.65 to 0.77 g/cm^3, respectively (Table 1).

Table 1. Summary of microthermometric data and their calculations of andradite, grossular, quartz, and calcite minerals from the different skarn zones from the study area.

Mineral		**Andradite**	**Grossular**		**Quartz**		**Calcite**
Stage		Prograde	Retrograde	Retrograde	Retrograde	Retrograde	Retrograde
Zone		Proximal	Proximal	Proximal	Intermediate	Distal	Proximal
Inclusion type (primary)		L–V	L–V	L–V	L–V	L–V	L–V
Eutectic temperature (°C)	Interval	From −37.8 to −12.1	From −8.8 to −1.6	From −26.7 to −0.8	From −26.6 to −1.7	From −8.6 to −1.9	From −17.8 to −1.1
	n	10	13	17	20	12	24
	Avg	−24.1	−4.9	−7.0	−11.9	−4.7	−6.3
Tm-ice (°C)	Interval	From −25.8 to −8.5	From −5.3 to −0.4	From −13.5 to −0.5	From −17.6 to −0.4	From −3.4 to −0.9	From −12.3 to −0.2
	n	10	13	17	20	12	24
	Avg	−17.6	−1.8	−3.7	−6.7	−1.8	−2.3
Th-tot (°C)	Interval	From 359.0 to 412.2	From 280.1 to 372.6	From 256.0 to 394.3	From 299.8 to 370	From 296.1 to 336.5	From 197.7 to 317.8
	n	10	13	17	20	12	24
	Avg	390.2	348.1	304.9	337.9	319.5	280.4
Salinity %NaCl eq.	Interval	From 12.3 to 26.1	From 0.7 to 8.3	From 0.9 to 17.3	From 0.7 to 20.7	From 1.6 to 5.6	From 0.4 to 16.2
	n	10	13	17	20	12	24
	Avg	26.1	2.9	5.7	9.4	3.1	3.6
Density (g/cm^{-3})	Interval	From 0.69 to 0.91	From 0.58 to 0.77	From 0.56 to 0.88	From 0.65 to 0.86	From 0.65 to 0.77	From 0.71 to 0.89
	n	10	13	17	20	12	24
	Avg	0.81	0.64	0.75	0.75	0.70	0.78
P_h (bar)	Interval	From 145.3 to 261.4	From 63.1 to 211.8	From 43.1 to 255.4	From 82.4 to 185.7	From 80.4 to 138.4	From 14.7 to 107.2
	n	10	13	17	20	12	24
	Avg	217.8	163.2	100.1	135.6	111.3	67.1

5.3. Stable Isotope Data

5.3.1. $\delta^{34}S$ Isotope of Sulfides

Stable isotope analysis was performed on 34 samples containing different sulfide minerals in order to ascertain the $\delta^{34}S$ isotopic composition (Table 2). The samples comprise six pyrite samples from the endoskarn zone, three pyrite samples and three chalcopyrite samples from the prograde proximal exoskarn zone, ten pyrite samples and three chalcopyrite samples from the retrograde proximal exoskarn zone, and seven pyrite samples and two chalcopyrite samples from the intermediate exoskarn zone (Table 2). All of these samples originated from the prograde and retrograde stages of mineralization, except for several pyrite minerals that formed alongside the Şaroluk granitic rock (see Figure 7). The $\delta^{34}S$ data of the two pyrite varieties from the endoskarn zone ranged from +0.27 to +0.57‰$_{VCDT}$ and from −4.56 to −3.98‰$_{VCDT}$. The pyrite and chalcopyrite acquired from the prograde proximal exoskarn zone displayed a narrow range of sulfur isotopic values. Specifically, the pyrite varied from −3.65 to −0.17‰$_{VCDT}$, while the chalcopyrite ranged from −4.23 to −0.25‰$_{VCDT}$ (Table 2). The sulfur isotopic values of pyrite and chalcopyrite in the retrograde proximal exoskarn zone ranged from −4.27 to −2.59‰$_{VCDT}$ for pyrite and from −3.84 to −2.67‰$_{VCDT}$ for chalcopyrite. In contrast, the pyrite and chalcopyrite detected in the intermediate exoskarn zone displayed a lower sulfur isotopic value. Specifically, the pyrite varied from −9.44 to −5.46‰$_{VCDT}$, while the chalcopyrite ranged from −8.88 to −5.79‰$_{VCDT}$ (Table 2). The $\delta^{34}S$ ratios of the ore-forming fluids were measured on the assumption that the sulfur included in these fluids was mostly in the form of H_2S. The calculations relied on fractionation equations established by Czamanske and Rye [41] and Ohmoto and Rye [42], which use the $\delta^{34}S$ values of sulfide minerals. The formational temperatures of the fluids were determined through the analysis of fluid inclusions. These analyses revealed that the average temperature of garnet minerals in the prograde zone was 390 °C, while the average temperature of the garnet, quartz, and calcite minerals in the retrograde zone was 323 °C. The measured $\delta^{34}S$ values of H_2S in the hydrothermal fluids varied between −5.47‰ and −0.34‰, with an average of −2.82‰ ± 0.18, and between −5.39‰ and −1.08‰, with an average of −4.24‰ ± 0.18, for the sulfides from the endoskarn and proximal zones, respectively. In contrast, these values ranged from −10.57‰ to −4.54‰, with an average of −7.54‰ ± 0.23, for the sulfides from the intermediate zone. These values represent the overall sulfur isotopic composition of the magmatic–hydrothermal fluid ($\delta^{34}S_{\Sigma S}$), as shown in Table 2.

Table 2. Sulfur isotope values of sulfides from the Yolindi Cu-Fe Skarn mineralization.

Sample ID	Mineral	Host	Metamorphic Grade	Measured $\delta^{34}S$ (‰)	at T °C	$\delta^{34}S_{H2S}$ Fluid (‰) (3)		Error Range
BG-220A	Pyrite	Endoskarn		−4.56	390 (1)	−5.47	±	0.18
BG-220B	Pyrite	Endoskarn		−4.11	390 (1)	−5.02	±	0.18
BG-220C	Pyrite	Endoskarn		−3.98	390 (1)	−4.89	±	0.18
BG-85A	Pyrite	Endoskarn		0.57	390 (1)	−0.34	±	0.18
BG-85B	Pyrite	Endoskarn		0.27	390 (1)	−0.64	±	0.18
BG-87	Pyrite	Endoskarn		0.33	390 (1)	−0.58	±	0.18
BG-102	Pyrite	Proximal	Prograde	−3.65	390 (1)	−4.56	±	0.18
BG-33A	Pyrite	Proximal	Prograde	−0.17	390 (1)	−1.08	±	0.18
BG-103	Pyrite	Proximal	Prograde	−2.73	390 (1)	−3.64	±	0.18
BG-102	Chalcopyrite	Proximal	Prograde	−4.13	390 (1)	−5.04	±	0.18
BG-102A	Chalcopyrite	Proximal	Prograde	−4.23	390 (1)	−5.14	±	0.18
BG-33A	Chalcopyrite	Proximal	Prograde	−0.25	390 (1)	−1.16	±	0.18
BG-237F2	Pyrite	Proximal	Retrograde	−3.54	323 (2)	−4.67	±	0.23
BG-237F2	Pyrite	Proximal	Retrograde	−3.49	323 (2)	−4.62	±	0.23
BG-31A	Pyrite	Proximal	Retrograde	−2.59	323 (2)	−3.72	±	0.23
BG-44	Pyrite	Proximal	Retrograde	−3.7	323 (2)	−4.83	±	0.23
BG-44	Pyrite	Proximal	Retrograde	−3.9	323 (2)	−5.03	±	0.23
BSD-1	Pyrite	Proximal	Retrograde	−3.64	323 (2)	−4.76	±	0.23
BSD-10	Pyrite	Proximal	Retrograde	−3.48	323 (2)	−4.61	±	0.23
BSD-11	Pyrite	Proximal	Retrograde	−4.08	323 (2)	−5.21	±	0.23
BSD-11	Pyrite	Proximal	Retrograde	−4.26	323 (2)	−5.39	±	0.23
BSD-8	Pyrite	Proximal	Retrograde	−3.01	323 (2)	−4.14	±	0.23
BG-31A	Chalcopyrite	Proximal	Retrograde	−2.67	323 (2)	−3.80	±	0.23
BSD-10	Chalcopyrite	Proximal	Retrograde	−3.84	323 (2)	−4.96	±	0.23
BSD-8	Chalcopyrite	Proximal	Retrograde	−3.41	323 (2)	−4.54	±	0.23
BG-100	Pyrite	Intermediate	Retrograde	−9.44	323 (2)	−10.57	±	0.23
BG-101	Pyrite	Intermediate	Retrograde	−8.92	323 (2)	−10.05	±	0.23

Table 2. *Cont.*

Sample ID	Mineral	Host	Metamorphic Grade	Measured $\delta^{34}S$ (‰)	at T °C	$\delta^{34}S_{H2S}$ Fluid (‰) (3)		Error Range
BG-205	Pyrite	Intermediate	Retrograde	−5.72	323 (2)	−6.85	±	0.23
BSS-4	Pyrite	Intermediate	Retrograde	−5.81	323 (2)	−6.94	±	0.23
BSS-5	Pyrite	Intermediate	Retrograde	−5.46	323 (2)	−6.58	±	0.23
BSS-6	Pyrite	Intermediate	Retrograde	−5.52	323 (2)	−6.65	±	0.23
BSS-6	Pyrite	Intermediate	Retrograde	−5.59	323 (2)	−6.72	±	0.23
BG-101	Chalcopyrite	Intermediate	Retrograde	−8.87	323 (2)	−10.00	±	0.23
BG-205	Chalcopyrite	Intermediate	Retrograde	−5.79	323 (2)	−6.92	±	0.23

(1) Based on the average homogenous temperature of the garnet minerals in the endoskarn and prograde zones; (2) Based on the average homogenous temperature of the garnet, quartz, and calcite minerals in the retrograde zone; (3) Calculated using the sulfur isotope fractionation equations outlined by Czamanske and Rye [41] and Ohmoto and Rye [42].

5.3.2. Carbon ($\delta^{13}C$) and Oxygen ($\delta^{18}O$) Isotope Studies

The carbon and oxygen isotopic compositions of calcite minerals obtained from marble (*n* = 5 samples), retrograde skarn (*n* = 9 samples), and prograde skarn (*n* = 11 samples) exhibited $\delta^{13}C$ values ranging from +1.89‰$_{VPDB}$ to +2.23‰$_{VPDB}$ (with an average of +2.06‰$_{VPDB}$), from −3.8‰$_{VPDB}$ to −2.25‰$_{VPDB}$ (with an average of −3.02‰$_{VPDB}$), and from −6.0‰$_{VPDB}$ to +0.09‰$_{VPDB}$ (with an average of −2.48‰$_{VPDB}$), respectively (Table 3). Additionally, the $\delta^{18}O$ values ranged from +21.61‰$_{VSMOW}$ to +21.73‰$_{VSMOW}$ (with an average of +21.67‰$_{VSMOW}$), from +0.94‰$_{VSMOW}$ to +3.62‰$_{VSMOW}$ (with an average of 2.45‰$_{VSMOW}$), and from +6.22‰$_{VSMOW}$ to +18.14‰$_{VSMOW}$ (with an average of +11.24‰$_{VSMOW}$) for the calcite minerals from marble, retrograde skarn, and prograde skarn, respectively (Table 3). Hence, the carbon isotope ratios ($\delta^{13}C$ values) of the samples generally exhibit a consistent pattern, showing little enrichment in the marble as compared to the prograde and retrograde calcite samples (Table 3). Nevertheless, the $\delta^{18}O$ values indicate a decrease in isotopic composition from the marble sample to the prograde and retrograde calcite samples. Moreover, the data provided relates to the declining carbon and oxygen isotopic compositions seen in calcite minerals throughout their transition from prograde skarn to retrograde skarn, which may be attributed to the elevated temperatures experienced in the prograde zone [43]. The fluid $\delta^{13}C_{CO2}$ values calculated at 318 °C (the highest homogeneous temperatures (T_h (°C)) for calcite minerals based on the fluid inclusion data of calcite minerals) using the calcite–CO_2 fractionation equation established by Scheele and Hoefs [44] exhibited a range from −4.92‰ to −4.58‰ (with an average of −4.75‰), from −10.61‰ to −9.06‰ (with an average of −9.83‰) and from −9.45‰ to −6.72‰ (with an average of −8.47‰) for calcite collected from marble, retrograde skarn, and prograde skarn, respectively (Table 3). The fluid $\delta^{18}O_{H2O}$ values were determined using the equation for oxygen isotopic fractionation of calcite–water developed by O'Neil et al. [45]. These values were calculated at the highest T_h of 318 °C (current fluid inclusion data) for calcite minerals. The range of values acquired for calcite from marble, retrograde skarn, and prograde skarn were from +20.88‰$_{VSMOW}$ to +21.00‰$_{VSMOW}$, from 0.91‰$_{VSMOW}$ to +3.49‰$_{VSMOW}$, and from +6.03‰$_{VSMOW}$ to +17.57‰$_{VSMOW}$, respectively (Table 3). In Figure 9a, the diagram illustrates the relationship between $\delta^{13}C_{PDB}$ and $\delta^{18}O_{SMOW}$. The marble samples fell within the marine carbonate field, indicating their composition. On the other hand, the samples of prograde calcite were plotted in the fields of granites, mantle-derived carbonatite, and xenoliths, as well as the transitional area between these fields and marine carbonate. This suggested that the oxygen isotope ratio increased in prograde calcite samples as a consequence of the dissolution and decarbonization of Torosan formation marble at the contact of the Şaroluk intrusion (Figure 9a). In contrast, the calcite samples collected from the retrograde skarn exhibited a reduction in oxygen isotope values, which may be attributed to the influence of meteoric water (see Figure 9a). The isotopic fractionation factor was calculated using the calcite–water equation proposed by O'Neil et al. [45]. The mixing curves for M–A were computed for several values of X (CO_2) and fluid rock ratios, including 0.01, 0.1, and 0.5, and 0.5, 2, 10, 40, 80, and 100, respectively (Figure 9b). The isotopic composition of $\delta^{13}C$ and $\delta^{18}O$ in the prograde and retrograde stage skarn calcites of the Yolindi skarn mineralization exhibited a depletion pattern that aligns with the curves of 0.01, 0.1, and, rarely, 0.5 X (CO_2). Isotopic depletion was observed in both the oxygen and carbon isotopes. The fluid–rock ratios of skarn calcites during the prograde stage exhibited the highest variability, with their values ranging from 10 to 80. In contrast, the fluid–rock ratios pertaining to retrograde-stage skarn calcites have yet to be identified.

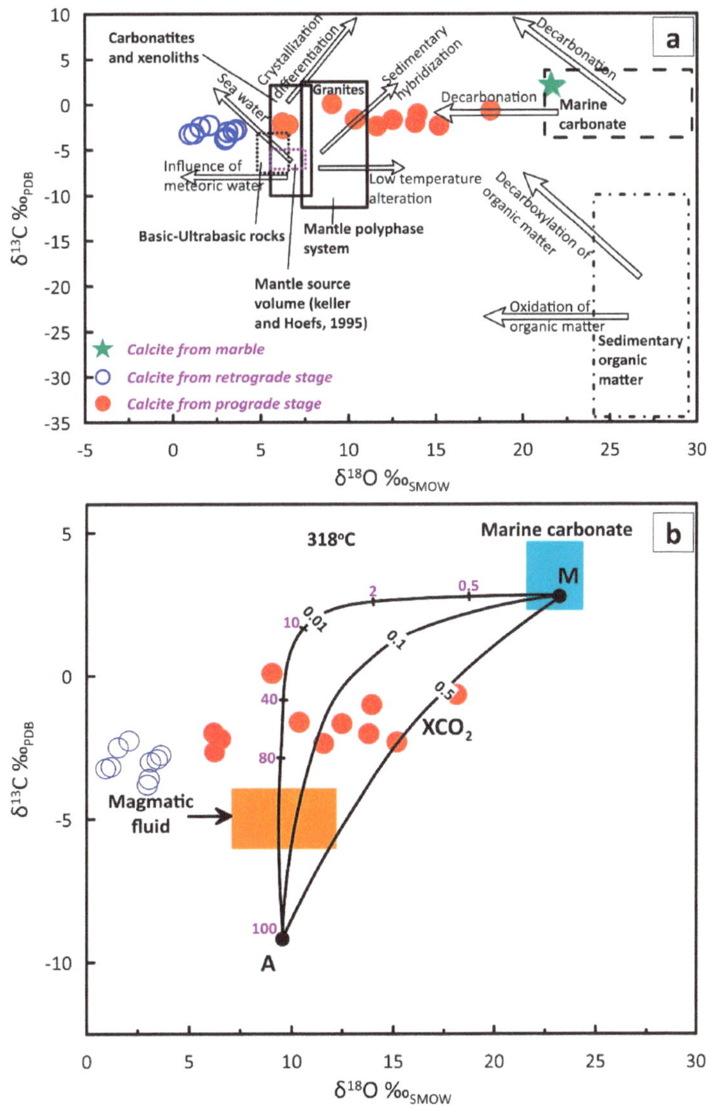

Figure 9. (**a**) $\delta^{13}C_{PDB}$ vs. $\delta^{18}O_{SMOW}$ binary diagram for calcites from marble, prograde, and retrograde skarn zones. These compositions were then compared with the $\delta^{13}C$ and $\delta^{18}O$ isotopic compositions

of other rock types, as documented by Sheppard [46] and Hoefs [47]. (**b**) Plots of $\delta^{13}C_{CO2}$ vs. $\delta^{18}O_{H2O}$ isotopes of skarn calcites on the fluid–rock interaction model of Taylor [48], with curves generated for various X (CO_2) and fluid–rock ratios of 0.01, 0.1, 0.5, 2, 10, 40, 80, and 100, respectively.

Table 3. Carbon ($\delta^{13}C$) and oxygen ($\delta^{18}O$) isotope analyses of the calcite minerals from the different zones and rocks in the Yolindi Cu-Fe Skarn mineralization.

Sample ID	Zone	$\delta^{13}C$ (‰VPDB)	$\delta^{18}O$ (‰SMOW)	$\delta^{13}C_{CO2}$ (‰) (1)	$\delta^{18}O_{H2O}$ (‰) (2)
BG-41	Prograde	−1.98	6.22	−8.79	6.07
BG-237e	Prograde	−2.64	6.26	−9.45	6.11
BG-237d2	Prograde	−2.19	6.55	−9.00	6.39
BSS-9	Prograde	0.09	9.05	−6.72	8.84
BG-204b	Prograde	−1.61	10.38	−8.42	10.14
BG-62	Prograde	−1	13.94	−7.81	13.61
BG-39A	Prograde	−2.3	15.18	−9.11	14.82
BG-50	Prograde	−0.65	18.14	−7.46	17.70
BG-39b	Prograde	−2.35	11.60	−9.16	11.32
BG-39c	Prograde	−2.01	13.80	−8.82	13.47
BG-39d	Prograde	−1.66	12.50	−8.47	12.20
BSD-10	Retrograde	−3.23	0.94	−10.04	0.92
BSD-1	Retrograde	−3.17	1.20	−9.98	1.17
BG-35A	Retrograde	−2.49	1.56	−9.30	1.52
BSD-14	Retrograde	−2.25	2.09	−9.06	2.04
BSS-12	Retrograde	−3.8	2.97	−10.61	2.90
BSS-13	Retrograde	−3.6	3.06	−10.41	2.99
BSS-14	Retrograde	−3	3.12	−9.81	3.05
BSS-15	Retrograde	−2.9	3.45	−9.71	3.37
BSS-16	Retrograde	−2.76	3.62	−9.57	3.53
BG-304	Marble	1.89	21.73	−4.92	21.21
BG-305	Marble	2.23	21.61	−4.58	21.10
BG-306	Marble	1.96	21.71	−4.85	21.19
BG-307	Marble	2.07	21.67	−4.74	21.15
BG-308	Marble	2.16	21.63	−4.65	21.12

(1) Equation of carbon isotope for calcite, as discussed by Scheele and Hoefs [44]. (2) Equation of oxygen isotopic fractionation of calcite-water, as outlined by O'Neil et al. [45].

6. Discussion

6.1. Sulfur Isotopic Compositions and Their Geological Implications

The skarn deposits exhibit geological significance in relation to geodynamic settings, crustal recycling and metallogenesis, paleoenvironmental conditions, ancient hydrothermal systems, mineral resource potential, and regional metamorphic history [49,50]. Skarn deposits often form in fractures, hydrothermal systems, and igneous–carbonate contact zones [49]. Similar to earlier hydrothermal systems and pathways, metal-rich hydrothermal fluids mix with surrounding rocks to generate skarn deposits [50]. According to Ault and Kulp [51], the variation in $\delta^{34}S$ values within an ore deposit offers insights on the conditions under which it was generated, namely whether the deposits originated from fluids originating from a deep-seated magmatic source or from a more restricted source in the upper crust. Furthermore, it is crucial to recognize that various origins of sulfur in fluids responsible for ore formation encompass unique $\delta^{34}S$ values. Firstly, it is common for a mantle or magmatic source to have a $\delta^{34}S$ value within the range of $0 \pm 3‰$ [52]. McCuaig and Kerrich [53] stated that magmatic sources are formed through the dissolution or desulfidation of magmatic sulfides, having $\delta^{34}S$ values that range from 0‰ to +9‰. A seawater source has been identified via a $\delta^{34}S$ value of about +20‰ [53]. Furthermore, sedimentary rocks that have large negative $\delta^{34}S$ values exhibit a notable decrease in sulfur concentration, as stated by Rollinson [54].

The sulfur isotopic composition ($\delta^{34}S$) values obtained from the Yolindi Cu-Fe Skarn mineralization exhibited distinct variations. In the endoskarn zone, two varieties of pyrite minerals occurred, with one having $\delta^{34}S$ values ranging from +0.27 to +0.57‰$_{VCDT}$ and the other ranging from −4.56 to −3.98‰$_{VCDT}$. Furthermore, the sulfide minerals in the exoskarn zones exhibited negative $\delta^{34}S$ values. The proximal zone showed a slight negative signature, ranging from −4.27 to −0.17‰$_{VCDT}$ for pyrite and from −4.23 to −0.25‰$_{VCDT}$ for chalcopyrite. On the other hand, the intermediate zone displayed significantly more negative values, ranging from −9.44 to −5.46‰$_{VCDT}$ for pyrite and from −8.88 to −5.59‰$_{VCDT}$ for chalcopyrite (see Table 2). These findings indicated the potential occurrence of a sulfur source that exhibits varying consistency and encompasses a number of characteristics with magmatic sources, as well as other sources, including sedimentary, biogenic, and/or metamorphic sources (Figure 10a). The $\delta^{34}S$ values of hydrogen sulfide (H_2S) in ore fluids for sulfide minerals in the endoskarn zone (from −5.47‰ to −0.34‰) and proximal zone (from −5.39‰ to −1.08‰) exhibited a slight negative pattern, which fell slightly within the range of sulfide minerals associated with the magmatic–hydrothermal signature described by Richards [55]. These findings indicated that the sulfur found in the pyrite and chalcopyrite minerals did not mainly originate from the Earth's mantle. However, the $\delta^{34}S_{fluid}$ values still remained within the range of sulfur derived from magmatic sources [47]. The occurrence of very low (large and negative) $\delta^{34}S_{fluid}$ values in the intermediate zone (from −10.57‰ to −4.54‰) indicates the potential influence of negative $\delta^{34}S$ reservoirs, such as organic-rich sources, as detailed by Chambers [56] and Strauss [57], or similarities to metamorphic sulfide, as reported by Hutchison et al. [58] (Figure 10b). In summary, the sulfide-bearing fluid may have originated from either a magmatic–hydrothermal fluid, from the leaching process in igneous rocks, or from contributions from the organic-rich layers and/or metamorphic sources of the host metamorphic rocks of the Torasan formation. The fluctuation in sulfur isotope ratio from endoskarn to exoskarn zones and from prograde to retrograde zones in the Yolindi skarn mineralization may provide valuable insights into their physical and chemical evolution, as well as the processes involved in the processing of these deposits. Our previous article [4] discussed the transition from endoskarns to exoskarns in the Yolindi area, which indicates a spatial and temporal zonation in mineral chemistry within a skarn system. This transition is marked by different combinations of minerals, such as massive, un-zoned andradite and augite with magnetite in the endoskarn and zoned andradite in the proximal zone, indicating a prograde character [4]. Epidote, actinolite, and Cl-rich scapolite are newly formed hydrous minerals in the exoskarn intermediate and distal zones, as part of the retrograde system [4]. One instance of scapolite occurred in exoskarns, which had a greater concentration of chlorine. This indicates the presence of a hydrothermal fluid with a high NaCl content, possibly originating from the country metasediment rock of the Torasan formation [59]. The sulfur isotopic compositions at the first stage of skarn mineralization showed that the sulfur originated from a magmatic source. The sulfur isotopic compositions ($\delta^{34}S$) varied from 0.27‰ to 0.57‰ and from −4.56‰ to −0.17‰ in the endoskarn zone (see Table 2 and Figure 10). As the system evolved from the prograde stage to the retrograde stage, the ($\delta^{34}S$) values changed from −9.44‰ to −2.59‰ (see Table 2 and Figure 10). During this transition, there was a gradual decrease in the $\delta^{18}O$ fluid values from 6.0‰ to 17.6‰ in the prograde zone and from 0.9‰ to 3.5‰ in the retrograde zone (see Table 3). This suggested that there was an increase in mixing with meteoric water [60]. Figures 10 and 11 depict the typical ranges of sulfur isotopic compositions displayed under different geological settings. The $\delta^{34}S$ values of sulfide minerals reported in several skarn deposits in Turkey and worldwide displayed a broad range of values, all of which were within the range of $\delta^{34}S$ values seen in granitic rocks (Figure 10a). When differentiating between ore deposits from sedimentary and igneous settings, it is important to approach the use of sulfur isotopes with caution due to the substantial overlap seen in $\delta^{34}S$ values (Figure 10a). This occurrence may be attributed to the significant deposition of sulfide and sulfate minerals from the hydrothermal fluids during the mineralization process [61]. Figure 11 revealed that the studied Yolindi Cu-Fe skarn mineralization shares similarities with various other deposits, namely the Çataltepe deposits in Turkey [62], Fırıncıkdere in Kalkım in Turkey [63], Mazraeh in NW Iran [60],

Ditto in Japan [64], Shenshan in NE China [65], and Velardeña in Mexico [66] (Figure 11). These deposits exhibited a significant reduction in $\delta^{34}S$, indicating a strong link between the sulfur isotopic composition of rock sulfur and ore sulfur in the magnetite series of granitoids. This also suggested that they may have developed via contact metasomatism, with their light sulfur likely originating from biogenic sulfur present in the sedimentary rocks [64,67].

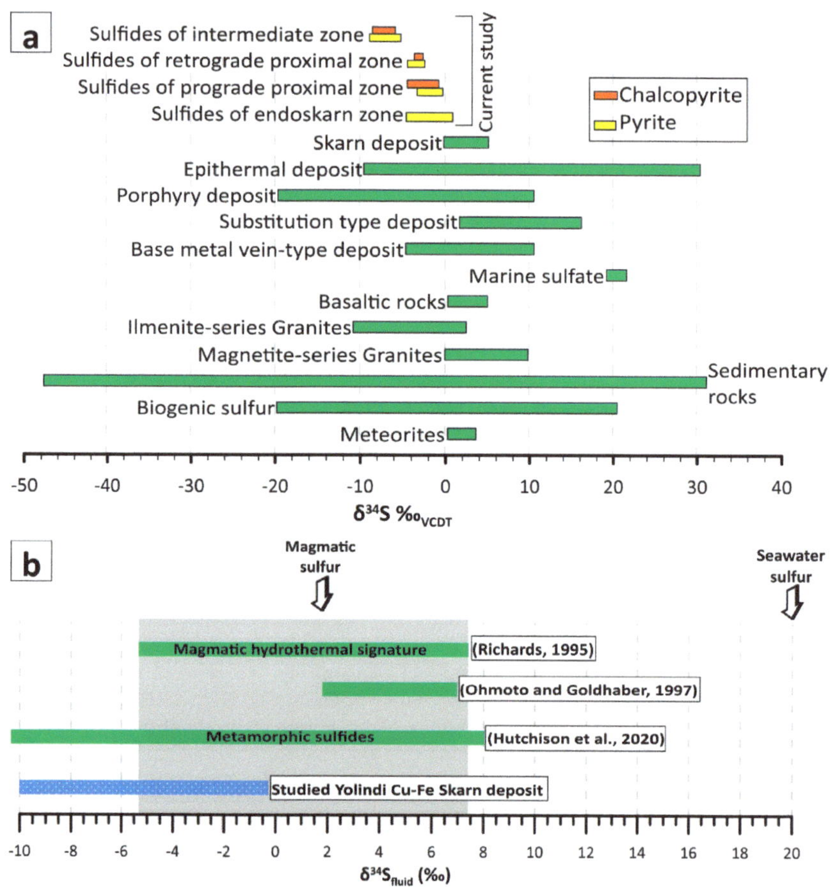

Figure 10. (**a**) $\delta^{34}S$ values of various geologic reservoirs, as reported by Seal [68], while skarn deposits were detailed by Taylor [69]. (**b**) $\delta^{34}S_{fluid}$ values of the Yolindi skarn mineralization and their comparison with the magmatic–hydrothermal deposits from Richards [55] and Ohmoto and Goldhaber [70] and metamorphic sulfide from Hutchison et al. [58].

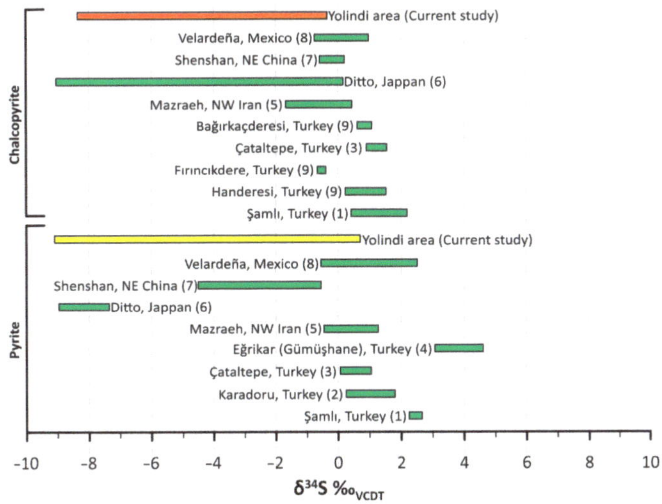

Figure 11. Sulfur isotopic composition diagram for sulfides from Yolindi area in the Biga Peninsula. Compilation of S isotope data: (1) Şamlı, Turkey [71]; (2) Karadoru, Turkey [72]; (3) Çataltepe, Turkey [62]; (4) Eğrikar (Gümüşhane), Turkey [73]; (5) Mazraeh, NW Iran [60]; (6) Ditto, Japan [64]; (7) Shenshan, NE China [65]; (8) Velardeña, Mexico [66]; and (9) Handeresi-Bağırkaçderesi-Fırıncıkdere in Kalkım, Turkey [63].

6.2. Source of Carbon

The $\delta^{13}C/\delta^{12}C$ isotope ratio has often been used for the purposes of determining the source of carbon dioxide (CO_2) derived from gas associated with magmatic–hydrothermal systems. In addition, it can be inferred that the computed $\delta^{13}C_{CO2}$ is almost considered the $\delta^{13}C_{\Sigma C}$ of the fluid [74]. This ratio is a valuable tool for distinguishing between various potential sources of fluid. These sources include: (i) gases originating from the Earth's mantle, which typically have $\delta^{13}C_{CO2}$ values ranging from $-7‰$ to $-3‰_{VPDB}$ [75,76]; (ii) the degradation of organic matter, which often produces significantly negative $\delta^{13}C_{CO2}$ values ($\leq -20‰_{VPDB}$; Hoefs [47] and O'Leary [77]); and (iii) thermal metamorphic processes associated with the decomposition of limestone, which are typically characterized by CO_2 with isotopically enriched values ($\delta^{13}C_{CO2}$: $>0 \pm 1‰_{VPDB}$; Sano and Marty [78], Tedesco and Scarsi [79], Minissale [80] and Clark [81]). Typically, the presence of intermediate $\delta^{13}C_{CO2}$ values has often been attributed to the amalgamation of carbon dioxide derived from several sources. Nevertheless, several studies have provided evidence indicating that the carbon isotopic composition of various gases is subjected to significant alterations due to the secondary processes of isotope fractionation. These processes include the interaction between gas and water, as demonstrated by Weinlich [82], Gilfillan et al. [83], Darrah et al. [84], and Güleç and Hilton [85]. Additionally, the precipitation or dissolution of carbonate minerals has been shown to have an impact on carbon isotopes in gases, as evidenced by Ohwada et al. [86], Barry et al. [87], and Venturi et al. [88]. Hence, it is important to be careful in interpreting $\delta^{13}C_{CO2}$ data, especially for gases derived from regions with restricted geological constraints [89]. The measured $\delta^{13}C_{CO2}$ values of prograde calcite varied between -9.45 and $-6.72‰_{VPDB}$ and demonstrated conformity with a magmatic source derived from the Earth's mantle, with an influence from the thermal degradation of organic materials (Figure 12a). In contrast, the $\delta^{13}C_{CO2}$ values of retrograde calcite exhibited a range from -10.61 to $-9.06‰_{VPDB}$, indicating that the CO_2 was solely derived by the thermal degradation of organic matter (Figure 12a).

6.3. Calcite-Forming Fluid Sources

Hydrothermal fluids from various sources have distinct carbon–oxygen isotopic compositions, as shown by Demény and Harangi [90], Demény et al. [91], Zheng and Chen [92], and Zhou et al. [93]. Particularly, carbon dioxide (CO_2) originating from the mantle or igneous carbonatite sources displays a variety of isotopic compositions. More precisely, it has a $\delta^{13}C_{PDB}$ value that varies between −8‰ and −4‰, and a $\delta^{18}O_{SMOW}$ value that ranges from 6‰ to 10‰ [91,94]. However, CO_2 that comes from marine carbonate sources has distinct isotopic characteristics varying between −4‰ and 4‰, whereas the $\delta^{18}O_{SMOW}$ value ranges from 20‰ to 30‰ [95]. In addition, CO_2 originating from sedimentary organic matter has unique isotopic compositions, having $\delta^{13}C_{PDB}$ values ranging from −30‰ to −15‰, whereas its $\delta^{18}O_{SMOW}$ value ranges from 24‰ to 30‰ [47]. Furthermore, the carbon isotopic composition of the hydrothermal fluids, when paired with the oxygen isotopic composition, offers more insight into the origins of the fluid in the hydrothermal system. However, the carbon isotope signature of the water source remained rather ambiguous. The $\delta^{18}O_{H2O}$ data in the Yolindi area exhibited a wide range of variability. These values for calcite from prograde skarns varied from +6.03‰ to +17.57‰, and for calcite from retrograde skarns, they ranged from +0.91‰ to +3.49‰. These findings indicate that the hydrothermal fluid that caused the formation of calcite crystals in the prograde skarn zone is a mixture of metamorphic and magmatic sources, perhaps originating from mantle or igneous origins (Figure 12b). The calcite minerals in the retrograde skarn zone were formed via the activity of hydrothermal fluids originating from meteoric sources, as shown in Figure 12b.

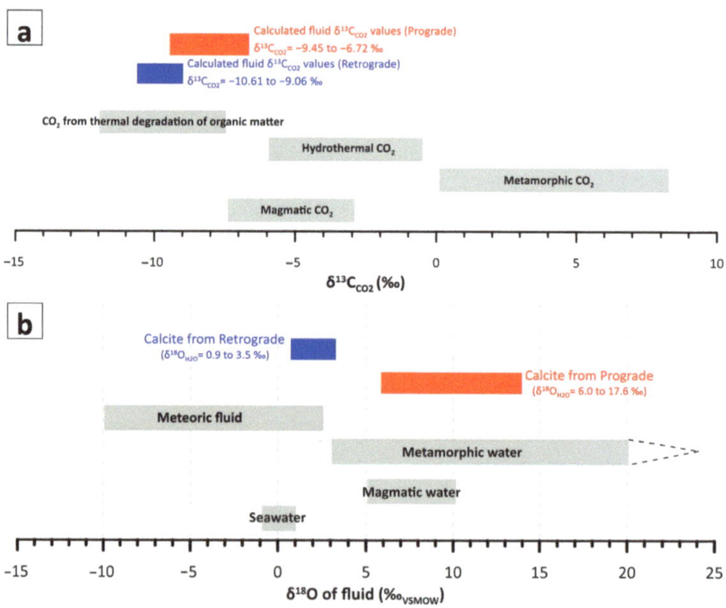

Figure 12. (**a**) The $\delta^{13}C_{CO2}$ values of calcite minerals from prograde skarn and retrograde skarn were compared to the natural $\delta^{13}C_{CO2}$ values from different sources, as reported by Schidlowski [96] and Planavsky et al. [97]. (**b**) The $\delta^{18}O_{H2O}$ values of calcite minerals from prograde skarn and retrograde skarn were compared to the natural $\delta^{18}O_{H2O}$ values from different sources, as detailed by Sheppard [98] and Rollinson and Pease [99].

6.4. Composition of Hydrothermal Solutions

Eutectic temperatures (Te) have been ascertained for different types and phases of inclusions occurring in andradite, grossular, quartz, and calcite minerals. The eutectic temperatures of the prograde proximal stage inclusions in andradite ranged from −37.80 to −12.10 °C. The temperatures mentioned are associated with the mixing of H_2O-$CaCl_2$-$MgCl_2$-NaCl salt solutions [33], rather than being specific to individual phases (Figure 13). Nevertheless, the eutectic temperatures of the retrograde stage inclusions of grossular, quartz, and calcite minerals in the proximal, intermediate, and distal zones were found to be frequently comparable with the eutectic temperature of the H_2O–NaCl system [33] (Figure 13). Considering these eutectic temperatures at all stages, the composition of the prograde early phases of inclusions was characterized by a combination of $CaCl_2$- and $MgCl_2$-dominated systems [33]. $CaCl_2$ and $MgCl_2$ are the main components of hydrothermal fluids, as stated by Roedder [32]. The salinity and temperature of this particular stage of inclusion exhibited greater values compared to the retrograde phases. The existence of $CaCl_2$ and $MgCl_2$ indicates a direct or indirect interaction with either hydrothermal fluid or marine carbonate in the nearby area [100,101]. The carbonaceous rocks that were found in the Torasan formation within the skarn contacts are the only plausible origin for the first-stage inclusions bearing $CaCl_2$ and $MgCl_2$ during the prograde stage. Thus, the existence of $CaCl_2$ and $MgCl_2$ in the fluid inclusions can be attributed to a variety of factors, including the evolution of hydrothermal systems, temperature and salinity fluctuations, fluid immiscibility (or boiling), and the presence of distinct types of fluid inclusions, such as liquid-rich and vapor-rich inclusions, in this stage [102,103]. Later on, it was discovered that the retrograde stage inclusions were mostly composed of NaCl, based on the very similar eutectic temperatures of the NaCl–H_2O system [33]. The transition from a composition mostly consisting of $CaCl_2$ and $MgCl_2$ to one dominated by NaCl in the fluid inclusions indicates that there were significant interactions between the hydrothermal solutions and carbonaceous host rocks during the first stages of skarnification. During the retrograde phase, a reduction in the interaction between the hydrothermal fluids and carbonates may have resulted in alterations that led to compositions dominated by NaCl.

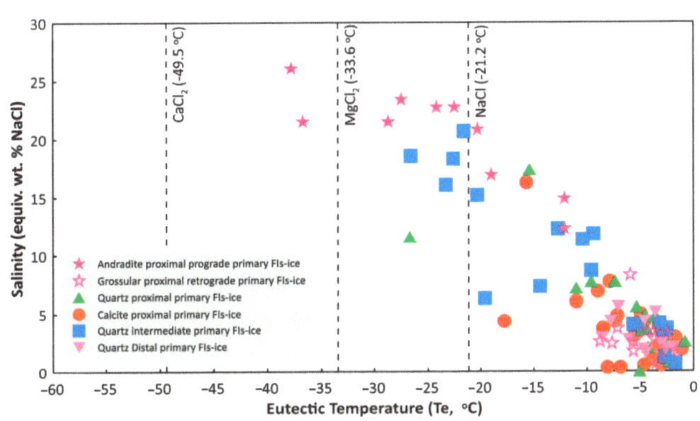

Figure 13. Comparing the measured eutectic temperatures of fluid inclusions from various mineral phases at different stages with the eutectic temperatures of certain salt solutions according to Shepherd et al. [33] and Wilkinson [104].

6.5. Evolution of Ore-Forming Fluids

The prograde-stage inclusions had higher salinity, reaching up to 26.07 wt.% NaCl equivalent, and higher Th temperatures, reaching up to 412.20 °C (Figure 14). In contrast, the fluid inclusions during the retrograde stage exhibited reduced salinity and Th temperatures. The salinity content of the minerals formed during the retrograde stage (i.e., grossular, quartz, and calcite) dropped as the temperature of the fluid inclusions decreased

(Figure 14). This decrease in salinity was ascribed to the amalgamation of meteoric water and hydrothermal solutions [104], as well as the interaction of the solution with the organic material within the host Torasan formation. The results unequivocally show that the salinity content of hydrothermal solutions was higher during the first phase (prograde) of skarn formation due to the interaction between the fluids originating from magma and carbonaceous rock units of the Torasan formation. During the later stages (retrograde), the addition of meteoric water resulted in a decrease in salinity levels and a reduction in fluid temperatures.

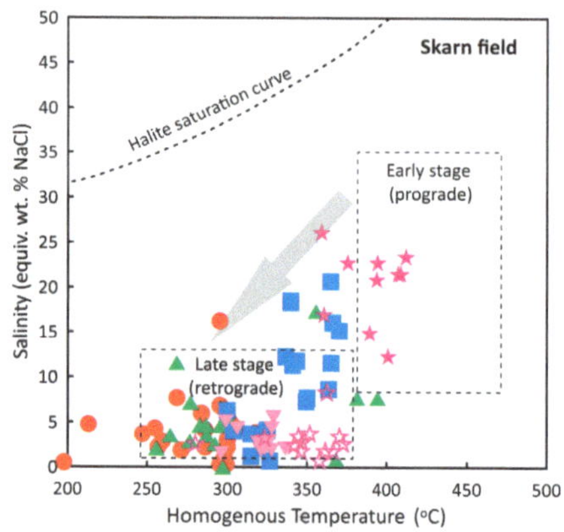

Figure 14. Salinity vs. total homogenization temperature plot of the fluid inclusions [105], the skarn field after Roedder [32], showing the fluid evolution in the Yolindi Cu-Fe skarn mineralization. The symbols are shown in Figure 13.

The ore-forming fluid evolution of the Yolindi Cu-Fe skarn mineralization began with the crystallization and cooling of the Şaroluk quartz monzonite and granodiorite in the shallow crust. The early skarn stage (prograde) is characterized by relatively moderate-to-high salinities, reaching up to 26 wt.% NaCl equivalents. These salinities were believed to be the result of a magmatic fluid component that underwent boiling [106] from the crystallizing Şaroluk quartz monzonite–granodiorite parental magma. It has been widely reported that the Şaroluk intrusion, which formed at shallow depths, contains volatile-rich magmas. This has been supported by various studies [27,107–111]. The presence of these magmas promotes the direct release of metalliferous and highly saline aqueous fluids [55]. In this particular setting, the first fluids that separate will possess salinities of moderate levels (6–8 wt.% NaCl equivalent; Bodnar) [106]. During the process of crystallization, the concentration of chlorine increases in the molten substance. This causes the fluids that are still separating to have progressively increasing levels of salinity. The andradite minerals in this early-stage (prograde) proximal exoskarn zone contain liquid-rich and vapor-rich two-phase (L+V) aqueous inclusions. These inclusions indicate that fluid boiling occurred during this early stage, resulting from the boiling process of the initial supercritical fluid. This fluid boiling was found to be directly associated with the earlier skarn stage (Figure 15). Therefore, the magnetite deposition in the proximal zone during the prograde stage was most likely a result of boiling, along with a reduction in temperature and/or pressure of the hydrothermal fluids.

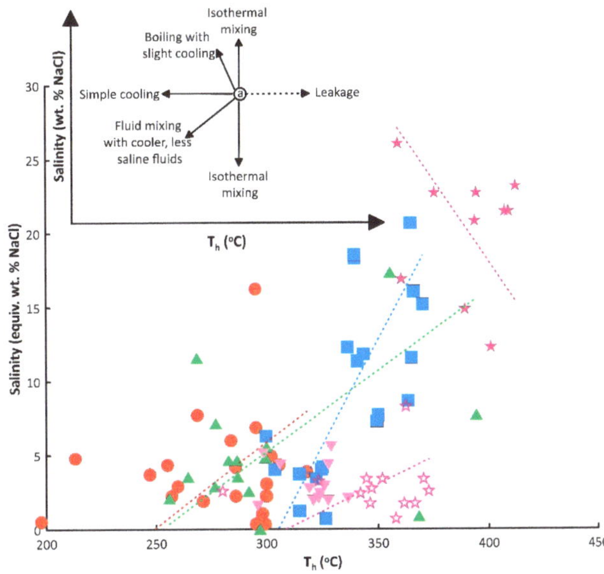

Figure 15. Homogenization temperature (Th°C) and salinity (wt.% NaCl) diagram for the Yolindi Cu-Fe skarn mineralization [104]. Shepherd et al. [33] modified the figure inset to depict temperature–salinity trends or fluid evolution pathways deriving from various geological events. The dotted lines indicate the line graphs for each sample type. The symbols are shown in Figure 13.

On the other hand, the cooling and reduction in pressure and salinity, as well as changes in fluid composition, during the transition from the prograde stage to the retrograde stage of skarnification can be elucidated via various processes [112]: (1) the infiltration of meteoric water and its mixing with a prevailing magmatic fluid; (2) the addition of basinal fluids from the nearby sedimentary host rocks; and/or (3) the liberation of $CaCl_2$ during the creation of calc-silicate minerals and skarn [113]. Several porphyry and skarn systems, such as those described by Roedder [32] and Lai et al. [114], have shown similar tendencies. The fluids responsible for forming ores during the later stage (retrograde) in the proximal, intermediate, and distal exoskarn zones at the Yolindi area were found to be colder and more diluted compared to the earlier stage. This is evident from the obvious linear patterns shown in Figure 15, indicating that the influence of mixing with meteoric water is considerable [1]. The ongoing influx and mixture of meteoric water, which is relatively cold and has low salinity, resulted in ore-forming fluids with a moderate–low temperature and moderate–low salinity. The occurrence of relatively higher iron grades, associated with retrograde skarn formations, suggested that magnetite and specular hematite were formed concurrently during this stage of skarn mineralization. This is in contrast to the earlier phases, where only magnetite was formed as the initial iron oxide. The principal cause of iron deposition seems to be the combination of magmatic-dominated fluids from the crystallizing Şaroluk intrusion with cooler fluids emanating from the surface. The sulfide-rich mineralization deposition during the latter phase of the retrograde stage in the intermediate and distal zones indicated that it may have taken place under skarns to porphyry–epithermal transitional conditions (?) e.g., Bouabdellah et al. [112]. The probable explanation of this phenomenon may be attributed to a drop in temperature, pressure, and salinity, together with the occurrence of vuggy quartz and the substitution of lattice-bladed calcite with quartz. The mineralization originated from fluids with a salinity of less than 15 wt.% NaCl equivalent. The formation occurred at temperatures ranging from 390 to 197 °C and pressures below 250 bars [112]. Therefore, it is very likely that the cooling and boiling of these fluids, together with the majority of fluid mixing, significantly influenced the creation of iron and sulfide minerals.

7. Conclusions

This comprehensive study of the Yolindi Cu-Fe skarn mineralization represents a significant advancement in understanding the complex interplay of geological processes responsible for skarn formation and mineralization. Utilizing a multidisciplinary approach, combining microthermometry of fluid inclusions with isotopic analysis ($\delta^{34}S$, $\delta^{13}C$, and $\delta^{18}O$), this study has elucidated the physiochemical conditions and fluid evolution within this skarn mineralization process. The findings highlight the significance of magmatic impacts, meteoric water interactions, and the role of the host rock matrix in shaping the mineralization process.

The isotopic data, particularly sulfur isotopes in sulfide minerals and carbon and oxygen isotopes in calcite, have provided key insights into the sources of sulfur and the composition and temperature of fluid–rock interactions. The variable $\delta^{34}S$ levels across different skarn zones highlight the diverse sulfur origins, ranging from magmatic impacts in endoskarn zones to organic-rich and potentially metamorphic origins in exoskarn zones. This isotopic variability reflects the complex dynamics of sulfur incorporation into the skarn system. Additionally, carbon and oxygen isotope evidence from calcites revealed a significant shift in fluid sources and interactions, transitioning from a dominantly magmatic CO_2 source in the prograde skarn to predominantly meteoric water and organic material during the retrograde stage.

The fluid inclusion studies further complemented these findings, delineating a clear temporal and compositional transition of the hydrothermal fluids. The early prograde stage was characterized by high salinity and boiling magmatic fluids, indicative of intense interactions with carbonaceous rocks. The retrograde phase, marked by lower salinity and temperatures, signified the increasing influence of meteoric water and organic-rich host rocks, culminating in a complex mineralization process. This transition from high-salinity, magmatic-dominated fluids to more diluted compositions reflects the intricate interplay between deep magmatic processes and surface water influences in the formation of the Yolindi skarn mineralization.

In summary, the Yolindi Cu-Fe skarn mineralization offers a quintessential example of skarn formation, where the convergence of magmatic and meteoric fluids, coupled with the geochemical characteristics of the host rock, plays a pivotal role in mineralization. The achievements of this study lie in its detailed characterization of these interactions and their influence on ore-forming processes, contributing to a deeper understanding of the formation and evolution of skarn deposits.

Author Contributions: Conceptualization, M.K. (Mustafa Kaya), M.K. (Mustafa Kumral), C.Y. and A.A.; methodology, M.K. (Mustafa Kaya) and A.A.; software, M.K. (Mustafa Kaya) and A.A.; validation, M.K. (Mustafa Kaya) and A.A.; formal analysis, M.K. (Mustafa Kaya), M.K. (Mustafa Kumral) and A.A.; investigation, M.K. (Mustafa Kaya) and A.A.; resources, M.K. (Mustafa Kaya) and A.A.; data curation, M.K. (Mustafa Kaya) and A.A.; writing—original draft preparation, M.K. (Mustafa Kaya) and A.A.; writing—review and editing, M.K. (Mustafa Kaya), M.K. (Mustafa Kumral) and A.A.; visualization, M.K. (Mustafa Kaya) and A.A.; supervision, M.K. (Mustafa Kumral) and A.A. All authors took part in the fieldwork. All authors have read and agreed to the published version of the manuscript.

Funding: This work was funded by the Scientific Research Project (BAP project ID: 44594) for graduate students at Istanbul Technical University (ITU, Turkey).

Data Availability Statement: The data are available in Tables 1–3.

Acknowledgments: This study is part of the thesis of the first author. The authors would like to express their gratitude to Lecturer Beril Tanç Kaya and Hatice Nur Bayram (ITU, Turkey) for their assistance in the laboratory work. They would also like to thank geologist Sercan Öztürk (ITU, Turkey) for his help during the fieldwork. Mustafa Selman Aydoğan (Balıkesir University, Turkey) and Yusuf Kağan Kadioğlu (Ankara University, Turkey) are appreciated for their contributions in the geologic map. Additionally, the contributions of the editor and anonymous reviewers are gratefully recognized for their valuable comments and recommendations, which significantly enhanced the quality of the manuscript.

Conflicts of Interest: The authors declare no conflict of interest.

References

1. Meinert, L.D.; Dipple, G.M.; Nicolescu, S.; Hedenquist, J.W.; Thompson, J.F.H.; Goldfarb, R.J.; Richards, J.P. World Skarn Deposits. In *Economic Geology 100th Anniversary Volume*; Society of Economic Geologists: Littleton, CO, USA, 2005; pp. 299–336.
2. Yigit, O. A prospective sector in the Tethyan Metallogenic Belt: Geology and geochronology of mineral deposits in the Biga Peninsula, NW Turkey. *Ore Geol. Rev.* **2012**, *46*, 118–148. [CrossRef]
3. Kuşcu, İ. Skarns and skarn deposits of Turkey. In *Mineral Resources of Turkey*; Pirajno, F., Ünlü, T., Dönmez, C., Şahin, M.B., Eds.; Modern Approaches in Solid Earth Sciences; Springer: Berlin/Heidelberg, Germany, 2019; Volume 16, pp. 283–336.
4. Kaya, M.; Kumral, M.; Yalçın, C.; Abdelnasser, A. Genesis and Evolution of the Yolindi Cu-Fe Skarn Deposit in the Biga Peninsula (NW Turkey): Insights from Genetic Relationships with Calc-Alkaline Magmatic Activity. *Minerals* **2023**, *13*, 1304. [CrossRef]
5. Kaya, M.; Kumral, M.; Abdelnasser, A.; Yalçın, C.; Öztürk, S.; Bayram, H.N.; Tanç-Kaya, B. Evolution of the hydrothermal fluids of the Yolindi Fe-Cu skarn deposit, Biga peninsula, NW Turkiye: Evidence from carbon-oxygen isotopic variations of calcite minerals. In Proceedings of the EGU General Assembly 2023, Vienna, Austria, 23–28 April 2023.
6. Kuşcu, İ.; Tosdal, R.M.; Gençalioğlu-Kuşcu, G. Porphyry-Cu Deposits of Turkey. In *Mineral Resources of Turkey*; Pirajno, F., Ünlü, T., Dönmez, C., Şahin, M.B., Eds.; Modern Approaches in Solid Earth Sciences; Springer: Berlin/Heidelberg, Germany, 2019; pp. 337–425.
7. Karaman, M.; Kumral, M.; Yildirim, D.K.; Doner, Z.; Afzal, P.; Abdelnasser, A. Delineation of the porphyry-skarn mineralized zones (NW Turkey) using concentration–volume fractal model. *Geochemistry* **2021**, *4*, 125802. [CrossRef]
8. Abdelnasser, A.; Kumral, M.; Zoheir, B.; Yilmaz, H. Evolution of the Tepeoba porphyry-skarn Cu-Mo-Au deposit, NW Turkey: New mineralogical and geochemical findings. *Ore Geol. Rev.* **2022**, *147*, 104967. [CrossRef]
9. Yildirim, D.K. Genesis of the Halılar Metasediment-Hosted Cu-Pb (±Zn) Mineralization, NW Turkey: Evidence from Mineralogy, Alteration, and Sulfur Isotope Geochemistry. *Minerals* **2022**, *12*, 991. [CrossRef]
10. Meinert, L.D. Skarns and Skarn Deposits. *Geosci. Can.* **1992**, *19*, 145–162.
11. Özden, S.; Över, S.; Poyraz, S.A.; Güneş, Y.; Pınar, A. Tectonic implications of the 2017 Ayvacık (Çanakkale) earthquakes, Biga Peninsula, NW Turkey. *J. Asian Earth Sci.* **2018**, *154*, 125–141. [CrossRef]
12. Okay, A.; Siyako, M.; Burkan, K. Geology and tectonic evolution of the Biga Peninsula, northwest Turkey. *Bull.-Tech. Univ. Istanb.* **1991**, *44*, 191–256.
13. Göncüoğlu, M.C. *Introduction to the Geology of Turkey: Geodynamic Evolution of the Pre-Alpine and Alpine Terranes*; ODTÜ Jeoloji Mühendisliği Bölümü: Ankara, Turkey, 2010; pp. 1–66.
14. Birkle, P.; Satir, M.; Erler, A.; Ercan, T.; Bingol, E.; Orcen, S. Dating, geochemistry and geodynamic significance of the Tertiary magmatism of the Biga Peninsula, NW Turkey. In *Geology of the Black Sea Region: Ankara, Turkey*; Erler, A., Ed.; Mineral Research and Exploration Institute of Turkey (MTA): Ankara, Turkey, 1995; pp. 171–180.
15. Yılmaz, Y.; Genç, Ş.; Karacık, Z.; Altunkaynak, Ş. Two contrasting magmatic associations of NW Anatolia and their tectonic significance. *J. Geodyn.* **2001**, *31*, 243–271. [CrossRef]
16. Özdamar, Ş.; Roden, M.F.; Zou, H.; Billor, M.Z.; Hames, W.; Georgiev, S.; Dunkl, I. Petrogenesis of oligocene plutonic rocks in western Anatolia (NW Turkey): Insights from mineral and rock chemistry, Sr-Nd isotopes, and U-Pb, Ar-Ar and (U-Th)/He geochronology. *Geochemistry* **2021**, *81*, 125747. [CrossRef]
17. Özdamar, Ş.; Zou, H.; Billor, M.Z.; Hames, W.E. Petrogenesis of mafic microgranular enclaves (MMEs) in the oligocene-miocene granitoid plutons from northwest Anatolia, Turkey. *Geochemistry* **2021**, *81*, 125713. [CrossRef]
18. Kamacı, Ö.; Altunkaynak, Ş. Petrological insights into connections between the S- and I-type magmatic associations in metamorphic core complexes: A case study of the Çataldağ metamorphic core complex (NW Turkey). *Lithos* **2023**, *464–465*, 107433. [CrossRef]
19. Özdamar, Ş.; Zou, H.; Billor, M.Z.; Hames, W.; Roden, M.F.; Sarıkaya, O.; Georgiev, S. Petrogenesis, geochronology and thermochronology of Oligocene to Miocene Western Anatolia granitoid plutons in Turkey. *Lithos* **2024**, *464–465*, 107430. [CrossRef]
20. Genç, Ş.; Altunkaynak, Ş. On the Eybek granite (Biga Peninsula, NW Anatolia): A new evaluation in the light of new geochemical data. *J. Earth Sci. Appl. Res. Cent. Hacet. Univ.* **2007**, *28*, 75–98.
21. Yılmaz, Y.; Karacık, Z. Geology of the northern side of the Gulf of Edremit and its tectonic significance for the development of the Aegean grabens. *Geodin. Acta* **2001**, *14*, 31–43. [CrossRef]
22. Yigit, O. Mineral deposits of Turkey in relation to Tethyan metallogeny: Implications for future mineral exploration. *Econ. Geol.* **2009**, *104*, 19–51. [CrossRef]
23. Duru, M.; Pehlivan, S.; Dönmez, M.; Ilgar, A.; Akçay, A. *Geological Map of The Balıkesir-i18 Quadrangle*; General Directorate of Mineral Research and Exploration: Ankara, Turkey, 2007.

24. Aysal, N. *Petrology of the Mesozoic–Tertiary Magmatism and Metamorphism in Northern Biga (Çanakkale)*; Istanbul University: Istanbul, Turkey, 2005.
25. Aysal, N.; Ustaomer, T.; Ongen, S.; Keskin, M.; Koksal, S.; Peytcheva, I.; Fanning, M. Origin of the Early-Middle Devonian magmatism in the Sakarya Zone, NW Turkey: Geochronology, geochemistry and isotope systematics. *J. Asian Earth Sci.* **2012**, *45*, 201–222. [CrossRef]
26. Aslan, Z.; Demir, H.; Altın, İ. U–Pb zircon geochronology and petrology of the early Miocene Göloba and Şaroluk plutons in the Biga Peninsula, NW Turkey: Implications for post-collisional magmatism and geodynamic evolution. *J. Afr. Earth Sci.* **2020**, *172*, 103998. [CrossRef]
27. Aysal, N. Mineral chemistry, crystallization conditions and geodynamic implications of the Oligo–Miocene granitoids in the Biga Peninsula, Northwest Turkey. *J. Asian Earth Sci.* **2015**, *105*, 68–84. [CrossRef]
28. Krushensky, R. Neogene calc-alkaline extrusive and intrusive rocks of the karalar-yesiller area: Northwest Anatolia, Turkey. *Bull. Volcanol.* **1975**, *39*, 336–360. [CrossRef]
29. Dönmez, M.; Akçay, A.E.; Genç, Ş.C.; Şükrü, A. Middle-late Eocene volcanism and marine ignimbrites in Biga peninsula (NW Anatolia-Turkey). *Bull. Miner. Res. Explor.* **2005**, *131*, 21.
30. Karacik, Z.; Yilmaz, Y.; Pearce, J.A. The Dikili-Çandarlı volcanics, Western Turkey: Magmatic interactions as recorded by petrographic and geochemical features. *Turk. J. Earth Sci.* **2007**, *16*, 493–522.
31. Meinert, L.D. Skarn zonation and fluid evolution in the Groundhog mine, Central mining district, New Mexico. *Econ. Geol.* **1987**, *82*, 523–545. [CrossRef]
32. Roedder, E. Volume 12: Fluid inclusions. *Rev. Mineral.* **1984**, *12*, 644.
33. Shepherd, T.; Rankin, A.; Alderton, D. *A Practical Guide to Fluid Inclusion Studies*; Blackie and Son Ltd.: Glasgow, UK, 1985; p. 239.
34. Van den Kerkhof, A.M.; Hein, U.F. Fluid inclusion petrography. *Lithos* **2001**, *55*, 27–47. [CrossRef]
35. Darling, R.S. An extended equation to calculate NaCl contents from final clathrate melting temperatures in H_2O-CO_2-NaCl fluid inclusions: Implications for PT isochore location. *Geochim. Cosmochim. Acta* **1991**, *55*, 3869–3871. [CrossRef]
36. Archer, D.G. Thermodynamic properties of the NaBr+ H2O system. *J. Phys. Chem. Ref. Data* **1991**, *20*, 509–555. [CrossRef]
37. Bodnar, R. Revised equation and table for determining the freezing point depression of H2O-NaCl solutions. *Geochim. Cosmochim. Acta* **1993**, *57*, 683–684. [CrossRef]
38. Steele-MacInnis, M.; Lecumberri-Sanchez, P.; Bodnar, R.J. Short note: HokieFlincs_H_2O-NaCl: A Microsoft Excel spreadsheet for interpreting microthermometric data from fluid inclusions based on the PVTX properties of H_2O-NaCl. *Comput. Geosci.* **2012**, *49*, 334–337. [CrossRef]
39. Bakker, R.J. Package FLUIDS 1. Computer programs for analysis of fluid inclusion data and for modelling bulk fluid properties. *Chem. Geol.* **2003**, *194*, 3–23. [CrossRef]
40. Zhang, Y.-G.; Frantz, J.D. Determination of the homogenization temperatures and densities of supercritical fluids in the system NaCl-KCl-$CaCl_2$-H_2O using synthetic fluid inclusions. *Chem. Geol.* **1987**, *64*, 335–350. [CrossRef]
41. Czamanske, G.K.; Rye, R.O. Experimentally determined sulfur isotope fractionations between sphalerite and galena in the temperature range 600 degrees to 275 degrees C. *Econ. Geol.* **1974**, *69*, 17–25. [CrossRef]
42. Ohmoto, H.; Rye, R. Isotopes of sulfur and carbon. In *Geochemistry of Hydrothermal Ore Deposits*; John Wiley & Sons, Inc.: Hoboken, NJ, USA, 1979; pp. 509–567.
43. Valley, J.W.; Taylor, H.P.; O'Neil, J.R. *Stable Isotopes in High Temperature Geological Processes*; Walter de Gruyter GmbH & Co. KG: Berlin, Germany, 2018; Volume 16.
44. Scheele, N.; Hoefs, J. Carbon isotope fractionation between calcite, graphite and CO2: An experimental study. *Contrib. Mineral. Petrol.* **1992**, *112*, 35–45. [CrossRef]
45. O'Neil, J.R.; Clayton, R.N.; Mayeda, T.K. *Oxygen Isotope Fractionation in Divalent Metal Carbonates*; University of Chicago: Chicago, IL, USA, 1969.
46. Sheppard, S.M. Isotope geothermometry. In *Thermométrie et Barométrie Géologiques*; Lagache, M., Ed.; French Society of Mineralogy and Crystallography: Paris, France, 1984; pp. 349–412.
47. Hoefs, J. *Stable Isotope Geochemistry*; Springer Science & Business Media: Cham, Switzerland, 2008.
48. Taylor, J. Water/rock interactions and the origin of H_2O in granitic batholiths: Thirtieth William Smith lecture. *J. Geol. Soc.* **1977**, *133*, 509–558. [CrossRef]
49. Chen, Y.-J.; Chen, H.-Y.; Zaw, K.; Pirajno, F.; Zhang, Z.-J. Geodynamic settings and tectonic model of skarn gold deposits in China: An overview. *Ore Geol. Rev.* **2007**, *31*, 139–169. [CrossRef]
50. Dawson, K.M.; Eckstrand, O.R.; Sinclair, W.D.; Thorpe, R.I. Skarn Deposits. In *Geology of Canadian Mineral Deposit Types*; Geological Society of America: Boulder, CO, USA, 1995; Volume 8, p. 650.
51. Ault, W.U.; Kulp, J.L. Sulfur isotopes and ore deposits. *Econ. Geol.* **1960**, *55*, 73–100. [CrossRef]
52. Chaussidon, M.; Lorand, J.-P. Sulphur isotope composition of orogenic spinel lherzolite massifs from Ariege (North-Eastern Pyrenees, France): An ion microprobe study. *Geochim. Cosmochim. Acta* **1990**, *54*, 2835–2846. [CrossRef]
53. McCuaig, T.C.; Kerrich, R. P—T—T—Deformation—Fluid characteristics of lode gold deposits: Evidence from alteration systematics. *Ore Geol. Rev.* **1998**, *12*, 381–453. [CrossRef]
54. Rollinson, H. *Using Geochemical Data*; Longman Scientific & Technical Publisher: Harlow, UK, 1993; p. 352.

55. Richards, J.P. Alkalic-type epithermal gold deposits—A review. *Magmas Fluids Ore Depos. Mineral. Assoc. Can. Short Course* **1995**, *23*, 367–400.
56. Chambers, L. Sulfur isotope study of a modern intertidal environment, and the interpretation of ancient sulfides. *Geochim. Cosmochim. Acta* **1982**, *46*, 721–728. [CrossRef]
57. Strauss, H. The isotopic composition of sedimentary sulfur through time. *Palaeogeogr. Palaeoclimatol. Palaeoecol.* **1997**, *132*, 97–118. [CrossRef]
58. Hutchison, W.; Finch, A.A.; Boyce, A.J. The sulfur isotope evolution of magmatic-hydrothermal fluids: Insights into ore-forming processes. *Geochim. Cosmochim. Acta* **2020**, *288*, 176–198. [CrossRef]
59. Zhu, Q.; Xie, G.; Mao, J.; Li, W.; Li, Y.; Wang, J.; Zhang, P. Mineralogical and sulfur isotopic evidence for the incursion of evaporites in the Jinshandian skarn Fe deposit, Edong district, Eastern China. *J. Asian Earth Sci.* **2015**, *113*, 1253–1267. [CrossRef]
60. Rajabpour, S.; Hassanpour, S.; Jiang, S.-Y. Physicochemical evolution and mechanism of a skarn system: Insights from the world-class Mazraeh Cu deposit, NW Iran. *GSA Bull.* **2023**, 1–20. [CrossRef]
61. Oyman, T.; Bozan, S.; Çiçek, M.; Chiaradia, M.; Kuşcu, İ.l. Metal source and the origin of the Darıderesi Pb-Zn (Ag) veins in the Balya Mining District, NW Türkiye: Constraints from ore mineral chemistry, fluid inclusions and S-Pb isotopic signatures. *All Earth* **2023**, *35*, 210–241. [CrossRef]
62. Demirale, G. *Geology and Genesis of the Çataltepe (Lapseki/Çanakkale) Pb–Zn ± Cu ± Ag Deposit*; Ankara University: Ankara, Turkey, 2011.
63. Akıska, S.; Demirela, G.; Sayili, S. Geology, mineralogy and the Pb, S isotope study of the Kalkim Pb-Zn+/-Cu deposits, Biga Peninsula, NW Turkey. *J. Geosci.* **2013**, *58*, 379–396. [CrossRef]
64. Shimazaki, H.; Yamamoto, M. Sulfur isotope ratios of some Japanese skarn deposits. *Geochem. J.* **1979**, *13*, 261–268. [CrossRef]
65. Ma, X.-L.; Wang, K.-Y.; Li, S.-D.; Shi, K.-T.; Wang, W.-Y.; Yang, H. Geology, fluid inclusion, and H–O–S–Pb isotopic study of the Shenshan skarn Fe–Cu deposit, Southern Great Xing'an Range, Northeast China. *J. Geochem. Explor.* **2019**, *200*, 167–180. [CrossRef]
66. Jiménez-Franco, A.; Canet, C.; Alfonso, P.; González-Partida, E.; Rajabi, A.; Escalante, E. The Velardeña Zn-(Pb-Cu) skarn-epithermal deposits, central-northern Mexico: New physical-chemical constraints on ore-forming processes. *Bol. Soc. Geol. Mex.* **2020**, *72*, A270719. [CrossRef]
67. Sasaki, A.; Ishihara, S. Sulfur isotopic composition of the magnetite-series and ilmenite-series granitoids in Japan. *Contrib. Mineral. Petrol.* **1979**, *68*, 107–115. [CrossRef]
68. Seal, R.R. Sulfur isotope geochemistry of sulfide minerals. *Rev. Mineral. Geochem.* **2006**, *61*, 633–677. [CrossRef]
69. Taylor, B. Stable isotope geochemistry of ore-forming fluids. *Mineral. Assoc. Can. Short Course Handb.* **1987**, *13*, 337–445.
70. Ohmoto, H.; Goldhaber, M. Sulfur and carbon isotopes. *Geochem. Hydrothermal Ore Depos.* **1997**, *3*, 517–611.
71. Yılmazer, E.; Güleç, N.; Kuşcu, İ.; Lentz, D.R. Geology, geochemistry, and geochronology of Fe-oxide Cu (±Au) mineralization associated with Şamlı pluton, western Turkey. *Ore Geol. Rev.* **2014**, *57*, 191–215. [CrossRef]
72. Bozan, S.; Uçar, İ. Geology, Geochemistry and isotope characteristics of Karadoru/Yenice (Çanakkale) Pb-Zn-Cu (Au-Ag) Deposit. In Proceedings of the 72nd Geological Congress of Turkey with International Participation, Ankara, Türkiye, 28 January–1 February 2019; pp. 411–413.
73. Sipahi, F.; Akpınar, İ.; Eker, Ç.S.; Kaygusuz, A.; Vural, A.; Yılmaz, M. Formation of the Eğrikar (Gümüşhane) Fe–Cu skarn type mineralization in NE Turkey: U–Pb zircon age, lithogeochemistry, mineral chemistry, fluid inclusion, and OHCS isotopic compositions. *J. Geochem. Explor.* **2017**, *182*, 32–52. [CrossRef]
74. Ohmoto, H. Systematics of sulfur and carbon isotopes in hydrothermal ore deposits. *Econ. Geol.* **1972**, *67*, 551–578. [CrossRef]
75. Gerlach, T.M.; Taylor, B.E. Carbon isotope constraints on degassing of carbon dioxide from Kilauea Volcano. *Geochim. Cosmochim. Acta* **1990**, *54*, 2051–2058. [CrossRef]
76. Poreda, R.; Craig, H.; Arnorsson, S.; Welhan, J. Helium isotopes in Icelandic geothermal systems: I. 3He, gas chemistry, and 13C relations. *Geochim. Cosmochim. Acta* **1992**, *56*, 4221–4228. [CrossRef]
77. O'Leary, M.H. Carbon isotopes in photosynthesis. *Bioscience* **1988**, *38*, 328–336. [CrossRef]
78. Sano, Y.; Marty, B. Origin of carbon in fumarolic gas from island arcs. *Chem. Geol.* **1995**, *119*, 265–274. [CrossRef]
79. Tedesco, D.; Scarsi, P. Long term variations of the 3He/4He ratio and other noble gas isotopes ratios at Vulcano island (southern Italy): How the volcano works. *J. Geophys. Res.* **1998**, *104*, 10499–10510. [CrossRef]
80. Minissale, A. Origin, transport and discharge of CO_2 in central Italy. *Earth-Sci. Rev.* **2004**, *66*, 89–141. [CrossRef]
81. Clark, I. *Groundwater Geochemistry and Isotopes*; CRC Press: Boca Raton, FL, USA, 2015.
82. Weinlich, F.H. Isotopically light carbon dioxide in nitrogen rich gases: The gas distribution pattern in the French Massif Central, the Eifel and the western Eger Rift. *Ann. Geophys.* **2005**, *48*, 19–31. [CrossRef]
83. Gilfillan, S.M.; Lollar, B.S.; Holland, G.; Blagburn, D.; Stevens, S.; Schoell, M.; Cassidy, M.; Ding, Z.; Zhou, Z.; Lacrampe-Couloume, G. Solubility trapping in formation water as dominant CO_2 sink in natural gas fields. *Nature* **2009**, *458*, 614–618. [CrossRef] [PubMed]
84. Darrah, T.H.; Tedesco, D.; Tassi, F.; Vaselli, O.; Cuoco, E.; Poreda, R.J. Gas chemistry of the Dallol region of the Danakil Depression in the Afar region of the northern-most East African Rift. *Chem. Geol.* **2013**, *339*, 16–29. [CrossRef]
85. Güleç, N.; Hilton, D.R. Turkish geothermal fields as natural analogues of CO_2 storage sites: Gas geochemistry and implications for CO_2 trapping mechanisms. *Geothermics* **2016**, *64*, 96–110. [CrossRef]

86. Ohwada, M.; Satake, H.; Nagao, K.; Kazahaya, K. Formation processes of thermal waters in Green Tuff: A geochemical study in the Hokuriku district, central Japan. *J. Volcanol. Geotherm. Res.* **2007**, *168*, 55–67. [CrossRef]
87. Barry, P.; Hilton, D.; Füri, E.; Halldórsson, S.; Grönvold, K. Carbon isotope and abundance systematics of Icelandic geothermal gases, fluids and subglacial basalts with implications for mantle plume-related CO_2 fluxes. *Geochim. Cosmochim. Acta* **2014**, *134*, 74–99. [CrossRef]
88. Venturi, S.; Tassi, F.; Bicocchi, G.; Cabassi, J.; Capecchiacci, F.; Capasso, G.; Vaselli, O.; Ricci, A.; Grassa, F. Fractionation processes affecting the stable carbon isotope signature of thermal waters from hydrothermal/volcanic systems: The examples of Campi Flegrei and Vulcano Island (southern Italy). *J. Volcanol. Geotherm. Res.* **2017**, *345*, 46–57. [CrossRef]
89. Tassi, F.; Feyzullayev, A.A.; Bonini, M.; Sani, F.; Aliyev, C.S.; Darrah, T.H.; Vaselli, O.; Baghirli, R.J. Mantle vs. crustal fluid sources in the gas discharges from Lesser Caucasus and Talysh Mountains (Azerbaijan) in relation to the regional geotectonic setting. *Appl. Geochem.* **2020**, *118*, 104643. [CrossRef]
90. Demény, A.; Harangi, S. Stable isotope studies and processes of carbonate formation in Hungarian alkali basalts and lamprophyres: Evolution of magmatic fluids and magma-sediment interactions. *Lithos* **1996**, *37*, 335–349. [CrossRef]
91. Demény, A.; Ahijado, A.; Casillas, R.; Vennemann, T.W. Crustal contamination and fluid/rock interaction in the carbonatites of Fuerteventura (Canary Islands, Spain): A C, O, H isotope study. *Lithos* **1998**, *44*, 101–115. [CrossRef]
92. Zheng, Y.; Chen, J. *Stable Isotope Geochemistry*; Science Press: Beijing, China, 2000; pp. 62–118. (In Chinese)
93. Zhou, J.-X.; Huang, Z.-L.; Lv, Z.-C.; Zhu, X.-K.; Gao, J.-G.; Mirnejad, H. Geology, isotope geochemistry and ore genesis of the Shanshulin carbonate-hosted Pb–Zn deposit, southwest China. *Ore Geol. Rev.* **2014**, *63*, 209–225. [CrossRef]
94. Taylor, H.P.; Frechen, J.; Degens, E.T. Oxygen and carbon isotope studies of carbonatites from the Laacher See District, West Germany and the Alnö District, Sweden. *Geochim. Cosmochim. Acta* **1967**, *31*, 407–430. [CrossRef]
95. Veizer, J.; Hoefs, J. The nature of O18/O16 and C13/C12 secular trends in sedimentary carbonate rocks. *Geochim. Cosmochim. Acta* **1976**, *40*, 1387–1395. [CrossRef]
96. Schidlowski, M. Carbon isotopes as biogeochemical recorders of life over 3.8 Ga of Earth history: Evolution of a concept. *Precambrian Res.* **2001**, *106*, 117–134. [CrossRef]
97. Planavsky, N.; Partin, C.; Bekker, A. Carbon Isotopes as a Geochemical Tracer. In *Encyclopedia of Astrobiology*; Gargaud, M., Irvine, W.M., Amils, R., Cleaves, H.J., Pinti, D.L., Quintanilla, J.C., Rouan, D., Spohn, T., Tirard, S., Viso, M., Eds.; Springer: Berlin/Heidelberg, Germany, 2015; pp. 366–371.
98. Sheppard, S.M. Characterization and isotopic variations in natural waters. *Rev. Mineral. Geochem.* **1986**, *16*, 165–183.
99. Rollinson, H.; Pease, V. (Eds.) Using Stable Isotope Data. In *Using Geochemical Data: To Understand Geological Processes*, 2nd ed.; Cambridge University Press: Cambridge, UK, 2021; pp. 219–285.
100. Neng, J.; Jiuhua, X.; Mianxin, S. Fluid inclusion characteristics of mesothermal gold deposits in the Xiaoqinling district, Shaanxi and Henan provinces, People's Republic of China. *Miner. Depos.* **1999**, *34*, 150–162. [CrossRef]
101. Germann, K.; Lüders, V.; Banks, D.A.; Simon, K.; Hoefs, J. Late Hercynian polymetallic vein-type base-metal mineralization in the Iberian Pyrite Belt: Fluid-inclusion and stable-isotope geochemistry (S-O-H-Cl). *Miner. Depos.* **2003**, *38*, 953–967. [CrossRef]
102. Massawe, R.J.R.; Lentz, D.R. Petrogenesis and U–Pb (titanite) age of Cu–Ag skarn mineralization in the McKenzie Gulch area, northern New Brunswick, Canada. *J. Geochem. Explor.* **2022**, *232*, 106902. [CrossRef]
103. Ren, T.; Zhong, H.; Zhang, X.C. Fluid inclusion and stable isotope (C, O and S) constraints on the genesis of the high-grade Langdu Cu skarn deposit in Yunnan, SW China. *Ore Geol. Rev.* **2020**, *118*, 103354. [CrossRef]
104. Wilkinson, J. Fluid inclusions in hydrothermal ore deposits. *Lithos* **2001**, *55*, 229–272. [CrossRef]
105. Niu, P.-P.; Jiang, S.-Y.; Xiong, S.-F.; Hu, Q.-S.; Xu, T.-l. Fluid Inclusions and H-O-C-S Isotopes of the Wushan Copper Polymetallic Deposit in the Suizao Area, Hubei Province: Implications for Ore Genesis. *Geofluids* **2019**, *2019*, 3431909. [CrossRef]
106. Bodnar, R. Fluid-inclusion evidence for a magmatic source for metals in porphyry copper deposits. In *Magmas, Fluids and Ore Deposits*; Thompson, J., Ed.; Short Course Series; Mineralogical Association of Canada: Quebec, QC, Canada, 1995; Volume 23, pp. 139–152.
107. Altunkaynak, Ş.; Yılmaz, Y. The Mount Kozak magmatic complex, Western Anatolia. *J. Volcanol. Geotherm. Res.* **1998**, *85*, 211–231. [CrossRef]
108. Erkül, S.T. Petrogenetic evolution of the Early Miocene Alaçamdağ volcano-plutonic complex, northwestern Turkey: Implications for the geodynamic framework of the Aegean region. *Int. J. Earth Sci.* **2012**, *101*, 197–219. [CrossRef]
109. Erkül, S.T.; Erkül, F. Comment on "Al-in-Hornblende Thermobarometry and Sr-Nd-O-Pb Isotopic Compositions of the Early Miocene Alaçam Granite in NW Anatolia (Turkey)". *Turk. J. Earth Sci.* **2013**, *22*, 354–358. [CrossRef]
110. Karacık, Z.; Yılmaz, Y. Geology of the ignimbrites and the associated volcano–plutonic complex of the Ezine area, northwestern Anatolia. *J. Volcanol. Geotherm. Res.* **1998**, *85*, 251–264. [CrossRef]
111. Hasözbek, A.; Erdoğan, B.; Satir, M.; Siebel, W.; Akay, E.; Doğan, G.D.; Taubald, H. Al-in-hornblende thermobarometry and Sr-Nd-O-Pb isotopic compositions of the early miocene Alaçam granite in NW Anatolia (Turkey). *Turk. J. Earth Sci.* **2012**, *21*, 37–52. [CrossRef]
112. Bouabdellah, M.; Jabrane, R.; Margoum, D.; Sadequi, M. Skarn to Porphyry-Epithermal Transition in the Ouixane Fe District, Northeast Morocco: Interplay of Meteoric Water and Magmatic-Hydrothermal Fluids. In *Mineral Deposits of North Africa*; Bouabdellah, M., Slack, J.F., Eds.; Springer International Publishing: Cham, Switzerland, 2016; pp. 201–225.

113. Kwak, T.; Tan, T.H. The importance of $CaCl_2$ in fluid composition trends; evidence from the King Island (Dolphin) skarn deposit. *Econ. Geol.* **1981**, *76*, 955–960. [CrossRef]
114. Lai, J.; Chi, G.; Peng, S.; Shao, Y.; Yang, B. Fluid evolution in the formation of the fenghuangshan Cu-Fe-Au deposit, Tongling, Anhui, China. *Econ. Geol.* **2007**, *102*, 949–970. [CrossRef]

minerals

Article

Fluid Evolution of Greisens from Krupka Sn-W Ore District, Bohemian Massif (Czech Republic)

Michaela Krejčí Kotlánová [1,2,3], Zdeněk Dolníček [4,*], Miloš René [5], Walter Prochaska [6], Jana Ulmanová [4], Jaroslav Kapusta [7], Vlastimil Mašek [†] and Kamil Kropáč [7]

1 Department of Geological Sciences, Masaryk University, Kotlářská 267/2, 611 37 Brno, Czech Republic; kotlmi@seznam.cz
2 Research Institute for Building Materials, Hněvkovského 30/65, 617 00 Brno, Czech Republic
3 BIC Brno Spol. s r.o., Technology Innovation Transfer Chamber, Purkyňova 648/125, 612 00 Brno, Czech Republic
4 Department of Mineralogy and Petrology, National Museum, Cirkusová 1740, 193 00 Praha, Czech Republic; jana.ulmanova@nm.cz
5 Institute of Rock Structure and Mechanics, Academy of Sciences of the Czech Republic, V Holešovičkách 94/41, 182 09 Praha, Czech Republic; rene@irsm.cas.cz
6 Österreichisches Archäologisches Institut, Franz Klein-Gasse 1, 1190 Wien, Austria; walter.prochaska@oeaw.ac.at
7 Department of Geology, Palacký University, 17. Listopadu 1192/12, 771 46 Olomouc, Czech Republic; jaroslav.kapusta@upol.cz (J.K.); kamil.kropac@upol.cz (K.K.)
* Correspondence: zdenek.dolnicek@nm.cz
† Deceased author.

Abstract: The Sn-W ore deposits in the Krupka surroundings are associated with greisens, which occur in the upper parts of Late Variscan granitoid intrusions. Fluid inclusions were studied in samples of quartz, cassiterite, apatite, fluorite, and topaz in greisenized granites, greisens, and hydrothermal veins with Sn-W mineralization. The greisenization process took place at temperatures 370–490 °C and pressures 155–371 bars, and associated fluids had predominantly low salinity and a low gas (CO_2, N_2 and CH_4) content. The post-greisenization stage was connected with the formation of (i) low-salinity (0–8 wt. % NaCl eq.) fluid inclusions with homogenization temperatures <120–295 °C and (ii) high-salinity (18 to >35 wt. % NaCl eq.) fluid inclusions with homogenization temperatures 140–370 °C, often containing trapped crystals of quartz, topaz, and sulfides, or daughter crystals of salts and carbonates, which were identified by microthermometric measurements, electron microprobe analysis, and Raman spectroscopy. Analyses of fluid inclusion leachates have shown that Na and Ca chlorides predominate in fluids. According to hydrogen stable isotopes, the source of greisenizing and post-greisenizing fluids was not only magmatogenic but also meteoric water or fluids derived from sedimentary rocks.

Keywords: Krupka; Bohemian Massif; greisen; Sn-W mineralization; fluid inclusions; stable isotopes of hydrogen; crush-leach analyses; Raman spectroscopy

Citation: Krejčí Kotlánová, M.; Dolníček, Z.; René, M.; Prochaska, W.; Ulmanová, J.; Kapusta, J.; Mašek, V.; Kropáč, K. Fluid Evolution of Greisens from Krupka Sn-W Ore District, Bohemian Massif (Czech Republic). *Minerals* **2024**, *14*, 86. https://doi.org/10.3390/min14010086

Academic Editors: Yuichi Morishita and Napoleon Q. Hammond

Received: 19 December 2023
Revised: 3 January 2024
Accepted: 8 January 2024
Published: 11 January 2024

1. Introduction

The Erzgebirge/Krušné hory mountains are rich in occurrences of Sn-W mineralization bound to greisens, which form the apical parts of Late Variscan granitoid intrusions. In the western part of the Saxothuringian Zone of the Bohemian Massif, these occurrences are more abundant and are represented by deposits in Krásno near Horní Slavkov, in Přebuz, Rolava, and Podlesí in the Blatná granite body. In the east, the most important localities, where tin and tungsten ores were historically mined, are Krupka and Cínovec.

Older (OIC) and younger intrusive complex (YIC), differing in age and rock chemistry, were distinguished in the Krušné hory/Erzgebirge Batholith [1]. Opinions on the formation of highly fractionated granites and associated greisens are evolving, and there are many supporters of both magmatic and metasomatic origins of fluids that form these rocks. [2] assumed that all YIC granites were formed by the hydrothermal transformation of granites belonging to OIC granites. The mentioned author developed the theory of autometamorphism, which explains the transformation of granites in the postmagmatic stage due to the activity of cogenetic hydrothermal solutions. Ref. [3] considered the zonal structure of the YIC granites, the theory of albitization, and other metasomatic processes. Many authors are inclined toward the metasomatic origin of fluids that can form YIC granites, e.g., [4,5].

An often-accepted theory of the formation of highly fractionated granites involves their primary magmatic origin, where enrichment in volatile components and incompatible elements occurs during fractional crystallization [6]. When the magma rises to the surface, gradually smaller volumes of more differentiated melts are enriched in incompatible components (Rb, Cs, Li and others), including H_2O and other volatile substances (F, B), eventually leading to their segregation [7]. Another clue that supports a magmatogenic origin of fluids is the finding of magmatic breccias at several localities, e.g., Krupka, Horní Slavkov, and Podlesí. These rocks are formed by fragments of gneisses or phyllites cemented by granite matrix [8–11]. Many authors accept the opinion that greisenization fluids have a magmatogenic origin—e.g., [12,13].

The research of stable isotopes of hydrogen and oxygen in altered (greisenized) granites from the locality Cínovec and Krásno-Hub Stock [14] showed that the origin of greisenizing fluids is not only magmatogenic but partly also in meteoric waters and fluids derived from sedimentary rocks.

From a mineralogical and petrological point of view, the greisens in the Czech part of the Krušné hory/Erzgebirge Mts. are relatively well described, but comprehensive information about the genesis and formation conditions of greisens is lacking. Investigations of fluid inclusions in the minerals of greisens from the Czech part of the Saxothuringian Zone of the Bohemian Massif were mainly carried out by [15–18]. A large amount of data on the topic of fluid inclusions in greisens and associated hydrothermal veins from the Krupka ore district was collected by [15]. The aim of this research was to collect new data on temperature and pressure conditions of the formation of greisens and associated hydrothermal veins and try to determine the origin and composition of greisenizing and younger fluids at the Krupka ore district. This article expands the topic with new knowledge obtained by modern research methods.

2. Geological Settings

2.1. The Krušné Hory/Erzgebirge Crystalline Complex

The Krušné hory/Erzgebirge Crystalline Complex occurs in the northwestern part of the Bohemian Massif and is formed by a diverse sequence of metamorphic rocks of various protolith ages ranging from the Precambrian to the Lower Paleozoic. The intensity of Variscan metamorphosis decreases to the northwest. In the northeast, the rocks are represented by Proterozoic paragneisses, and in the southwest, by Ordovician phyllites and quartzites. The central part of the Saxothuringian unit is formed by Cambrian mica schists. In the north and northwest, Saxothuringian units sink beneath Permo-Carboniferous, Mesozoic, Tertiary, and Quaternary sediments [19,20]. Crystalline units are intruded by granitic plutons of the Variscan age (Figure 1).

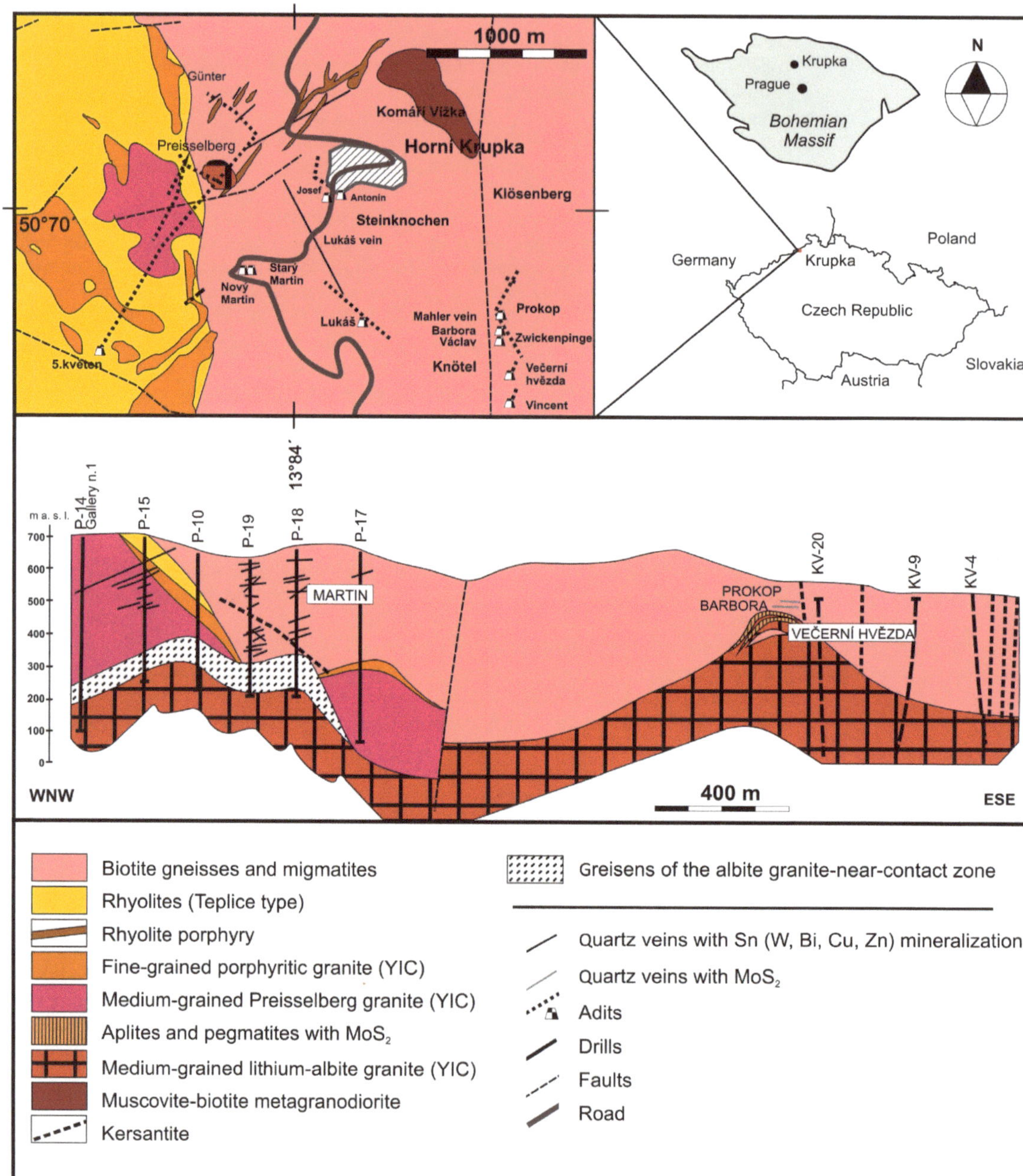

Figure 1. Geological map of the Krupka ore district with the location of the main ore fields and abandoned mine workings (modified after [21]) and the WNW–ESE geological cross-section through the Krupka surroundings (modified after [22]).

2.2. *Krušné Hory/Erzgebirge Batholith*

The Krušné hory/Erzgebirge Batholith was considered by the first geologists to be a single body [23,24], but it was later found that in space and time, there are several intrusions with their own evolution and different ages [25,26]. The Karlovy Vary, Nejdek-Eibenstock, Kirchberg, and Bergen plutons are described in the west of the Saxothuringian Zone of the Bohemian Massif. From the central part, the Geyer granite body, the Hora Svaté Kateřiny, and Hora Svatého Šebestiána bodies are known. In the east, the Telnice granite body and the Altenberg caldera occur [26–28].

As mentioned in the introduction, two types of granites were distinguished within the Krušné hory/Erzgebirge Batholith, differing in their age and chemical composition. The low-F granites of OIC included porphyritic biotite granites, granodiorites, and granites. The high-F granites of YIC were represented by coarse-grained, sometimes porphyritic, muscovite-biotite or tourmaline-biotite granites and Li-mica granites, which are Si-rich, reduced, and strongly peraluminous and are often affected by processes of autometamorphism and have increased contents of incompatible elements [1]. The Li-mica granites have crustal isotopic signatures [26]. The transitional types are sometimes present between OIC and YIC. According to chemistry, two subtypes of OIC and YIC granites—weakly and strongly peraluminous—are further distinguished [25]. Strongly peraluminous granites correspond to S-type granites, are enriched in P, and have relatively low content of HFSE and HREE. The weakly peraluminous granites correspond to A- and I-type granites and have a very low content of P and increased amounts of HFSE and HREE. The granites in the Krupka ore district belong to the younger generation of slightly peraluminous P-poor granites [25]. Compared to the rest of European Variscides, the Krušné hory/Erzgebirge Batholith exhibits a relatively high abundance of Li-mica granites. However, the reason is still not completely resolved because the isotopic composition of Li-mica granites does not differ significantly from peraluminous granites from other regions. However, the underlying gneisses and metasediments have high contents of incompatible elements, and this can also be an explanation for the different composition of granites in the Krušné hory/Erzgebige Batholith [26]. Granites of OIC are of the Upper Viséan–Westphalian age (340 Ma to 310 Ma) [20,29]. The age of the granites of YIC was estimated at 285 to 329 Ma [30–33]. Granites of YIC are often affected by greisenization [13,17,18] and connected with economically important accumulations of Sn, W, and Li.

2.3. Geological Situation of the Krupka Ore District

The geological situation in the Krupka ore district is relatively complicated. Two geological units meet there: (i) the gneiss complex, composed of Freiberg orthogneisses in the east and in the west with biotite paragneisses, and (ii) the Teplice rhyolite complex with different porphyritic rocks. The Teplice rhyolite complex is, in places, crosscut by granite porphyries and intruded by the granites of the younger intrusive complex (see Figure 1).

There are three ore deposits around Krupka town surroundings—Preisselberg, Knötel, and Komáří Vížka. The Preisselberg body consists mainly of fine-grained porphyritic biotite granite. In the central part, a drill hole reached lithium albitized apogranite with quartz-topaz-(zinnwaldite) greisen in its apical part. Two stages of greisenization are described here. The older one is connected with the formation of dark W-bearing mica-quartz greisen, whereas the younger one resulted in the formation of quartz-topaz-mica greisen. Ore mineralization consists of wolframite, cassiterite, scheelite, and sulfides. In the northeast, there is present an important hydrothermal vein, Lukáš, with a thickness of up to 0.5 m and abundant Sn-W mineralization. In the vicinity of the vein, the greisenization of rocks connected with the formation of Li-Fe micas and topaz is evident. The marginal parts of the Preisselberg granite are lined with pegmatite, composed mainly of quartz and K-feldspar [22]. At Komáří Vížka, gneisses are intersected by felsic quartz porphyries affected by greisenization. Greisens are quartz-Li-mica-topaz rocks with cassiterite and abundant amounts of sulfides. The Knötel deposit is the largest district in this area and is located in the north of Krupka. Under the mantle gneisses, which are sometimes crosscut by aplite veins and composed of quartz, topaz, micas, and feldspars, hidden bodies of lithium apogranite have been found by drill holes [34]. These rocks are greisenized in places. At Prokop Stock, which forms the apical part of the granite body, there is an occurrence of quartz greisen and quartz veins with abundant molybdenite and fluorite in the top parts of the body. Refs. [35,36] refer to these light rocks as quartz greisens, hydrothermal quartzites, or quartzite-like rocks. These rocks are formed mainly by quartz in association with molybdenite, fluorite, and accessory topaz, mica, clay, and ore minerals. The above-mentioned authors describe nests/pockets of massive quartz with molybdenite. Quartzite-like rocks contain gneiss xenoliths or form the matrix of breccias with fragments of surrounding gneisses and pegmatites. The lower parts of the body are formed by pegmatite, which was encountered by the adits Barbora and Večerní hvězda and consists mainly of K-feldspar, coarse-grained quartz, and biotite. Pegmatite is crosscut by aplite veins, quartz veins, and vein-type (crack-type) greisens. The mineral composition of greisens ranges from zinnwaldite-quartz greisen to light gray quartz greisen with topaz [22]. Ref. [37] distinguished four stages of the formation of ore mineralization in the Krupka ore district—albitization, greisenization, the formation of sulfides, and the formation of fluorite and carbonates.

3. Materials and Methods

Some samples for the study were collected from dump material occurring in the Krupka ore district. Another part of the samples was borrowed from the collections of the National Museum in Prague and Mgr. Jakub Mysliveček from Czech Geological Survey.

The doubly polished thin sections were prepared for the study of the fluid inclusions in quartz, topaz, cassiterite, fluorite, and apatite. The sections were examined using a polarizing microscope in reflected light and subsequently also in backscattered electrons (BSE) on electron microprobe Cameca SX-100 (AMETEK, Inc., Berwyn, PA, USA) in the National Museum in Prague.

Petrography of fluid inclusions was studied using polarizing microscopes Olympus BX50 (Olympus Co., Tokyo, Japan) and Nikon Eclipse LV100ND (Nikon Co., Tokyo, Japan). Then, microthermometry of fluid inclusions was performed using an Olympus BX51 microscope equipped with a Linkam THMSG 600 heating-freezing microthermometric chamber (Linkam Scientific Instruments, Surrey, UK) at the Department of Geology, Palacký University in Olomouc, Czech Republic. Primary, pseudosecondary, and secondary inclusions were studied in order to characterize all stages of fluid evolution associated with Sn-W mineralization, greisenization, and post-greisenization stages. The following parameters were measured: freezing temperature (Tf), melting temperature of the last crystal of ice (Tm_{ice}), homogenization temperature to vapor (Th_V), homogenization temperature to liquid (Th_L), homogenization temperature critical (Th_C), eutectic temperature (Te), clathrate melting temperature (Tm_{cla}), and dissolution temperature of halite (Tds_h). The measurement was calibrated between −56.6 and 374.1 °C with inorganic standards and natural fluid inclusions with known temperatures of phase transitions. The reproducibility of measurement is within 0.1 °C for temperatures between −56.6 and 0 °C, and within 1 °C for temperature of 374.1 °C.

The salinity of inclusions enclosing the H_2O-NaCl system (with Tm_{ice} from 0 to −21.3 °C) was calculated according to [38]. The composition, density, and molar volume of volatile gas-bearing FI were calculated using the programs ICE and BULK according to [39,40]. For fluid inclusions trapped from a heterogeneous fluid, the bulk density and composition of the inclusions were calculated in the BULK program according to [41], and then the pressure conditions of inclusion entrapment were calculated in the ISOC program according to [42,43] for H_2O-CO_2-NaCl fluids and [44,45] for H_2O-NaCl fluids. Programs ICE, BULK, and ISOC are included in the FLUIDS 1 and FLUIDS 2 program packages, available on the website of the University of Leoben [46].

Fluid inclusion leachates were analyzed in selected samples of quartz from greisens and veins. Samples were first crushed to a fraction of 0.3–1.1 mm and subsequently boiled in distilled water. Thereafter, the samples were dried, and other mineral phases and impurities were handpicked under a binocular microscope. A weighed quantity of quartz (1 g) was washed in deionized water and subsequently dried at 50 °C. One gram of the dried sample was ground in an agate mortar together with 5 mL of deionized water. Subsequently, the suspension was filtered through a 0.2 μm nylon filter. Analysis of the filtrate was performed using an ion chromatograph Dionex DX-500 (Thermo Fisher Scientific, Inc., Waltham, MA, USA). Detection limits are as follows: Cl (10 ppb), Br (2 ppb), F (5 ppb), I (0.1 ppb), and SO_4 (10 ppb).

In selected samples, Raman analysis was used to identify volatiles and solid phases enclosed in fluid inclusions. The analyses were carried out at the Institute of Molecular and Translational Medicine, Faculty of Medicine, Palacký University, Olomouc (Czech Republic). The samples were examined on the WITec Confocal Raman Imaging Microscope System alpha300 R+ spectrometer (Oxford Instruments, Abingdon, Oxfordshire, UK) with an excitation of 532 nm (25 mW power incident on the sample, lens 50× /NA 0.8, spectrum acquisition time 1 min). Part of the Raman analysis was carried out at the National Museum in Prague (Czech Republic) on the DXR dispersive Raman Spectrometer (Thermo Fisher Scientific, Inc., Waltham, MA, USA) mounted on a confocal Olympus microscope. The Raman spectra were collected in the range 4000–30 cm^{-1}; however, for our purposes, the unnecessary part of the spectrum was subsequently cut off. The instrument was set up by a software-controlled calibration procedure using multiple neon emission lines (wavelength calibration), multiple polystyrene Raman bands (laser-frequency calibration) and standardized white-light sources (intensity calibration). The Raman signal was excited by an unpolarized green 532 nm solid state, diode-pumped laser, and detected by a CCD detector. The experimental parameters: 100× objective, 5 s exposure time, 100 exposure, 50 μm pin spectrograph aperture, and 10 mW laser power level. The spectra were repeatedly acquired from different places/focus depths in order to obtain a representative spectrum with the best signal-to-noise ratio. The eventual thermal damage of the measured point on solid phases was excluded by visual inspection of the excited surface after measurement, by observation of possible decay of spectral features at the start of excitation and by checking for thermal downshift of Raman lines. Spectral manipulations were performed using the Omnic 9 software (Thermo Scientific). Publications of [47,48] were used for the determination of individual gaseous components, and the identification of solid phases was carried out according to the data contained in [49]. The quantification of gases in the non-aqueous phase of fluid inclusions was based on the measurement of peak areas [47] and calibration using the natural fluid inclusions with known compositions.

From a sample of vein quartz containing fluid inclusions with solid phases, a piece of approx. 0.5 × 0.5 cm was chipped off. Subsequently, the fresh surface of the piece was coated with a thin layer of carbon and then studied using a Jeol-JXA 8600 electron microprobe (JEOL Ltd., Akišima, Japan) at the Department of Geology of the Palacký University in Olomouc. Backscattered electron (BSE) microphotographs of selected opened fluid inclusions were taken at an acceleration voltage of 15 kV and beam current of 10 nA, and individual phases were identified by means of EDS spectra.

Hydrogen isotopes were analyzed in selected micas and clay minerals. The impurities were manually separated from the samples under a stereoscopic binocular microscope. Then, the selected minerals were ground in an agate mortar to obtain 5–10 mg of the sample powder. Samples were analyzed following the method by [50] at Eberhard Karls Universität, Tübingen, Germany. The samples were degassed in a vacuum line at 110 °C overnight and then heated until fusion to liberate all the water. Any potentially produced H_2 was oxidized to water with a CuO trap. Water and CO_2 were separated using liquid N_2 and ethanol cold traps. Water was then reduced by Zn at approximately 400 °C, and the isotopic composition of the resulting H_2 was measured on a Finnigan MAT 252 mass spectrometer (Finnigan MAT GmbH, Bremen, Germany). The results of isotope analysis are given in δD values concerning the V-SMOW standard. The measurement error is ±2‰. The published fractionation factors were used to calculate the H isotopic composition of the fluids [51,52].

Clay minerals from the vein cavity and from the greisen surface were scraped off, dried at 50 °C for approximately 24 h and subsequently ground in an agate mortar. The powdered samples were placed on a Si wafer and dripped with isopropyl alcohol. The qualitative XRD analysis of the samples was performed on the X-ray diffractometer Bruker

D8 Advance (Bruker Co., Billerica, MA, USA) with Cu anode with variable divergent aperture screens at Θ-Θ reflective Bragg-Brentano parafocusation geometry on Research Institute for Building Materials in Brno, Czech Republic. The XRD pattern was measured in the range $2\Theta = 5–100^\circ$, but subsequently, a part of the spectra was cut off due to the fact that the main peaks of clay minerals are located at low 2Θ angles. The step of the measurement was 0.020° (2Θ), and the step time was 188 s.

4. Results

4.1. Sample Types and Mineralogy

About 50 samples, including greisenized gneisses, greisenized granites, greisens, and hydrothermal veins connected with Sn-W mineralization from the Krupka ore district, were collected or borrowed to study.

Most samples were collected in the Knötel ore deposit. On the Prokop Stock, the greisenization mainly affects mantle gneisses and granites of the younger intrusive complex. Samples collected there were predominantly greisenized gneisses and quartz greisens with abundant molybdenite (Figure 2a), fluorite (Figure 2b), accessory topaz (Figure 2c), cassiterite (Figure 2d), wolframite, and apatite (Figure 2e1,e2). Greisens are, in many cases, crosscut by hydrothermal veins composed of quartz, fluorite, accessory wolframite, and sulfides. A clay mineral corresponding to dickite and less illite (confirmed by XRD analysis; Figure 3) is present in the greisens and vein cavities. Greisens from the dump material of adit Sedmi spáčů are rich in sulfides, especially chalcopyrite, which form inclusions or nests in the rock (Figure 2f). In the Preisselberg area, there are Preisselberg-type granites, which are biotite granites in places affected by greisenization (Figure 2g). Samples of greisenized granites and greisens were collected from this area. Greisens are sometimes crosscut by quartz veins up to 0.5 cm thick (Figure 2g). Many samples come from the Lukáš ore vein. Very well to perfectly limited quartz crystals (Figure 2h) reach a size up to 7 × 3 cm and have a zonal structure that is often visible to the naked eye.

Figure 4a shows a fine-grained matrix of greisen from the Preisselberg area, which is cut by a quartz vein with quartz rich in fluid inclusions. Cassiterite often shows a zonal structure (Figure 4b), showing different interference colors and pleochroism in individual zones. In samples from Knötel ore deposits this mineral encloses wolframite (Figure 4c). Micas (corresponding to protolithionite to zinnwaldite by their chemistry), enclosed in quartz, also show a zonal structure and enclose small grains of cassiterite (Figure 4d). Inclusions of wolframite often occur in cassiterite (Figure 4e). Apatite (Figure 4f) and a TiO_2 phase with a distinctly zonal structure were found in association with ore minerals (Figure 4f. The present study is based on a detailed laboratory investigation of 15 representative samples, including greisenized granites, greisens, and hydrothermal veins connected with Sn-W mineralization of the Krupka ore district (Table 1).

4.2. Petrography of Fluid Inclusions

Fluid inclusions were studied in quartz, cassiterite, topaz, fluorite, and apatite from greisens, greisenized granites, and hydrothermal veins. Fluid inclusions in quartz are very abundant. All genetic types—primary (PFI), pseudosecondary (PSFI), and secondary (SFI) fluid inclusions occur in most of the samples. In the case of topaz and apatite, fluid inclusions occur less frequently. Primary inclusions are two-phase, containing gas and aqueous components (Figure 5a). In greisens and greisenized granites, fluid inclusions (PFI1) usually have a lower degree of the filling (LVR—liquid/vapor ratio: 0.05 to 0.6) than inclusions from hydrothermal veins (0.4–0.95; PFI2). The shape of the inclusions is irregular in most cases, isometric, or less often, the inclusions have the shape of a negative crystal (Figure 5b). PFI are predominantly distributed as solitary inclusions (Figure 5c) or form small clusters. In the case of vein crystals of quartz, which have a visible growth zonation, the fluid inclusions sometimes follow the growth zones. This trend is also observed in some of the cassiterite crystals. The size of the primary inclusions varies 15–120 μm; the largest inclusions were found in quartz from greisens, and the smallest in fluorite from veins.

Figure 2. The texture and mineral composition of studied samples from the Krupka ore district. (**a**) Greisen sample with molybdenite from Prokop Stock, sample KR-11; (**b1**) Quartz greisen with hydrothermal vein with a cavity filled with fluorite, sample KR-12; (**b2**) Detail of cavity with fluorite crystals; (**c**) Topaz on greisen sample from Prokop Stock; sample JM-1; (**d**) Greisen sample with cassiterite from Knötel area, sample KR-10; (**e1**) Violet fluorite and light blue apatite on greisen sample from adit Martin, sample MA-3; (**e2**) Detail of apatite from (**e**); (**f**) Greisen sample with an abundant amount of sulfides, predominantly chalcopyrite from dump material from adit Sedmi spáčů, Knötel area, sample KR-3; (**g**) Sample of greisenized granite with dark greisen zones around quartz vein from Knötel area, adit Preisselberg II—sample PRK-1; (**h**) Perfectly limited crystal of quartz from Steinknochen area, Martin adit—sample KNM-3.

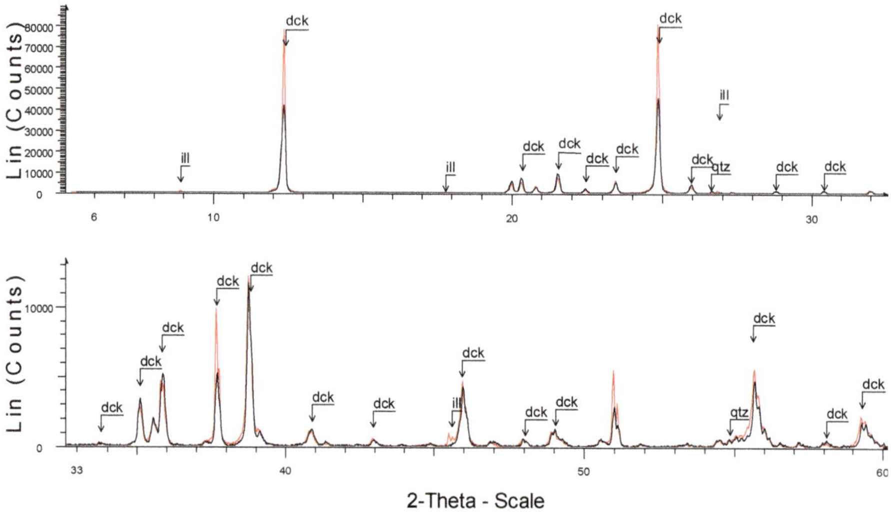

Figure 3. XRD records of clay minerals from greisen sample KR-4 (black spectrum) and hydrothermal vein sample KR-12 (red spectrum) from Knötel area; dck—dickite, ill—illite, qtz—quartz.

Table 1. List of samples from the Krupka ore district used for this study.

Sample	Locality	Type of Sample	Mineral Assemblages
KNM-1	Knötel, Prokop Stock	greisen	quartz, Li-mica, wolframite, molybdenite
KNM-2	Knötel	greisen	quartz, Li-mica, apatite, cassiterite, TiO_2 phase, wolframite
KNM-3	Steinknochen, Martin adit	hydrothermal vein	quartz, oxi-hydroxides of Fe
KNM-4	Knötel	greisen	quartz, Li-mica, wolframite
KR-17	Knötel	greisenized granite	quartz, micas, K-feldspar, plagioclase, cassiterite,
KR-3	Knötel, adit Sedmi spáčů	greisen	quartz, Li-mica, cassiterite, wolframite, chalcopyrite, pyrite, galena, arsenopyrite
KR-4	Knötel	greisen	quartz, Li-mica, wolframite, dickite, illite, fluorite
KR-5	Knötel	greisen	quartz, protolithionite/zinnwaldite, cassiterite
KR-7	Knötel	greisen	quartz, phengite/Li-phengite, molybdenite, fluorite, woframite
KR-8	Knötel, Prokop Stock	greisen	quartz, molybdenite, wolframite, fluorite, dickite
KR-10	Knötel, Prokop Stock	greisen	quartz, Li-mica, cassiterite, wolframite
KR-11	Knötel, Prokop Stock	greisen	quartz, mica, molybdenite, Bi-minerals
KR-12	Knötel, Prokop Stock	greisen	quartz, wolframite, fluorite, clay minerals
KR-12	Knötel, Prokop Stock	hydrothermal vein	quartz, fluorite, clay mineral
PRK-1	Preisselberg	greisenized granite	quartz, plagioclase, K-feldspar, Li-mica, cassiterite, zircon
PRK-1	Preisselberg	greisen	quartz, Li-mica, zinnwaldite, muscovite, cassiterite, topaz
PRK-1	Preisselberg	hydrothermal vein	quartz

Pseudosecondary inclusions (PSFI) occur mainly in quartz from greisenized granites, greisen, and veins. They are found less frequently in cassiterite and fluorite (Figure 5d) and were not detected in topaz and apatite. PSFIs are two-phase, gas-liquid with LVR 0.3 to 0.95 and reach a maximum size of 45 µm. Two types of pseudosecondary inclusions were distinguished according to their degree of filling (PSFI1-LVR: 0.3–0.7; PSFI2-LVR: 0.7–0.95). PSFIs often have an irregular or oval shape. Inclusions with a negative crystal shape and drop-like-shaped inclusions occur less often. The distribution of inclusions is as short trails not passing through the whole grains or forming small clusters. Inclusions often have varying degrees of filling, especially in quartz from greisens (Figure 5e).

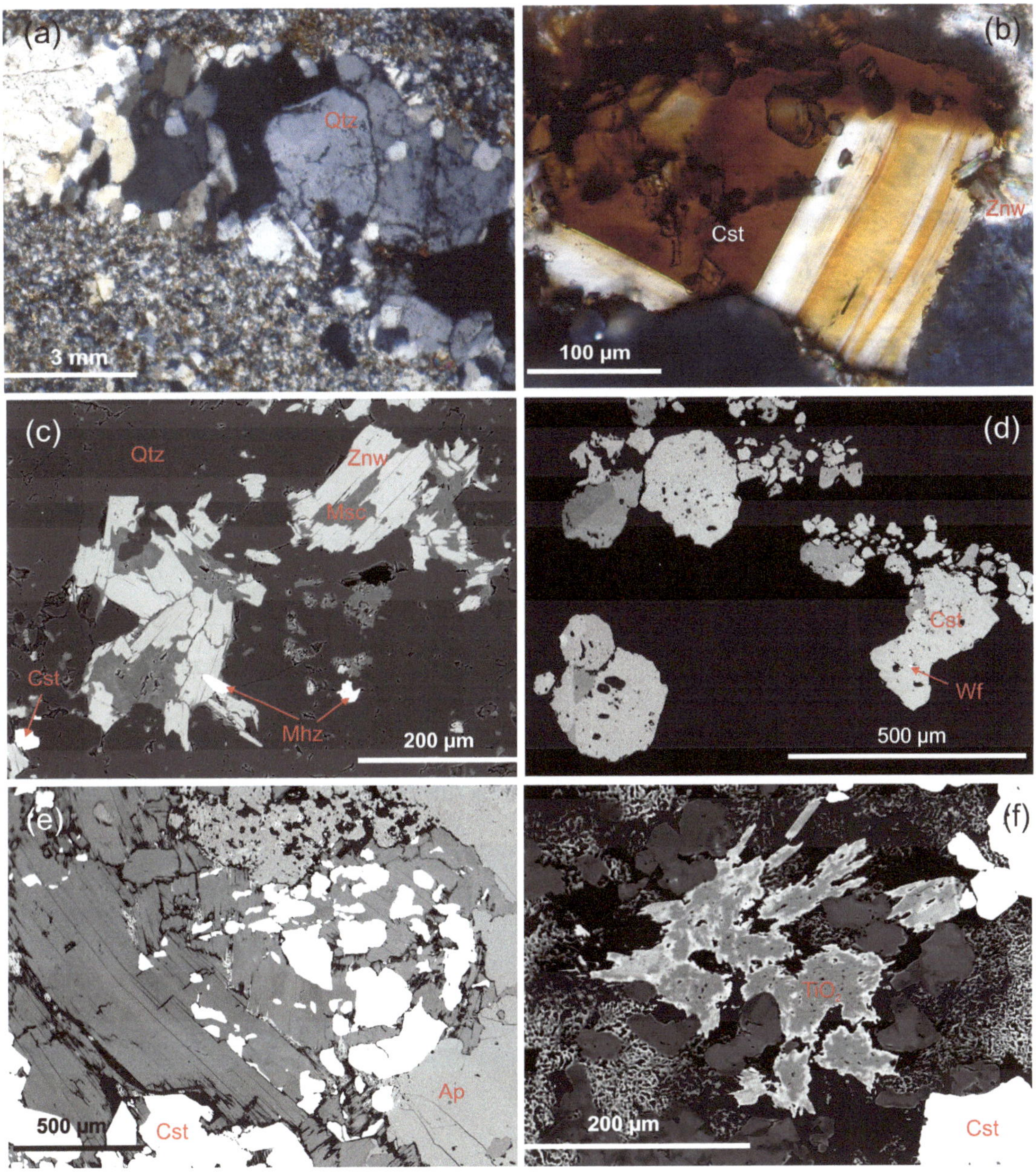

Figure 4. Mineral assemblages and texture of the samples. (**a**) Greisen from Preisselberg (sample PRK-1) cut by quartz vein (Qtz), polarized light, XPL; (**b**) Grain of cassiterite (Cst) with zonal structure and zinnwaldite (Znw) fan-like crystals in greisenized granite from Preisselberg (PRK-1), polarized light, XPL; (**c**) Zonal tabular crystals of micas corresponding to muscovite (Msc) and zinnwaldite (Znw) enclosed in quartz (Qtz) from greisen (PRK-1), BSE image; (**d**) Wolframite grains enclosed in cassiterite, BSE image; (**e**) Cassiterite (Cst) in mica interlayers and in association with apatite (Ap) from greisen from Knötel–Prokop Stock (KNM-2), BSE image; (**f**) Zonal TiO_2 phase in association with quartz and cassiterite (Cst) in greisen from Knötel–Prokop Stock (KNM-2), BSE image.

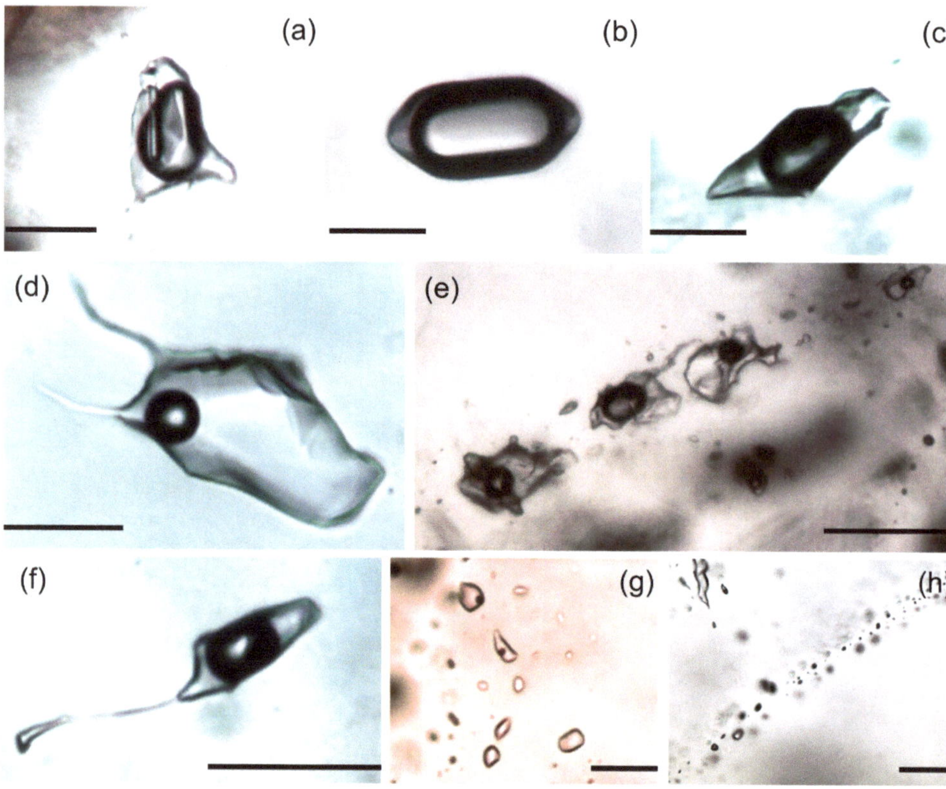

Figure 5. Appearance of fluid inclusions in minerals from the Krupka ore district in polarized light (PPL) at room temperature: (**a**) Primary inclusion from topaz from greisen sample KNM-2; (**b**) Vapor-rich negative crystal-shaped primary fluid inclusions from quartz from greisen sample KR-11 from Knötel; (**c**) Solitary primary fluid inclusion from quartz from vein sample KNM-3; (**d**) Pseudosecondary inclusion from fluorite from vein sample KR-12; (**e**) Short trail of pseudosecondary inclusions from quartz from greisen with variable LVR, sample KR-12; (**f**) Secondary inclusion from quartz from greisenized granite affected by necking-down, sample PRK-1; (**g**) Group of secondary liquid and liquid-vapor fluid inclusions from greisen, sample KR-12; (**h**) Trail of secondary inclusions from vein quartz; sample PRK-1. Scale bar is 20 μm for all microphotographs.

For secondary inclusions, three types were distinguished, differing in the degree of filling and the presence of solid phases enclosed in inclusions. Secondary inclusions of the first type (SFI1) occur in quartz and less often in cassiterite from greisen and greisenized granites. They have a lower LVR (0.4–0.95) than inclusions of the second type (SFI2; LVR 0.7–1.0), which occur predominantly in vein minerals. SFI1 inclusions are situated on cracks and form trails that mostly cross the whole grains; sometimes, they continue through the neighboring grain. Secondary inclusions in quartz and cassiterite are often affected by the necking-down (Figure 5f). Stretched inclusions with varying degrees of filling often occur side by side, with some inclusions containing liquid phase. SFI2 inclusions are two-phase but quite often are also single-phase, containing only the liquid phase. SFI2 inclusions are usually smaller than SFI1, and the shape of the inclusions is oval, spherical, or irregular. Inclusions form small groups on cracks (Figure 5g,h) and occur predominantly in quartz, less often in cassiterite and fluorite. The youngest generation of SFI2 are fluid inclusions that intersect whole grains and sometimes also several grains, have a high degree of filling, and often are all-liquid. Opaque SFI1 or PSFI1 were found in some samples of cassiterite. The third type of secondary inclusions (SFI3) contains, in addition to the gaseous and aqueous phase, one or more solid phases (Figure 6a–j) and is only found in quartz (Figure 6a–i) and cassiterite (Figure 6j). The amount of vapor phase in this type of inclusion does not exceed 50 vol. %. The shape of the inclusions is irregular, or the inclusions have the shape of a negative crystal (Figure 6a,i). Inclusions reach up to 70 μm in size. SFI3 inclusions occur on trails crossing the grains. Some multiphase inclusions occur as solitary inclusions, and some may also perhaps be classified as pseudosecondary inclusions. SFI3 forming short trails composed of only a few inclusions often exhibit decreasing sizes of inclusions toward the grain core, indicating the healing of small cracks in the minerals. In a few cases of FI on cracks, these inclusions have different degrees of filling, with inclusions toward the grain edge having a higher content of the gaseous phase than inclusions toward the grain core. As far as the age of SFI3, it seems most likely that the solitary or small group inclusions represent the older generation of SFI3, followed by inclusions healing small cracks or forming trails intersecting whole grains. Older solitary multiphase inclusions have a large size and often contain more solid phases than inclusions on trails.

Identification of some solids in SFI3 inclusions was performed by microthermometric measurements; other phases were identified by Raman spectroscopy and EDS analysis. Quartz, carbonates (calcite, siderite), and an unspecified sulfate were identified by Raman spectroscopy. The Raman spectrum of one of the analyzed carbonates is shown in Figure 7. An intense peak at 1085 cm^{-1} and a less intense one at 736 cm^{-1} correspond with a high probability of siderite, according to [49]. An unspecified sulfate has the shape of needles with dimensions of approximately 9 × 1 μm. It was found in only one quartz-hosted secondary inclusion from the vein sample KR-12. The Raman spectrum of this phase is characterized by peaks at 449, 1001, and 1064 cm^{-1}. It is quite difficult to identify chlorides using Raman spectroscopy, as they often have only a weak signal and in combination with strong fluorescence, identification is often impossible. A vein quartz sample from Prokop Stock was studied using an electron microprobe focusing on fluid inclusions containing solid phases. The EPMA revealed the presence of quartz (Figure 6k), topaz (Figure 6l), galena, and an unspecified Na-Al silicate. Multiphase inclusions enclosing quartz reach a size of up to 30 μm and form a short trail. Quartz is present in all inclusions in the trail (Figure 6k). Topaz forms a columnar, perfectly confined crystal (Figure 6l). A Na-Al silicate is in the form of needles up to 25 μm long. More than 10 needles of this phase occur in the inclusion.

4.3. Microthermometry of Fluid Inclusions

The results of microthermometric measurements are shown in Figure 8a–d and Table 2. The homogenization to liquid or vapor and, in some cases, also critical homogenization was observed in primary fluid inclusions. The latter type of homogenization was frequently observed in quartz from greisens and also, in a few cases, in apatite, cassiterite, and topaz. The highest homogenization temperatures were measured in PFI1 hosted by cassiterite, quartz, and topaz (Figure 8a), with higher temperatures recorded for homogenization to vapor (356–498 °C) than in case of critical homogenization (389–421 °C) and homogenization to liquid (336–441 °C) (Table 2). Pseudosecondary and secondary inclusions homogenize to liquid in many cases; only a few inclusions, which have a low degree of filling, homogenize to vapor. The lowest homogenization temperatures were measured for fluorite (Table 2, Figure 8a). The dark opaque pseudosecondary/secondary inclusions in cassiterite did not react to heating or cooling, as no phase transitions were visible. Halite in PSFI/SFI3 dissolves in a wide range of temperatures from 120 to 335 °C, and the total homogenization temperature of SFI3 inclusions with halite is in the range 141 to 369 °C (Figure 9, Table 3). In the Th vs. Tds_H plot, the data points are located below and above the diagonal reference 1:1 line (Figure 9). The SFI3 from greisens shows a wider range of Tds_H values than inclusions from greisenized granites and veins and, consequently, also has a wider range of salinities (Figure 9). Halite-bearing SFI3 from hydrothermal veins homogenizes on average at lower temperatures than SFI3 from greisens and greisenized granites. In a few cases of multiphase SFI3 inclusions, decrepitation occurred before the inclusion was homogenized. The phase transitions during heating of a multiphase inclusion from vein quartz of sample KNM-3 are illustrated in Figure 10. At room temperature, the inclusion contains a gas bubble, aqueous liquid, and four solid phases, one of which is halite, which disappears at a temperature of 268 °C and then the inclusion homogenizes into a liquid at a temperature of 370 °C. Other solid phases dissolve at very low temperatures, up to 60 °C (Figure 10).

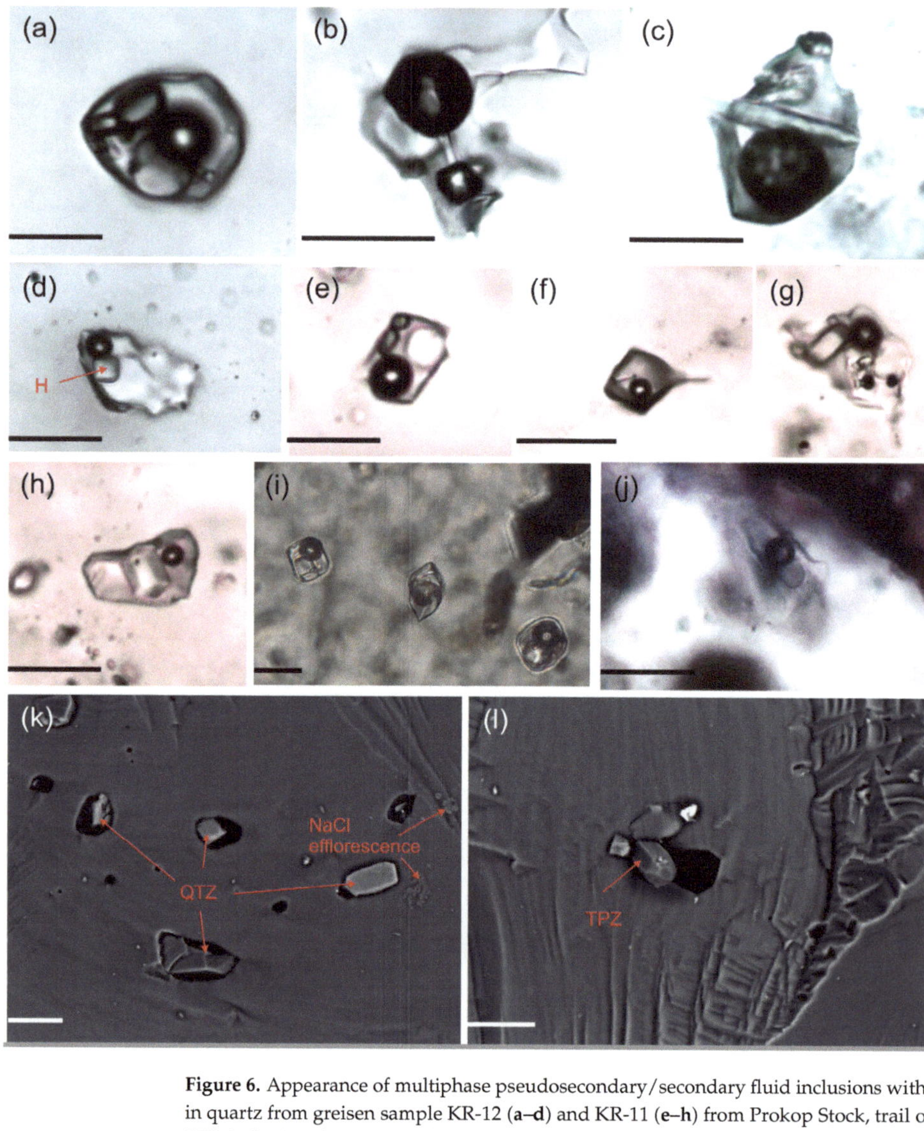

Figure 6. Appearance of multiphase pseudosecondary/secondary fluid inclusions with solid phases in quartz from greisen sample KR-12 (**a**–**d**) and KR-11 (**e**–**h**) from Prokop Stock, trail of multiphase SFI3 inclusions from quartz from vein sample KNM-3 (younger generation of SFI3); (**i**) from Steinknochen (Martin adit) and cassiterite from greisenized granite (sample PRK-1) from Preisselberg (**j**) under the microscope in polarized light at room temperature (PPL), H—halite; fluid inclusions in quartz from greisen sample KR-12 in BSE: (**k**) Group of opened fluid inclusions in quartz from greisen sample KR-12 from Prokop Stock with grains of quartz (QTZ); (**l**) Fluid inclusion in quartz from greisen sample KR-12 from Prokop Stock encloses crystal of topaz (TPZ); scale bar is 20 µm for all microphotographs.

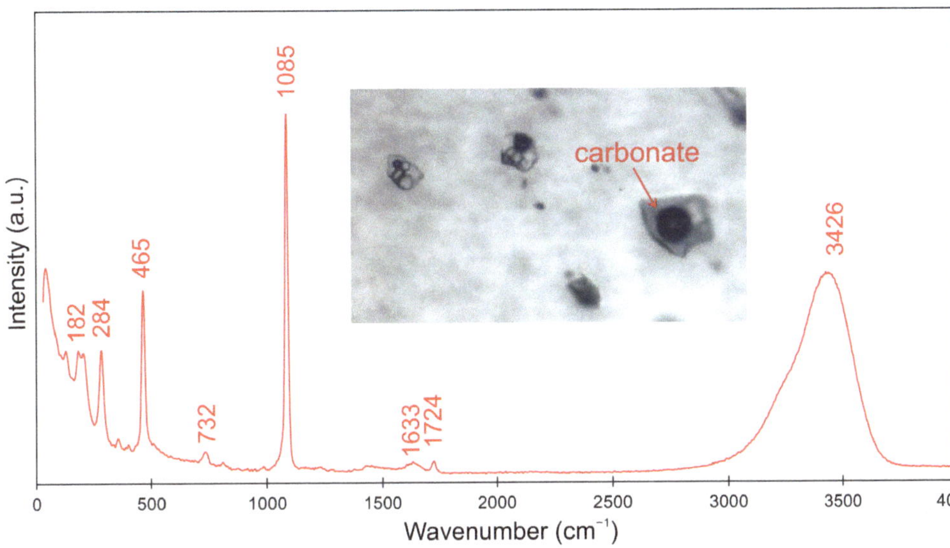

Figure 7. Raman spectrum of a solid carbonate enclosed in a multiphase SFI3 from vein quartz from Krupka (sample KMN-3). The size of the studied inclusion is 27 μm.

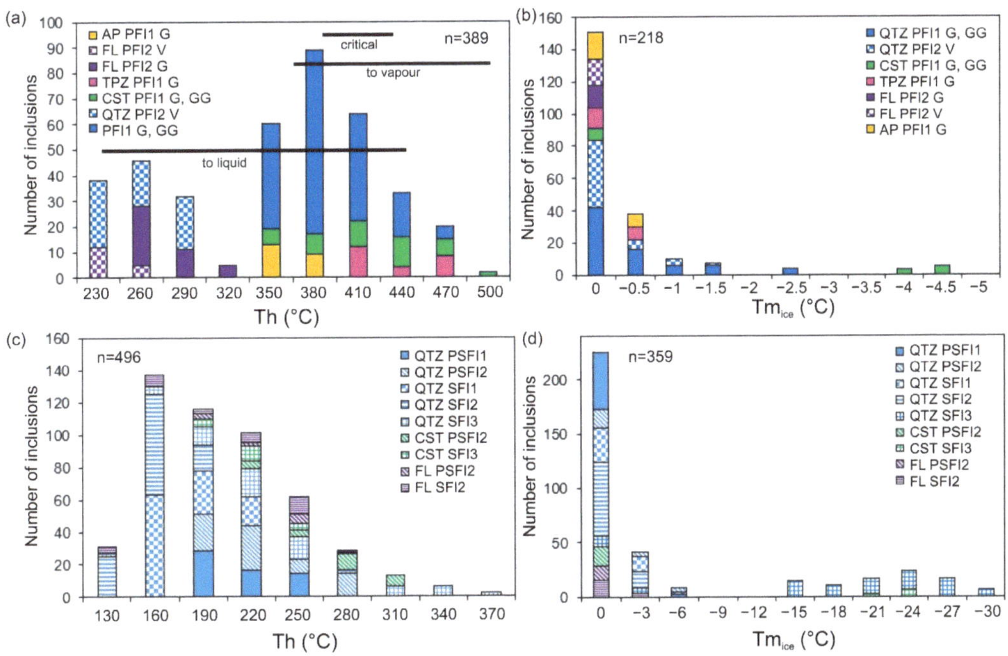

Figure 8. Histograms of homogenization temperatures and ice melting temperatures of fluid inclusions from Krupka, G—greisens, GG—greisenized granites, V—veins: (**a**) Homogenization temperatures of primary inclusions, bars indicate the ranges for individual modes of homogenization; (**b**) Ice melting temperatures of primary inclusions; (**c**) Homogenization temperatures of pseudosecondary and secondary fluid inclusions; (**d**) Ice melting temperatures of pseudosecondary and secondary inclusions.

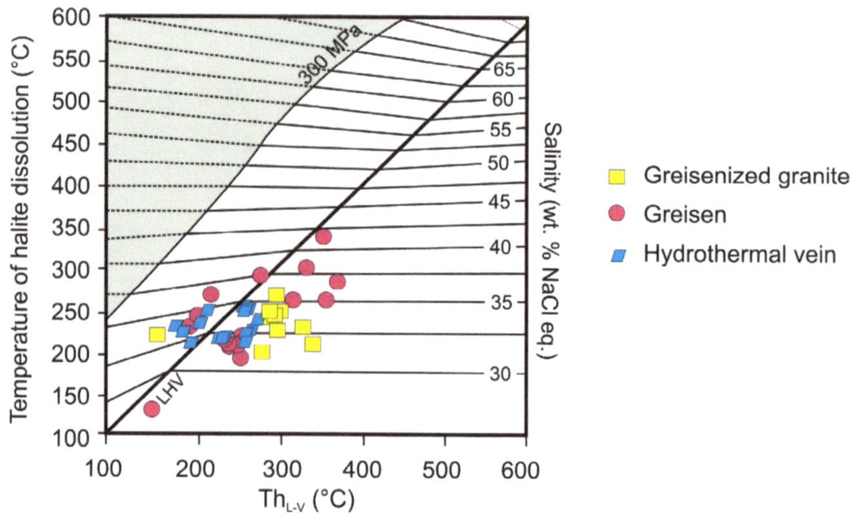

Figure 9. Plot Th vs. Tds$_H$ vs. salinity (model) after [53] with data points of SFI3 from the Krupka ore district.

Table 2. Results of microthermometric measurements of fluid inclusions from the Krupka ore district.

Sample	Mineral	No. of FI	Genesis	Phase Composition	LVR	Th_L (°C)	Th_V (°C)	Th_C (°C)	Tm_{ice} (°C)	Te (°C)
					Greisenized granite					
PRK-1	qtz	17	PFI1	L+V	0.05–0.6	351–441	392–456	-	−2.5 to 0	−22.8 to −23.0
PRK-1	qtz	15	PSFI1	L+V	0.4–0.7	236–252	247–265	-	−1.5 to 0	-
PRK-1	qtz	14	PSFI2	L+V	0.7–0.95	178–295	-	-	−5.2 to 0	-
PRK-1	qtz	13	SFI1	L+V	0.4–0.95	156–234	-	-	−3.5 to 0	-
PRK-1	qtz	14	SFI2	L+V	0.8–0.95	115–205	-	-	−0.6 to 0	-
PRK-1	cst	14	PFI1	L+V	0.1–0.5	362–423	396–498	-	−0.2 to 0	-
PRK-1	cst	9	PSFI2	L+V	0.7–0.95	208–311	-	-	−1.5 to 0	-
PRK-1	cst	-	SFI1/PSFI1	V (?)	-	-	-	-		-
KR-17	qtz	13	PFI1	L+V	0.05–0.5	342–416	382–421	-	−2.5 to 0	−22.8
KR-17	qtz	11	PSFI2	L+V	0.7–0.95	182–235	-	-	−0.8 to 0	−23.0 to −23.1
KR-17	qtz	8	SFI1	L+V	0.4–0.95	165–228	-	-	−3.3 to 0	-
KR-17	qtz	12	SFI2	L+V	0.7–0.95	117–192	-	-	−0.3 to 0	-
KR-17	cst	9	PFI1	L+V	0.2–0.5	378–435	363–465	-	−4.6 to −0.1	
					Greisen					
PRK-1	qtz	24	PFI1	L+V	0.1–0.6	377–401	392–425	389–412	−2.3 to −0.1	-
PRK-1	qtz	13	PSFI1	L+V	0.3–0.6	205–249	-	-	−1.3 to 0	-
PRK-1	qtz	8	PSFI2	L+V	0.7–0.95	231–282	-	-	−3.6 to 0	-
PRK-1	qtz	16	SFI1	L+V	0.5–0.9	149–227	-	-	−4.0 to 0	-
PRK-1	qtz	15	SFI2	L, L+V	0.7–1.0	121–199	-	-	−2.6 to −0.1	−23
PRK-1	cst	16	PFI1	L+V	0.1–0.5	362–425	383–406	-	−0.2 to 0	-
PRK-1	cst	8	SFI2	L+V	0.7–0.95	145–201	-	-		
PRK-1	tpz	16	PFI1	L+V	0.05–0.6	393–423	408–456	401–416	−0.7 to 0	-
KR-3	qtz	16	PFI1	L+V	0.2–0.5	367–402	398–431	406, 413	−2.1 to 0	
KR-4	qtz	14	PFI1	L+V	0.1–0.5	345–387	376–469		−0.8 to −0.1	
KR-4	fl	9	PFI2	L+V	0.7–0.9	224–289	-	-	−0.2 to 0	-
KR-5	qtz	10	PFI1	L+V	0.2–0.5	354–421	382–439		−2.0 to 0	
KR-7	qtz	10	PFI1	L+V	0.2–0.6	372–418	390–447		−1.5 to 0	
KR-8	qtz	12	PFI1	L+V	0.1–0.5	345–421	364–473		−2.5 to −0.2	
KR-8	fl	11	PFI2	L+V	0.7–0.95	254–296	-	-	−0.1 to 0	-
KR-11	qtz	31	PFI1	L+V	0.1–0.5	357–406	370–428	397–409	−0.5 to 0	
KR-12	qtz	27	PFI1	L+V	0.1–0.5	348–411	381–489	397–421	−2.1 to −0.2	−22.8
KNM-1	qtz	18	PFI1	L+V	0.1–0.5	383–396	356–402		−2.3 to 0	
KNM-4	qtz	15	PFI1	L+V	0.2–0.5	361–380	377–421		−1.7 to −0.1	
KR-11	qtz	8	PSFI1	L+V	0.5–0.7	208–232	-	-	−1.0 to 0	-

Table 2. *Cont.*

Sample	Mineral	No. of FI	Genesis	Phase Composition	LVR	Th_L (°C)	Th_V (°C)	Th_C (°C)	Tm_{ice} (°C)	Te (°C)
KR-12	qtz	12	PSFI1	L+V	0.5–0.7	203–212	-	-	−0.5 to 0	-
KNM-1	qtz	6	PSFI1	L+V	0.5–0.7	216–262	-	-	−0.3 to 0	-
KR-12	qtz	11	PSFI2	L+V	0.8–0.95	201–232	-	-	−3.1 to 0	-
KR-11	qtz	21	SFI1	L+V	0.6–0.8	190–213	-	-	−2.5 to 0	-
KR-12	qtz	28	SFI1	L+V	0.5–0.8	175–209	-	-	−3.2 to 0	-
KNM-1	qtz	7	SFI1	L+V	0.6–0.7	169–193	-	-	−4.0 to −0.2	-
KR-5	qtz	13	SFI2	L+V	0.8–0.95	116–186	-	-	−4.2 to 0	−37.7
KR-11	qtz	21	SFI2	L, L+V	0.8–1.0	145–174	-	-	−2.5 to −0.3	−38.4
KR-12	qtz	28	SFI2	L, L+V	0.8–1.0	121–206	-	-	−5.8 to 0	−37.3, −37.5
KNM-1	qtz	9	SFI2	L, L+V	0.8–1.0	139–152	-	-	−0.4 to 0	
KNM-4	qtz	11	SFI2	L, L+V	0.8–1.0	119–128	-	-	−0.2 to 0	
KR-5	cst	10	PFI1	L+V	0.1–0.5	352–440	376–462	-	−0.2 to 0	-
KR-10	cst	14	PFI1	L+V	0.2–0.6	336–412	399–403	-	−0.1 to 0	-
KR-5	cst	7	PSFI2	L+V	0.8–0.95	246–289	316–322	-	−2.8 to 0	-
KR-10	cst	-	SFI1	L+V, V (?)	0–0.05 (?)	-	-	-		-
KNM-2	tpz	8	PFI1	L+V	0.2–0.5	398–431	405–416	412–423	−0.4 to 0	-
KNM-2	ap	21	PFI1	L+V	0.1–0.5	342–390	-	-	−0.5 to 0	−22.8 to −22.9
KR-11	fl	12	PFI2	L+V	0.6–0.9	216–301	-	-	−0.2 to 0	-
KR-11	fl	6	PSFI2	L+V	0.8–0.95	176–241	-	-	−1.3 to 0	-
KR-11	fl	13	SFI2	L+V	0.7–0.95	121–247	-	-	−2.1 to 0	-
					Hydrothermal vein					
KNM-3	qtz	23	PFI2	L+V	0.5–0.9	218–291	-	-	−1.1 to 0	
KNM-3	qtz	4	PSFI1	L+V	0.5–0.7	213–266	-	-	−1.2 to 0	-
KNM-3	qtz	12	PSFI2	L+V	0.8–0.95	177–242	-	-	−6.5 to 0	-
KNM-3	qtz	13	SFI1	L+V	0.4–0.9	165–230	-	-	−3.5 to 0	-
KNM-3	qtz	21	SFI2	L+V	0.7–0.95	<50–201	-	-	−2.6 to 0	-
KR-12	qtz	42	PFI2	L+V	0.4–0.7	256–302	-	-	−0.4 to 0	−22.9
KR-12	qtz	17	PSFI2	L+V	0.8–0.95	199–245	-	-	−1.5 to 0	-
KR-12	qtz	21	SFI1	L+V	0.7–0.9	160–202	-	-	−3.5 to 0	-
KR-12	qtz	32	SFI2	L, L+V	0.8–0.95	137–182	-	-	−4.5 to 0	−36.3 to −37.9
KR-12	fl	14	PFI2	L+V	0.7–0.95	247–276	-	-	−0.2 to 0	-
KR-12	fl	16	SFI2	L+V	0.95	115–204	-	-	−3.8 to 0	-
KR-8	fl	6	PFI2/PSFI1	L+V	0.7–0.95	191–213			−0.3 to 0	
PRK-1	qtz	26	PFI2	L+V	0.5–0.8	259–279	-	-	−1.2 to 0	-
PRK-1	qtz	8	SFI1	L+V	0.7–0.9	148–223	-	-	−2.8 to 0	-
PRK-1	qtz	14	SFI2	L, L+V	0.9–1.0	<50–162	-	-	−1.3 to 0	-

Table 3. Results of microthermometric measurements of multiphase inclusions from the Krupka ore district.

Sample	Mineral	Genesis	No. of FI	Phase Composition	LVR	Th_L (°C)	Tm_{ice} (°C)	Tds_h (°C)
				Greisenized granite				
PRK-1	qtz	PSFI/SFI3	6	L+V+S	0.6–0.95	151–303	−31.0 to −14.1	162–202
PRK-1	cst	PSFI/SFI3	8	L+V+S	0.5–0.9	293–335	−25.5 to −19.5	178–224
				Greisen				
PRK-1	qtz	PSFI/SFI3	12	L+V+S_{1-3}	0.7–0.95	263–369	−26.1 to −15.2	189–314
KR-3	cst	PSFI/SFI3	6	L+V+S_{1-3}	0.5–0.95	141–327	−23.2 to −21.7	145–335
KR-12	qtz	PSFI/SFI3	15	L+V+S_{1-6}	0.5–0.95	196–242	−27.6 to −16.0	182–281
KNM-1	qtz	PSFI/SFI3	9	L+V+S_{1-6}	0.6–0.95	189–221	−28.1 to −18.3	178–204
KR-11	qtz	PSFI/SFI3	17	L+V+S_{1-5}	0.5–0.95	212–248	−25.8 to −20.4	210–270
KR-11	qtz	SFI3	4	L+V+S	0.8–0.95	182–311	−1.3 to −0.2	-
KR-10	cst	PSFI/SFI3	5	L+V+S_{1-2}	0.7–0.95	192–245	−24.5 to −19.7	214–275
				Hydrothermal vein				
KNM-3	qtz	PSFI/SFI3	16	L+V+S_{1-5}	0.6–0.95	164–213	−26.5 to −17.2	201–251
KR-12	qtz	SFI3	8	L+V+S_{1-2}	0.7–0.95	126–362	−3.0 to 0	-
KR-12	qtz	PSFI/SFI3	14	L+V+S_{1-6}	0.6–0.95	182–265	−30.8 to −14.3	192–231

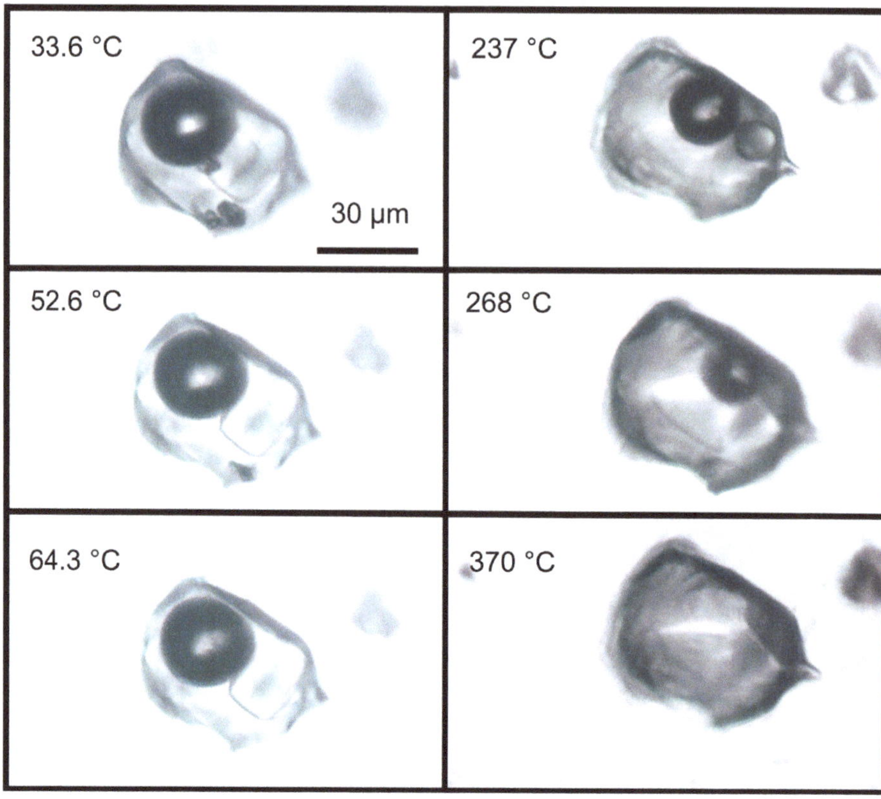

Figure 10. Example of phase transitions during heating of a PSFI/SFI3 fluid inclusion from vein quartz from sample KNM-3 (Lukáš vein, Martin adit).

In the cooling mode, the freezing temperatures of FI range from −18 to −62 °C. In several cases of PFI in quartz samples from greisens with low LVR and in some PSFI/SFI3 with solid phases, the inclusions did not freeze even when cooled to −120 °C. During heating of the frozen inclusions, the eutectic temperature was occasionally measured, which ranged from −22.8 to −23.2 °C in most cases, and for a few secondary inclusions in quartz from greisens and veins, it varies from −36.3 to −38.4 °C (Table 2). The last ice crystal melts in most primary inclusions at temperatures −1.5 to 0 °C, rarely down to −2.6 °C (for quartz) and −4.6 °C (for cassiterite), which corresponds to salinity 0 to 7.3 wt. % NaCl eq. The last ice crystal in pseudosecondary and secondary inclusions also melts mostly in the range of −3 to 0 °C, less often down to −6 °C for quartz (Figure 8d). Exceptions are PSFI/SFI3 from quartz and cassiterite with solid phases that showed Tm_{ice} values at much lower temperatures (down to −31 °C; Table 3, Figure 8d). In some PSFI multiphase inclusions, where the ice melted at low temperatures, the melting of hydrohalite was observed at temperatures around 2–5 °C in quartz from sample KNM-3. SFI3, which does not contain crystals of halite but usually contains one other phase (topaz, quartz, sulfides or other opaque minerals) present in several samples of quartz from greisens and veins. These inclusions show a melting temperature of the last ice crystal from −3 °C to 0 °C (Table 3). Solitary multiphase PSFI mostly shows lower temperatures of ice melting than inclusions on trails.

The system H_2O-gases-salts was recognized only in primary and pseudosecondary inclusions of a few samples of quartz, cassiterite, and topaz from greisens. Inclusions with a low degree of filling, in most cases, homogenize to vapor; inclusions with a dominant liquid phase usually homogenize to liquid. In several cases, critical homogenization has also occurred. Fluid inclusions containing gases are two-phase at room temperature, and the condensation of the liquid carbonic phase was not observed during cooling runs. In some inclusions, however, the presence of gases was identified by the crystallization of clathrate. Clathrate dissociates at temperatures between 4.2 and 10.1 °C, with higher temperatures measured for cassiterite from sample PRK-1 (Table 4). The total homogenization temperatures of clathrate-bearing inclusions do not differ significantly from the homogenization temperatures of inclusions without detectable gases.

Table 4. Results of microthermometry of clathrate-bearing inclusions from Krupka ore district.

Sample	Mineral	Genesis	No. of FI	Phase Composition	LVR	Th_L (°C)	Th_V (°C)	Thc (°C)	Tm_{ice} (°C)	Tm_{cla} (°C)
PRK-1	cst	PFI1/PSFI1	4	L+V	0.1–0.4	-	399–456		−0.2 to 0	5.6–10.1
KR-5	cst	PFI1	2	L+V	0.1–0.3	-	396, 402		−0.1 to 0	4.8, 6.1
KR-12	qtz	PFI1	2	L+V	0.1–0.3	-	392	408	−2.1 to 0	4.2, 4.7
KNM-1	qtz	PFI1	2	L+V	0.2–0.4	-	422,438		−0.8 to 0	5.8, 7.3
KNM-2	qtz	PFI1	5	L+V	0.1–0.5	389, 401	406–419		−1.4 to −0.1	4.2–7.3
KNM-2	tpz	PFI1	7	L+V	0.2–0.5	399–413	403–408	401	−0.6 to 0	4.4–8.4

4.4. Composition of Vapor Phase

Raman spectroscopy of gaseous FI in quartz from greisen (KR-12) and vein (KNM-3) samples was performed. Only CO_2 was identified by Raman analysis in a primary inclusion. In another PFI from greisen, trace amounts of N_2 (1.7 mol. %) and CH_4 (2.1 mol. %) were recorded in addition to CO_2 (96.2 mol. %). An example of the Raman spectrum of gaseous phases in fluid inclusions hosted by quartz from greisen is shown in Figure 11. In the quartz sample from the Lukáš vein, gaseous phases were analyzed in three pseudosecondary/secondary fluid inclusions. A higher CH_4 content (28.8 and 100 mol. %) was recorded in two inclusions; the nitrogen content is either none or negligible (0.1 mol. %). The rest of the non-aqueous phase is CO_2. The results are presented in Table 5.

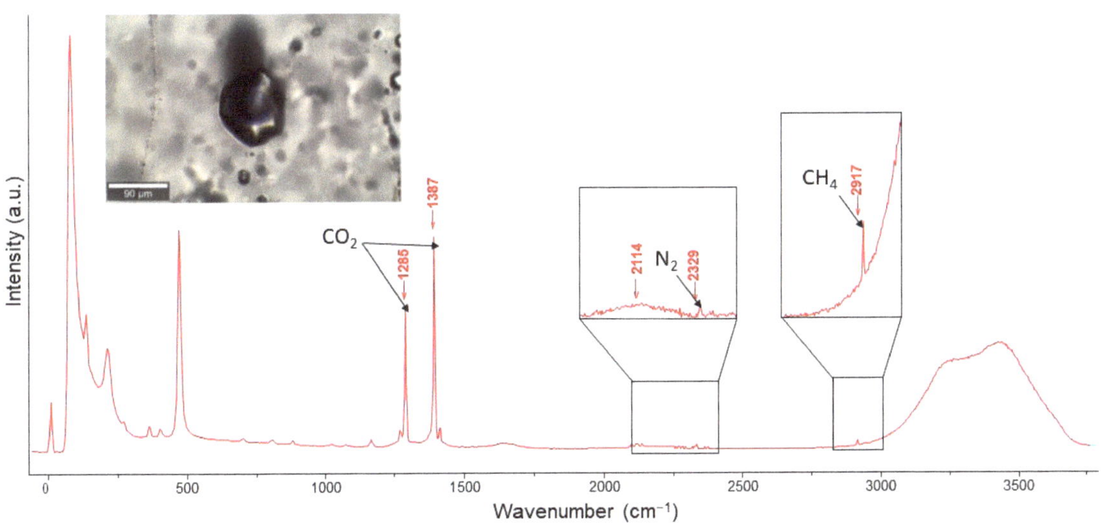

Figure 11. Raman spectrum of gaseous phase enclosed in a primary fluid inclusion from quartz in greisen from Krupka (KR-12).

Table 5. Composition of non-aqueous phase in fluid inclusions from the Krupka ore district.

			CH_4	CO_2	N_2
Sample	**Mineral**	**Genesis**		**mol. %**	
KNM-3	qtz	PFI2/PSFI1	28.8	71.1	0.1
KNM-3	qtz	PFI2/PSFI1	1.3	98.7	0
KNM-3	qtz	PFI2/PSFI1	100	0	0
KR-12	qtz	PFI1	0	100	0
KR-12	qtz	PFI1	2.1	96.2	1.7

4.5. Crush-Leach Analyses of Fluid Inclusions

Fluid inclusion leachates of three quartz samples were analyzed: one from greisen with fluorite, the second from greisen with molybdenite, and the last from a hydrothermal vein. The dominant cation is Na^+ in leachate from quartz from greisen with fluorite and vein quartz, while Ca predominates in leachate from greisen with molybdenite. The Na/K ratio is lowest for greisen with molybdenite and highest for quartz from a vein. The highest Mg contents are present in the quartz sample from greisen with molybdenite. Cl^- is the dominating anion in all leachates. The content of nitrates in greisen with molybdenite is also relatively high and more than 4.5 times higher than in leachate from vein quartz (Table 6). The contents of Li, Br, I, and F are relatively low (Table 6). The $Br/Cl \times 10^3$ varies from 3.7 to 4.8, and $I/Cl \times 10^6$ from 12 in vein quartz to 66 in greisen with fluorite. The charge balance (Q^+/Q^-) of the quartz sample from the hydrothermal vein is equal to 1. The Q^+/Q^- of the quartz from the greisen sample with fluorite is also relatively balanced, but the greisen sample with molybdenite has a charge balance of 2.3, which means that another not-analyzed anion must be present or either some solid phases enclosed in FI or a possibly present contaminating mineral phase dissolved during leachate preparation.

Table 6. Crush-leach analyses of fluid inclusions in quartz from Krupka (values in ppb).

Sample	KR-11 (Greisen)	KR-12 (Greisen)	KR-12 (Vein)
Li^+	27	30	32
Na^+	4496	4250	4055
K^+	2582	2124	1677
Mg^{2+}	602	173	94
Ca^{2+}	10,198	1878	952
F^-	81	219	250
Cl^-	9866	9179	8564
Br^-	47	38	31
I^-	1.4	0.6	0.1
NO_3^-	3829	1573	845
SO_4^{2-}	777	186	191
$Br/Cl \times 10^3$	4.8	4.2	3.7
Na/Br	95	111	129
$I/Cl \times 10^6$	15	66	12
Na/K	1.7	2.0	2.4
K/Na	0.6	0.5	0.4
Cl/SO_4	13	49	45
Q^+/Q^-	2.3	1.2	1.0

4.6. Hydrogen Isotopes

The δD values and water contents were determined for two mica samples (KR-5 and KR-7) and two clay minerals (KR-4 and KR-8). Greisen samples with different mineral compositions (see Table 1) and degrees of autometamorphism from the Knötel deposit were chosen so that the hydrogen isotope analyses covered both the greisenization and post-greisenization stages. A wide range of δD values was found, ranging between −54.6 and 33.6‰ V-SMOW (Table 7).

Table 7. The determined $\delta D_{mineral}$ values, contents of water and calculated δD_{fluid} for given temperatures in clay minerals and micas.

Mineral	Sample	$\delta D_{mineral}$ (‰ V-SMOW)	δD_{fluid} (‰ V-SMOW)	Temperature (°C)	Content of Water (%)
Dickite	KR-4	−22.2	−45.7 to −37.9	100–250	9.79
Dickite	KR-8	−51.2	−74.7 to −66.9	100–250	14.31
Phengite/Li-phengite	KR-7	−54.6	−92.4 to −77.8	350–400	2.34
Protolithionite/zinnwaldite	KR-5	33.6	−20.6 to −6.0	350–450	0.92

5. Discussion

5.1. Chemical Composition of Fluids

As is evident from microthermometric measurements, two fluid systems were involved in the formation of greisenized granites, greisens, and associated Sn-W hydrothermal veins in the Krupka ore district. The H_2O-salt fluid system is most frequent and occurs in all samples, all minerals, and all genetic types of inclusions of studied rocks and veins. Microthermometric data from fluid inclusions with the H_2O-salts fluid system were compared with already published data from other localities within the Czech and German parts of the Bohemian Massif. The data for the primary inclusions are very similar to the data from the Hub Stock [17] and Zinnwald [54], but the above-mentioned papers report a narrower range of homogenization temperatures than cover our data from Krupka (Figure 12). The melting temperature of the last ice crystal is in the range of −5 to 0 °C at all compared localities. The last crystal of ice in secondary inclusions from Krupka containing salt crystals melts at lower temperatures (down to −31 °C). The measured positive temperatures of melting of hydrohalite (2–5 °C) indicate metastable behavior [55]. Ref. [17] also reported high-salinity secondary inclusions in topaz and cassiterite from the Hub Stock, but most of these inclusions displayed metastable behavior during cooling runs. The mentioned paper reported the dissolution temperature of halite in inclusions in topaz in the range of 176–215 °C. All these inclusions homogenize to a liquid at temperatures of 468–512 °C. Multiphase inclusions in cassiterite homogenize into a liquid at temperatures of 485–498 °C, and halite in them dissolves at 158–167 °C [17]. Ref. [15] describes inclusions with solid phases in quartz from Krupka and in topaz from Cínovec but reported the salinity (35 wt. % NaCl eq.) from only a single inclusion containing crystals of salt. The above-mentioned paper also published the dissolution temperature of halite in the range 200–315 °C, whereas halite in inclusions measured in this work dissolves at temperatures between 120 and 335 °C (Figure 9). In a number of cases, the halite dissolves first and then disappears the vapor bubble, but in a few cases, halite dissolves after the bubble disappearance.

Less frequent is the H_2O-gases-salt fluid system, where a variable amount of CO_2 and trace amounts of N_2 and CH_4 were identified by microthermometric measurements (Section 4.3) and by Raman spectroscopy (Section 4.4). The non-aqueous phase is dominated by CO_2 in most cases, but in one case, pure CH_4 was found. Clathrate-bearing inclusions have low salinity (0–4.1 wt. % NaCl eq.; [40,41]). Compared to other localities with Sn-W or W mineralization, the non-aqueous phase in the fluid inclusions from Krupka shows a wider compositional range compared to the W mineralization from Tanvald granite from Jablonec nad Nisou [56], where CO_2 is always the dominant phase, and fluids have similar salinities (0–4.6 wt. % NaCl eq.). Carbon dioxide also dominates in the vapor phase from Cínovec ([57], unpublished data of authors) (Figure 13). The composition of the non-aqueous phase in the fluid inclusions suggests two possible trends, although the amount of data is small. One of these trends is already considered in [56] for W mineralization in Tanvald granite. In the primary inclusions from Krupka and from other Sn-W and W localities in the Bohemian Massif, a CO_2-rich fluid is probably mixed with fluids with very similar CH_4:N_2 ratios. The second trend for PSFI/SFI from Krupka seems to be characterized by a mixing of CO_2-rich fluids and CH_4-rich fluids with negligible N_2 contents (Figure 13). CH_4 in the fluids is a sign of a more reducing environment when the fluids are captured. The contents of other gases (CH_4 and N_2) in addition to CO_2 were neglected during calculations of bulk fluid composition since the contents of those gases are very low or zero in PFI inclusions studied by Raman spectroscopy. Compositions and molar volumes of representative clathrate-bearing inclusions are listed in Table 8. The bulk molar volume of fluids in FI with volatiles ranges from 37.3 to 84.9 cm^3/mol, while the highest values are reached in quartz from greisen in sample KR-12 (Table 9). The calculated amount of CO_2 in inclusions ranges from 2.6 to 16.5 mol. % (Table 9).

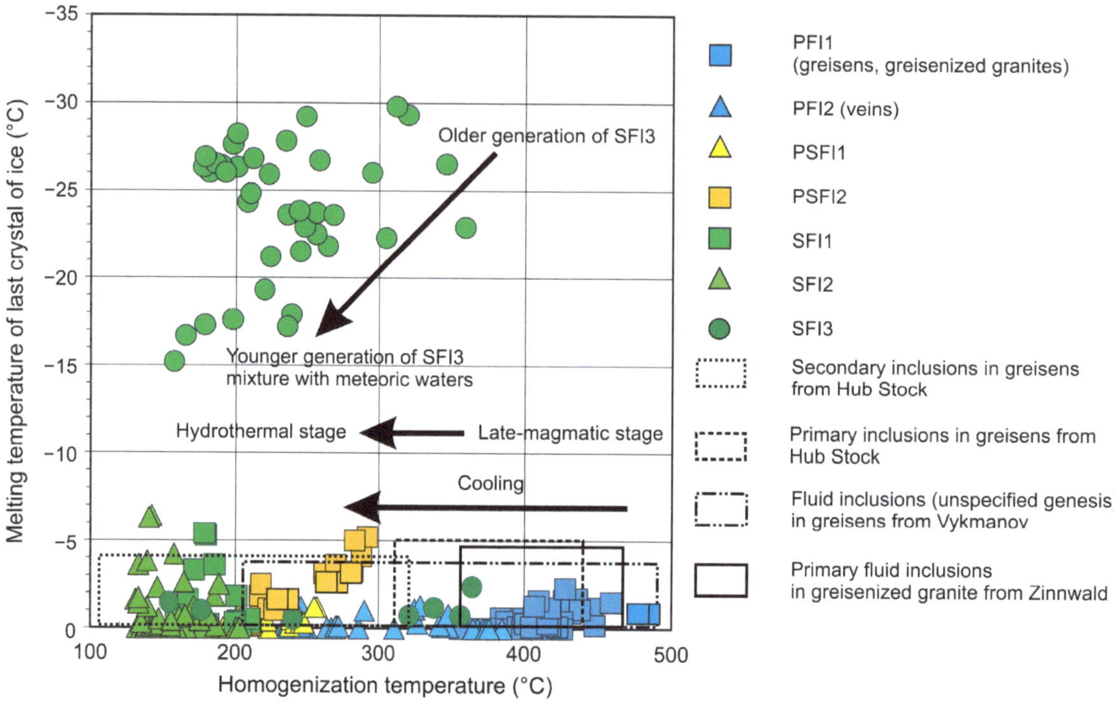

Figure 12. Plot of temperature of homogenization vs. ice melting temperature for fluid inclusions from Krupka and comparison with localities Vykmanov [16], Horní Slavkov–Hub Stock [17], and Zinnwald [54].

Table 8. Composition and molar volumes of representative clathrate-bearing inclusions from the Krupka ore district.

Sample	KNM-2	KMN-2	KNM-2	KNM-1	KR-5	KR-12
Mineral	Tpz	Tpz	Qtz	Qtz	Cst	Qtz
Genesis	PFI1	PFI1	PFI1	PFI1	PFI1	PFI1
Th_v (°C)	408	403	412	422	396	392
Tm_{ice} (°C)	−0.2	−0.1	−0.7	−0.8	−0.2	−2.1
Tm_{cla} (°C)	4.6	6.1	5.1	5.8	5.6	4.2
LVR	0.2	0.3	0.2	0.4	0.3	0.2
$X(H_2O)$	0.89	0.90	0.87	0.91	0.90	0.88
$X(CO_2)$	0.11	0.09	0.10	0.07	0.08	0.10
X(NaCl)	0.00	0.01	0.02	0.02	0.02	0.02
Molar volume (cm^3/mol)	83.0	56.8	79.5	42.9	58.5	82.5

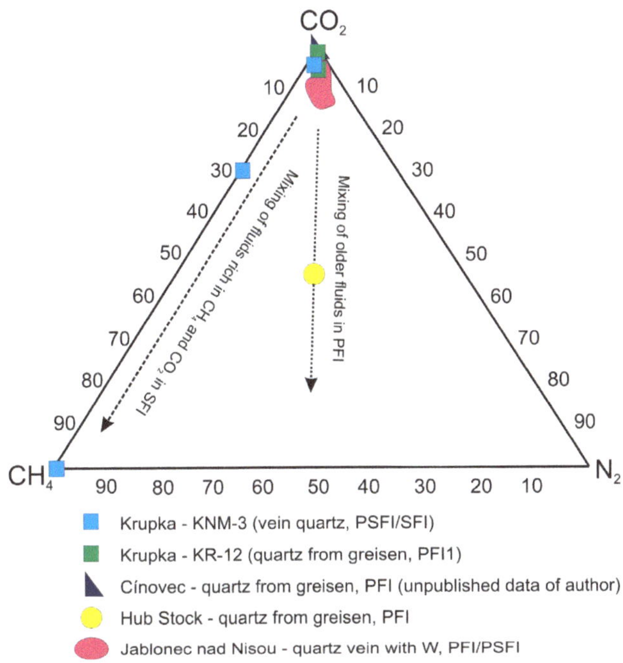

Figure 13. Results of Raman analyses of non-aqueous phases of fluid inclusions from Krupka in ternary plot CO_2-CH_4-N_2 and comparison with samples from other localities in the Bohemian Massif, W-mineralization from Jablonec nad Nisou [56] and greisens from Hub Stock [17].

Table 9. Ranges of composition and molar volumes of clathrate-bearing fluid inclusions from the Krupka ore district.

Sample	Mineral	Genesis	Salinity (wt. % NaCl eq.)	Molar Volume (cm^3/mol)	mol. % CO_2
PRK-1	Cst	PFI1/PSFI1	0.0–2.8	79.3–83.4	12.6–14.6
KR-5	Cst	PFI1	3.1–4.0	56.4–55.8	8.1–9.4
KR-12	Qtz	PFI1	3.2–3.4	81.6–84.9	10.7–11.3
KNM-1	Qtz	PFI1	1.3–3.3	55.9–81.7	2.6–8.9
KNM-2	Qtz	PFI1	1.5–3.9	58.6–81.6	7.7–14.8
KNM-2	Tpz	PFI1	0.5–4.1	56.6–80.2	7.6–16.5

5.2. P-T Conditions of Greisenization and Post-Greisenization Processes

Considering the facts that the primary inclusions with and also without CO_2 in some samples of greisens homogenize in three different modes (to vapor, to liquid, and in a critical way) and the homogenization temperatures differ only slightly, it is likely that these inclusions were trapped from a heterogeneous fluid [49]. In such a situation, the minimum recorded homogenization temperatures would correspond to the actual temperatures of fluid entrapment. For vapor-homogenizing inclusions, it was possible to calculate the pressure at the time of fluid entrapment. The results even with bulk molar volume and mol. % of CO_2 in representative fluid inclusions are summarized in Table 10, and results of P-T calculations are visualized in Figure 14. The results show that the greisenization process took place at temperatures of 370 to 490 °C at low pressures from 155 to 371 bar. The fluids with CO_2 were trapped at higher pressures (322 to 371 bar) than fluids of system H_2O-NaCl. Fluids without volatile components were captured at pressures of 155–320 bar, with the highest pressures calculated for inclusions in quartz from greisens from samples KR-11 and KR-12. The pressure and temperature conditions of greisenization given from the Hub

Stock [17] correspond to temperatures of 350 to 450 °C and 250 to 530 bar. Greisenization in the western part of the Erzgebirge took place in a wider temperature and pressure interval (Figure 14).

Table 10. Calculated bulk molar volumes of vapor-homogenizing fluid inclusions trapped from heterogenous fluid and calculated pressure of their entrapment.

Fluid System	H_2O-CO_2-(NaCl)		H_2O-NaCl					
Sample Mineral	KNM-2 tpz	KR-12 qtz	PRK-1 qtz	PRK-1 tpz	KR-3 qtz	KNM-2 tpz	KR-11 qtz	KR-12 qtz
Th_v (°C)	403, 408	392	392–425	408–456	398–431	405–416	379–428	381–456
Tm_{ice} (°C)	−0.1,−0.2	−2.1	−2.3 to −0.2	−0.3 to −0.1	−2.1 to −0.3	−0.4 to 0	−0.5 to 0	−2.1 to −0.2
Tm_{cla} (°C)	4.6, 6.1	4.2						
LVR	0.2, 0.3	0.2	0.2–0.4	0.2–0.3	0.2–0.3	0.2–0.3	0.1–0.3	0.1–0.4
Mol. % CO_2	10.7, 8.9	9.8						
Bulk molar volume (cm^3/mol)	56.8, 83.0	82.4	38.3–46.7	47.0–95.1	39.0–52.9	45.2–52.2	37.5–60.2	33.7–80.54
Pressure (bar)	352, 371	322	173–223	197–268	182–224	193–209	155–320	157–313

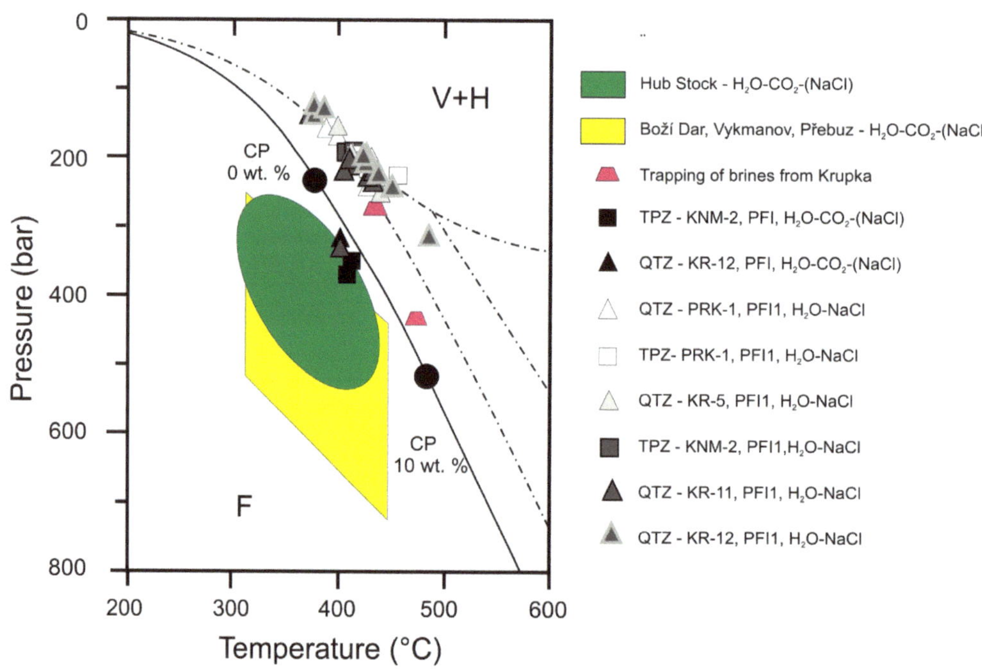

Figure 14. Pressure and temperature conditions of trapping of PFI connected with process of greisenization from Krupka and comparison with other localities, Horní Slavkov–Hub Stock [17], Boží Dar, Vykmanov, Přebuz [16] and trapping of brines from Krupka [15]. Diagram modified after [58].

In the vast majority, PSFI1,2 and SFI1,2 homogenize only to liquid, and conditions of capture of such homogeneous fluid cannot be determined in the absence of results from independent geothermometers or geobarometers. The post-greisenization stage, characterized by the formation of hydrothermal veins with fluorite, clay minerals (dickite, illite), and Fe oxides and oxyhydroxides, took place at low temperatures because secondary fluid inclusions that contain only a liquid phase indicate the capture of fluid inclusions at temperatures less than 100 °C [49]. However, the temperature values of the fluids captured by PSFI1, SFI1, and SFI3 will be much higher, as shown by the measured homogenization temperatures reaching up to 370 °C. Hence, the available data from these late fluid inclusions likely cover the whole cooling history of the host magmatic rock. Ref. [15] reports the temperature and pressure conditions for two episodes of trapping of high-salinity fluids in quartz from Krupka (Figure 14) and states values of 430–440 °C and 280 bar and 480–490 °C and 460 bar.

5.3. Origin of Fluids

As in the fluids from the Hub Stock [17], sodium chloride dominates in the fluids from Krupka according to the recorded eutectic temperatures of the fluid inclusions [59]. Mg and Fe chlorides are likely present, depending on the lower values of the eutectic temperatures (see Section 4.3), in addition to NaCl in some secondary inclusions. The leachates of fluid inclusions showed the presence of nitrates and sulfates in relatively significant amounts in addition to chlorides (see Table 3). Sulfates are present as part of the brines in SFI3, as verified by Raman spectroscopy. Nitrates can be leached from overlying sedimentary rocks or basins and infiltrated along fractures into the granitoid massif. Among the cations, Na^+ usually dominates, but in one sample of quartz from greisen from the Prokop Stock, calcium is predominant. The high content of Ca in fluids can be explained by the albitization of plagioclase when the rock is depleted of Ca, which can then be present in greisenizing solutions [60]. Another explanation may be the dissolution of calcite that is present as a solid in SFI3 inclusions. Calcium-rich fluid was also found in topaz and cassiterite from the Hub Stock [17] (Figure 15). The I/Cl values (12, 15, and 66 $\times 10^{-6}$; Table 3) and Br/Cl (3.7, 4.2 and 4.8 $\times 10^{-3}$) in fluids from Krupka are comparable with shield brines from the Canadian and Baltic Shields and also fall into the field of post-Variscan high-salinity fluids of the Bohemian Massif (Figure 16). The charge balance (Q^+/Q^-) of one greisen sample from Krupka is not balanced (Table 3) and indicates the presence of other anions in the fluids. The presence of carbonates in fluids seems to be the most probable, as calcite and siderite solids were determined in fluid inclusions by means of Raman spectroscopy (Figure 11). The Na/K in fluids from Knötel–Prokop Stock is higher than in fluids from Martin adit referred by [15]. This is very likely caused by a higher degree of argillitization of samples from Prokop Stock. As stated [61,62], the argillitization process begins around 200 °C and is associated with the activity of low-salinity and acid solutions poor in sulfate ions.

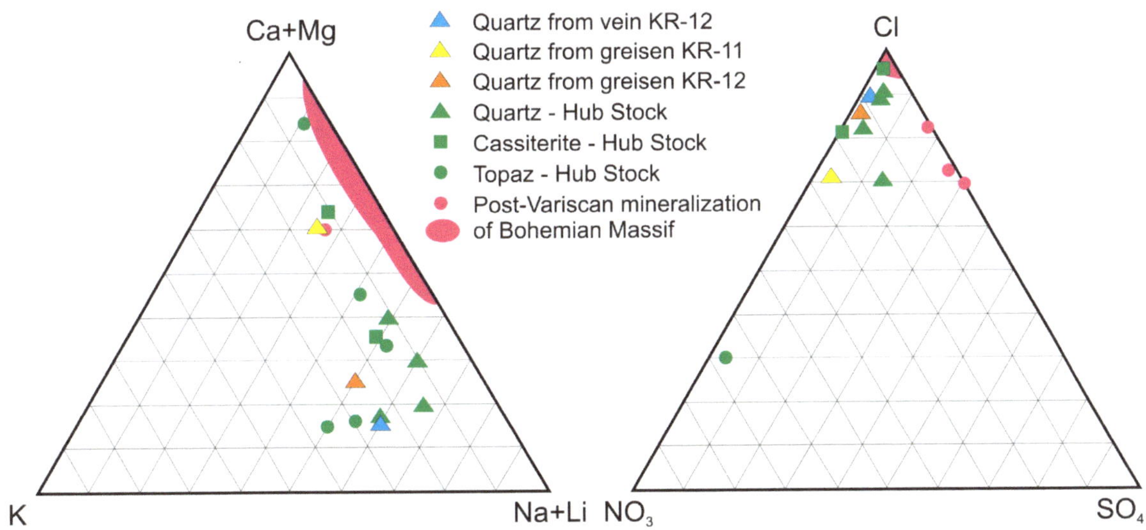

Figure 15. Ternary diagrams NO_3^--SO_4^{2-}-Cl^- and K^+-(Na^++Li^+)-(Ca^{2+}+Mg^{2+}) in leachates of fluid inclusions from Krupka and comparison with leachates from Horní Slavkov–Hub Stock [17] and post-Variscan hydrothermal mineralization from the Bohemian Massif [63].

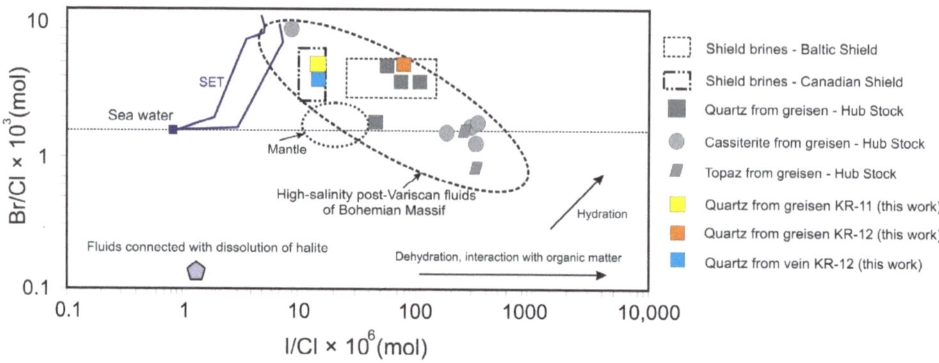

Figure 16. I/Cl vs. Br/Cl plot for leachates of fluid inclusions from Krupka and comparison with other localities—Hub Stock [17], high-salinity post-Variscan fluids of Bohemian Massif [63,64] and various other sources—sea water, mantle; SET—seawater evaporation trajectory. Plot modified after [17].

High-salinity PSFI/SFI3 are probably connected with fluids derived from brines during the secondary boiling of magma. When the magma is initially enriched in CO_2 and other volatile components, fluids with different compositions can be exsolved during the secondary boiling [65]. Refs. [54,66,67] deal with the research of fluid and melt inclusions in altered granites, greisens, and associated hydrothermal veins from the German part of Erzgebirge. According to these studies, granitic melt at great depths and at high temperatures is undersaturated with H_2O. During melt-fluid separation at the interface of the magmatic and hydrothermal phases, high-temperature brines are often exsolved, which can concentrate metals and produce economically important deposits. In the case of the Krupka ore district, high-salinity fluids and low-salinity steam were also exsolved. Some solid phases (quartz and sulfides) occurring in the high-salinity fluid inclusions do not dissolve even at high temperatures. As already mentioned above, multiphase inclusions probably belong to at least two generations, but they do not differ in principle either in terms of homogenization temperature. However, in some inclusions (apparently in the younger generation), there is a decrease in salinity, probably due to interaction with meteoric waters. Very low salinities have multiphase inclusions without salts, containing only captured crystals of other mineral phases (mainly topaz and quartz). Evidence of brines in greisens is also known from other localities in the Bohemian Massif, including the Hub Stock near Horní Slavkov, Krupka, and Cínovec [15,17].

Some clues on the origin of the fluids also offer analyses of stable isotopes of hydrogen in micas and clay minerals. According to δD_{fluid} values ranging between −92.4 and −6.0‰ V-SMOW, calculated from $\delta D_{mineral}$ values and estimated formation temperatures (Table 7), the fluids were most likely derived from a magmatic source, but some contribution of another source (sedimentary or metamorphic fluid) cannot be excluded (Figure 17). The calculated δD values of greisenizing fluids (sample KR-7) approximately correspond with the results from other localities in Bohemian Massif and other world Sn-W deposits (Figure 17). The δD_{fluid} for protolithionite/zinnwaldite from greisen sample KR-5 has only a slightly negative value, which can be the consequence of the redistribution of isotopes during phase separation associated with boiling [68] of greisenizing fluids. Samples KR-4 and KR-8 originated from strongly altered greisens with fluorite and clay minerals. The δD_{fluid} values of these samples most likely correspond to an origin in meteoric waters. The δD_{fluid} values for samples KR-4 and KR-5 are very similar to the δD values of altered granites from the Bohemian Massif (Figure 17). Based on oxygen isotopes, the participation of both magmatic and meteoric fluids during the formation of quartz veins with Sn-W mineralization was also reported [69] from the Takatori deposit in Japan.

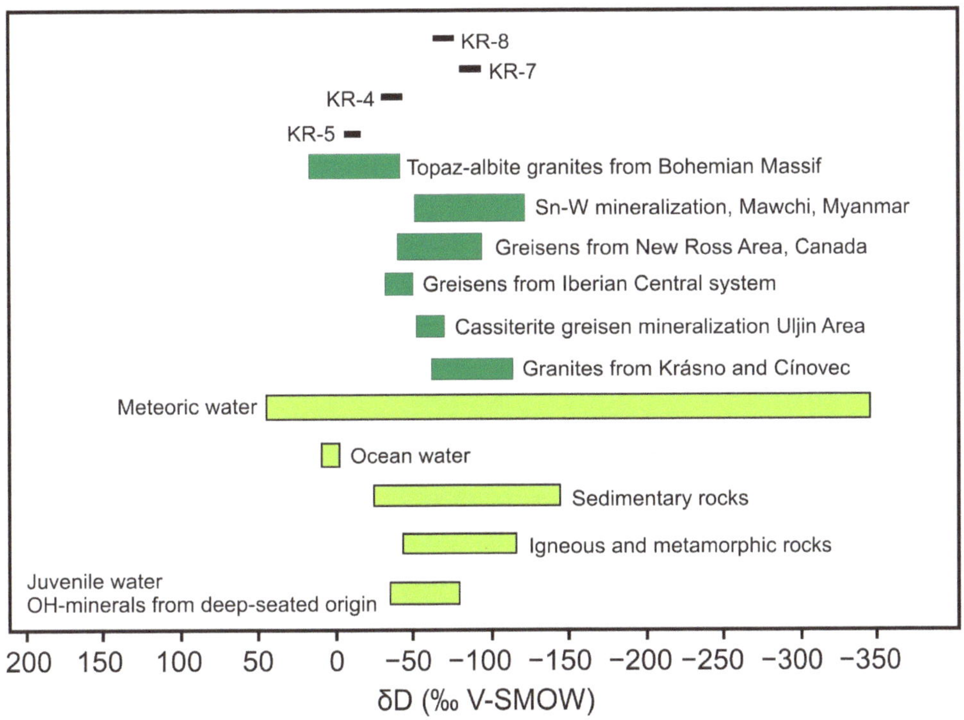

Figure 17. The calculated δD values of parent fluids from samples from Krupka ore district and their comparison with geologically important reservoirs [68], granites from Horní Slavkov–Hub Stock [17] and Cínovec [14], greisens from Iberian Central system [70], greisen cassiterite mineralization from Uljin Area [71], greisens from New Ross area in Canada [72], Sn-W mineralization, Mawchi, Myanmar [73], and unpublished data by M. René for topaz-albite granites of the Bohemian Massif.

Low salinity and the presence of only single-phase secondary inclusions in a number of greisen and vein samples suggest that the hydrothermal stage was associated with meteoric waters that infiltrated the hydrothermal system [48].

Based on the evidence obtained in this work, it is possible to construct a simplified model of fluid evolution. Figure 18 illustrates the evolution of fluids in the Knötel area. The initial stages of the greisenization of granites involved H_2O-NaCl-gas (CO_2, CH_4, and N_2) fluids with low salinity. As the temperature decreased and the magma rose closer to the surface, the exsolution of high-salinity fluids (up to 31 wt. % NaCl eq.) occurred, which are trapped in multiphase inclusions interpreted as PSFI or as an older generation of SFI3. These late inclusions have a relatively low degree of filling. Further cooling led to the infiltration of external fluids originating in either meteoric water or overlying sedimentary rocks or sedimentary basins. These fluids had low salinity and often healed cracks in greisens and hydrothermal veins. The degree of filling of FI is still relatively low. In the same stage of evolution, multiphase inclusions could also have been trapped, which represent a mix of magmatogenic brines and fluids originating from meteoric waters. The salinity of fluids trapped in SFI3 inclusions of the younger generation is slightly lower than that of fluids of the older generation. In the final stages of the hydrothermal stage, only low-salinity, low-temperature fluids dominated by surface waters healed fractures in greisens and newly formed veins (Figure 18).

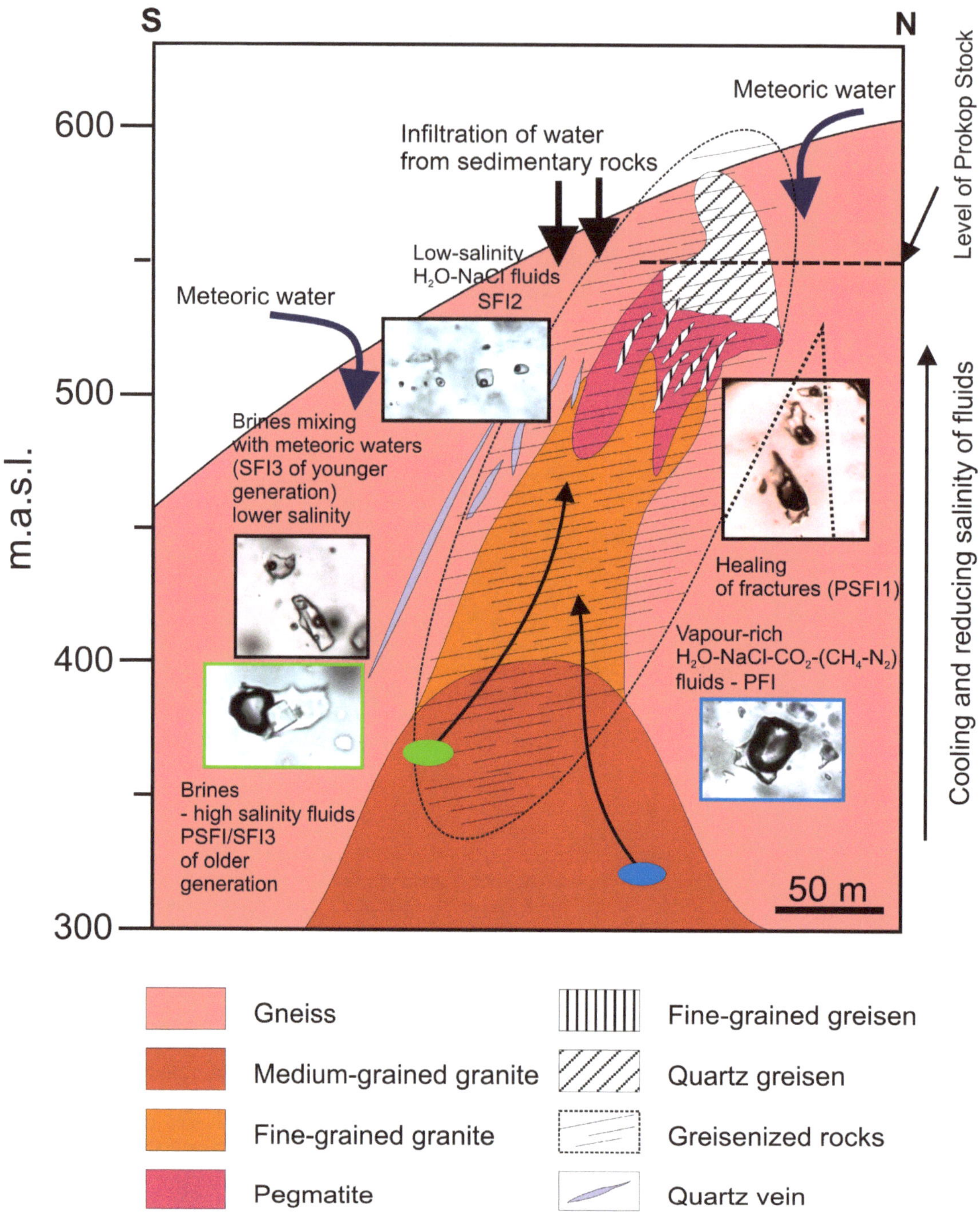

Figure 18. Simplified model of fluid evolution in the Knötel Stock.

6. Conclusions

The results of the study of greisenized granites, greisens, and associated hydrothermal veins from the Krupka ore district showed that the greisenization process took place at temperatures of 370–490 °C and pressures of 155–371 bar. The greisenization process was related to the activity of near-critical low-salinity aqueous fluids with low gas content (CO_2, N_2, CH_4). The post-greisenization processes, including the formation of some ore veins and the crystallization of clay and other secondary minerals, were associated with the activity of low-salinity and high-salinity fluids with temperatures of <120–370 °C and containing captured or daughter crystals, predominantly of salts and carbonates. According to the isotopic composition of H of fluids, the source of fluids is partly magmatogenic, but post-greisenizing fluids also originated in meteoric waters and fluids derived from sedimentary rocks.

Author Contributions: Conceptualization, M.K.K., Z.D., M.R., W.P., J.U., J.K., and K.K.; methodology, M.K.K., Z.D., V.M., J.U., and J.K.; investigation, M.K.K. and Z.D.; resources, M.K.K.; writing—original draft preparation, M.K.K.; writing—review and editing, M.K.K., Z.D., M.R., W.P., J.U., J.K., and K.K. All authors have read and agreed to the published version of the manuscript.

Funding: The study was financially supported by the Ministry of Culture of the Czech Republic (long-term project DKRVO 2024-2028/1.I.a; National Museum, 00023272) to Z.D. and J.U, DKRVO, long-term conceptual development of a research organization Research Institute for Building Materials project 2023–2027 and EXPRO 2019 project of the Czech Science Foundation (No. 19-29124X) to M.K.K.

Data Availability Statement: All data are contained in this paper.

Acknowledgments: H. Taubald (University of Tübingen) is thanked for analyses of hydrogen isotopes. Constructive comments by two referees helped to improve the original draft of the manuscript.

Conflicts of Interest: The authors declare no conflicts of interest. The funder had no role in the design of this study, in the collection, analysis, or interpretation of data, in the writing of this manuscript, or in the decision to publish these results.

References

1. Lange, H.; Tischendorf, G.; Palchen, W.; Klemm, I.; Ossenkopf, W. Zur Petrographie und Geochemie der Granite des Erzgebirges. *Geologie* **1972**, *21*, 457–494.
2. Teuscher, E.O. Primäre Bildungen des granitischen Magmas und seine Restlösungen im Massif von Eibenstock-Neudeck. *Miner. Petr. Mitt.* **1936**, *19*, 211–266.
3. Beus, A.A.; Zalashkova, N.E. High-temperature postmagmatic metasomatic processes in grantioids. *Izv. Akad. Nauk. SSSR Ser. Geol.* **1962**, *4*, 13–31.
4. Štemprok, M. Petrology and the vertical extent of mineralization in the Cínovec (Zinnwald) granite cupola. *Sbor. Geol. Věd Lož. Geol. Miner.* **1965**, *5*, 7–106.
5. Tischendorf, G.; Förster, H.J. Hercynian granite magmatism and related metallogenesis in the Erzgebirge. *Monogr. Ser. Miner. Depos.* **1994**, *31*, 5–23.
6. Kovalenko, V. *Petrology and Geochemistry of Rare Metal Granites*; Nauka: Novosibirsk, Russia, 1977; p. 206.
7. Černý, P.; Blevin, P.L.; Cuney, M.; London, D. Granite-Related Ore Deposits. *Econ. Geol.* **2006**, *107*, 337–370. [CrossRef]
8. Jarchovský, T.; Pavlů, D. Albite-topaz microgranite from Horní Slavkov (Slavkovský les Mts.) NW Bohemia. *Věst. Ústř. Úst. Geol.* **1991**, *66*, 13–22. [CrossRef]
9. Eisenreich, M.; Breiter, K. Krupka Sn-W-Mo deposits in Eastern Krušné hory Mts. *Věst. Čes. Geol. Úst.* **1993**, *63*, 3.
10. Breiter, K. From explosive breccia to unidirectional solidification textures: Magmatic evolution of a phosphorus- and fluorine-rich granite system (Podlesí, Krušné hory Mts., Czech Republic). *Bull. Geosci.* **2002**, *77*, 67–92.
11. René, M.; Škoda, R. Nb-Ta-Ti oxides fractionation in rare-metal granites: Krásno-Horní Slavkov ore district, Czech Republic. *Miner. Petrol.* **2011**, *103*, 37–48. [CrossRef]
12. Thomas, R. Estimation of the viscosity and the water content of silicate melts from melt inclusion data. *Eur. J. Miner.* **1994**, *6*, 511–535. [CrossRef]
13. Jarchovský, T. The nature and genesis of greisen stock at Krásno, Slavkovský les area—Western Bohemia, Czech Republic. *J. Czech Geol. Soc.* **2006**, *51*, 201–216. [CrossRef]
14. Dolejš, D.; Štemprok, M. Magmatic and hydrothermal evolution of Li-F granites: Cínovec and Krásno intrusions, Krušné hory batholith, Czech Republic. *Bull. Geosci.* **2001**, *76*, 77–79.

15. Ďurišová, J.; Charoy, B.; Weisbrod, A. Fluid inclusion studies in minerals from tin and tungsten deposits in the Krušné Hory Mountains (Czechoslovakia). *Bull. Minéral.* **1979**, *102*, 665–675. [CrossRef]
16. Ďurišová, J. Formation conditions of greisen parageneses of the western Erzgebirge Mountains. *Věst. Ústř. Úst. Geol.* **1984**, *59*, 141–152. (In Czech)
17. Dolníček, Z.; René, M.; Prochaska, W.; Kovář, M. Fluid evolution of the Hub Stock, Horní Slavkov–Krásno Sn–W ore district, Bohemian Massif, Czech Republic. *Miner. Depos.* **2012**, *47*, 821–833. [CrossRef]
18. Štemprok, M.; Pivec, E.; Langrová, A. The petrogenesis of a wolframite-bearing greisen in the Vykmanov granite stock, Western Krušné hory pluton (Czech Republic). *Bull. Geosci.* **2005**, *80*, 165–184.
19. Matte, P.; Maluski, H.; Rajlich, P.; Franke, W. Terrane boundaries in the Bohemian Massif: Result of large-scale Variscan shearing. *Tectonophysics* **1990**, *177*, 151–170. [CrossRef]
20. Klomínský, J.; Jarchovský, T.; Rajpot, G.S. *Atlas of Plutonic Rocks and Orthogneisses in the Bohemian Massif, Saxothuringicum*; Czech Geological Survey: Prague, Czech Republic, 2010; p. 94.
21. Sejkora, J.; Breiter, K. Historical Krupka ore district, Erzgebirge Mountains. *Bull. Miner. Petrol. Odd. Nár. Muz.* **1999**, *7*, 29. (In Czech)
22. Janečka, J.; Malásek, F.; Štemprok, M.; Tischendorf, G.; Zoubek, V. *Metallogeny of Tin and Tungsten in the Krušné Hory—Erzgebirge*; International Geological Correlation Programme—Excursion Guide; Ústřední Ústav Geologický: Prague, Czech Republic, 1974.
23. Schumacher, F. Die Erzgebirgishe Metallprovinz und ihre Genesis. *Erzmetall* **1933**, *30*, 161–166.
24. Watznauer, A. Die erzgebirgischen Granitintrusionen. *Geologie* **1954**, *6–7*, 688–706.
25. Breiter, K.; Förster, H.-J.; Seltmann, R. Variscan silicic magmatism and related tin-tungsten mineralization in the Erzgebirge-Slavkovský les metallogenic province. *Miner. Depos.* **1999**, *34*, 505–521. [CrossRef]
26. Forster, H.-J.; Tischendorf, G.; Trumbull, R.B.; Gottesmann, B. Late-Collisional Granites in the Variscan Erzgebirge, Germany. *J. Pet.* **1999**, *40*, 1613–1645. [CrossRef]
27. Škvor, V. *Geology of the Czech Part of the Erzgebirge Mountains and Smrčiny*; Ústřední Ústav Geologický: Prague, Czech Republic, 1975; p. 48. (In Czech)
28. Škvor, V. The Krušné Hory Mountains pluton and its interpretation. *Věst. Ústř. Úst. Geol.* **1986**, *61*, 65–71. (In Czech)
29. Siebel, W.; Trzebski, R.; Stettner, G.; Hecht, L.; Casten, U.; Höhndorf, A.; Műller, P. Granitoid magmatism of the NW Bohemian massif revealed: Gravity data, composition, age relations and phase concept. *Geol. Rund.* **1997**, *86*, 545–563. [CrossRef]
30. Romer, R.L.; Thomas, R.; Stein, H.J.; Rhede, D. Dating multiply overprinted Sn-mineralized granites—Examples from the Erzgebirge, Germany. *Miner. Depos.* **2007**, *42*, 337–359. [CrossRef]
31. Tichomirova, M.; Leonhardt, D. New age determination (Pb/Pb zircon evaporation, Rb/Sr) on the granites from Aus-Schwarzenberg and Eibenstock, Western Erzgebirge, Germany. *Z. Geol. Wiss.* **2010**, *38*, 99–123.
32. Štemprok, M.; Blecha, V. Variscan Sn–W–Mo metallogeny in the gravity picture of the Krušné hory/Erzgebirge granite batholith (Central Europe). *Ore Geol. Rev.* **2015**, *69*, 285–300. [CrossRef]
33. Richter, P.; Stettner, G. Geochemische und petrographishe Untersuchungen der Fichtelgebirgsgranite. *Geologica Bavar.* **1979**, *78*, 144.
34. Štemprok, M.; Šulcek, Z. Geochemical profile through an ore-bearing lithium granite. *Econ. Geol.* **1969**, *64*, 392–404. [CrossRef]
35. Peterková, T.; Dolejš, D. Magmatic-hydrothermal evolution of highly evolved granite stock Knöttel near Krupka in Erzgebirge. *Zpr. Geol. Výzk.* **2017**, *50*, 189–194. [CrossRef]
36. Peterková, T.; Dolejš, D. Magmatic-hydrothermal transition of Mo-W-mineralized granite-pegmatite-greisen system recorded by trace elements in quartz: Krupka district, Eastern Krušné hory/Erzgebirge. *Chem. Geol.* **2019**, *523*, 179–202. [CrossRef]
37. Žák, L. Origin of the molybdenite and feldspar deposit of Krupka. II. Paragenetic relations. *Acta Univ. Carol. Geol.* **1966**, *22*, 167–195.
38. Bodnar, R. Revised equation and table for determining the freezing point depression of H_2O-NaCl solutions. *Geochim. Cosmochim. Acta* **1993**, *57*, 683–684. [CrossRef]
39. Duan, Z.; Møller, N.; Weare, J.H. An equation of state for the CH_4-CO_2-H_2O system: I. Pure systems from 0 to 1000 °C and 0 to 8000 bar. *Geochim. Cosmochim. Acta* **1992**, *56*, 2605–2617. [CrossRef]
40. Duan, Z.; Møller, N.; Weare, J.H. An equation of state for the CH_4-CO_2-H_2O system: II. Mixtures from 50 to 1000 °C and 0 to 1000 bar. *Geochim. Cosmochim. Acta* **1992**, *56*, 2619–2631. [CrossRef]
41. Bowers, T.S.; Helgeson, H.C. Calculation of the thermodynamic and geochemical consequences of nonideal mixing in the system H_2O-CO_2-NaCl fluids at high pressures and temperatures. *Geochim. Cosmochim. Acta* **1983**, *47*, 1247–1275. [CrossRef]
42. Zhang, Y.-G.; Frantz, J.D. Determination of the homogenization temperatures and densities of supercritical fluids in the system NaCl–KCl–$CaCl_2$–H_2O using synthetic fluid inclusions. *Chem. Geol.* **1987**, *64*, 335–345. [CrossRef]
43. Bakker, R.J. Adaptation of the Bowers and Helgeson (1983) equation of state to the H_2O–CO_2–CH_4–N_2–NaCl system. *Chem. Geol.* **1999**, *154*, 225–236. [CrossRef]
44. Available online: https://fluids.unileoben.ac.at/Computer.html (accessed on 17 December 2023).
45. Bodnar, R.J.; Vityk, M.O. Interpretation of microthermometric data for H_2O-NaCl fluid inclusions. In *Fluid Inclusions in Minerals: Methods and Applications*; De Vivo, B., Frezzotii, M.L., Eds.; Short Course IMA: Pontignano-Siena, Italy, 1994; pp. 117–130.
46. Knight, C.; Bodnar, R. Synthetic fluid inclusions: IX. Critical PVTX properties of NaCl-H_2O solutions. *Geochim. Cosmochim. Acta* **1989**, *53*, 3–8. [CrossRef]
47. Frezzotti, M.L.; Tecce, F.; Casagli, A. Raman spectroscopy for fluid inclusion analysis. *J. Geochem. Explor.* **2012**, *112*, 1–20. [CrossRef]
48. Burke, E.A.J. Raman microspectrometry of fluid inclusions. *Lithos* **2001**, *55*, 139–158. [CrossRef]

49. Hurai, V.; Huraiová, M.; Slobodník, M.; Thomas, R. *Geofluids Development in Microthermometry, Spectroscopy, Thermodynamics, and Stable Isotopes*, 1st ed.; Elsevier Inc.: Amsterdam, The Netherlands, 2015. [CrossRef]
50. Vennemann, T.V.; O´Neil, J.R. A simple and inexpensive method of hydrogen isotope and water analyses of minerals and rocks based on zinc reagent. *Chem. Geol.* **1993**, *103*, 227–234. [CrossRef]
51. Suzuoki, T.; Epstein, S. Hydrogen isotope fractionation between OH-bearing minerals and water. *Geochim. Cosmochim. Acta* **1976**, *40*, 1229–1240. [CrossRef]
52. Gilg, H.A.; Sheppard, S.M. Hydrogen isotope fractionation between kaolinite and water revisited. *Geochim. Cosmochim. Acta* **1996**, *60*, 529–533. [CrossRef]
53. Lecumberri-Sanchez, P.; Steele-MacInnis, M.; Bodnar, R.J. A numerical model to estimate trapping conditions of fluid inclusions that homogenize by halite disappearance. *Geochim. Cosmochim. Acta* **2012**, *92*, 14–22. [CrossRef]
54. Webster, J.; Thomas, R.; Förster, H.J.; Seltman, R.; Tappen, C. Geochemical evolution of halogen-enriched granite magmas and mineralizing fluids of the Zinnwald tin-tungsten mining district, Erzgebirge, Germany. *Miner. Depos.* **2004**, *39*, 452–472. [CrossRef]
55. Zwart, E.W.; Touret, J.L.R. Melting behavior and composition of aqueous fluid inclusions on fluorite and calcite: Application within system H_2O-$CaCl_2$-NaCl. *Eur. J. Miner.* **1994**, *6*, 773–786. [CrossRef]
56. Ulmanová, J.; Dolníček, Z. New occurrence of vein W-mineralization in Tanvald granite from Jablonec nad Nisou—Mineralogy, chemical compositions of minerals and fluid inclusions. *Bull. Min. Petrolog.* **2019**, *27*, 193–204. (In Czech)
57. Nasdala, L.; Smith, D.C.; Kaindl, R.; Ziemann, M.A. Raman spectroscopy: Analytical perspectives in mineralogical research. *EMU Notes Miner.* **2004**, *6*, 20. [CrossRef]
58. Tsuruoka, S. The Evolution of Hydrothermal Fluids from the Deep Porphyry Environment to the Shallow Epithermal Environment. Ph.D. Thesis, Colorado School of Mines, Faculty and the Board of Trustees, Golden, CO, USA, 2017. Volume 197. p. 17.
59. Borisenko, A.S. Izučenije solevogo sostava rastvorov gazovožidkich vklučenij v mineralach metodom kriometrii. *Geolog. Geofiz.* **1977**, *8*, 16–27.
60. Klemm, W. Chemical evolution of hydrothermal solutions during Variscan and post-Variscan mineralization in Erzgebirge, Germany. In *Metallogeny of Collisional Orogens: Focused on the Erzgebirge and Comparable Metallogenic Settings*; Seltmann, R., Kämpf, H., Möller, P., Eds.; Czech Geological Survey: Prague, Czech Republic, 1994; pp. 150–158.
61. Charoy, B. Ploemeur kaolin deposit (Brittany), an example of hydrothermal alteration. *Pétrologie* **1975**, *1*, 253–266.
62. Charoy, B. *Definition of Importance des Phenomenes Deuteriques et des Fluids Associes dans les Granites. Consequences Metallogeniques*; Sciences de la Terre: Paris, France, 1979; p. 364.
63. Dolníček, Z.; Fojt, B.; Prochaska, W.; Kučera, J.; Sulovský, P. Origin of the Zálesí U–Ni–Co–As–Ag/Bi deposit, Bohemian Massif, Czech Republic: Fluid inclusion and stable isotope constraints. *Miner. Depos.* **2009**, *44*, 81–97. [CrossRef]
64. Kotlánová, M.; Dolníček, Z. Origin and chemical composition of fluids of post-Variscan hydrothermal mineralization at locality Zlatý důl near Hlubočky (Lower Carboniferous of the Nízký Jeseník Upland). *Geol. Výzk. Mor. Slez.* **2016**, *23*, 74–80. (In Czech) [CrossRef]
65. Edmonds, M.; Woods, A.W. Exsolved volatiles in magma reservoirs. *J. Volcanol. Geotherm. Res.* **2018**, *368*, 13–30. [CrossRef]
66. Thomas, R. Fluid evolution in relation to the emplacement of the Variscan granites in the Erzgebirge region: A review of the melt and fluid inclusion evidence. In *Metallogeny of Collisional Orogens*; Seltmann, R., Kämpf, H., Möller, P., Eds.; Czech Geological Survey: Prague, Czech Republic, 1994; pp. 70–81.
67. Thomas, R.; Förster, H.J.; Appel, K.; Webster, J. Formation of extremely F-rich hydrous melt fractions and hydrothermal fluids during differentiation of highly evolved tin-granite magmas: A melt/fluid-inclusion study. *Contrib. Miner. Pet.* **2005**, *148*, 582–601. [CrossRef]
68. Hoefs, J. *Stable Isotope Geochemistry*, 6th ed.; Springer: Berlin/Heidelberg, Germany, 2009. [CrossRef]
69. Morishita, Y.; Nishio, Y. Ore Genesis of the Takatori Tungsten–Quartz Vein Deposit, Japan: Chemical and Isotopic Evidence. *Minerals* **2021**, *11*, 765. [CrossRef]
70. Tornos, F.; Delgado, A.; Casquet, C.; Galindo, C. 300 Million years of episodic hydrothermal activity: Stable isotope evidence from hydrothermal rocks of the Eastern Iberian Central System. *Miner. Depos.* **2000**, *35*, 551–569. [CrossRef]
71. Moon, S.H.; Park, H.-I.; Ripley, E.M.; Lee, I. Mineralogic and stable isotope studies of cassiterite greisen mineralization in the Uljin area, Korea. *Econ. Geol.* **1996**, *91*, 916–933. [CrossRef]
72. Carruzzo, S.; Kontak, D.J.; Clarke, D.B.; Kyser, T.K. An integrated fluid-mineral stable-isotope study of the granite-hosted mineral deposits of the New Ross area, South Mountain Batholith, Nova Scotia, Canada: Evidence for multiple reservois. *Canad. Miner.* **2004**, *42*, 1425–1441. [CrossRef]
73. Myint, A.Z.; Yonezu, K.; Boyce, A.J.; Selby, D.; Scherstén, A.; Tindell, T.; Watanabe, K.; Swe, Y.M. Stable isotope and geochronological study of the Mawchi Sn-W deposit, Myanmar: Implications for timing of mineralization and ore genesis. *Ore Geol. Rev.* **2018**, *95*, 663–679. [CrossRef]

minerals

MDPI

Article

The Solubility of Antimony (Sb) in Liquid Hydrocarbons and Its Implication for the Ore-Forming Process of Orogenic Antimony-Gold Deposits in Southern Tibet

Yue Su [1], Xiaoming Sun [1,2,3,*] and Zhengpeng Ding [1,3]

1 School of Earth Sciences and Engineering, Guangdong Provincial Key Laboratory of Geodynamics and Geohazards, Sun Yat-Sen University, Zhuhai 519082, China; suyue6@mail2.sysu.edu.cn (Y.S.); dingzhp@mail2.sysu.edu.cn (Z.D.)
2 School of Marine Sciences, Guangdong Provincial Key Laboratory of Marine Resources and Coastal Engineering, Sun Yat-Sen University, Zhuhai 519082, China
3 CSIRO Mineral Resources, Australian Resources Research Centre, Kensington, WA 6151, Australia
* Correspondence: eessxm@mail.sysu.edu.cn

Abstract: Orogenic antimony-gold deposits contribute significantly to the global antimony resource base. China's orogenic antimony-gold deposits are primarily in southern Tibet. Investigations indicate that antimony combines and migrates with sulfur hydroxides in ore-forming fluids. Previous research on fluid inclusions in orogenic antimony-gold ores with sedimentary rock accommodation revealed the presence of organic inclusions, including liquid hydrocarbons, alongside common components such as CO_2, H_2O, CH_4, and NaCl. However, the impact of liquid hydrocarbons on antimony migration and mineralization is still debatable. To investigate the transportability of antimony by liquid hydrocarbons in orogenic antimony ores, we selected n-dodecanethiol and n-dodecane as the subjects. We measured the solubility and occurrence form of antimony in these compounds at various temperatures and durations. The results indicate that after 5 and 10 days of reaction at 100 °C, the antimony concentrations in the n-dodecanethiol and n-dodecane groups were 67.44 ± 7.62 ppm, 75.15 ± 16.74 ppm, 1.40 ± 1.02 ppm, and 3.02 ± 3.09 ppm, respectively. At 150 °C for 5 and 10 days, the respective concentrations were 50.58 ± 5.39 ppm, 77.26 ± 45.20 ppm, 2.66 ± 3.08 ppm, and 2.41 ± 2.03 ppm. At 200 °C for 5 and 10 days, the corresponding concentrations were 339.76 ± 71.94 ppm, 218.97 ± 25.03 ppm and 6.53 ± 7.17 ppm, 2.27 ± 0.82 ppm (n = 3). The measured solubility of antimony in the n-dodecanethiol group increased gradually with rising temperature. The solubility of antimony in the n-dodecane group was low and notably inferior to that observed in the n-dodecanethiol group. X-ray photoelectron spectroscopy (XPS) analysis demonstrated a distinct thiol (R-SH) peak at 163.31 eV and compound peaks of antimony reacting with thiols at 162.06 and 160.87 eV. This suggests that antimony predominantly forms complexes with thiols for migration. Our findings suggest that specific liquid hydrocarbon components, predominantly thiols, can interact with antimony at metallogenic temperatures and persist in ore-forming fluids, facilitating migration and mineral enrichment. Earlier experimental studies on gold and crude oil have indicated that liquid hydrocarbons also play an essential role in the transportation and enrichment of gold during the formation of gold deposits, thus indicating that liquid hydrocarbons possess the considerable potential to act as an ore-forming fluid during orogenic antimony-gold deposit formation in southern Tibet.

Keywords: antimony; liquid hydrocarbons; thiol; ore-forming fluid; orogenic antimony-gold deposits; southern Tibet

Citation: Su, Y.; Sun, X.; Ding, Z. The Solubility of Antimony (Sb) in Liquid Hydrocarbons and Its Implication for the Ore-Forming Process of Orogenic Antimony-Gold Deposits in Southern Tibet. *Minerals* **2024**, *14*, 141. https://doi.org/10.3390/min14020141

Academic Editors: Yuichi Morishita and Napoleon Q. Hammond

Received: 2 January 2024
Revised: 22 January 2024
Accepted: 23 January 2024
Published: 27 January 2024

1. Introduction

Regions such as South China, the Qinling Mountains, Southern Tibet, and Xinjiang Tianshan are known for their abundant antimony-gold resources [1]. Situated within the

Tethys Himalayan region, the antimony polymetallic mineralization belt in southern Tibet is positioned between the Indus-Yarlung Zangbo Suture Zone (IYZSZ) and the South Tibetan Detachment System (STDS) (Figure 1) [2]. The region's stratigraphy is extensively exposed, exhibiting intricate tectonic features, and antimony single deposits and antimony polymetallic deposits are widely distributed due to multiple phases of collision and accretion between the Indian plate and the Eurasian plate [3]. Antimony (gold) deposits within the extensive South Tibetan metallogenic belt are predominantly associated with sedimentary rocks as the main ore-forming wall rocks, exhibiting a wide distribution range, significant production outputs, and substantial metal reserves. The near east–west and north–south fault structures control the spatial distribution of these antimony (gold) deposits (Figure 1), and the majority of them occur in Triassic, Jurassic, and Cretaceous siltstones, feldspathic sandstones, and carbonaceous slates, as well as alkali-rich intermediate-basic intrusive rock masses. The organic stuff is abundant in the ore-forming wall rocks.

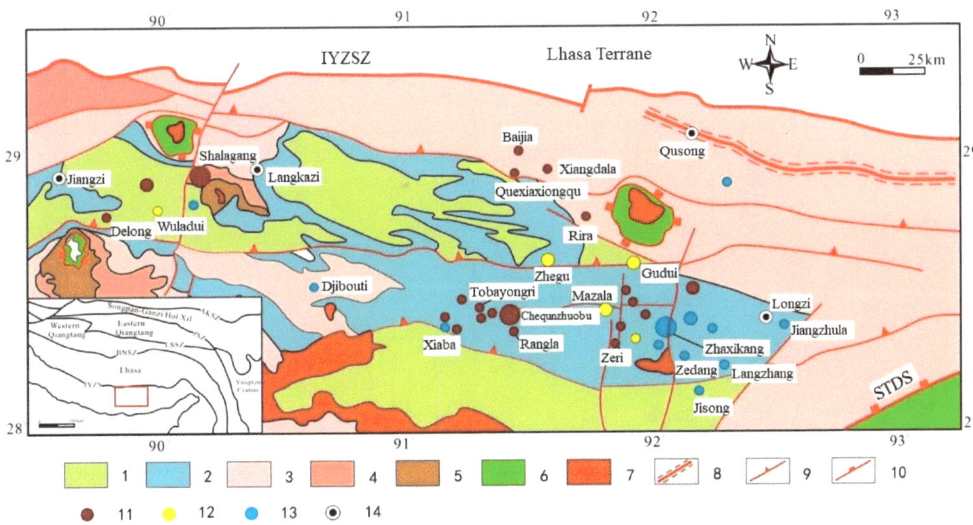

Figure 1. Metallogenic map of antimony ore belt in south Tibet (modified from [4,5]). IYZSZ = Indus-Yarlung Zangbo Suture Zone, STDS = South Tibet Detachment System. 1. Cretaceous: clastic and carbonate rocks, interlayered with volcanic rocks; 2. Jurassic: shelf-facies shale (slate), sandstone, limestone, interlayered with volcanic rocks; 3. Upper Triassic: deep-water shelf-facies sandstone shale (slate), interlayered with volcanic rocks; 4. Lower-middle Triassic: shallow-marine sandstone and shale (slate); 5. Permian: conglomerate, sandstone, marble, and silty slate; 6. Precambrian metamorphic rocks; 7. Cenozoic granitoids; 8. Ductile shear zone; 9. Thrust fault; 10. Detachment fault; 11. Antimony deposit; 12. Gold-antimony deposit; 13. Antimony polymetallic ore; 14. City.

Organic matter plays a crucial role in the migration and enrichment processes of mineralizing elements, particularly in the formation of metallic sedimentary deposits [6–8]. Significant quantities of liquid hydrocarbons can be generated through the thermal decomposition of organic matter during both diagenesis and deep diagenesis [9,10]. These liquid hydrocarbons, formed during the evolution of sedimentary organic matter, have been identified in various metal deposits, such as orogenic antimony-gold deposits [11–14], Witwatersrand-type Au-U deposits in South Africa [15], sediment-hosted U deposits [16,17], and Mississippi Valley-type Pb-Zn deposits [18–20].

Numerous antimony-only deposits and antimony-gold deposits are closely associated with liquid hydrocarbons. Antimony single deposits, such as the DaChang antimony deposit in Qinglong, Guizhou, contain liquid hydrocarbons or organic inclusions that are present within the ore and oil source rocks in the marginal basins, and there are multi-layered asphalt layers (ancient oil deposits), with asphalt reserves totaling 368,400 tons. Additionally, there is a close spatial relationship between the antimony mineralization layer and asphalt layers [21,22]. Furthermore, isotopic analyses of ore samples from the Meiduo antimony deposit in Northern Tibet indicate that organic matter was involved in the mineralization process, and Raman laser and micro-infrared spectroscopic analyses reveal that the mineralized fluids contain components including CH_4, C_6H_6, and CH_2 [23]. In the fluid inclusions of the mineralized ores, antimony gold deposits, such as the gold and antimony ore deposits in Lannigou, Guizhou, contain saturated hydrocarbons, unsaturated hydrocarbons, aromatic hydrocarbons, and other organic substances, and are intimately associated with the oil layer [24,25]. Moreover, Zhai et al. [14] employed petrographic observation, Raman spectroscopy, and fluid inclusion micro thermometry to detect gas-liquid two-phase or one-phase organic inclusions at room temperature in the vein mineral quartz and the ore mineral phengite, both of which formed during the metallogenic phase of the Shalagang antimony deposit and the Mazala antimony-gold deposit in southern Tibet (Figure 2a). Raman spectroscopy analysis indicated that the organic inclusions consisted mainly of alkanes and polycyclic aromatic hydrocarbons (Figure 2b). Levine [26] proposed that the presence of a considerable quantity of liquid-phase and gas-liquid two-phase organic inclusions in the ore indicated the involvement of liquid hydrocarbons in the transport of ore-forming materials, and the occurrence of liquid organic inclusions indicated the maturation of hydrocarbons.

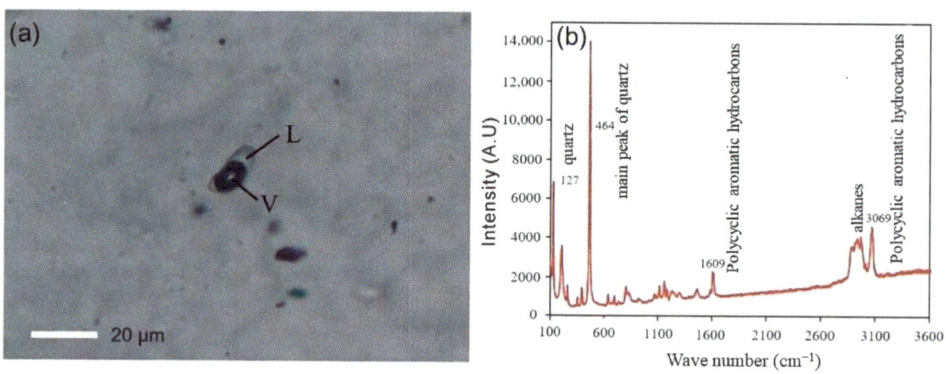

Figure 2. Photomicrograph of typical organic fluid inclusions from Mazala deposits (modified from [14]). L: Liquid; V: Vapor. (**a**) Isolated two-phase hydrocarbon inclusion in quartz from Mazala. (**b**) Laser Raman spectra of liquid-phase hydrocarbon in two-phase inclusions.

The origin of the ore-forming fluids in orogenic antimony-gold deposits continues to be a subject of debate. For instance, Yang et al. [2] determined that the ore-forming fluids in the Shalagang antimony deposit are dominated by circulating atmospheric water, and the ore-forming fluids in the Mazala antimony-gold deposit are a mixture of magmatic and atmospheric water. Nie et al. [27] concluded that the mineralization of the Mazala antimony-gold deposit was related to the tectonic-magmatic activity of the late Yanshan–early Himalayan period, with magmatic-hydrothermal fluids extracting metallic elements from the surrounding rocks, leading to mineralization. Zhai et al. [14] reported that the Shalagang antimony deposit and the Mazala antimony-gold deposit include a mixture of constructed and metamorphic water in the ore-forming fluids. Nonetheless, the current research on the ore-forming fluids of orogenic antimony-gold deposits primarily emphasizes the investigation of their inorganic components. The liquid hydrocarbon components found in the fluid inclusions of the mineralized ores associated with the Shalagang antimony deposit and the Mazala antimony-gold deposit suggest that liquid hydrocarbons may be able to play an active role in the antimony mineralization process. Therefore, we start from the organic components in the ore-forming fluids to investigate whether the liquid hydrocarbons can participate in transporting antimony to mineralization. Recent experimental studies by various scholars have shown that natural liquid hydrocarbons have the capacity to transport sufficient mineralizing materials to support the mineralization process. These studies have specifically investigated the solubility of zinc, uranium, palladium, and other metals in liquid hydrocarbons [28–33]. N-dodecane and n-dodecanethiol are typical constituents of liquid hydrocarbons formed through the pyrolysis of organic matter, representing the wide range of organic groups present in such hydrocarbons. These two organic compounds are suitable for this experiment because of their physical characteristics, including their melting and boiling temperatures, as well as their stability under various redox and temperature conditions [34]. Therefore, in our experimental investigation to assess the potential contribution of liquid hydrocarbons in ore-forming fluids to the transport of antimony, we utilized n-dodecane and n-dodecanethiol.

We conducted experiments involving the dissolution of antimony metal in n-dodecane and n-dodecanethiol. The experiments were conducted at temperatures of 100, 150, and 200 °C, with varying reaction durations ranging from 5 to 10 days. Additionally, we performed an X-ray photoelectron spectroscopy (XPS) analysis to investigate the surface composition of the antimony lumps after the reaction. This analysis enabled us to investigate the potential efficiency of n-dodecane and n-dodecanethiol in transporting antimony and to determine if liquid hydrocarbons can participate in the transportation of antimony in the ore-forming fluids. This research provides new insights into the transport mechanism of antimony in orogenic antimony-gold deposits located in southern Tibet and hosted by sedimentary rocks.

2. Materials and Methods

2.1. Sample Description

The organic reagents n-dodecane and n-dodecanethiol were obtained from Wengjiang Reagent and produced in the Guanchang Industrial Zone, Guandu Town, Wengyuan County, Shaoguan, Guangdong. The properties of n-dodecane and n-dodecanethiol are displayed in Table 1.

Table 1. Characterization of n-dodecane and n-dodecanethiol properties.

Parameters	N-Dodecane	N-Dodecanethiol
Concentration	AR, 98.0%	AR, 98.0%
Density (25 °C)	0.75 g/mL	0.845 g/mL
Molecular formula	$C_{12}H_{26}$	$C_{12}H_{26}S$
Molecular weight	170.34	202.40
Boiling point	215–217 °C	266–283 °C
Melting point	−9.6 °C	−7 °C

2.2. High-Temperature Experiments

The experiments were conducted in a sealed reactor consisting of a larger tube placed over a smaller tube, with the larger tube made of high borosilicate glass test tubes (OD = 12 mm, ID = 10 mm) and the smaller tube made of quartz test tubes (OD = 9 mm, ID = 7 mm) (Figure 3). The decision to conduct tests at 100, 150, and 200 °C was based on the temperature ranges identified through micro-thermometry of fluid inclusions in orogenic antimony deposits, as well as the physical and chemical properties of n-dodecane and n-dodecanethiol. To minimize the risk of thermal degradation of n-dodecane and n-dodecanethiol and ensure sufficient time for equilibrium to be reached, the lowest temperature within this range was chosen.

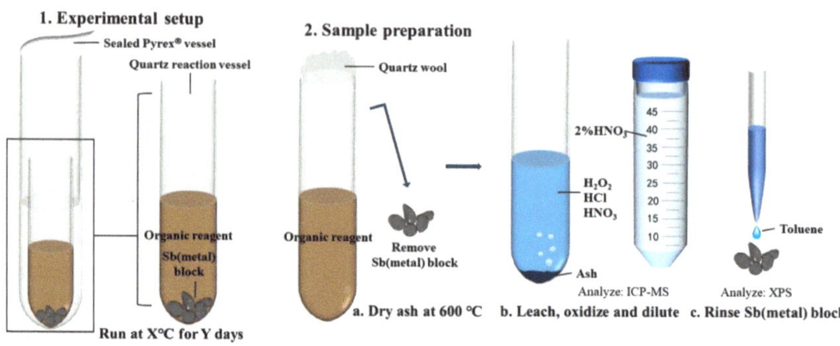

Figure 3. The experimental protocol employed for the investigation of antimony solubility and speciation in liquid hydrocarbons at 100, 150, and 200 °C.

Prior to the experiment, the borosilicate glass and quartz tubes were cleaned with nitric acid (~75% HNO_3) for 24 h, rinsed with ultrapure water, and dried at 100 °C for 2 h. The antimony was introduced into the quartz tubes in the form of pure metal blocks with a controlled mass of approximately 0.50 g. Two experimental groups were established: one group received a mixture of 0.25 mL n-dodecane and 0.25 mL n-dodecanethiol in the quartz test tubes, while the other group received only 0.25 mL of n-dodecane. After the addition of the reagents as well as the antimony blocks, the quartz test tubes were carefully introduced into the borosilicate glass test tubes and then sealed with a high-temperature flame. All the weights, that is, the weights of the quartz test tubes before and after the addition of the antimony metal blocks, as well as the n-dodecane and n-dodecanethiol, were carefully determined during the experiment using a high-precision analytical balance. Sanz-Robinson et al. [31,32] conducted kinetic experiments using a range of crude oil samples and a controlled temperature gradient to evaluate the solubility of palladium and zinc in crude oil and determine the equilibrium time. Palladium and zinc wires were reacted with equal volumes of crude oil for 5, 10, 15, and 30 days. The concentration of palladium in the crude oil started to stabilize after approximately five days, while zinc reached a steady concentration in less than 15 days and, in certain instances, within five days. Therefore, based on the findings of previous studies, reaction durations of 5 and 10 days were selected to ensure the attainment of steady-state concentrations of antimony within a reasonable timeframe.

After completing the weighing process, the sealed reactor was placed inside a table-top muffle furnace oven preheated to the required temperature. The temperature was maintained within a ±1 °C range. The experimental setup utilized a muffle furnace manufactured by Shanghai Jinping Instrumentation Co., Ltd, Shanghai, China. The furnace had dimensions of 500 mm in length, 300 mm in width, and 200 mm in height. It operated at a rated voltage of 380 V, a rated temperature of 1200 °C, and a power output of 12 kW, incorporating an electric heating wire as the heating element.

After the experiment, the reactors were opened using a diamond cutter, and the quartz test tubes were carefully extracted using disposable tweezers. Clean quartz wool was used to plug the openings of the quartz test tubes, ensuring no material loss due to volatilization and facilitating subsequent analytical tests.

2.3. XPS Analysis

Organic reagents that had not been completely reacted were evaluated using a method similar to that described by Sugiyama and Williams-Jones [35]. The crude oils were ashed using a combination of thermal combustion and chemical oxidation. The procedure involved transferring the remaining organic chemicals into a clean quartz test tube, sealing the tube's opening with clean quartz cotton, and subjecting the tube to cauterization at 600 °C for 12 h. After cauterization, the quartz test tube was treated with a mixture of 0.25 mL HNO_3 (75%), 0.25 mL HCl (35%), and 0.5 mL H_2O_2 (30%) and left to leach for 24 h, ensuring complete dissolution of any remaining ash. The leaching solution was diluted 1000 times with 2% HNO_3 solution before analyzing and determining the amount of dissolved antimony in the solution using an ICP-MS quadruple rod plasma mass spectrometer (Agilent 7700x, Santa Clara, CA, USA) with Y as the internal Standard. Incompletely reacted antimony blocks were removed with disposable tweezers, washed with toluene, and vacuum-dried for 24 h at room temperature. Subsequently, the surface composition of the antimony blocks was analyzed and studied using X-ray Photoelectron Spectroscopy (XPS) on a Thermo Fisher Scientific Kα spectrometer. The measurement employed Al Kα radiation (1486.6 eV, 12 kV, 720 W) with an X-ray spot size of 400 μm. The scanning mode was set to CAE, with the lens mode as Standard. The full-spectrum scanning fluence energy was 150 eV, the narrow-spectrum scanning fluence energy was 40 eV, and the resolution was $\leq$0.45 eV. Energy correction was conducted using the surface contamination C1s peak (284.8 eV). The resulting data were fitted and analyzed using XPS Peak41 software (version 4.0) and background removal through the Shirley-type method.

3. Results

3.1. Results from the Antimony Solubility Experiments

The dissolution of antimony blocks in n-dodecane and n-dodecanethiol reagents at 100, 150, and 200 °C is reported in Table 2 and Figure 4.

Table 2. A summary of the experimentally determined solubility of antimony in n-dodecane and n-dodecanethiol at 100, 150, and 200 °C (ppm).

Temp. / Times	N-Dodecane: N-Dodecanethiol = 1:1		Only N-Dodecane	
	5 Days	**10 Days**	**5 Days**	**10 Days**
100 °C	67.44 ± 7.62	75.15 ± 16.74	1.40 ± 1.02	3.02 ± 3.09
150 °C	50.58 ± 5.39	77.26 ± 45.20	2.66 ± 3.08	2.41 ± 2.03
200 °C	339.76 ± 71.94	218.97 ± 25.03	6.53 ± 7.17	2.27 ± 0.82

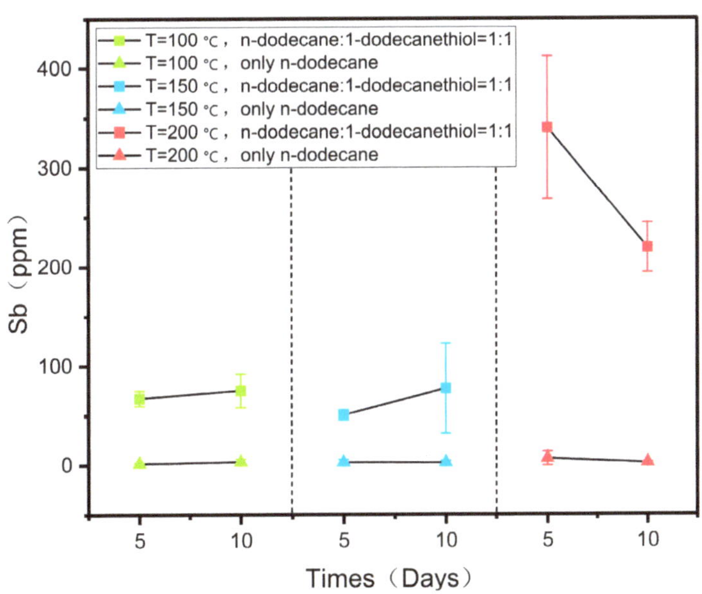

Figure 4. Concentration of antimony in both groups of reagents at 100, 150, and 200 °C as a function of the duration of the experiments.

The reaction was divided into two groups, with three experiments performed under each set of conditions. One group contained a 1:1 mixture of n-dodecane and n-dodecanethiol, while the other group used n-dodecane alone as the reagent (Table 2). The dissolution of antimony was investigated by comparing the two groups. The solubility of antimony in the 1:1 mixture of n-dodecane and n-dodecanethiol increased with temperature, reaching a maximum of 339.76 ppm at 200 °C, as indicated in Table 2 and Figure 4. Moreover, with an increase in reaction time at 100 and 150 °C, the solubility of antimony increased, reaching values of 75.15 ppm and 77.26 ppm on the 10th day, respectively. However, at 200 °C, the solubility of antimony decreased, and on the 5th day, it was higher than on the 10th day, possibly due to slight pyrolysis of the organic reagents during prolonged heating. The solubility of antimony did not increase with temperature in reagents consisting solely of n-dodecane. Only a small amount of antimony was dissolved in n-dodecane, and its solubility did not significantly change at different temperatures or reaction durations, remaining in the range of a few ppm (Table 2, Figure 4). These findings demonstrate that the thiol content plays a crucial role in determining the dissolution of antimony.

3.2. Results of XPS Analyses

XPS was employed to measure the spectra of antimony blocks after the reaction in a 1:1 combination of n-dodecane and n-dodecanethiol at 200 °C, as illustrated in Figure 5. Figure 5a illustrates the examination of reagent residues on the surface of the antimony blocks. It was determined that sulfur accounts for approximately 21% of the surface of the antimony metal block. The sulfur (S2p) peaks in the green and red lines correspond to sulfur bound to antimony and have binding energies of 162.06 eV and 160.87 eV (Figure 5a), respectively. The prominent sulfur (S2p) peak in the blue line corresponds to the unbound free thiol functional group, with a binding energy of 163.31 eV (Figure 5a) [36].

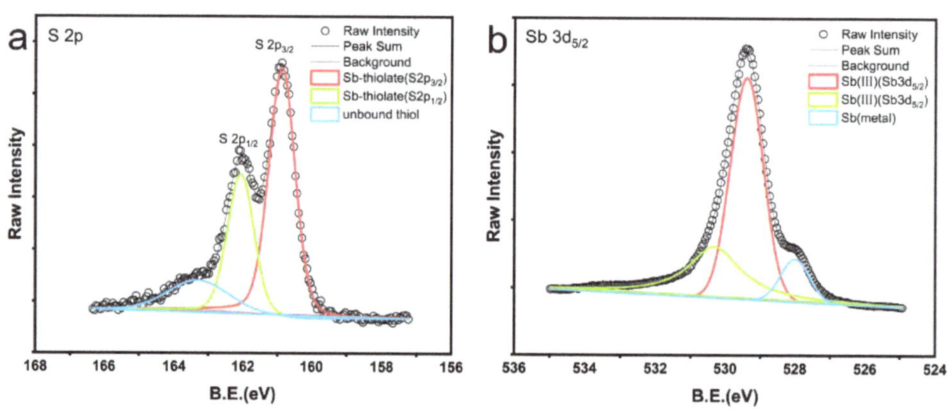

Figure 5. XPS spectra obtained by X-ray photoelectron spectroscopy of antimony blocks after reaction with a 1:1 mixture of n-dodecane and n-dodecanethiol reagents at 200 °C. (**a**) shows the XPS spectrum of the residual reagent sulfur form on the antimony block after the reaction. The thiol peak was separated into an unbound thiol peak in blue (163.31 eV), while the sulfur peaks bound to antimony are represented by the green (162.06 eV) and red (160.87 eV) peaks. (**b**) displays the XPS spectrum of the antimony block surface after the reaction and rinsing with toluene. The green and red peaks represent the compound antimony, while the blue peak represents the antimony metal.

XPS tests indicate that most sulfur is chemically bonded to antimony, while a portion exists as unbound free thiols. Additionally, Figure 5b demonstrates that XPS measurements of the toluene-rinsed surface of the antimony block reveal the presence of a blend of antimony metal and antimony compound (III). The antimony (Sb3d) peaks are depicted by the green and red lines, with binding energies of 530.30 eV and 529.39 eV (Figure 5b), respectively, corresponding to antimony compounds formed through thiol reactions. During the reaction with the mixed reagents, the antimony metal undergoes oxidation while the thiols undergo reduction.

4. Discussion

4.1. Factors Determining the Concentration and Speciation of Antimony in Liquid Hydrocarbons

Previous experimental studies have demonstrated that temperature exerts a significant influence on the solubility of antimony in solution. In a study by Kolpakova [37], the solubility of stibnite ores in H_2S-H_2O solution was examined across temperatures ranging from 100 to 300 °C. The results revealed a substantial alteration in stibnite solubility due to temperature variations, with the maximum solubility observed at 200 °C. He et al. [38] conducted water–rock leaching experiments on the S- and Cl-containing stratum of antimony in a medium-low temperature hydrothermal system. Their findings revealed that the antimony leaching rate increased with rising temperature within the range of 100–200 °C, reaching its peak at 200 °C. William-Jones and Norman [39] provided evidence of the substantial impact of temperature on antimony precipitation in hydrothermal fluids, highlighting that lower temperatures were more favorable for the formation of stibnite through antimony precipitation. This experimental study on the dissolution of antimony was conducted at 100, 150, and 200 °C temperatures, and the results showed that temperature is an important parameter affecting the solubility of antimony in organic reagents. Figure 4 illustrates the relationship between antimony solubility and temperature. In the 1:1 mixture of n-dodecane and n-dodecanethiol reagent group, the solubility of antimony exhibits a gradual increase with temperature, reaching its peak at 200 °C (Figure 4). Research indicates that the solubility of most minerals increases as temperature rises. Within a specific range, higher temperatures enhance the extraction of mineralized metals from surrounding mineralized rocks by ore-forming fluids, leading to a significant elevation in the concentration of mineralized components in the fluids and facilitating mineralization enrichment. Additionally, temperature impacts the density of the ore-forming fluid. As temperature increases, the density of the fluid decreases, prompting its upward flow toward faults, fracture zones, and other regions of the ore that are conducive to mineralization. This, in turn, affects the mineralization process.

The findings of our study indicate that the thiol group can dissolve substantial quantities of antimony, with 339 ppm of antimony dissolved at 200 °C. Additionally, Figure 4 illustrates that the solubility of antimony in the thiol group is significantly greater compared to the n-dodecane group (Figure 4). Based on the XPS characterization results, unbound thiols were found to persist on the surface of the metallic antimony block after the completion of the reaction (Figure 5). Additionally, the surface of the antimony block exhibited a higher presence of sulfur-bound compounds, providing evidence for the strong affinity between antimony and the reduced sulfur in the reagent mixture. The experimental results reveal that thiol is the critical parameter governing the dissolution of antimony in organic reagents. Metals can form ligands by interacting with organic compounds, and the stability constants of organic-metal complexes are significantly greater than those of inorganic complexes. Therefore, as long as low concentrations of organic ligands are present in mineralized hydrothermal fluids, organic-metal complexes are dominant [40]. The ability of specific sulfur compounds, especially thiols, to create organometallic complexes with chalcophile elements is widely acknowledged [41,42]. In contrast, antimony is a chalcophile element that effortlessly combines with sulfur. Furthermore, unlike n-dodecane, which is a soft base, n-dodecanethiol has polarity due to the presence of sulfur-based functional groups, and antimony metal is a junction acid; hence, the thiol reacts more readily with antimony, enabling its solubilization in the reagent. Therefore, liquid hydrocarbons with a high concentration of thiols can play a role in the transportation of antimony during the mineralization process. The amount of hydrogen sulfide in the mineralized layer affects the concentration of thiols in liquid hydrocarbons. In the deep source area of antimony-gold mines in southern Tibet, hydrocarbons and sulfates undergo thermochemical reduction reactions, resulting in the generation of substantial amounts of hydrogen sulfide. This hydrogen sulfide can react with organic compounds present in crude oil and natural gas, leading to an increase in the concentration of thiols and the subsequent production of liquid hydrocarbons enriched with thiols [43–46].

4.2. Formation of Liquid Hydrocarbons in Orogenic Antimony-Gold Ores and Their Role in the Mineralization Process

The collision between the Indian and Eurasian plates occurred approximately 65 million years ago [47–49], resulting in a collisional orogenic process that can be divided into two distinct phases: early collision-extrusion and late orogenic intracontinental extension. Throughout this process, various geological features emerged in southern Tibet, including fold tectonics, reverse faults, metamorphic core complexes, and near east–west and near north–south fault systems (Figure 1) [2,50–54]. These geological features have played a role in constraining the mineralization of antimony (gold) deposits. Numerous antimony-only deposits and antimony-gold deposits are present within the Southern Tibetan Plateau, of which the Shalagang and Mazala deposits represent examples of antimony-only and antimony-gold types, respectively, both hosted in sedimentary rocks. The Triassic, Jurassic, and Cretaceous sedimentary units, where these two deposits are located, are mostly represented by siltstone, sandstone, quartz sandstone, carbonaceous slate, and marl. These units contain a large amount of sedimentary organic matter and often constitute the ore-forming wall rocks of the deposits [27]. Previous studies have revealed that the Shalagang antimony deposit and the Mazala antimony-gold deposit are shallow, low-temperature deposits. Additionally, fluid inclusions within the relevant mineralized ores of these two deposits have been found to contain liquid hydrocarbon components (Figure 2) [14,55].

The results of our tests indicate that liquid hydrocarbons play a crucial role in antimony mineralization transport. Previous research by Migdisov et al. [28] examined the solubility of gold in a variety of natural crude oils at temperatures ranging from 100 to 300 °C, yielding results comparable to those presented here. Their experiments demonstrated that liquid hydrocarbons may have significantly contributed to the formation of the deposits. Liquid hydrocarbons can extract gold from their host rocks, transport it in appreciable concentrations, and deposit it in economically exploitable amounts. Crede et al. [29,30] performed an in situ XAS experiment at 25–250 °C to determine the morphology and structural properties of gold complexes in aqueous and oil-based fluids (S-free n-dodecane, S-bearing 1-dodecanethiol). Since sulfur and organothiol compounds are ubiquitous and abundant components in natural oils, the experimental results show that natural oils can potentially enrich gold from coexisting gold-bearing brines, demonstrating the importance of liquid hydrocarbons in conveying and enriching gold mineralization. As a result, we may conclude that liquid hydrocarbons can play an essential role in both antimony transport and gold mineralization.

This also provides a possible explanation mechanism for the co-occurrence of antimony and gold ores in southern Tibet from the ore-forming fluid. Regional metamorphic tectonic events impact the organic-rich rocks around the sediment-hosted orogenic antimony-gold deposits, leading to the liberation of organic components, including liquid hydrocarbons. These components are subsequently enriched in the ore-bearing fluids, facilitating the entry of the active and more active antimony and gold elements into the reservoir in complex and other forms, in conjunction with water and oil. Ductile shear zones and faults create favorable conditions for the upwelling of ore-bearing fluids. These fluids are subsequently influenced by changes in temperature and pressure, leading to their migration, precipitation in favorable tectonic sites, and the eventual formation of mineralized bodies [2,11].

4.3. Evaluation of the Potential of Liquid Hydrocarbons as Ore-Forming Fluids in Orogenic Antimony-Gold Ores

When evaluating the potential of liquid hydrocarbons as ore-forming fluids, it is crucial to compare antimony solubility in them to antimony solubility in hydrothermal fluids. The solubility characteristics of stibnite indicate that solutions of different properties can dissolve and transport high concentrations of antimony. Elevated temperatures and mildly acidic or alkaline conditions are particularly favorable for the dissolution and transport of antimony [56]. Extensive research data on the antimony concentration in modern geothermal fluids have shown significant variability, with the majority having antimony contents around 0.1 ppm [56,57]. In contrast, fluids in New Zealand's geothermal areas exhibit antimony contents as high as 84.42–238.25 ppm [58], demonstrating the presence of fluids with elevated antimony levels in nature. Nevertheless, compared to the antimony solubility in this experiment, it is clear that the antimony level in modern geothermal fluids is far from saturation. In addition, studies have shown that hydrothermal solutions forming epithermal gold deposits contain only a few tens of ppb of gold [59], which is within the solubility range of gold in crude oils identified in the study by Migdisov et al. [28]. Hence, we conclude that liquid hydrocarbons have great potential as mineralizing fluids for transporting antimony and gold mineralization.

Additionally, the antimony concentration and extractability of the host rock are crucial factors in the formation of antimony ore-forming fluids. The average antimony content in magmatic rocks is generally lower than Clark's value, with ultramafic rocks having the lowest antimony content (0.1 ppb), basaltic and acidic rocks exhibiting similar antimony levels (0.2 ppb), and sedimentary shales and mudstone rocks containing the highest antimony content (1.5 to 2 ppm) [60]. The prevalence of sedimentary rocks in southern Tibet, along with the abundant organic matter in the surrounding rocks, creates a favorable metallogenic environment conducive to hydrocarbon reservoir formation. When hydrocarbons come into contact with sulfate minerals at temperatures exceeding 100 °C, they trigger reactions like thermal sulfate reduction (TSR), leading to a substantial increase in thiol levels [61]. Based on the findings of this experimental work, thiols readily react and combine with antimony, causing antimony oxidation and thiol reduction, effectively inhibiting antimony migration during mineralization. Hydrocarbons with a high thiol content significantly enhance the extraction of antimony metal from ore-forming wall rocks. Consequently, in sediment-hosted antimony deposits, organic fluids, rather than aqueous fluids, primarily facilitate antimony transfer. The hydrocarbons found in the inclusions are likely the remaining organic components devoid of thiol after their involvement in metal transportation for mineralization.

This experiment's XPS characterization results show that antimony is readily bonded to thiols (Figure 5). However, we are unsure of the exact manner in which antimony is transported in liquid hydrocarbons in this instance. Additional ligands or functional groups present in liquid hydrocarbons may also play a significant role in facilitating the transport of antimony. In addition to the formation of organic complexes for metal transportation during mineralization, liquid hydrocarbons can exhibit other forms. An example of this is the synthesis of noble metal nanoparticles (NMNPs) to explain the enrichment and transportation of gold and platinum group elements (PGE) within naturally occurring organic (hydrocarbon) systems [33]. Furthermore, it has been found that liquid hydrocarbons are more efficient in transporting precious metals when they are in the form of nanoparticles rather than molecules. Hence, a detailed investigation into the specific form of liquid hydrocarbons involved in the transportation of antimony during mineralization is required.

5. Conclusions

(1) Antimony exhibits a pronounced chemical affinity towards reduced sulfur components. The solubility of antimony in organic reagents with a high thiol concentration surpasses its solubility in fluids previously identified as antimony-rich mineralized.

(2) The thermochemical sulfate reaction that occurs in the surrounding rocks of sediment-hosted orogenic antimony-gold deposits in a specific temperature range produces high concentrations of hydrogen sulfide. This, in turn, leads to the formation of a liquid hydrocarbon rich in thiols. This liquid hydrocarbon with a high thiol content can serve as a carrier within the ore-forming fluids for antimony mineralization.

(3) There is a clear positive correlation between the solubility of antimony and thiols, suggesting that thiols are determinants of antimony solubilization in liquid hydrocarbons. Nevertheless, we cannot rule out the possibility that other organic ligands, such as porphyrins and carboxylic acids, may also play a role in the dissolution and transport of antimony.

Author Contributions: Y.S. and Z.D. conceived and designed the experiments; Y.S. analyzed the data and wrote this paper; Z.D. and X.S. revised the original manuscript. All authors have read and agreed to the published version of the manuscript.

Funding: This research was financially supported by the National Natural Science Foundation of China (91855213, 41972070, 41672071, and U1302233), the National Key Research and Development Program of China (2018YFC0603603 and 2018YFA0702605), and the International Program for Candidates, Sun Yat-Sen University (SYSU-32110-20230907-0002).

Data Availability Statement: Data are contained within the article.

Acknowledgments: We would like to thank editors and anonymous reviewers for providing valuable remarks and constructive comments that have enhanced the overall quality of this paper.

Conflicts of Interest: The authors declare that they have no known competing financial interests or personal relationships that could appear to have influenced the work reported.

References

1. Li, F.; Zhang, G.Y.; Deng, Z.; Zhang, J.F.; Qin, Z.P.; Chen, X.; Lan, Z.W. Mineralization process of the Mazhala Au-Sb deposit, South Tibet: Constraints from fluid inclusions, H-O-S isotopes, and trace elements. *Ore Geol. Rev.* **2023**, *160*, 105595. [CrossRef]
2. Yang, Z.S.; Hou, Z.Q.; Gao, W.; Wang, H.P.; Li, Z.Q.; Meng, X.J.; Qu, X.M. Metallogenic characteristics and genetic model of antimony and gold deposits in south Tibetan detachment system. *Acta Geol. Sin.* **2006**, *80*, 1377–1391.
3. Feng, X.L.; Du, G.S. The distribution, mineralization types and prospecting and exploration of the gold deposits in Xizang. *Tethyan Geol.* **1999**, *23*, 31–38.
4. Sun, X.; Zheng, Y.Y.; Wang, C.M.; Zhao, Z.Y.; Geng, X.B. Identifying geochemical anomalies associated with Sb-Au-Pb-Zn-Ag mineralization in North Himalaya, southern Tibet. *Ore Geol. Rev.* **2016**, *73*, 1–12. [CrossRef]
5. Li, Y.; Song, Y.; Tang, J.X.; Chen, W.; Sun, H. Study on the distribution, type and metallogenic regularity of antimony deposit in Tibet. *Geol. China* **2023**, 1–34.
6. Fu, J.M.; Peng, P.A.; Lin, Q.; Liu, D.H.; Jia, R.F.; Shi, J.X.; Lu, J.L. A few issues in the study of organic geochemistry of stratified mineral deposits. *Adv. Earth Sci.* **1990**, *5*, 43–49.
7. Zhuang, H.P.; Lu, J.L. Metallogenic deposits associated with organic matter. *Geology-Geochemistry* **1996**, *4*, 6–11.
8. Zhuang, H.P.; Lu, J.L.; Wen, H.J.; Mao, H.H. Organic matter in hydrothermal ore-forming fluid. *Geology-Geochemistry* **1997**, *1*, 85–91.
9. Wan, C.L.; Zhai, Q.L.; Jin, Q. Interaction between source rocks and igneous rocks—A preliminary investigation of dissolution of igneous rocks by organic acids and catalysis of transition metals in the generation of hydrocarbons in the source rocks. *Geology-Geochemistry* **2001**, *29*, 46–51.
10. Miao, Y.N.; Li, X.F.; Wang, X.Z.; Chen, Y.; Zhou, Y.J.; Liu, F.W.; Xia, J. Review on hydrocarbon generation, pores formation and its methane adsorption mechanism in shale kerogen. *Sci. Sin. Phys. Mech. Astron.* **2017**, *47*, 114604. [CrossRef]
11. Mo, R.W.; Sun, X.M.; Zhai, W.; Zhou, F.; Liang, Y.H. Ore-forming fluid geochemistry and metallogenic mechanism from Mazhala gold-antimony deposit in southern Tibet, China. *Acta Petrol. Sin.* **2013**, *29*, 1427–1438.
12. Hu, S.Y.; Evans, K.; Craw, D.; Rempel, K.; Bourdet, J.; Dick, J.; Grice, K. Raman characterization of carbonaceous material in the Macraes orogenic gold deposit and metasedimentary host rocks, New Zealand. *Ore Geol. Rev.* **2015**, *70*, 80–95. [CrossRef]
13. Mirasol-Robert, A.; Grotheer, H.; Bourdet, J.; Suvorova, A.; Grice, K.; McCuaig, T.C.; Greenwood, P.F. Evidence and origin of different types of sedimentary organic matter from a Paleoproterozoic orogenic Au deposit. *Precambrian Res.* **2017**, *299*, 319–338. [CrossRef]
14. Zhai, W.; Zheng, S.Q.; Sun, X.M.; Wei, H.X.; Mo, R.W.; Zhang, L.Y.; Zhou, F.; Yi, J.Z. He-Ar isotope compositions of orogenic Mazhala Au-Sb and Shalagang Sb deposits in Himalayan orogeny, southern Tibet: Constrains to ore-forming fluid origin. *Acta Petrol. Sin.* **2018**, *34*, 3525–3538, (In Chinese with English Abstract).
15. Fuchs, S.; Williams-Jones, A.E.; Jackson, S.E.; Przybylowicz, W.J. Metal distribution in pyrobitumen of the Carbon Leader Reef, Witwatersrand Supergroup, South Africa: Evidence for liquid hydrocarbon ore fluids. *Chem. Geol.* **2016**, *426*, 45–59. [CrossRef]
16. Landais, P. Organic geochemistry of sedimentary uranium ore deposits. *Ore Geol. Rev.* **1996**, *11*, 33–51. [CrossRef]
17. Spirakis, C.S. The roles of organic matter in the formation of uranium deposits in sedimentary rocks. *Ore Geol. Rev.* **1996**, *11*, 53–69. [CrossRef]
18. Parnell, J. Metal enrichments in solid bitumens—A review. *Miner. Depos.* **1988**, *23*, 191–199. [CrossRef]

19. Gize, A.; Barnes, H. Organic processes in Mississippi Valley Type ore genesis. In Proceedings of the 28th International Geological Congress Abstracts, Washington, DC, USA, 9–19 July 1989; pp. 557–558.
20. Kesler, S.; Jones, H.; Furman, F.; Sassen, R.; Anderson, W.; Kyle, J. Role of crude oil in the genesis of Mississippi Valley Type deposits: Evidence from the Cincinnati arch. *Geology* **1994**, *22*, 609–612. [CrossRef]
21. Ye, Z.Y.; Shi, J.X.; Hu, R.Z. Organic matter and its role in mineralization in DaChang antimony deposit, Guizhou, China. *Acta Mineral. Sin.* **1997**, *3*, 310–315.
22. Wang, P.P. Relationship between organic matters and mineralization in the Qinglong antimony deposit, Guizhou province. *Value Eng.* **2019**, *38*, 243–245.
23. Yan, S.H.; Yu, J.J.; Zhao, Y.X.; Xu, Z.Z.; Wang, A.J.; Teng, R.L. Geology and Geochemistry of the Meiduo Antimony Ore Belt in Northern Tibet: Its Origin and Geodynamic Setting. *Acta Geosci. Sin.* **2004**, *25*, 541–548.
24. Zhang, Z.J.; Zhang, W.H. The study of organic ore-forming fluids in the Lannigou gold (mercury, antimony) deposit, Guizhou Province. *Miner. Depos.* **1998**, *17*, 354–361.
25. Yin, F.G.; Wan, F.; Tang, W.Q. The disseminated gold deposits in southwestern Guizhou: Mineralization model and its correlation with oil-generation theories. *Sediment. Geol. Tethyan Geol.* **2000**, *20*, 1–8.
26. Levine, R. Occurrence of fracture hosted imponite and petroleum fluid inclusion, Quebec city region, Canada. *AAPG Bull.* **1991**, *75*, 139–153.
27. Nie, F.J.; Hu, P.; Jiang, S.H.; Li, Z.Q.; Liu, Y.; Zhou, Y.Z. Type and temporal-spatial distribution of gold and antimony deposits (prospects) in southern Tibet, China. *Acta Geol. Sin.* **2005**, *3*, 373–385.
28. Migdisov, A.A.; Guo, X.; Xu, H.; Williams-Jones, A.E.; Sun, C.J.; Vasyukova, O.; Sugiyama, I.; Fuchs, S.; Pearce, K.; Roback, R. Hydrocarbons as ore fluids. *Geochem. Perspect. Lett.* **2017**, *5*, 47–52. [CrossRef]
29. Crede, L.S.; Evans, K.A.; Rempel, K.U.; Grice, K.; Sugiyama, I. Gold partitioning between 1-dodecanethiol and brine at elevated temperatures: Implications of Au transport in hydrocarbons for oil-brine ore systems. *Chem. Geol.* **2019**, *504*, 28–37. [CrossRef]
30. Crede, L.S.; Liu, W.; Evans, K.A.; Rempel, K.U.; Testemale, D.; Brugger, J. Crude oils as ore fluids: An experimental in-situ XAS study of gold partitioning between brine and organic fluid from 25 to 250 °C. *Geochim. Cosmochim. Acta* **2019**, *244*, 352–365. [CrossRef]
31. Sanz-Robinson, J.; Williams-Jones, A.E. Zinc solubility, speciation and deposition: A role for liquid hydrocarbons as ore fluids for Mississippi Valley Type Zn-Pb deposits. *Chem. Geol.* **2019**, *520*, 60–68. [CrossRef]
32. Sanz-Robinson, J.; Sugiyama, I.; Williams-Jones, A.E. The solubility of palladium (Pd) in crude oil at 150, 200 and 250 °C and its application to ore genesis. *Chem. Geol.* **2020**, *531*, 119320. [CrossRef]
33. Kubrakova, I.V.; Nabiullina, S.N.; Pryazhnikov, D.V.; Kiseleva, M.S. Organic matter as a forming and transporting agent in transfer processes of PGE and gold. *Geochem. Int.* **2022**, *60*, 748–756. [CrossRef]
34. Crede, L.S.; Rempel, K.U.; Hu, S.Y.; Evans, K.A. An experimental method for gold partitioning between two immiscible fluids: Brine and n-dodecane. *Chem. Geol.* **2018**, *501*, 35–50. [CrossRef]
35. Sugiyama, I.; Williams-Jones, A.E. An approach to determining nickel, vanadium and other metal concentrations in crude oil. *Anal. Chim. Acta* **2018**, *1002*, 18–25. [CrossRef]
36. Castner, D.G.; Hinds, K.; Grainger, D.W. X-ray photoelectron spectroscopy sulfur 2p study of organic thiol and disulfide binding interactions with gold surfaces. *Langmuir* **1996**, *12*, 5083–5086. [CrossRef]
37. Kolpakova, H.H.; Mapucharyants, B.O.; Zhang, F.X. Physicochemical conditions responsible for antimony and gold-antimony mineralization. *Geology-Geochemistry* **1993**, *5*, 17–22.
38. He, J.; Ma, D.S.; Liu, Y.J. Geochemistry of mineralization in the Zhazixi antimony ore belt on the margin of the Jiangnan old land. *Miner. Depos.* **1996**, *15*, 41–52.
39. Williams-Jones, A.E.; Norman, C. Controls of mineral parageneses in the system Fe-Sb-S-O. *Econ. Geol.* **1997**, *92*, 308–324. [CrossRef]
40. Lu, J.L.; Yuan, Z.Q. Experimental studies of organic-Zn complexes and their stability. *Chin. J. Geochem.* **1986**, *5*, 369–380.
41. Lewan, M. Factors controlling the proportionality of vanadium to nickel in crude oils. *Geochim. Cosmochim. Acta* **1984**, *48*, 2231–2238. [CrossRef]
42. Giordano, T.H. Metal transport in ore fluids by organic ligand complexation. In *Organic Acids in Geological Processes*; Springer: Berlin, Heidelberg, 1994; pp. 319–354.
43. Ho, T.Y.; Rogers, M.A.; Drushel, H.V.; Koons, C.B. Evolution of sulfur compounds in crude oils. *AAPG Bull.* **1974**, *58*, 2338–2348.
44. Wei, Z.; Mankiewicz, P.; Walters, C.; Qian, K.; Phan, N.T.; Madincea, M.E.; Nguyen, P.T. Natural occurrence of higher thiadiamondoids and diamondoidthiols in a deep petroleum reservoir in the Mobile Bay gas field. *Org. Geochem.* **2011**, *42*, 121–133. [CrossRef]
45. Cai, C.; Worden, R.H.; Bottrell, S.H.; Wang, L.; Yang, C. Thermochemical sulphate reduction and the generation of hydrogen sulphide and thiols (mercaptans) in Triassic carbonate reservoirs from the Sichuan Basin, China. *Chem. Geol.* **2003**, *202*, 39–57. [CrossRef]
46. Nguyen, V.P.; Burklé-Vitzthum, V.; Marquaire, P.M.; Michels, R. Thermal reactions between alkanes and H_2S or thiols at high pressure. *J. Anal. Appl. Pyrolysis* **2013**, *103*, 307–319. [CrossRef]

47. Kaneko, Y.; Katayama, I.; Yamamoto, H.; Misawa, K.; Ishikawa, M.; Rehman, H.U.; Kausar, A.B.; Shiraishi, K. Timing of Himalayan ultrahigh-pressure metamorphism: Sinking rate and subduction angle of the Indian continental crust beneath Asia. *J. Metamorph. Geol.* **2003**, *21*, 589–599. [CrossRef]
48. Leech, M.; Singh, S.; Jain, A.; Klemperer, S.; Manickavasagam, R. The onset of India- Asia continental collision: Early, steep subduction required by the timing of UHP metamorphism in the western Himalaya. *Earth Planet. Sci. Lett.* **2005**, *234*, 83–97. [CrossRef]
49. Hou, Z.Q.; Qu, X.M.; Yang, Z.S.; Meng, X.J.; Li, Z.Q.; Yang, Z.M.; Zheng, M.P.; Zheng, Y.Y.; Nie, F.J.; Gao, Y.F.; et al. Metallogenesis in Tibetan collisional orogenic belt: III. Mineralization in post-collisional extension setting. *Miner. Depos.* **2006**, *25*, 629–651.
50. Burchfie, B.C.; Chen, Z.; Hodges, K.V.; Liu, Y.P.; Royden, L.H.; Deng, C.R.; Xu, J.N. The south Tibetan detachment system, Himalayan orogen: Extension contemporaneous with and parallel to shortening in a collisional mountain belt. *Geol. Soc. Am. Spec. Paper* **1992**, *269*, 1–41.
51. Schares, E.; Xu, R.H.; Allegere, C.J. U-Pb geochronology of the Gangdese (Transhimalaya) plutonism in the Lhasa-Xizang region, Tibet. *Earth Planet. Sci. Lett.* **1984**, *69*, 311–320. [CrossRef]
52. Harrison, T.M.; McKeegan, K.D.; LeFort, P. Detection of inherited monazite in the Manaslu leucogranite by $^{208}Pb/^{232}Th$ ion microprobe dating: Crystallization age and tectonic implications. *Earth Planet. Sci. Lett.* **1995**, *133*, 271–282. [CrossRef]
53. Hou, Z.Q.; Gao, Y.F.; Qu, X.M.; Rui, Z.Y.; Mo, X.X. Origin of adakitic intrusives generated during mid-Miocene east-west extension in southern Tibet. *Earth Planet. Sci. Lett.* **2004**, *220*, 139–155. [CrossRef]
54. Zheng, Y.Y.; Wang, D.; Yi, J.Z.; Yu, Z.Z.; Jiang, Z.Y.; Li, X.X.; Shi, G.W.; Xu, J.; Liang, Y.C.; Dou, X.F.; et al. Antimony mineralization and prospecting orientation in the north Himalayan metallogenic belt, Tibet. *Earth Sci. Front.* **2022**, *29*, 200–230.
55. Zhai, W.; Sun, X.M.; Yi, J.Z.; Zhang, X.G.; Mo, R.W.; Zhou, F.; Wei, H.X.; Zeng, Q.G. Geology, geochemistry, and genesis of orogenic gold-antimony mineralization in the Himalayan Orogen, South Tibet, China. *Ore Geol. Rev.* **2014**, *58*, 68–90. [CrossRef]
56. Hu, X.W. Characteristics and discussion of stibnite solubility in different solutions. *Miner. Resour.* **1994**, *201*, 33–42.
57. Zhang, T.Y.; Li, C.Y.; Sun, S.J.; Hao, X.L. Geochemical characteristics of antimony and genesis of antimony deposits in South China. *Acta Petrol. Sin.* **2020**, *36*, 44–54, (In Chinese with English Abstract).
58. Wood, S.A.; Crerar, D.A.; Borcsik, M.P. Solubility of the assemblage pyrite-pyrrhotite-magnetite-sphalerite-galena-gold-stibnite-bismuthinite-argen-tite-molybdenite in H_2O-NaCl-CO_2 solutions from 200-350 °C. *Econ. Geol.* **1987**, *82*, 1864–1887. [CrossRef]
59. Williams-Jones, A.E.; Bowell, R.J.; Migdisov, A.A. Gold in solution. *Elements* **2009**, *5*, 281–287. [CrossRef]
60. Yi, J.B. A preliminary study on the basic features of global antimony mineralization and the background of mega antimony deposits in the world. *Geotecton. Metallog.* **1994**, *3*, 199–208.
61. Machel, H.G. Bacterial and thermochemical sulfate reduction in diagenetic settings—Old and new insights. *Sediment. Geol.* **2001**, *140*, 143–175. [CrossRef]

minerals

MDPI

Article

Cu and Au Mineralization of the Tolparovo Ore Occurrence: Evidence for the Formation of Redbed Copper Occurrences in Neoproterozoic Deposits of the Southern Urals

Sergey Vasilievich Michurin * and **Gulfiya Mavletovna Kazbulatova**

Institute of Geology, Ufa Federal Research Centre, Russian Academy of Sciences, Karl Marx Street 16/2, 450077 Ufa, Russia; kazbulatova@mail.ru
* Correspondence: s_michurin@mail.ru

Abstract: The mineralization and geochemical features of the Tolparovo ore occurrence are studied to reveal the contribution of diagenetic and epigenetic processes in the formation of copper mineralization in redbed deposits. The methods of electron microscopy, ICP MS, X-ray fluorescence, X-ray phase, atomic absorption, thermogravimetric analyses, and C and O isotopic composition were used. The ore is confined to the fault zone and feathering dolomite–calcite veins, having formed at temperatures of ~330–200 °C and pressures of 0.8–0.2 kbar. Similar to other copper redbed occurrences of the Bashkirian uplift, the Tolparovo copper ore occurrence is confined to basal Neoproterozoic deposits. Siliciclastic and carbonate deposits of this level were accumulated at low paleoequator latitudes in arid climates of continental and coastal–marine environments close to evaporite ones. Rocks of this stratigraphic interval demonstrate increased background concentrations of copper (~2–5 times exceeding the clark), which explains the stratification of redbed copper ore occurrences, indicating a predominantly sedimentary copper source. However, most of the redbed copper ore occurrences of the Bashkirian uplift are located in tectonic zones and are associated with the dikes of the Inzer gabbrodolerite complex. This connects the generation of ore occurrences with the formation of the Southern Urals Arsha Large Igneous Province (707–732 Ma) and the activity of postmagmatic fluids. Magmatic processes were presumably activated due to the collapse of the Rodinia supercontinent. It is shown that copper mineralization in stratiform deposits may result from a joint manifestation of dia- and epigenesis processes.

Keywords: sediment-hosted stratiform copper deposits; redbed copper ore occurrences; Neoproterozoic; Southern Urals; copper sulfides; native gold; tellurides

Citation: Michurin, S.V.; Kazbulatova, G.M. Cu and Au Mineralization of the Tolparovo Ore Occurrence: Evidence for the Formation of Redbed Copper Occurrences in Neoproterozoic Deposits of the Southern Urals. *Minerals* **2024**, *14*, 148. https://doi.org/10.3390/min14020148

Academic Editors: Yuichi Morishita and Napoleon Q. Hammond

Received: 30 November 2023
Revised: 25 January 2024
Accepted: 27 January 2024
Published: 29 January 2024

1. Introduction

The western slope of the Southern Urals is recognized as a promising region for the search and detection of copper in industrial concentrations [1–7]. Numerous magmatic (Cu–Ni, Cu–Ti–Fe), volcanogenic–hydrothermal (Cu–porphyry), and hydrothermal–metamorphogenic (Cu–sulfide) small deposits and ore occurrences of copper, known in this region, are located mainly in the central part of the Bashkirian uplift. Here, they form a wide strip in the submeridional and northeastern direction (Figure 1a, deposits and ore occurrences are indicated by numbers 1–23). In the vast majority of cases, the deposits are spatially and genetically related to various heterochronous magmatic complexes, localized in Mesoproterozoic deposits [4].

At the same time, in the western part of the Bashkirian uplift (Inzer synclinorium and Alatau anticlinorium), stratiform ore occurrences of copper have been found in Neoproterozoic redbed deposits (see Figure 1a,b, ore occurrences are indicated by numbers 24–27,29), which are referred to as sedimentary–diagenetic formations [2,4,5]. At present, stratiform copper deposits play an important economic role [8–11]. They contain about 20% of the world's copper reserves [12]. One significant group of stratiform deposits is represented by the so-called redbed copper deposits, which are widely represented globally [13–21].

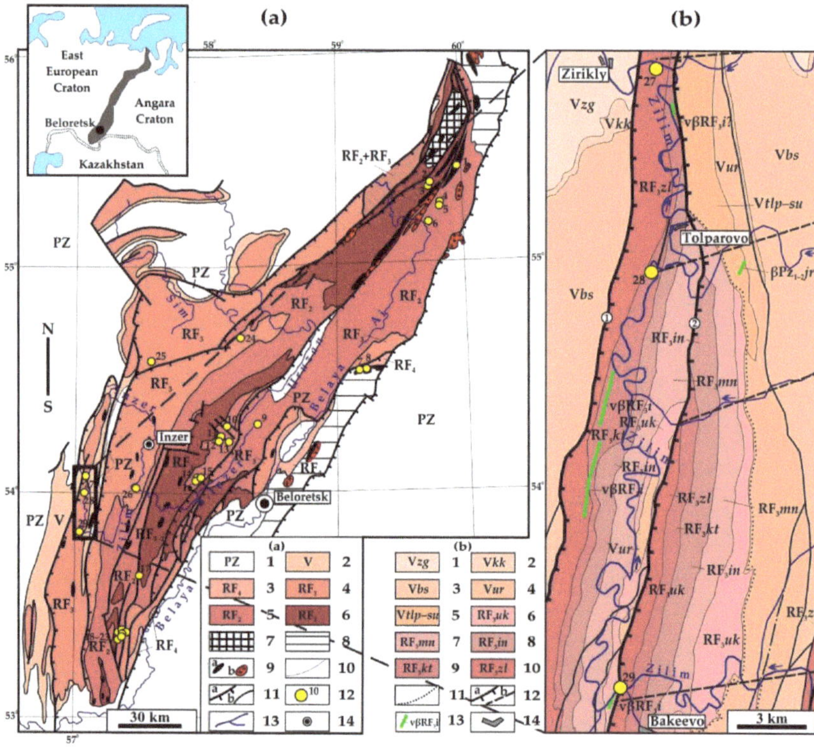

Figure 1. (**a**) Schematic geological map of the Bashkirian uplift according to ([22]) and (**b**) a fragment of the Alatau anticlinorium geological map (according to [23] with additions) with the location of copper deposits and ore occurrences (according to [1,2,5,6,24–26] and materials of the West Bashkirian complex geological expedition). Legend. For the map (**a**): 1–6—undivided deposits: 1—Paleozoic (PZ), 2—Vend (V), 3—final (RF_4), 4—upper (RF_3), 5—middle (RF_2), 6—lower (RF_1) Riphean; 7—Taratash metamorphic complex; 8—Uraltau and Ufaley metamorphic complexes; 9—igneous rocks: gabbro (a) and granites (b); 10—geological boundaries; 11—main tectonic disturbances: a—thrusts; b—faults; 12—copper deposits and ore occurrences (1—Fofanovka, 2—Andreevka, 3—Nikolaevka, 4—Evgrafovka, 5—Nadezhdinka, 6—Urenga, 7—Barsinka, 8—Kiryabinka, 9—Medvezhka, 10—Kazavda, 11—Yamantau, 12—Kuzelga, 13—Kalpak, 14—Kaintyube, 15—Vasilyevskaya Polyana, 16—Arepkulovo, 17—Bogryashka, 18—Aktash-1, 19—Bezymyannoye, 20—Aktash-2, 21—Bolshye Polyany, 22—Yuzhny Aktash, 23—Berezovye kashi, 24—Katav, 25—Lemesa, 26—Bolshoy Inzer, 27—Zirikly, 28—Tolparovo, 29—Bakeevo); 13—rivers; 14—settlements. For the map (**b**): 1–5—Vendian deposits, formations: 1—zigan (V*zg*), 2—kukkarauk (V*kk*), 3—basa (V*bs*), 4—uryuk (V*ur*), 5—undivided tolparovo and suirovo (V*tlp-su*); 6–10—Upper Riphean deposits, formations: 6—uk (RF_3uk), 7—minyar (RF_3mn), 8—inzer (RF_3in), 9—katav (RF_3kt), 10—zilmerdak (RF_3zl); 11—boundary of deposit disconformity; 12—tectonic disturbances: a—thrusts (1—Avdyrdak-Alatau, 2—Western Kalinsk), b—probable; 13—gabbrodolerite dikes ($\nu\beta RF_3i$—Inzer complex, $\beta Pz_{1-2}jr$—Yurmaty complex); 14—settlements.

Despite the relatively low concentrations of Cu (<1.0 wt %), the redbed copper ore occurrences of the Bashkirian uplift are promising in terms of identifying industrial copper deposits on the western slope of the Southern Urals, where large-scale mineralization is predicted [6,7]. The copper–sulfide mineralization here is confined to the Zilmerdak and Katav formations (Figure 2) with an age of approximately ~1000–800 Ma according to [27–29]. The deposits of the formations have a characteristic red color due to the presence of diagenetic hematite and magnetite [30,31].

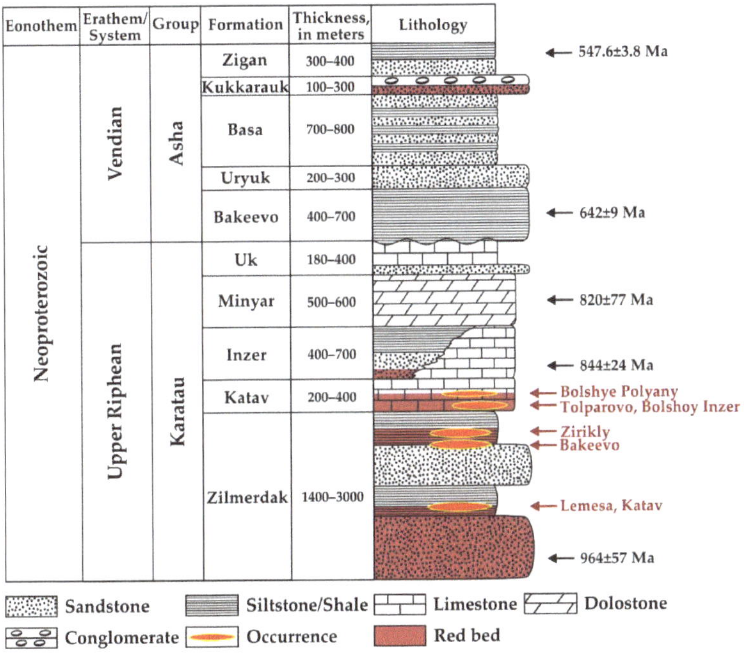

Figure 2. Generalized stratigraphic column of Neoproterozoic strata of the Southern Urals according to [22,32] with current geochronological constraints and locations of redbed copper occurrences. Geochronological data are provided according to [27,28,33,34].

However, no consensus has been reached so far on the time of copper sulfide formation in the stratiform deposits (whether it is diagenetic or epigenetic), as well as on the duration of the sulfide formation processes [35–40]. For example, the genesis of copper belt deposits in Central Africa remains controversial. Some researchers believe [18,41,42] the copper sulfides in these deposits to have formed at the stage of diagenesis; hence, their age comprises ~800 Ma and is consistent with the age of host dolomitic shales, formed in evaporite depositional environments. Other researchers claim [43,44] that disseminated and vein mineralization in the copper deposits of Central Africa has an epigenetic nature, having formed in the synorogenic time from 609 to 473 Ma. However, the existence of an earlier syndiagenetic copper mineralization remains possible. In the Dongchuan Copper Deposits of Southern China, studies [45,46] found that the mineralization of copper in the Paleoproterozoic host rocks was non-diagenetic despite its formation ~780–700 Ma ago in connection with the pulses of magmatic activity during the collapse of the Rodinia supercontinent. At the same time, according to [47], the time of the copper–sulfide mineral formation in Southern China deposits comprised ~1700 Ma and was close to the age of its hosting dolostones (1742 ± 13 Ma). It is widely believed that copper mineralization in stratiform deposits is multistage and is due to the implementation of various geological processes throughout the sedimentary–tectonic evolution of host basins [8,9,13,36,38,42–44,48,49].

To date, no detailed mineralogical studies of redbed copper ore occurrences on the western slope of the Southern Urals have been carried out. The works [2,5–7] provide only a macroscopic description of the ore mineralization in the Katav and Lemesa ore occurrences of the Zilmerdak formation. Meanwhile, in 2022–2023, in the Alatau anticlinorium near the Tolparovo village, we found a new ore occurrence of copper in the redbed limestones of the Katav formation (see Figure 1a,b, the ore occurrence is indicated by the number 28). In this study, we analyze the geological, mineralogical, and geochemical features of the Tolparovo ore occurrence, as well as other redbed copper ore occurrences in the western slope of the Southern Urals. The aim was to identify the participation of diagenetic or epigenetic processes in the formation or transformation of the copper mineralization on the basis of mineralogical and geochemical indicators, as well as to determine the conditions of its formation in the Neoproterozoic redbed deposits of the Alatau anticlinorium.

2. Geological Setting

The Alatau anticlinorium is located in the western part of the Bashkirian uplift. It covers an area of about 240 km × 25 km. The western and eastern borders of the anticlinorium pass along the Alatau and Zilmerdak thrusts, respectively. It consists of three overthrusts, moving upon each other from the east, with a thickness varying from 1–2 to 3–4 km and an output width ranging from 4 to 12 km [50]. Accordingly, from west to east, a threefold repetition of a stratigraphic sequence for the anticlinorium sedimentary section is recorded in general terms. The overthrust boundaries include the Avdyrdak-Alatau and Western Kalinsk submeridional thrusts (see Figure 1b). In their frontal parts, a significant deformation of rocks is noted with an extensive formation of steep anticlinal folds and tectonic faults [50]. The anticlinorium is divided into a series of submeridionally elongated anticlines (Avdyrdak, Alatau, etc.) and synclines (Yashkurt, Kulgunino, etc.) [51].

Neoproterozoic deposits of the Alatau anticlinorium are represented by the Karatau and Asha series, whose ages are approximately ~1000–540 Ma (see Figure 2) [27,28,33,34]. The Karatau series includes the Zilmerdak, Katav, Inzer, Minyar, and Uk formations [22,52]. The Zilmerdak formation, which is the basal formation for the Karatau series, unites (from below) the Biryan, Nugush, Lemesa, and Bederysh subformations [52]. Redbed sandstones with interlayers of gravelstones, conglomerates, and rare subordinate layers of siltstones, mudstones, and dolostones prevail in the Biryan subformation. The Nugush subformation is composed of sandstones, siltstones, mudstones, and clay dolostones. The Lemesa subformation is represented by light gray sandstones with subordinate layers of siltstones and mudstones. The Bederysh subformation is composed of sandstones, siltstones, mudstones, and clay dolostones. The rocks of the Nugush and Bederysh subformations in the lower part are red and red–brown, an in the upper part—light and green–gray [2,5,6]. The Katav formation is composed of red, red–brown (lower subformation), and gray (upper subformation) clay limestones [22,52]. The Inzer formation is represented by sandstones, siltstones, limestones, and mudstones. The Minyar formation is composed of dark gray, gray dolostones, and dolomitic limestones with chert interlayers and lenses in the upper part. The Uk formation is represented by sandstones with stromatolites, limestones, siltstones, and mudstones.

The magmatic formations of the Alatau anticlinorium are represented by gabbrodolerite dikes of the Inzer complex [50,51]. According to [51], the U–Pb age of the Inzer complex comprises 721 ± 6.9 Ma. The gabbrodolerite dikes are spatially biased to the Avdyrdak-Alatau and Western Kalinsk thrusts, where they break through the deposits of the Zilmerdak and Katav formations (see Figure 1b). Dikes can be traced along the strike for 6–7 km. In addition, thin dikes of gabbrodolerites, secant to the Uryuk formation deposits, are rarely observed in the anticlinorium (see Figure 1b). The geochronological data for these dikes are absent. They can be presumably attributed to the Jurmatin Early-Middle Paleozoic complex [23,50].

The Tolparovo ore occurrence is located in the southern part of the Avdyrdak anticline on the left bank of the river Zilim, ~1.5 km south–west of Tolparovo village (see Figure 1b). Here, red–brown and light gray, fine- and medium-laminated, micrograined clay limestones of the Katav formation are exposed (Figure 3a–g). The rocks are contorted into a synclinal fold (see Figure 3b). The rock stratification of its southeastern and northwestern limbs is inclined to the south–west (207–244° ∠27–73°) and to the east (65–95° ∠35–80°), respectively, with a dip azimuth of the fold bend and dip angle equal to 145–150° and 20°, respectively.

Figure 3. Geological conditions of gray and red limestones of the Katav formation on the left bank of the river Zilim ~1.5 km south–west of Tolparovo village, location of the mineralized ore zone and the relationship of mineralized dolomite–calcite veins with late barren calcite veins. Legend: (**a**)—general view of the exposure; arrows show the fold bend and reverse fault (dashed line) locations in the section; (**b**)—synclinal fold bend; (**c**)—dashed lines denote the zone of the mineralized postfold low-amplitude reverse fault in the northwestern limb of the synclinal fold; (**d**,**e**)—relationship of mineralized dolomite–calcite veins, filling shear fractures ((**d**), number 1 in circles) and tension joints ((**e**), number 1 in circles), complementary to the reverse fault, with late barren milky-white calcite veins ((**d**,**e**), number 2 in circles) in gray (**d**) and red (**e**) layers; (**f**)—late barren calcite veins in redbed limestone (number 2 in circles); (**g**)—red and gray subordinated layers in limestones.

In the limestones of the northwestern limb near the fold bend, a postfolded low-amplitude reverse fault is observed with the fault plane inclined to the west by 265–283° at an angle of 64–67° (see Figure 3c). The mineralized fault zone with a width of ~12–20 cm is filled with poorly cemented, ferruginated, brecciated, boudinaged fragments of dolomite–calcite rocks with slickensides. The reverse fault is feathered by low-thick (0.5–2.5 cm) yellow–brown, frequently ferruginated, mineralized ore dolomite–calcite veins, filling fault-associated shear fractures (275–288°∠50–80°, see Figure 3d) and tension joints (170–215° ∠46–85°, see Figure 3e).

In addition, late barren milky-white calcite veins, secant for both brecciated rocks in the fault zone and feathering dolomite–calcite veins, are formed in the exposure (see Figure 3d–f). Here, they have a zonal distribution and are most often found in the southeastern limb of the fold near the bend in an area about 15 m long. In this section part, their thickness reaches 10 cm with a length of several meters. In other parts of the section, late calcite veins are observed less often with their thicknesses not exceeding 0.3–1.0 cm. The bedding elements of late calcite veins in various limbs of the fold are similar: their dip azimuth in the southeastern limb equals 296–348° with a dip angle of 39–70°; in the northwestern limb, these values comprise 298–334° and 28–50°, respectively. This indicates a postfold formation of late calcite veins.

The mineralized ore reverse fault zone is traced in the submeridional direction along the azimuth of 10–15° for a distance of about 2.5 km (see Figure 1b). A low-amplitude reverse fault was recorded [22] north of the considered ore occurrence in the section of the Katav formation near Tolparovo village with a vertical displacement amplitude of the fault wall equal to ~3 m. Although no detailed geochemical study of this section was performed, single samples of limestones also revealed increased Cu content (up to 0.04 wt %).

3. Materials and Methods

A total of 95 samples were selected from the Katav formation, including 61 samples directly from the Tolparovo ore occurrence and 34 samples from Katav limestones at a considerable distance therefrom. In the ore occurrence, samples were taken from host rocks (24 samples), fault zone (17 samples), mineralized ore dolomite–calcite veins (14 samples), and late barren calcite veins (6 samples). The concentrations of petrogenic oxides and rare elements in rocks and veins (SiO_2, TiO_2, Al_2O_3, Fe_2O_3, MnO, CaO, mgO, Na_2O, K_2O, S_{total}, P_2O_5, V, Cr, Co, Ni, Cu, Zn, Rb, Sr, Y, Zr, Ba) were determined by X-ray fluorescence analysis using an spectrometer XRF (X-Calibur, Xenemetrix LTD., Migdal HaEmek, Israel) (Ag-anode, 15–50 kV, 60–1000 mA). Calibration curves were constructed using state reference materials of limestones (KN, KN-2, KN-3, OSO No. 469-11), dolomitic limestones (SI-2), and dolomites (SI-1, SI-3). The detection limit while measuring petrogenic oxides and rare elements was 0.01–0.2 wt % and 3–10 g/t, respectively.

The concentrations of Au, Pb, and As were determined by atomic absorption analysis using a Persee A3 spectrometer (PG General Instrument Co., Beijing, China). In order to concentrate and isolate gold from interfering elements, it was extracted with butyl acetate from a hydrochloric acid solution (1/7) at an aqueous to organic phase ratio of 10:1. The concentration of As was determined using a WHG-630 A hydride generator (Xi'an Yima Optoelec Co., Ltd., Xi'an, China). The detection limit in the measurement of Au, Pb, and As comprised 0.2 ppm.

The FeO concentration of bulk samples was established by the volumetric bichromatic titrimetric method. In addition, the fluorine content was determined in the Katav formation samples by the photometric method through the formation of lanthanum fluoride alizarin complexonate according to a procedure described in [53]. Measurements were carried out using a KFK-3-01 photometer (Zagorsk Optical-Mechanical Plant, Sergiyev Posad, Russia) with two state reference materials in each single measurement at a detection limit of 0.005 wt %.

In addition, for determining the copper background concentration in the rocks of the Alatau anticlinorium, samples were taken from Zilmerdak (48 samples), Inzer (34 samples),

Minyar (18 samples), and Uk (5 samples) formations. The concentrations of Cu in the rocks of these formations were determined by X-ray fluorescence analysis.

The mineralogical composition of rocks and veins was determined by thermogravimetric and X-ray diffraction analyses (n = 35). Differential thermal analysis was performed by analyst T.I. Chernikova using a Q-1500 derivatograph (Mom, Budapest, Hungary) through heating in air from 20 to 1000 °C at a rate of 10 °C/min. The sample mass was about 500 mg. Radiographic studies were carried out by analyst G.S. Sitdikova using a DRON-4 diffractometer in powder samples with a mass of 0.5–1 g. The analysis was performed in Cu Kα radiation (40 kV, 40 mA) in a 2Θ region of 4–60° with a step of 0.02° and a counting time equal to 10 s.

The composition study of minerals in the host limestones (n = 4) and veins (n = 5) of the ore occurrence was carried out in polished thin sections by analyst S.S. Kovalev using a Tescan electron microscope (Vega 4 Compact, Tescan Brno s.r.o., Brno, Czech Republic) equipped with an Explorer 15 energy dispersion analyzer (Oxford Instruments, Oxford, UK). Spectra were processed automatically in the AzTec One software package using the TrueQ procedure (version 5.1, Oxford Instruments, Oxford, UK). During the analysis, the following parameters were used: an acceleration voltage of 20 kV, a probe current of 4 nA, 60 s spectrum point accumulation time in the Point&ID mode, a beam diameter of ~3 µm.

Determination of trace and rare earth elements (REE) (Li, Be, Sc, V, Cr, Co, Ni, Cu, Zn, As, Se, Rb, Sr, Y, Zr, Nb, Mo, Ag, Cd, Sn, Sb, Te, Cs, Ba, La, Ce, Pr, Nd, Sm, Eu, Gd, Tb, Dy, Ho, Er, Tm, Yb, Lu, Hf, Ta, W, Tl, Pb, Bi, Th, U) in host limestones (n = 3) and veins (n = 6) was carried out at the Institute of Problems of Microelectronics Technology and Highly Pure Materials (Chernogolovka, Russia, Analytical Group Head V.K. Karandashev) by mass spectrometry with inductively coupled plasma using an X-7 mass spectrometer (Thermo Electron Corporation, Warwickshire, UK). The resulting patterns for REE were normalized against NASC [54]. In addition, the ICP MS method was applied to analyze two samples of an insoluble residue after treating the samples with cold 5% HCl for 16 h.

The isotopic composition of C and O in six ore vein samples was determined at the Geoscience Center of the Institute of Geology of the Komi Scientific Center UB RAS (Syktyvkar, Group Head V.L., Andreichev). The digestion of samples to CO_2 in orthophosphoric acid at a temperature of 80 °C and the measurement of the isotopic composition of C and O in the continuous helium flow mode was carried out using an analytical complex, including the Gas Bench II sample preparation and introduction system, connected to a DELTA V Advantage mass spectrometer (Thermo Scientific, Karlsruhe, Germany). International reference materials NBS18 and NBS19 were used for calibration. The measurement reproducibility of $\delta^{13}C$ and $\delta^{18}O$ was within ±0.2 ‰.

4. Results

4.1. Geochemistry

4.1.1. Trace Elements

The average chemical compositions and average contents of trace elements in the host limestones and veins of the Tolparovo ore occurrence, as well as in the background limestones of the Katav formation, located outside the ore occurrence, according to the results of an X-ray fluorescence and atomic absorption analysis, are given in Table 1. The most significant variations in the chemical composition and distribution of trace elements are characteristic of rocks from the fault zone. In addition, they have the highest concentrations of copper, ranging from 97 to 2981 ppm. In dolomite–calcite veins, filling shear fractures, copper concentrations vary from 35 to 193 ppm, while, for tension joints, this value ranges across 31 to 134 ppm. Late milky-white calcite veins are characterized by the lowest concentrations of copper, from 4 to 22 ppm.

Table 1. Average chemical compositions (wt %) and average concentrations of trace elements (ppm) in the host limestones and veins of the Tolparovo ore occurrence, as well as in the background limestones of the Katav formation outside the ore occurrence.

Oxide, Element	1 (n = 34)	2 (n = 24)	3 (n = 17)	4 (n = 9)	5 (n = 5)	6 (n = 6)
SiO_2	10.96	10.24	14.33	6.95	4.32	3.72
TiO_2	0.15	0.12	0.41	0.08	0.08	0.04
Al_2O_3	2.59	2.01	5.24	1.57	1.14	0.86
FeO	0.29	0.32	0.26	0.39	0.33	0.25
Fe_2O_3	0.96	0.71	3.14	0.29	2.15	0.15
MnO	0.04	0.04	0.37	0.05	0.38	0.04
CaO	46.45	47.42	39.23	48.76	48.57	52.11
MgO	1.00	1.47	3.00	1.72	2.74	1.06
Na_2O	0.36	0.31	0.24	0.27	0.21	0.16
K_2O	0.64	0.33	1.30	0.21	0.18	0.09
P_2O_5	0.07	0.06	0.06	0.05	0.04	0.04
S_{total}	0.02	0.03	0.04	0.03	0.02	0.02
ППП	36.41	37.17	32.05	39.33	40.30	40.96
Σ	99.92	100.23	99.65	99.69	100.42	99.52
F	248	203	478	532	315	281
V	21	18	65	23	30	25
Cr	24	11	37	15	13	24
Co	3	3	8	3	10	4
Ni	11	24	18	18	20	12
Cu	19	12	1427	122	81	8
Zn	18	10	24	10	10	7
As	0.6	0.1	15	0.4	0.2	1
Rb	19	16	28	12	11	10
Sr	115	125	65	135	92	140
Y	9	7	9	5	6	4
Zr	41	46	43	38	31	29
Ba	132	68	336	116	413	188
Au	–	<0.2	0.4	0.5	<0.2	<0.2
Pb	3	2	8	2	0.1	4

1—background limestones of the Katav formation outside the Tolparovo ore occurrence; 2–6—Tolparovo ore occurrence: 2—host limestones; 3—mineralized reverse fault zone; 4–5—mineralized dolomite–calcite veins, filling shear fractures (4) and tension joints (5), complementary to the reverse fault; 6—late barren milky-white calcite veins. n—number of samples. Dash—n/a.

Copper in the ore occurrence reveals weakly expressed positive geochemical relationships with Si, Ti, Al, Fe, Mn, K, V, Cr, Ni, As, Rb, Au, and Pb (Figure 4). The coefficients of correlation with these elements are at the level of 0.2–0.4. The absence of a positive relationship between the copper and sulfur contents is explained by the development of supergenic processes in the ore occurrence and the replacement of sulfides with copper carbonates. Significant correlation coefficients (~0.5) are demonstrated by copper only with mg and Ba in mineralized dolomite–calcite veins. Along with these elements, the content of Fe, Mn, V, and Ni increases noticeably in the veins that fill tension joints. Host limestones near the veins are similarly enriched in mg.

Au contents of the host rocks are below 0.2 ppm. In rocks of the fault zone and dolomite–calcite veins from shear fractures, the average gold concentrations comprise 0.4–0.5 ppm (see Table 1). In mineralized veins, Au forms weak positive geochemical relationships only with Cu and As with correlation coefficients of ~0.3–0.4.

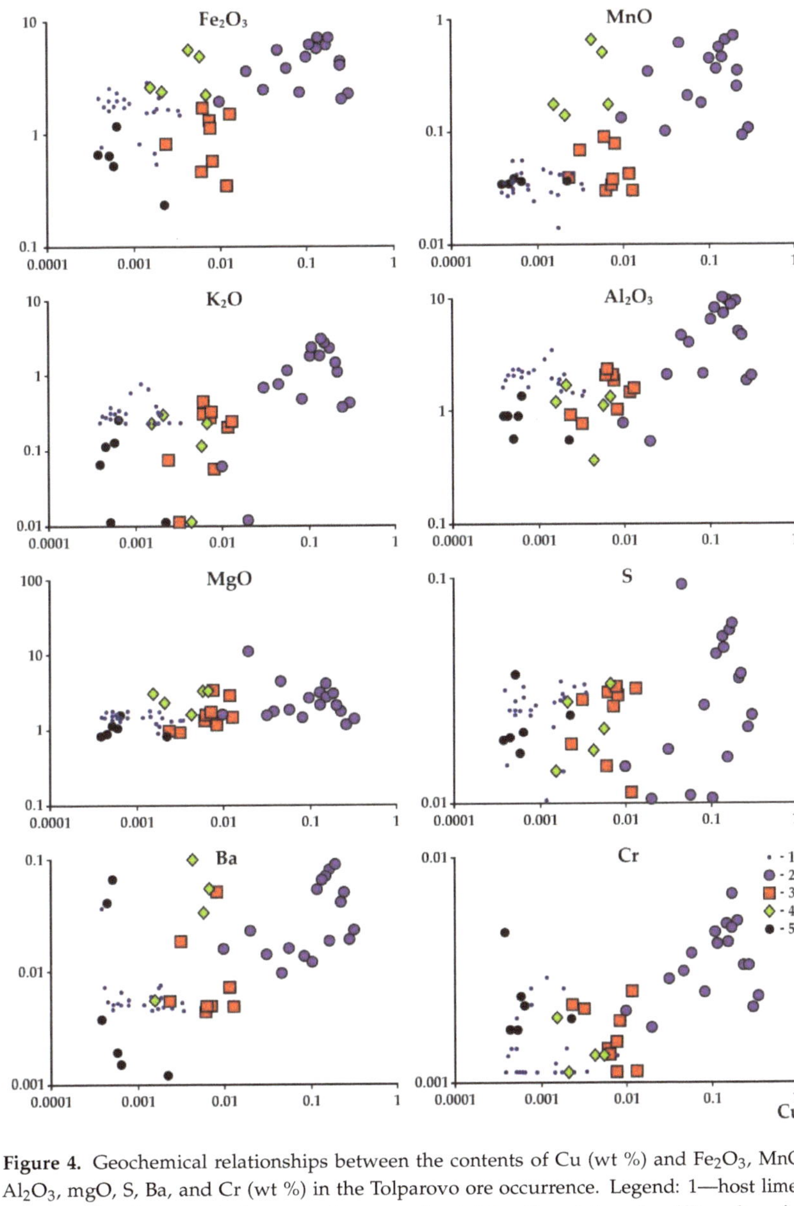

Figure 4. Geochemical relationships between the contents of Cu (wt %) and Fe_2O_3, MnO, K_2O, Al_2O_3, mgO, S, Ba, and Cr (wt %) in the Tolparovo ore occurrence. Legend: 1—host limestones; 2—mineralized reverse fault zone; 3–4—mineralized dolomite–calcite veins, filling shear fractures (3) and tension joints (4), complementary to the reverse fault; 5—late barren milky-white calcite veins.

The results of an X-ray fluorescence analysis were generally consistent with the ICP MS data of the mineralized dolomite–calcite veins (Table 2). In the veins, concentrations of V, Co, Ni, As, Se, Rb, Cs, Ba, Tl, Bi, and U increase significantly compared with the "background" limestones of the Katav formation. At the same time, the contents of most elements are similar in dolomite–calcite veins and "background" limestones. The concentration of REE in dolomite–calcite veins is equal to ~25 ppm. This is even slightly lower than in the "background" limestones of the Katav formation with a REE concentration of ~35 ppm.

Table 2. Average concentrations of trace elements (ppm) in the limestones of the Katav formation and mineralized dolomite–calcite veins of the Tolparovo ore occurrence.

Element	1 (n = 3)	2 (n = 3)	3 (n = 3)	4 (n = 1)	5 (n = 1)
Li	6.36	4.16	3.12	50.20	30.40
Be	0.99	0.25	0.11	1.60	1.10
Sc	1.93	2.90	1.35	9.30	13.10
V	bdl	8.40	8.32	90.40	80.50
Cr	5.61	7.35	7.21	70.90	70.15
Co	3.03	7.14	4.16	19.10	13.35
Ni	10.61	81.60	59.74	40.40	31.65
Cu	1.65	92.70	61.80	600.60	600.30
Zn	22.98	13.26	11.44	130.15	110.20
Ga	1.38	1.62	1.80	21.35	18.10
As	bdl	0.85	0.96	10.70	9.60
Se	bdl	0.51	0.05	0.41	0.48
Rb	0.34	10.20	8.32	77.60	80.50
Sr	231.36	171.70	103.00	18.20	40.10
Y	7.67	5.05	5.15	8.30	9.40
Zr	12.10	8.16	11.44	130.10	110.70
Nb	0.98	0.17	0.55	14.10	11.15
Mo	bdl	0.07	0.12	0.80	1.20
Ag	0.02	0.07	0.06	0.74	0.57
Cd	0.10	0.04	0.02	0.19	0.15
Sn	0.36	0.27	0.27	3.00	2.70
Sb	0.24	0.09	0.10	1.30	1.00
Te	bdl	0.02	0.01	0.13	0.06
Cs	0.03	0.31	0.32	3.70	3.20
Ba	11.02	28.56	206.00	131.70	2003.20
La	7.84	5.10	6.12	5.15	8.10
Ce	13.20	9.18	10.20	6.20	11.30
Pr	1.57	1.20	1.10	0.65	1.05
Nd	6.33	5.10	5.20	2.10	3.50
Sm	1.27	1.00	1.05	0.50	0.70
Eu	0.24	0.19	0.20	0.21	2.40
Gd	1.24	0.86	0.92	0.62	1.10
Tb	0.19	0.13	0.12	0.14	0.15
Dy	1.10	0.70	0.61	1.20	1.20
Ho	0.22	0.15	0.14	0.31	0.30
Er	0.67	0.42	0.35	1.10	1.10
Tm	0.09	0.06	0.05	0.19	0.18
Yb	0.56	0.40	0.35	1.40	1.30
Lu	0.09	0.06	0.05	0.22	0.21
Hf	0.43	0.23	0.27	4.15	4.05
Ta	0.07	0.00	0.03	1.00	0.80
W	0.17	0.03	0.19	2.60	1.90
Tl	0.01	0.03	0.05	0.50	0.50
Pb	1.95	4.08	1.05	9.00	2.00
Bi	0.02	0.09	0.06	0.90	0.60
Th	1.17	0.90	0.92	7.05	9.15
U	0.21	1.10	0.41	5.20	2.10

1—host limestones; 2–3—mineralized dolomite–calcite veins, filling shear fractures (2) and tension joints (3); 4–5—insoluble residue of dolomite–calcite veins, filling shear fractures (4) and tension joints (5); n—number of samples; bdl—below detection limit.

4.1.2. Isotopic Composition of C and O

In mineralized dolomite–calcite veins, the δ^{13}C comprises 1.2–1.8‰ (VPDB), while δ^{18}O ranges from −8.4 to −8.0 ‰ (VPDB) (Table 3). The carbon and oxygen isotopic compositions of the veins vary within very narrow limits and have a weak negative correlation with each other (K_{corr} = −0.38).

Table 3. Isotopic composition of C and O (‰, VPDB) in mineralized dolomite–calcite veins of the Tolparovo ore occurrence.

	Sample No.					
	b8	m2612a	m2620	m3059	m3062	m3063
$\delta^{13}C$	1.5	1.8	1.7	1.3	1.2	1.3
$\delta^{18}O$	–8.3	–8.4	–8.2	–8.4	–8.2	–8.0

4.2. *Mineralogy*

4.2.1. Host Rocks

The content of calcite in the host limestones of the Katav formation averages 81–86% (Table 4). In variable amounts from 1 to 8%, quartz, muscovite, albite, chlorite, and hematite are present in the rocks. For host rocks, extremely low contents of dolomite and K-feldspar are typical, averaging less than 1% and being, in most cases, below the detection limit of thermogravimetric and X-ray diffraction analyses.

Table 4. Average mineralogical compositions of Katav formation background limestones, as well as host rocks and veins of the Tolparovo ore occurrence.

Mineral	1 (n = 21)	2 (n = 5)	3 (n = 4)	4 (n = 3)	5 (n = 2)
Calcite	81	86	68	77	68
Dolomite	<1	<1	10	6	19
Quartz	8	5	6	6	2
Muscovite	3	2	7	4	5
Chlorite	3	3	5	4	3
Albite	3	2	2	2	1
K-feldspar	<1	<1	1	<1	<1
Hematite	1	<1	<1	<1	<1

1—background limestones of the Katav formation outside the Tolparovo ore occurrence; 2–5—Tolparovo ore occurrence: 2—host limestones; 3—mineralized reverse fault zone; 4–5—mineralized dolomite–calcite veins, filling shear fractures (4) and tension joints (5), complementary to the reverse fault; n—number of samples.

4.2.2. Ore Veins

In the ore occurrence, calcite, quartz, muscovite, albite, chlorite, and hematite are also the main forming minerals of rocks from the fault zone and their feathering ore veins. However, in contrast to the host rocks, these rocks are associated with significant dolomite contents, averaging 6–19% (see Table 4). Vein dolomite is more frequently represented by individual rhombohedral crystals and their segregations, and less often by vein-shaped secretions. Generally, dolomite in veins can be associated with chlorite and ore minerals (Figure 5a,b). The content of mgO and FeO in vein calcite varies from 0.21 to 0.53 and 0.14 to 0.33 wt %, respectively, with the MnO concentration reaching 0.19 wt %. In the dolomite of ore dolomite–calcite veins, the concentrations of FeO and MnO comprise 0.35–5.20 and 0.15–0.50 wt %, respectively.

Vein chlorite forms "pedate" precipitations near the rhombohedral crystals of dolomite and copper sulfides (see Figure 5a,b,i). The chemical composition of chlorite mainly corresponds to pycnochlorite and diabantite in single analytical procedures (Table 5, Figure 6). Chlorites are characterized by slight fluctuations in iron content with x(Fe) values ranging within 0.32–0.37.

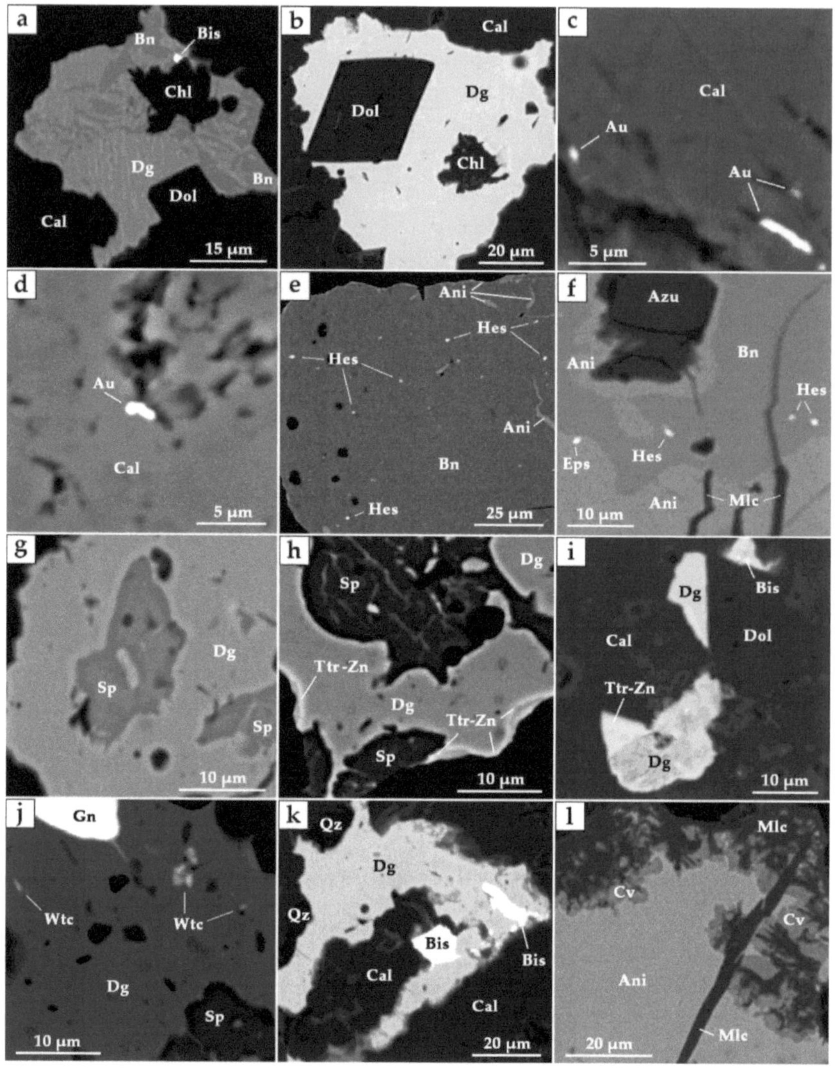

Figure 5. BSE-images of minerals from the dolomite–calcite veins of the Tolparovo ore occurrence. Legend: (**a**)—micrographic texture of the bornite and digenite intergrowth in association with dolomite and chlorite; (**b**)—digenite in association with dolomite and chlorite; (**c**,**d**)—secretions of native gold; (**e**)—inclusions of hessite in bornite; bornite is substituted by anilite along the fractures; (**f**)—inclusions of hessite and empressite in bornite; bornite is substituted by anilite, azurite, and malachite; (**g**)—intergrowth of digenite and sphalerite; (**h**)—disseminated texture of digenite and sphalerite intergrowth; secretions of tetrahedrite-(Zn) in digenite at the sphalerite bondary; (**i**)—intergrowth of digenite and tetrahedrite-(Zn); (**j**)—inclusions of galena and wittichenite in digenite, digenite and sphalerite intergrowths; (**k**)—inclusions of bismite in digenite, digenite associated with quartz; (**l**)—bornite is substituted with anilite, covellite, and malachite. Ani—anilite; Au—native gold; Azu—azurite; Bis—bismite; Bn—bornite; Cal—calcite; Chl—chlorite; Cv—covellite; Dg—digenite; Dol—dolomite; Eps—empressite; Gn—galena; Hes—hessite; Mlc—malachite; Qz—quartz; Sp—sphalerite; Ttr-Zn—tetrahedrite-(Zn); Wtc—wittichenite.

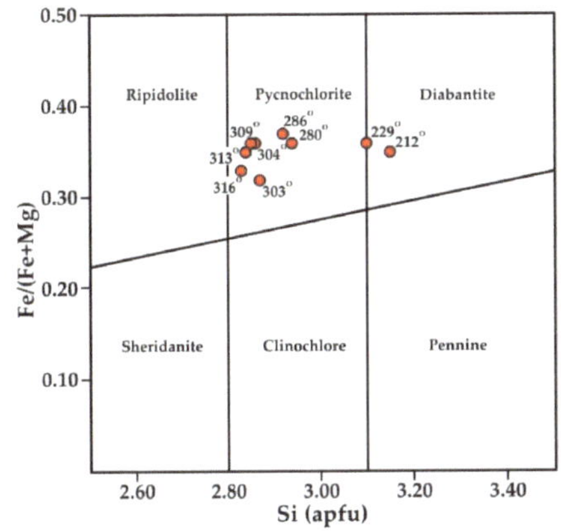

Figure 6. Chlorite composition on the classification diagram of [56]. Numbers show the formation temperatures calculated using a chlorite geothermometer [55] (Table 5).

Table 5. Chemical composition of chlorite (wt %) and calculated temperatures of its formation.

	Grain No.								
	1	2	3	4	5	6	7	8	9
SiO_2	27.30	28.00	27.17	27.92	27.34	27.38	30.07	29.40	27.74
TiO_2	0.00	0.00	0.00	0.00	0.00	0.00	0.00	1.17	0.00
Al_2O_3	21.80	21.53	21.64	20.73	22.27	21.59	21.31	19.61	21.40
Cr_2O_3	0.00	0.00	0.00	0.00	0.00	0.00	0.00	0.00	0.00
FeO	18.21	17.66	18.19	19.30	17.83	17.17	16.99	17.97	16.35
MnO	0.00	0.00	0.00	0.00	0.00	0.00	0.00	0.00	0.00
MgO	17.95	17.86	17.83	18.06	18.74	19.98	17.55	17.56	19.75
CaO	0.39	0.45	0.87	0.27	0.00	0.00	0.00	0.00	0.39
Na_2O	0.00	0.00	0.00	0.00	0.00	0.00	0.00	0.00	0.00
K_2O	0.00	0.05	0.00	0.00	0.00	0.00	0.00	0.30	0.24
Total	85.65	85.55	85.70	86.28	86.18	86.12	85.92	86.01	85.87
Crystallochemical coefficients in the chlorite formula (calculation for 20 cations)									
Si	5.73	5.88	5.70	5.84	5.67	5.65	6.30	6.19	5.74
Ti	0.00	0.00	0.00	0.00	0.00	0.00	0.00	0.19	0.00
Al	5.39	5.33	5.35	5.11	5.45	5.25	5.26	4.87	5.21
Cr	0.00	0.00	0.00	0.00	0.00	0.00	0.00	0.00	0.00
Fe‴	0.00	0.00	0.00	0.00	0.00	0.00	0.00	0.00	0.00
Fe″	3.19	3.09	3.19	3.37	3.09	2.96	2.97	3.16	2.82
Mn	0.00	0.00	0.00	0.00	0.00	0.00	0.00	0.00	0.00
Mg	5.61	5.59	5.57	5.63	5.79	6.14	5.47	5.51	6.08
Ca	0.09	0.10	0.20	0.06	0.00	0.00	0.00	0.00	0.09
Na	0.00	0.00	0.00	0.00	0.00	0.00	0.00	0.00	0.00
K	0.00	0.01	0.00	0.00	0.00	0.00	0.00	0.08	0.06
x(Mg)	0.64	0.64	0.64	0.63	0.65	0.67	0.65	0.64	0.68
Al(IV)	2.27	2.12	2.30	2.16	2.33	2.35	1.70	1.81	2.26
Al(VI)	3.12	3.21	3.05	2.95	3.12	2.90	3.56	3.06	2.95
x(Fe)	0.36	0.36	0.36	0.37	0.35	0.33	0.35	0.36	0.32
T, °C	304	280	309	286	313	316	212	229	303

In calculating the crystallochemical formula, Al(IV) supplements Si position to 8 cations and Al(VI) accounts for the remaining part of the total calculated amount of Al. The temperatures were calculated using a chlorite geothermometer [55].

Barite in dolomite–calcite veins is found in the form of xenomorphic secretions and crystals of a rectangular cross-section up to 80 μm in size, often in association with quartz. In the chemical composition of barite, up to 1.62 wt % of Sr impurity is recorded.

Ore mineralization in the veins is represented by chalcopyrite, pyrite, native gold, bornite, digenite, sphalerite, galena, tetrahedrite-(Zn), wittichenite, hessite, empressite, bismite, anilite, geerite, covellite, azurite, malachite, and barite. Chalcopyrite and pyrite form a rare thin dissemination in dolomite–calcite veins from a system of shear fractures. The size of the chalcopyrite secretions is 3–60 μm, and cubic pyrite crystals reach 0.5 mm in size. Cu (up to 1.72 wt %) and Co (up to 0.45 wt %) impurities are observed in pyrite (Table 6). In one case, an Mo impurity (2.37 wt %) was found in the chalcopyrite.

Table 6. Chemical composition of iron, copper, and bismuth sulfides (wt %).

Mineral	Pyrite	Chalcopyrite	Bornite	Digenite	Anilite	Geerite	Covellite	Wittichenite
n	7	15	15	35	20	4	12	9
				Chemical composition of minerals (wt %)				
Cu	1.19	33.38	62.94	77.93	76.44	74.78	67.06	38.76
Fe	46.61	30.43	10.00	0.01	0.11	0.43	0.05	bdl
Co	0.11	Bdl	bdl	Bdl	bdl	bdl	bdl	bdl
Bi	bdl	Bdl	bdl	Bdl	bdl	bdl	bdl	42.59
S	51.96	35.42	26.25	21.73	22.71	23.71	31.79	19.29
Total	99.76	99.23	99.19	99.67	99.21	98.89	98.90	100.63
				Crystallochemical coefficients in the minerals formula (a.p.f.u)				
Cu	0.02	0.97	4.98	9.02	6.92	7.95	1.03	3.02
Fe	1.01	1.00	0.90	0.00	0.01	0.05	0.00	0.00
Co	0.00	0.00	0.00	0.00	0.00	0.00	0.00	0.00
Bi	0.00	0.00	0.00	0.00	0.00	0.00	0.00	1.01
S	1.96	2.03	4.12	4.98	4.07	5.00	0.97	2.97
Cu/S	–	–	–	1.81	1.70	1.59	1.06	–

bdl—below detection limit, *n*—number of analyses.

In dolomite–calcite veins, native gold is associated with pyrite and chalcopyrite. Gold is found in calcite in the form of isometric and lamellar secretions with a size of 1–5 μm (Figure 5c,d). The composition of gold includes significant and variable amounts of copper (35.45–51.69 wt %), a small amount of silver (4.11–5.08 wt %), and often an impurity of zinc (0.64–1.19 wt %) (Table 7). The chemical composition of gold is variable. It varies from an almost stoichiometric composition of auricuprid $Cu_{3.02}(Au_{0.84}Ag_{0.14})_{0.98}$ to Cu-containing gold $(Cu_{3.07}Zn_{0.05})_{3.12}(Au_{1.65}Ag_{0.23})_{1.88}$ and $(Cu_{1.96}Zn_{0.06})_{2.02}(Au_{0.84}Ag_{0.14})_{0.98}$.

Table 7. Chemical composition of native gold and silver tellurides (wt %).

	Sample No.						
	b8-1	b8-2	b8-3	b8-4	m2609-1	m2609-2	m2609-3
n	3	3	1	2	4	6	2
				Chemical composition of minerals (wt %)			
Au	58.99	53.64	53.22	44.66	bdl	bdl	bdl
Ag	4.43	4.78	5.08	4.11	62.60	61.20	41.20
Cu	35.45	40.42	39.85	51.69	bdl	bdl	bdl
Zn	0.64	0.97	1.19	bdl	bdl	bdl	bdl
Te	bdl	bdl	bdl	bdl	37.70	37.81	57.54
Total	99.50	99.82	99.34	100.45	100.30	99.01	98.74
				Crystallochemical coefficients in the minerals formula (a.p.f.u)			
Au	1.65	0.84	0.84	0.84	0.00	0.00	0.00
Ag	0.23	0.14	0.15	0.14	1.99	1.97	0.92
Cu	3.07	1.97	1.95	3.02	0.00	0.00	0.00
Zn	0.05	0.05	0.06	0.00	0.00	0.00	0.00
Te	0.00	0.00	0.00	0.00	1.01	1.03	1.08

bdl—below detection limit, *n*—number of analyses.

Bornite stands out both in the form of independent grains and in the form of intergrowths with digenite. The size of bornite grains ranges from 5 μm × 15 μm to 150 μm × 300 μm. In polished sections, their typical shape is elongated without clear boundaries and with large parallelogram grains. Intergrowths of bornite and digenite form micrographic (myrmecite) textures (see Figure 5a), which indicate the simultaneous formation of these minerals. The typical size of bornite–digenite secretions of an isometric or slightly elongated rectangular shape is 50–70 μm. In the chemical composition of bornite ($Cu_{4.98}Fe_{0.90}S_{4.12}$), a slight iron deficiency and some excess sulfur contents are recorded (see Table 6).

Bornite contain numerous small (2–3 μm) inclusions of silver tellurides—mainly hessite and, less often, empressite (see Figure 5e,f). The chemical composition of hessite ($Ag_{1.97–1.99}Te_{1.01–1.03}$) does not significantly deviate from the stoichiometric composition of this mineral (see Table 7). In empressite ($Ag_{0.92}Te_{1.08}$), a slight excess of tellurium is observed (see Table 7).

Digenite is most common in the ore veins of copper sulfides. In addition to micrographic intergrowths with bornite, it forms disseminating intergrowths with sphalerite (see Figure 5g,h,j), as well as independent secretions up to 70 μm in size (see Figure 5b,k). In digenite, associated with sphalerite, tetrahedrite-(Zn) secretions (see Figure 5h,i), galena, and small (1–3 μm) inclusions of wittichenite (see Figure 5j) are noted. In many cases, isometric or xenomorphic bismite secretions up to 20 μm in size are noted along the edges of digenite grains (see Figure 5i,k). The ratio of copper and sulfur (1.81) in digenite ($Cu_{9.02}S_{4.98}$) is close to stoichiometric (see Table 6). Rarely, an insignificant impurity of Fe, reaching 0.13 wt %, is noted in this copper sulfide.

Sphalerite is established only in intergrowths with digenite. Small isometric inclusions of galena are sometimes noted in sphalerite. Various grains of sphalerite exhibit minor variations in the chemical composition ($(Zn_{0.96–0.98}Cu_{0.01–0.02}Fe_{0.00–0.01})_{0.97–1.00}S_{1.00–1.03}$, which corresponds to the almost ideal stoichiometric composition of this mineral (Table 8). Sphalerite is characterized by a low Fe content (0.18–0.45 wt %) and relatively high Cu concentrations (0.50–1.07 wt %).

Table 8. Chemical composition of sphalerite (wt %).

	Grain No.									
	1	2	3	4	5	6	7	8	9	10
Zn	65.27	64.43	64.51	65.44	65.68	65.42	66.34	64.70	65.33	66.01
Fe	0.21	0.45	0.20	0.29	0.20	0.29	0.20	0.18	0.33	0.28
Cu	0.50	0.87	0.81	0.80	0.82	0.80	0.75	1.07	0.83	0.68
S	34.65	33.15	32.93	33.59	33.53	33.58	34.49	33.08	33.07	33.17
Total	100.63	98.90	98.45	100.12	100.23	100.09	100.78	99.03	99.56	100.14
Crystallochemical coefficients in the minerals formula (a.p.f.u)										
Zn	0.96	0.97	0.97	0.97	0.98	0.97	0.98	0.97	0.98	0.98
Fe	0.00	0.01	0.00	0.01	0.00	0.01	0.00	0.00	0.01	0.01
Cu	0.01	0.01	0.01	0.01	0.01	0.01	0.01	0.02	0.01	0.01
S	1.03	1.01	1.01	1.01	1.01	1.01	1.01	1.01	1.00	1.00
FeS, mol %	0.36	0.79	0.35	0.50	0.35	0.50	0.34	0.32	0.57	0.49
CuS, mol %	0.76	1.35	1.25	1.22	1.25	1.22	1.14	1.65	1.28	1.04

Tetrahedrite-(Zn) forms thin (1–2 μm) coronas in digenite (see Figure 5h) and intergrowths with this mineral along smooth boundaries (see Figure 5i). Tetrahedrite-(Zn) has a variable chemical composition (Table 9). Fluctuations are noted in the content of Zn (5.46–8.84 wt %), As (9.64–13.87 wt %), Sb (9.61–14.55 wt %), and Bi, which is found in relatively high concentrations (1.34–2.25 wt %) only in some tetrahedrites-(Zn). Fe concentrations vary slightly from 0.52 to 0.70 wt %. The As/(As + Sb) ratio ranges within 0.40–0.57, compared with Zn/Fe ranging within 9.84–12.91 (see Table 9).

Table 9. Chemical composition of tetrahedrite-(Zn) (wt %).

	Grain No.									
	1	2	3	4	5	6	7	8	9	10
Cu	41.65	42.34	41.57	43.74	42.78	41.30	42.26	41.16	41.21	38.89
Zn	5.69	6.10	5.80	5.46	6.64	6.66	6.89	6.89	7.23	8.84
Fe	0.53	0.56	0.54	0.52	0.61	0.54	0.70	0.58	0.56	0.70
As	12.59	11.50	13.05	11.92	13.61	9.86	13.87	10.84	9.64	11.02
Sb	9.61	10.08	11.20	10.56	10.30	13.40	10.30	13.87	14.55	14.28
Bi	2.03	2.25	1.34	1.77	bdl	bdl	bdl	bdl	bdl	bdl
S	25.95	26.86	26.91	27.26	27.17	26.77	26.86	26.39	26.96	25.42
Total	98.05	99.69	100.41	101.23	101.11	98.53	100.88	99.73	100.15	99.15
	Crystallochemical coefficients in the minerals formula (a.p.f.u)(calculation for 29 atoms)									
Cu	10.46	10.42	10.18	10.59	10.28	10.25	10.20	10.18	10.12	9.77
Zn	1.39	1.46	1.38	1.28	1.55	1.61	1.62	1.66	1.73	2.16
Fe	0.15	0.16	0.15	0.14	0.17	0.15	0.19	0.16	0.16	0.20
As	2.68	2.40	2.71	2.45	2.77	2.08	2.84	2.27	2.01	2.35
Sb	1.26	1.29	1.43	1.33	1.29	1.74	1.30	1.79	1.86	1.87
Bi	0.15	0.17	0.10	0.13	0.00	0.00	0.00	0.00	0.00	0.00
S	12.91	13.10	13.05	13.08	12.94	13.17	12.85	12.94	13.12	12.65
As/(As + Sb)	0.57	0.53	0.54	0.53	0.57	0.42	0.57	0.44	0.40	0.44
ln(Zn/Fe)	2.38	2.39	2.36	2.36	2.39	2.51	2.29	2.48	2.56	2.54

bdl—below detection limit.

Wittichenite occurs as inclusions (1–3 μm) in both independent grains of digenite and in digenite–sphalerite associations (see Figure 5j). The chemical composition of wittichenite ($Cu_{3.02}Bi_{1.01}S_{2.97}$) is close to the stoichiometric one (see Table 6).

Anilite typically forms after bornite in association with azurite and malachite (see Figure 5f,l). Individual small (up to 20 μm) xenomorphic grains, completely composed of anilite, are less common. In the chemical composition of anilite ($Cu_{6.92}Fe_{0.01}S_{4.07}$), as well as in bornite, a slight iron deficiency and some excess sulfur contents are recorded. Accordingly, the ratio of copper and sulfur (1.70) in this mineral is less than in stoichiometric anilite (see Table 6). In anilite, up to 0.27 wt % of an Fe impurity is sometimes recorded.

Geerite and covellite typically form after anilite in association with azurite and malachite (see Figure 5k,l). Geerite corresponds to an almost ideal stoichiometric composition ($Cu_{7.95}Fe_{0.05}S_{5.00}$) with a Cu/S ratio of 1.59 (see Table 6). Covellite ($Cu_{1.03}S_{0.97}$) forms skeletons around malachite secretions. Impurities of Fe up to 1.28 and up to 0.14 wt % are occasionally noted in geerite and covellite, respectively.

5. Discussion

5.1. Reconstruction of Physico-Chemical Parameters of Ore Formation

5.1.1. Calcite, Dolomite, and Chlorite

According to the distribution of mgO, FeO, and MnO in the dolomites and calcites from ore veins, it is possible to calculate the temperatures and pressures of their formation using a geothermobarometer [57]. The calculation of the temperature and pressure is carried out according to the formulas $mg'_{Cal} = mg_{Cal}(1 + 11.5Fe_{Cal} + 5Fe^2{}_{Cal} + 550Fe^3{}_{Cal} + 8Mn_{Cal} - 50Mn^2{}_{Cal})$, $C_{Fe} = Fe_{Cal}/(Fe_{Dol} + Mn_{Dol})$, where mg_{Cal}, Fe_{Cal}, Fe_{Dol}, Mn_{Cal}, and Mn_{Dol} are mole fractions of mg, Fe, and Mn in calcite and dolomite. Using this geothermobarometer, e.g., for gold deposits of the Southern Urals [58], demonstrates a good correspondence between the calculated formation temperatures of dolomite–calcite veins and the homogenization temperatures of their fluid inclusions. The calculated values of the temperature and pressure for dolomite–calcite paragenesis formation in the Tolparovo ore occurrence comprise 198–330 °C and 0.2–0.8 Kbar (20–80 MPa), respectively (Figure 7). These temperatures agree well with the temperatures of chlorite formation, ranging between 212 and 316 °C and calculated using a chlorite thermometer [55].

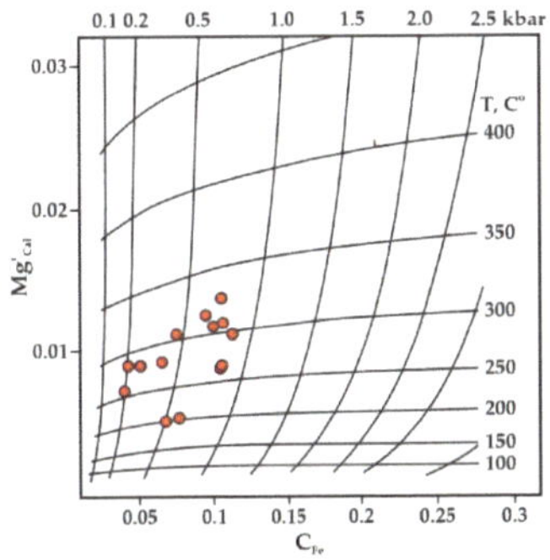

Figure 7. Composition of dolomite–calcite paragenesis in mineralized veins of the Tolparovo ore occurrence on the dolomite–calcite geothermobarometer diagram according to [57].

5.1.2. Pyrite, Chalcopyrite, and Native Gold

In the ore occurrence, the native gold of the dolomite–calcite veins associates with pyrite and chalcopyrite. In the Au–Cu–Ag diagram (Figure 8a), the figurative points of the chemical composition for the auricuprid of the Tolparovo ore occurrence are located near the liquidus isotherm of 300 °C, and the figurative points of other gold compounds are about 400 °C. The mutual solubility of the components in solid solutions of Au–Cu–Ag decreases with a temperature drop from 700 to 350 °C [59]. At a temperature of <350 °C, the formed gold phases with a low silver or copper content become metastable [60,61]. Experimental studies [62] show that under hypergene conditions with water/rock ratios equal to 200–250, oxidative processes can greatly affect the stability of intermetallic compounds in the Au–Cu–Ag system. Under such conditions, auricuprid decomposes according to the following scheme: $AuCu_3 \rightarrow AuCu \rightarrow Au_3Cu \rightarrow$ disordered solid solution $\rightarrow$ copper oxides. According to the published data, summarized in [62], the stability temperature of the cubic and tetragonal auricuprid is more than 390 and 285 °C, respectively. On this basis and the data presented in the diagram (Figure 8a), it can be concluded that the temperature of the auricuprid formation in the Tolparovo ore occurrence and, accordingly, its associated pyrite and chalcopyrite, is not less than 285 °C.

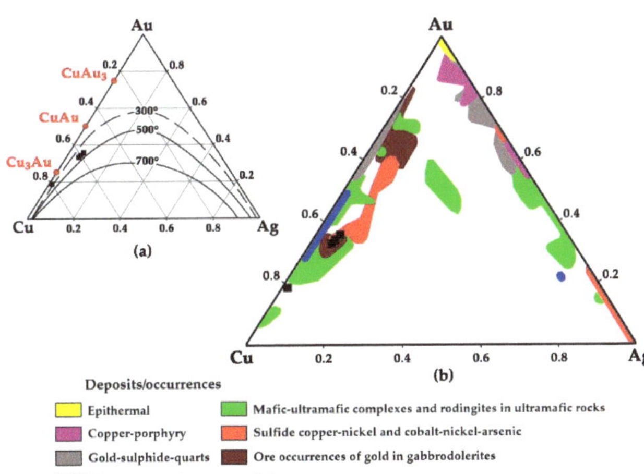

Figure 8. (**a**) Au–Ag–Cu diagrams with liquidus isotherms (according to [62,63]) and (**b**) compositions of Cu-containing native gold and silver (at %) from deposits/ore occurrences of various types (compiled according to [64–70]).

5.1.3. Sphalerite, Digenite, and Tetrahedrite-(Zn)

Intergrowths of sphalerite, digenite, and tetrahedrite-(Zn), represented by disseminated textures (see Figure 5h) in the Tolparovo ore occurrence, indicate the conditions of their joint crystallization. It should be noted that the considered mineral association is rather rare in stratiform copper deposits. Here, sphalerite tends to form intergrowths with chalcopyrite and galena, as well as with tennantite, pyrite, and, rarely, with chalcocite [49]. According to these studies, the temperature of the sphalerite–chalcopyrite paragenesis formation ranges between 230 and 350 °C, sometimes reaching up to 400 °C. Despite the sphalerite + digenite association being rare, it was obtained experimentally through a hydrothermal synthesis at a pressure of 1 Kbar and a temperature of 400 °C [71]. In addition, the micrographic intergrowths of bornite, digenite, sphalerite, and galena with high concentrations of gold and silver are described as the main minerals in the deposits of the modern subaerial geothermal system of Reykjanes in Iceland [72]. In this geothermal system, fluids rise from a depth of 2.5 km, having a temperature of 270–315 °C. The precipitation of bornite, digenite, sphalerite, and galena occurs at the outlet of the pipes under the conditions of boiling fluids due to a sharp pressure drop to 22–37 bar at a temperature of about 250 °C.

In order to calculate the temperature of the sphalerite and tetrahedrite-(Zn) formation in the Tolparovo ore occurrence, we used a geothermometer, developed for the paragenesis of permanent sphalerite with the tetrahedrite and tennantite of variable compositions [73]. The authors of the geothermometer assume an isothermal deposition of zonal faded ore, when each successive zone crystallizes under conditions of local osmotic equilibrium, fixed by the distribution of Fe and Zn in sphalerites and tetraedrites–tenantites. The temperature is calculated according to the equation $T(K) = -((664 \pm 30)/m)$, where T(K) is the kelvin temperature and m is the slope ratio of the regression line for the figurative points of tetrahedritic tennantite compositions on the As/(As + Sb)–ln(Zn/Fe) diagram [73]. The temperatures for the formation of coexisting sphalerite and tetrahedrite-(Zn) (195–280 °C), calculated using the geothermometer, e.g., for the Teremki gold deposit in the Eastern Transbaikalia, correspond well to the temperatures of the homogenization of fluid inclusions (216–298 °C) in the productive stage quartz [74].

In the dolomite–calcite veins of the Tolparovo ore occurrence with a variable composition of tetrahedrite-(Zn), the Fe content in sphalerite slightly varied from 0.32 to 0.79 mol % FeS (see Table 8). Therefore, we can assume that with the crystallization of sphalerite and coexisting tetrahedrite-(Zn), an osmotic equilibrium of the Fe and Zn exchange was achieved. Between the values of As/(As + Sb) and Zn/Fe ratios in tetrahedrites-(Zn), a significant inverse relationship is recorded with a correlation coefficient equal to −0.85. The slope ratio of the regression line for the figurative points of tetrahedrite-(Zn) compositions on As/(As + Sb)–ln(Zn/Fe) diagram equals to −1.222 (Figure 9a) with the calculated temperature of its formation being 270 ± 25 °C.

Taking into account data on the contents of FeS and CuS in sphalerite (see Table 8) and the calculated crystallization temperature of coexisting sphalerite and tetrahedrite-(Zn), it is possible to estimate the fugacity of sulfur in the ore-forming system according to the equation from [75]: $\log_{10} f\,S_2 = 11.01 - 9.49(1000/K) + [0.187 - 0.252(1000/K)] \times$ (mol % FeS in sphalerite) + $[0.35 - 0.2(1000/K)] \times$ (mol % CuS in sphalerite), where K is the kelvin temperature. In sphalerite of the Tolparovo ore occurrence, a relatively high copper content (0.76–1.65 mol % CuS) (see Table 8) is revealed, which, however, does not exceed its solubility limit. According to the studies in [76], the solubility limit of CuS in sphalerite is 2.4 mol % at temperatures of 300–500 °C without a correlation between the contents of Fe and Cu in sphalerite ($K_{corr} = 0.02$), which excludes the presence of the microinclusions of chalcopyrite in sphalerite. The calculations carried out using the equation from [75] show that the deposition of sphalerite and tetrahedrite-(Zn) in mineralized veins occur at a high sulfur fugacity, ranging from 10–7.6 to 10–5.8 (Figure 9b).

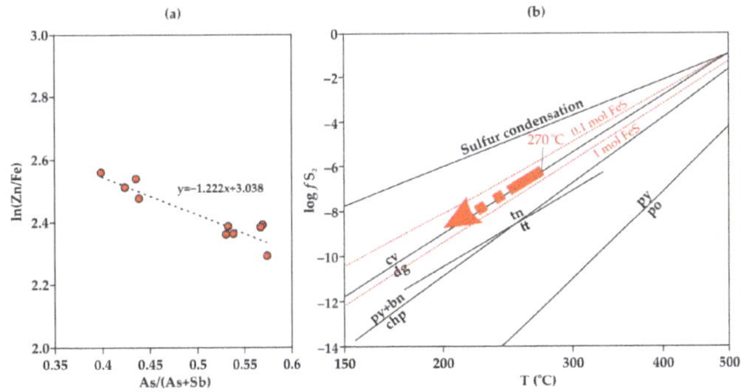

Figure 9. (**a**) Tetrahedrite-(Zn) composition on the As/(As + Sb)–ln(Zn/Fe) diagram and (**b**) sulfur fugacity diagram (log *f* S_2)—temperature (T°C) (from [73,77]). The dotted line shows the isoplets of FeS content (mol %) in sphalerite. Legend: bn—bornite; cv—covellite; chp—chalcopyrite; dg—digenite; po—pyrrhotite; py—pyrite; tn—tennantite; tt—tetrahedrite.

5.1.4. Hessite, Empressite

According to [78], the temperatures of telluride formation are typically less than 250 °C and do not exceed 350 °C. According to the experimental data and thermodynamic calculations [79,80], hessite and empressite are stable at temperatures below 500 °C (α-Ag_2Te-β-Ag_2Te's phase transformation temperature is about 150 °C) and 192 °C, respectively. Empressite is less common than hessite. Together, these minerals are described, e.g., in the Sandaowanzi epithermal Au–Ag–Te deposit (Aihui District, Heihe, Heilongjiang, China), where they were deposited in the final stages of the hydrothermal process under oxidative conditions at a temperature of about 240 °C [81]. Hessite and empressite were also noted as part of the late gold–telluride–palladium mineralization in the gabbrodolerites of the Pay-Khoy Ridge (Yugorsky Peninsula, Nenets Autonomous Okrug, Russia) [67]. The formation temperature of silver tellurides in this ore occurrence does not exceed 260 °C.

Hessite is generally one of the first formed telluride minerals. For the Fancha gold deposit (Xiaoqinling mining district, Lingbao City, Sanmenxia, Henan, China) and San Luis Alta Au–Ag–Te–(Bi–Se) deposit (Rio Grande district, Condesuyos Province, Arequipa, Peru), this is due to the fact that this mineral requires a relatively low fugacity of Te for its formation [82,83].

The calculated values of the sulfur fugacity during the formation of a sphalerite–digenite–tetrahedrite mineral association allow the tellurium fugacity in the Tolparovo ore occurrence to be estimated. In the diagram log *f* S_2–log *f* Te_2 (Figure 10), the hessite stability field, limited by the Au equilibrium line, is located between the tellurium fugacity values of $10^{-10.7}$ and $10^{-7.8}$.

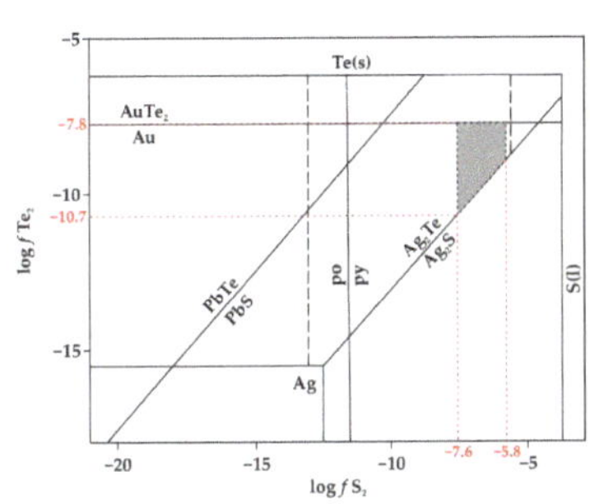

Figure 10. Diagram log *f* S_2–log *f* Te_2 at 300 °C (from [84]) and stability field for silver telluride minerals of the Tolparovo ore occurrence.

5.1.5. Sequence of Mineral Formation in the Tolparovo Ore Occurrence

The average temperatures of mineral formation in the Tolparovo ore occurrence according to dolomite–calcite, chlorite, and sphalerite–tetrahedrite geothermometers are 277 ± 39°, 284 ± 38°, and 270 ± 25°C, respectively. The similarity of temperatures, obtained using three independent geothermometers, indicates the reliability of the values obtained. At the initial stage of mineral formation in the ore-forming system, the pressure reached 0.7–0.8 Kbar (dolomite–calcite geothermobarometer estimates). The formation of ore minerals proceeded with a decrease in temperatures and pressures to <200 °C and 0.2 Kbar, respectively. At this stage, pyrite, chalcopyrite, auricuprid, bornite, digenite, sphalerite, tetrahedrite-(Zn), galena, wittichenite, hessite, and empressite were deposited in association with dolomite, chlorite, and barite (Figure 11). Mineral deposition occurred at a high sulfur fugacity, typical of bornite-containing ores [19,84,85], equal to $10^{-7.6}$–$10^{-5.8}$. The tellurium fugacity was $10^{-10.7}$–$10^{-7.8}$. Mineral formation probably occurred with a sequential change in oxidative conditions to alkaline ones. The calculations of the physicochemical parameters of ore formation in stratiform copper deposits show that the deposition of the pyrite → chalcopyrite → bornite → chalcocite association should be accompanied by a change in the pH of the medium from acidic to alkaline [19]. The initial oxidative conditions in the Tolparovo ore occurrence can be indicated by the presence of barite in the mineralized veins, where Cu directly correlates with Ba contents (see Figure 4).

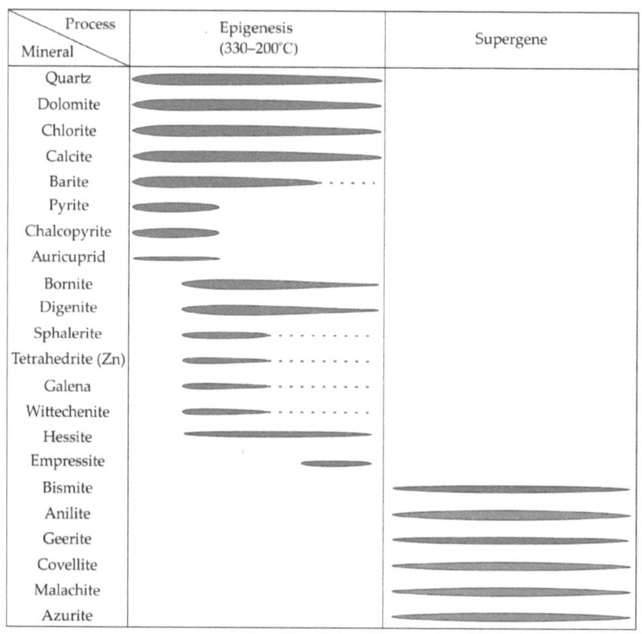

Figure 11. Paragenetic succession of the Tolparovo ore occurrence.

Ore mineralization in dolomite–calcite veins has small sizes, typically not exceeding 0.1 mm and reaching 0.3–0.5 mm in rare cases. This is probably not accidental and may be due to the deposition of minerals from supersaturated solutions under the conditions of an extremely rapid decrease in temperature and pressure. Fast cooling products from supersaturated solutions are typically characterized by fine-grained intergrowths of bornite, digenite, and sphalerite, observed in both hydrothermal experiments [71,85,86] and natural objects [72].

Anilite, geerite, and covellite in the ore occurrence develop in close association with copper carbonates—azurite and malachite (see Figure 5e,f,k,l). All minerals of this asso ciation are involved in the substitution of bornite and digenite. High-copper minerals of the chalcocite–digenite series are unstable in an exogenous environment [13,49,85], as referenced in these works. The presence of even a small amount of oxygen in the ore-forming system leads to the oxidation and removal of copper. Experimental studies show that the dissolution of digenite via an acidic solution of iron sulfate leads to the formation of anilite, followed by metastable polymorphs of geerite, spioncopite, yarrowite, and covellite, whose structure is based on the cubic closest packing of sulfur atoms [87]. Anilite is characterized by a low temperature stability <76 °C, while covellite has two stability fields, i.e., <75 °C and about 507 °C [13,49,85], as referenced in these works. Based on the above materials and the observed textural features, it can be concluded that anilite, geerite, covellite, azurite, and malachite in the Tolparovo ore occurrence formed when bornite and digenite were substituted under hypergene conditions (see Figure 11) at low temperatures and, probably, increased acidity. The formation of bismite also appears to have been associated with supergenic processes. This bismuth oxide may have been formed by the oxidation of wittichenite, included in digenite.

5.2. Sources of Ore Mineralization

5.2.1. REE

The behavior of REE in stratiform copper deposits is used to identify the diagenetic or hydrothermal epigenetic origin of ore minerals, as well as to determine the degree of their epigenetic alterations (for example, [88–90]). The ores of some stratiform copper deposits demonstrate a relationship between the content of Cu and the amount of REE. This is explained either by hydrothermal activity with an REE input (for example, in the Paratoo deposit in South Australia [88]), or by the diagenetic processes of ore mineral formation and their subsequent oxidation (for example, the Kupferschiefer deposit in Poland [89] and the Nahand-Ivand Area deposits in Iran [90]).

In the Tolparovo ore occurrence, the distribution of REE in the dolomite–calcite veins and background limestones of the Katav formation is highly similar (Figure 12a). Both veins and background limestones exhibit the same subhorizontal distribution trends with a characteristic slight enrichment of MREE in relation to LREE and HREE, as well as with mild negative anomalies of Eu and Ce. Similar distribution trends show that the source of REE in the mineralized dolomite–calcite veins were the host limestones of the Katav formation. At the same time, the contents of Cu and REE have no geochemical connection with each other. In the insoluble mineral residues of the veins, compared with their original composition (see Table 2), the content of Cu increases by 7–10 times, while the amount of REE (~20–33 ppm) remains at the same level as in the bulk vein samples (~25 ppm). This indicates the absence of a genetic link between REE and ore mineralization, representing an indirect argument in favor of the fact that the source of Cu was not only the host rocks.

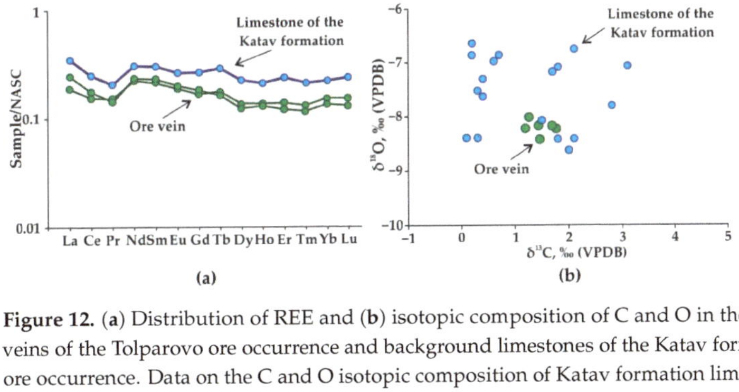

Figure 12. (**a**) Distribution of REE and (**b**) isotopic composition of C and O in the dolomite–calcite veins of the Tolparovo ore occurrence and background limestones of the Katav formation outside the ore occurrence. Data on the C and O isotopic composition of Katav formation limestones in the area of Tolparovo village are taken from [91].

5.2.2. C and O Isotopic Composition of Vein Calcite

According to [91], $\delta^{13}C$ and $\delta^{18}O$ in the limestones of the Katav formation in the vicinity of Tovparovo village range from 0.1 to 3.1 and from −8.7 to −6.8‰ (VPDB), respectively. According to these studies, the C- and O-isotope signals of the Katav limestones were insignificantly rearranged during the epigenesis, since the geochemical characteristics, typical of unaltered rocks, are preserved in limestones. They demonstrate low Mn/Sr ratios (<3.5) and an absence of correlations between the values of Mn/Sr, Fe/Sr, and $\delta^{18}O$, on the one hand, and $\delta^{13}C$, on the other [91]. In terms of C and O isotopic compositions, the dolomite–calcite veins of the Tolparovo ore occurrence are similar to the "background" limestones of the Katav formation (Figure 12b). This indicates that the source of carbon during the formation of the mineralized veins in the ore occurrence was host limestones.

5.2.3. Metal Sources

A comparative analysis of copper background concentrations in the deposits of the Alatau anticlinorium indicates (Figure 13) that, among five formations of the Karatau series, only redbed rocks of the Zilmerdak and Katav formations are enriched in copper. The average Cu content of the Zilmerdak sandstones and Katav limestones (19 ppm, see Table 1) is 1.9–4.7 times higher when compared with the clark of sandstones and carbonate rocks (respectively, $n \times 10^{-4}$ and 4 ppm according to [92]), and 1.2–1.3 times higher compared with the average Cu content of the upper continental crust (14.3 ppm according to [93]). At the same time, the concentrations of copper in the graybed carbonate rocks of the Inzer, Minyar, and Uk formations are at the near-clark level. It follows that a certain copper source in the Tolparovo ore occurrence could be the redbed deposits of the Zilmerdak and Katav formations.

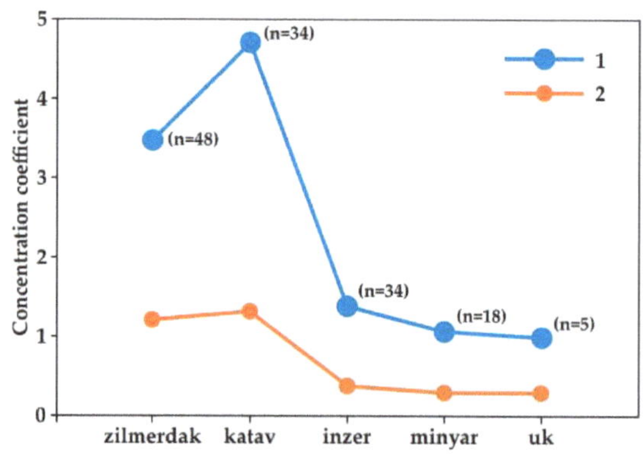

Figure 13. Concentration coefficients of average Cu contents in the rocks of the Karatau series of the Alatau anticlinorium. Legend: 1—relative to the Cu clark of [92]; 2—relative to the average Cu content in the upper continental crust of [93]. *n*—number of samples.

In the fault zone and mineralized veins, contents of TiO_2, Al_2O_3, Fe_2O_3, K_2O, F, V, Cr, Cu, Zn, As, Rb, Ba, and Pb in ore occurrences increase 2–5 times compared with the host rocks (see Table 1). At the same time, the host rocks of the ore occurrence are depleted of the listed elements in comparison with the background limestones of the Katav formation. Consequently, these elements, along with Cu, came from the host rocks to the fault zone. At the same time, among the listed elements, only the Cu and As contents in the fault zone increased significantly, by more than 100 times, compared with the host rocks. This may indicate a partial introduction of Cu and As from an additional source.

In order to determine the source of gold, data on its chemical composition can be used. The ratio of the main components in gold and the content of certain impurity elements may reflect the conditions of its formation and its typomorphic features [64,66,70,94]. Some phases of Cu-containing gold in the Tolparovo ore occurrence are close to the composition of Cu_3Au_2. The existence of this gold compound as an independent mineral has not been confirmed so far. In many cases, XRD studies show that Cu_3Au_2 compounds are copper-enriched cuproauride (cubic CuAu) or tetra-auricuprid (tetragonal CuAu) [64,95]. However, Cu-containing gold of such a chemical composition is often found in igneous deposits and the ore occurrences of gold [64,67,68,70]. It should be noted that native gold from the Tolparovo ore occurrence is chemically similar to that of rhodingites in mafite–ultramafite complexes and to gold, associated with gabbrodolerites (see Figure 8b). For example, similar gold compounds have recently been established as a part of a gold–telluride–palladium mineralization in gabbrodolerites of the Pai-Khoi ridge (Russia) [67].

In the Bashkirian uplift, Cu-containing gold of a similar composition was previously found [69] in the endo- and exocontacts of the gabbrodolerite dikes, breaking through the deposits of the neo- and mesoproterozoic in the area of the Aktash ore zone and the Suran sellaite–fluorite deposit. The geological, mineralogical, petrochemical, and geochemical characteristics of gabbrodolerites indicate their belonging to the Inzer gabbrodolerite complex [69]. In both ore zones, gold associates with chalcopyrite and is characterized by a specific composition, where close and significant amounts of Cu (29.8–36.4 wt %) and relatively low Ag contents (4.8–7.8 wt %) are recorded. The constantly present similar and significant Zn impurity (0.8–1.3 wt %) is particularly indicative. Zn represents a rather rare impurity in native gold [64,66,94]. According to the data [94], Zn, along with Pt and Pd, is more common in native gold, associated with relatively low-temperature chloride systems. Geological and geochemical features indicate the genetic relationship of Cu-containing gold in the Western Aktash zone and the Suran deposit with the Inzer gabbrodolerite complex and the activity of postmagmatic fluids [69]. According to [51], the U–Pb age of the Inzer complex is 721 ± 6.9 Ma.

In the Tolparovo ore occurrence, the composition of considered phases is characterized by close quantities of Cu (35.45–40.42 wt %), Ag (4.43–5.08 wt %), and similar contents of Zn (0.64–1.19 wt %) (see Table 7). The ore occurrence is located on the strike line of the Inzer complex gabbrodolerite dikes (see Figure 1b). In our opinion, this fact and similar specific features of the chemical composition for the native gold of the Tolparovo ore occurrence, the Aktash ore zone, and the Suran deposit indicate a single magmatic source of Au in these ore zones.

In the compared ore zones, the common geochemical features of host rocks involve their enrichment in Cu, As, Ba, and Zn. In the Aktash ore zone, at the contact with the dike complex, six copper ore occurrences are localized in Middle Riphean sandstones (see Figure 1a, ore occurrences are indicated by numbers 18–23) with Cu (up to 6.2 wt %), As and Sb (up to 1.0 wt %), Ba (up to 0.5 wt %), Zn (up to 0.3 wt %), and Au (up to 1.0 ppm). An enrichment of ore zones with the same elements, in particular Cu and As, suggests that their additional source in the Tolparovo ore occurrence could be postmagmatic fluids.

It can also be assumed that Ag, Te, and Bi were brought along with Au postmagmatic fluids. This assumption is consistent with the fact that hessite and empressite are not typical minerals in sediment-hosted stratiform copper deposits. Silver tellurides are characteristic of many epithermal gold–telluride deposits [81,96–98] and are often found in deposits of other types (volcanic-hosted massive sulfide (VHMS), intrusion-related gold deposits (IRGD), etc.) e.g., [67,82–84,99,100]. In metamorphogenic–hydrothermal and skarn deposits, gold and silver tellurides, in association with bismuth minerals (wittichenite and Bi-tellurides), are often noted as inclusions in hypogenic bornite [101]. In the ores of almost all sedimentary stratiform copper deposits, silver is a thorough and constantly present element, yet it typically manifests in its native form, its sulfides and sulfosalts, or it amalgams, e.g., [36,49].

Bismuth minerals in the ores of Cu formations similarly belong to relatively rare minerals [10,49]. Wittichenite is most often found among the bismuth minerals of copper deposits, usually intergrowing with bornite and chalcocite and often forming larger nests and flasers in quartz–calcite veins [49]. Bismutite forms in the oxidation zone after wittichenite. Among other bismuth minerals, copper sulfides sometimes associate with native bismuth, bismuthin, which, similar to wittichenite, gravitates towards the bornite zone.

5.3. General Features of Redbed Copper Ore Occurrences of the Bashkirian Uplift

5.3.1. Stratigraphy, Paleogeography, and Paleoclimate

Ore occurrences, localized in the Neoproterozoic deposits on the western slope of the Southern Urals, are characterized by some common features that are typical of stratiform redbed sedimentary–diagenetic copper deposits. First, all known ore occurrences (Lemesa, Zirikly, Katav, Bakeevo, Bolshoy Inzer, Tolparovo, Bolshye Polyany) are stratified. In remote areas of the Bashkirian uplift, they are confined to three close stratigraphic horizons, which form a rather narrow stratigraphic interval in the Neoproterozoic deposits as a whole (see Figure 2). Katav and Lemesa ore occurrences are bedded in sandstones and clay dolostones in the middle part of the Nugush subformation of the Zilmerdak formation [5,6]. The host rocks of Zirikly and Bakeevo ore occurrences are the sandstones of the Bederysh subformation of the Zilmerdak formation. The ore occurrences of Bolshoy Inzer, Tolparovo, and Bolshye Polyany are bedded in red and gray limestones in the lower and middle parts of the Katav formation [1,2]; this article.

Secondly, ore occurrences are found in redbed deposits (see Figure 2, Table 10), where they are typically confined to the boundary of redbed and graybed rocks. At the same time, the highest concentrations of ore elements, for example, in the Bakeevo ore occurrence, are noted mainly in redbed rocks. Bakeevo is the most productive of the considered copper ore occurrences. According to the data of the West Bashkirian complex geological expedition, several ore intervals with a thickness of 1.5–10 m are distinguished herein with a Cu content of 1 wt %.

Table 10. Characteristics of copper ore occurrences in the Neoproterozoic deposits on the western slope of the Southern Urals (compiled according to [1,2,5–7] and the materials of this article).

Occurrence	Stratigraphy	Host Rocks	Mineralization	Inzer Complex Gabbrodolerites
Lemesa	Zl$_2$	red and gray dolosiltstones, gray dolostones	Chp, Cc, Bn, Cv	spatial relationship
Katav	Zl$_2$	gray sandstones and dolostones	Py, Chp, Sp	not determined
Bakeevo	Zl$_4$	red and gray sandstones	no data	established
Zirikly	Zl$_4$	red and gray dolosiltstones, gray dolostones	Chp, Cc, Bn, Cv	spatial relationship
Bolshoy Inzer	Kt$_1$	red and gray limestones	Py, Chp, Mrc, Gn, Brt	spatial relationship
Tolparovo	Kt$_1$	red and gray limestones	Py, Chp, Bn, Dg, Sp, Gn, Ttr-Zn, Wtc, Au, Hes, Eps, Ani, Gee, Cv, Mlc, Azu, Bis, Brt	spatial relationship
Bolshye Polyany	Kt$_{1-2}$	gray limestones	Bn, Flr	established

Ani—anilite; Au—native gold; Azu—azurite; Brt—baryte; Bis—bismite; Bn—bornite; Cc—chalcocite; Chp—chalcopyrite; Cv—covellite; Dg—digenite; Eps—empressite; Flr—fluorite; Gee—geerite; Gn—Galena; Hes—hessite; Mlc—malachite; Mrc—marcasite; Py—pyrite; Sp—sphalerite; Ttr-Zn—tetrahedrite-(Zn); Wtc—wittichenite.

Thirdly, a common feature of the host red rocks in the Zilmerdak and Katav formations is the presence of relics of evaporite minerals therein. In redbed sandstones, siltstones and clay shales of the Zilmerdak formation, halite pseudomorphs and numerous shrinkage cracks are found, while, in the redbed clay limestones of the Katav formation, halite and gypsum pseudomorphs are observed [102]. According to the studies [102–104], siliciclastic and carbonate deposits of Zilmerdak and Katav formations were accumulated in arid climates, and in continental and coastal–marine environments, close to evaporite ones.

Most of the world's stratiform redbed copper deposits are found in sedimentary continental and coastal–marine deposits, accumulated in the arid climates of the low (~20–30°) northern latitudes of the paleoequator [14,105]. The calculated paleo-latitude values for the deposits of the Zilmerdak and Katav formations using the data of [29,30] comprise ~6–18°, respectively, which corresponds to the data reported in [14,105]. The similarity of paleogeography determines the proximity of the paleoclimate and sedimentation conditions. With regards to the basal deposits of the Karatau series in the Bashkirian uplift, this similarity allows us to explain the possible cause of increased background Cu concentrations in the rocks of the Zilmerdak and Katav formations (see Figure 13), as well as the confinement of copper ore occurrences to this stratigraphic interval.

In terms of age and sedimentary conditions, the deposits of the Zilmerdak and Katav formations of the Bashkirian uplift are similar to the host rocks of well-known large stratiform deposits of the Central African Copperbelt, which occur in the redbed deposits of the Roan Group with an age of 880–727 Ma [18]. Siliciclastic and carbonate rocks of the Roan group are characterized by the presence of collapse breccias and dissolved relics of evaporite minerals—gypsum and anhydrite pseudomorphs.

According to [8,9,12,14], stratiform copper deposits are associated with intracraton basins, where basal synrift strata of redbed rocks are overlapped by marine and/or lake sediments and by thick evaporite strata in basins located in low latitudes. The presence of redbed rocks and overlapping evaporites form favorable conditions for the formation of the redbed copper deposits [8,9,106,107]. The main source of copper here is redbed hematite-containing layers of rocks with thicknesses typically measured from 100 to 1000 m. According to [8,9], the presence of evaporites leads to the formation of high-saline brines, which can descend into the lower part of sedimentary strata and leach metals from oxidized redbed rocks. During halokinesis, induced by basin inversion, mineralized brines can be discharged in reducing barriers at higher stratigraphic levels [8,9].

At the same time, the studies [106,107] have demonstrated that the amount of pore water, involved in the in situ oxidation of ferrous minerals in redbed formations, is too insignificant for copper transportation. Strictly limited Eh-pH levels of the medium are required for the formation of early-diagenetic copper-containing ore-forming solutions. This researcher believes that these levels can be achieved only in the case of continuous streams of descending oxygen-rich meteoric waters, passing through the evaporite strata. Such solutions can leach copper from the redbed rocks and carry it with the subsequent deposition on reduction barriers.

According to the study of thermodynamic equilibria in redbed copper deposits [19], the process of mineral formation is mainly controlled by the pH of the medium and is weakly dependent on the temperature. At the same time, mineralogical zonality is found in many redbed copper deposits, consisting in the sequential deposition of Fe and Cu sulfides from the edge parts of the ore zones to the central ones in the following order: pyrite $\rightarrow$ chalcopyrite $\rightarrow$ bornite $\rightarrow$ chalcocite [8,14,16,19]. In the Tolparovo ore occurrence, as well as in other ore occurrences of the Bashkirian uplift, the zonal deposition of ore minerals is either absent or rather weakly expressed. According to [7], the copper–sulfide mineral formations in the ore occurrences of Zirikly, Lemesa, and Katav are divided into two types: evaporite and coppery sandstones. The evaporite type (Zyrikly, Lemesa) is characterized by the azonal scattered distribution of chalcopyrite, bornite, chalcocite, and low Cu content (up to 0.1 wt %) in the host rocks. The type of coppery sandstones (Katav) is characterized by the preferential development of chalcopyrite in the rocks, the increased bituminousity of the host rocks, and higher concentrations of Cu (up to 0.6 wt %) [6]. The zonal distribution of copper sulfides and the gradual lateral alteration of the monomineral chalcopyrite mineralization by chalcocite–bornite, then chalcocite, and, finally, sulfide-free zones are assumed only for the type of copper sandstones [7].

5.3.2. Tectonic Control of Mineral Formation and Connection with Intrusive Magmatism

According to the ideas of V.V. Radchenko and G.S. Senchenko [3], copper ore occurrences in the southern part of the Bashkirian uplift (ore occurrences of the Aktash ore zone (see Figure 1a, indicated by numbers 18–23) and some others) represent hydrothermal-metasomatic formations that are localized in the zones of sublatitudinal faults and, especially, at their intersections with submeridional faults. According to these researchers, in these particular zones, the processes of dolomitization and muscovitization are intensively manifested, acting as indicators of the ore process. This conclusion is confirmed by our studies of the Tolparovo ore occurrence, where ore mineral formation is in paragenetic association with dolomite.

The researchers in [3] believe that the latitudinal faults of the Bashkirian uplift are the deepest and longest-lived. Their formation is associated with the processes of riftogenesis/tafrogenesis, which are accompanied by subalkaline highly titanic intrusive and effusive volcanism. Indeed, many copper deposits and ore occurrences of the Bashkirian uplift are controlled by sublatitudinal fault zones, traced by gabbrodolerite dikes. For example, Bogryashka copper–gold–sulfide ore occurrence, localized in Lower Riphean deposits (see Figure 1a), is located in the zone of a large sublatitudinal fault [26,58], considered as a reflected foundation fault [50]. Within the area of this occurrence, magmatic formations are represented by a series of heterochronous gabbrodolerite dikes.

The most famous, on the western slope of the Southern Urals' Kiryabinsk and Barsinsk copper–sulfide deposits (see Figure 1a) of a hydrothermal–metasomatic vein type, are confined to a large sublatitudinal tectonic zone [24,108]. The primary ores of these deposits were bedded in the contact zone of the Kiryabinsk peridotite–pyroxenite–gabbro complex with an age of 680 $\pm$ 3.4 Ma [109]. This differentiated intrusion is considered a fragment of the concealed Large Igneous Province (LIP) [110].

Eventually, all copper ore occurrences in the Alatau anticlinorium, including the considered Tolparovo occurrence, gravitate to the zones of supposed latitudinal faults. Among these, the Bakeevo ore occurrence is located at the intersection of the latitudinal fault and the Western Kalinsk submeridional thrust (see Figure 1b). At the same time, almost all ore occurrences of copper in the Neoproterozoic deposits of the Bashkirian uplift are associated with dikes of gabbrodolerites of the Inzer complex (see Table 10). They are determined within the area of Bakeevo and Bolshye Polyany ore occurrences [1]. The Bolshoy Inzer and Lemesa ore occurrences are located near the gabbrodolerites of the Inzer complex, while the Tolparovo and Zirikly ore occurrences are on the line of their strike (see Figure 1b). According to our data [69], all copper ore occurrences in the Aktash ore zone are spatially associated with this complex in the southern part of the Bashkirian uplift. These facts allow us to link the copper mineral formation in the ore occurrences of the Alatau anticlinorium with the introduction of the Inzer complex gabbrodolerites and the activity of postmagmatic fluids. The formation time of this complex coincides with the age of Arsha LIP, which was formed at the turn of 707–732 Ma due the collapse of the Rodinia supercontinent [111,112]. According to V.N. Puchkov [110], Kiryabinsk and Arsha LIPs represent relics of mantle plumes, which, along with a number of other plume events in the Urals, largely determine its metallogeny. At the same time, the main source of copper in the ore occurrences, bedded in the Neoproterozoic deposits of the Bashkirian uplift, apparently, were the host rocks of the Zilmerdak and Katav formations, enriched with Cu at the stage of diagenesis. However, the process of copper leaching from host rocks probably occurred at the stage of epigenesis due to orogenic events and magmatic activity, as demonstrated, for example, for the Central African Copperbelt [43,44]. A similar metallogenic model, linking the copper mineral formation with magmatic activity during the collapse of the Rodinia supercontinent is also proposed for copper deposits in Southern China [45,46].

6. Conclusions

Similar to the world's other stratiform redbed copper deposits, the Bashkirian uplift occurrences are confined to sedimentary deposits that accumulated in an arid climate in near evaporitic continental and coastal–marine settings in the low northern latitudes of the paleoequator. This explains the elevated background concentrations of copper in the host redbed rocks of the Zilmerdak and Katav formations and indicates a predominantly sedimentary source of copper in the Bashkirian uplift.

At the same time, tectonic control is manifested in the location of copper ore occurrences of the Alatau anticlinorium. These deposits are located in the zones of assumed latitudinal faults or at their intersections with the Western Kalinsk submeridional thrust.

Almost all copper ore occurrences in the Neoproterozoic deposits of the Bashkirian uplift are associated with gabbrodolerite dikes of the Inzer complex, which makes it possible to associate the formation of ore occurrences with the activity of postmagmatic fluids. Au, partially Cu and As, and possibly Te, Ag, Bi have a magmatic source. This is indicated by similar features of the chemical composition of native gold from the Tolparovo ore occurrence and other ore zones of the Bashkirian uplift, where it has a similar composition in the gabbroderites of the Inzer complex. The formation time of the Inzer intrusive complex coincides with the age of the Arsha LIP, which represents a relic of the mantle plume and was formed at the turn of 707–732 Ma, in connection with the collapse of the Rodinia supercontinent.

Our study has confirmed the predictions of previous studies concerning the prospects of identifying industrial copper deposits on the western slope of the Southern Urals. The results have both methodological and practical significance in forming the basis for future investigations and prospecting for copper mineralization in Neoproterozoic sediments of the Alatau anticlinorium. The most probable locations of ore accumulations involve fluid discharge zones at the intersections of latitudinal and submeridional faults. Dolomitization and baritization are the indicators of the ore process.

Author Contributions: Conceptualization, validation, writing—original draft preparation, writing-review and editing, formal analysis, supervision, and funding acquisition, S.V.M.; fieldwork, methodology, investigation, data curation, and visualization, S.V.M. and G.M.K. All authors have read and agreed to the published version of the manuscript.

Funding: This study was carried out as part of the IG UFRC RAS State assignment (state registration No. FMRS-2022-0015).

Data Availability Statement: The data presented in this study are mainly contained within the article and available in the references listed. To a minor degree, the data presented are not publicly available due to privacy and are available on request from the corresponding author.

Acknowledgments: The authors express their sincere appreciation to S.A. Danilenko for critical comments and for the fruitful discussion of research materials; V.S. Michurina, N.D. Sergeeva, S.A. Dyakova, and N.P. Parfiriev for their assistance in conducting field studies; and also to I.V. Golovanova and K.N. Danukalov for their assistance in the processing of paleomagnetic data.

Conflicts of Interest: The authors declare no conflicts of interest.

References

1. Radchenko, V.V. About sulfide mineralization in the southern part of the Bashkir uplift. In *Materials of the Second Scientific Conference of Young Scientists (Geology)*; BBAS USSR: Ufa, Russia, 1971; pp. 81–82.
2. Filippov, V.A.; Akbashev, R.M. To the findings of stratified copper and barite mineralization in the Karatau series of the western slope of the Southern Urals. In *Geology and Conditions of Formation of Copper Deposits in the Southern Urals*; BBAS USSR: Ufa, Russia, 1975; pp. 144–148.
3. Radchenko, V.V.; Senchenko, G.S. Features of tectonics and magmatism of the southern part of the Bashkir uplift in connection with the prospects of ore bearing. In *Geology and Formation of Copper Deposits in the Southern Urals*; BBAS USSR: Ufa, Russia, 1975; pp. 149–152.
4. Alekseev, A.A. Genetic types of copper mineralization and prospects of copper-bearing capacity of the western slope of the Southern Urals. In *Criteria and Methods of Searching for Hidden Pyrite Deposits in the Urals*; BBAS USSR: Ufa, Russia, 1979; pp. 45–47.
5. Filippov, V.A. Prospects for the identification of sedimentary copper deposits in the Upper Riphean deposits of the western slope of the Southern Urals. *Ores Met.* **2009**, *2*, 23–27.
6. Filippov, V.A.; Kutovoy, A.K.; Ryabinin, V.F. A new discovery of copper sulfides in the Zilmerdak formation of the Neoproterozoic on the western slope of the Southern Urals. *Ores Met.* **2015**, *2*, 50–55.
7. Filippov, V.A. Ufa tectonocenter: Features of development in the Neoproterozoic and signs of influence on the migration of hydrocarbons and ore genesis. *Ural. Geol. J.* **2020**, *5*, 35–45. [CrossRef]
8. Hitzman, M.W.; Kirkham, R.; Broughton, D.; Thorson, J.; Selley, D. The sediment-hosted stratiform copper ore system. In *100th Anniversary Volume*; Society of Economic Geologists: Littleton, CO, USA, 2005; pp. 609–642. [CrossRef]
9. Hitzman, M.W.; Selley, D.; Bull, S. Formation of sedimentary rock-hosted stratiform copper deposits through Earth history. *Econ. Geol.* **2010**, *105*, 627–639. [CrossRef]

10. Duczmal-Czernikiewicz, A.; Baibatsha, A.; Bekbotayeva, A.; Omarova, G.; Baisalova, A. Ore minerals and metal distribution in tailings of sediment-hosted stratiform copper deposits from Poland and Kazakhstan. *Minerals* **2021**, *11*, 752. [CrossRef]
11. Aurelien, N.; Ousmane, S.; Pitiya, R.P. Zambia's Copperbelt area and copper mining: A review. *J. Geosci. Environ. Prot.* **2022**, *10*, 67–75. [CrossRef]
12. Kirkham, R.V.; Carriere, J.J.; Laramee, R.M.; Garson, D.F. Global distribution of sediment-hosted stratiform copper deposits and occurrences. *Geol. Surv. Can.* **1995**, *2915*, 256.
13. Gablina, I.F. *Copper Accumulation Conditions in Red Beds Continental Formations*; Nauka: Moscow, Russia, 1983; 112 p.
14. Kirkham, R.V. Distribution, settings, and genesis of sediment-hosted stratiform copper deposits in Boyle sediment-hosted stratiform copper deposits. *Geol. Assoc. Can.* **1989**, *36*, 3–38.
15. Chartrand, F.M.; Brown, A.C.; Kirkham, R.V. Diagenesis, sulphides and metal zoning in the Redstone copper deposit, Northwest Territories. *Geol. Assoc. Can.* **1989**, *36*, 189–206.
16. Cox, D.P.; Lindsey, D.A.; Singer, D.A.; Diggles, M.F. Sediment hosted copper deposits of the world: Deposit model sand database. Open-file report. *US Geol. Surv.* **2003**, *107*, 1–53. [CrossRef]
17. Durieux, C.G.; Brown, A.C. Geological context, mineralization, and timing of the Juramento sediment-hosted stratiform copper-silver deposit. Salta district. Northwestern Argentina. *Miner. Depos.* **2007**, *42*, 879–899. [CrossRef]
18. Zientek, M.L.; Bliss, J.D.; Broughton, D.W.; Christie, M.; Denning, P.D.; Hayes, T.S.; Hitzman, M.W.; Horton, J.D.; Frost-Killian, S.; Jack, D.J.; et al. *Sediment-Hosted Stratabound Copper Assessment of the Neoproterozoic Roan Group, Central African Copperbelt, Katanga Basin, Democratic Republic of the Congo and Zambia*; Scientific Investigations Report 2010–5090–T; US Geological Survey: Reston, VA, USA, 2014; 178p. [CrossRef]
19. Zhang, Y.; Han, R.; Wen, S.; Wei, P. Physiochemical restrictions of mineral zoning of sediment-hosted stratiform copper deposit. *Geofluids* **2018**, *2018*, 4261795. [CrossRef]
20. Khankahdani, K.N.; Karimi, M. Sediment-hosted copper mineralization in Bavanat region. Southern Sanandaj-Sirjan. *J. Sci. Islam. Repub. Iran* **2019**, *30*, 61–75. [CrossRef]
21. Ghasemlou, A.; Kouhestani, H.; Mokhtari, M.A.A.; Zohdi, A. Hamzelou Cu deposit: Redbed type sediment-hosted copper mineralization in the Upper red formation. NW Zanjan. *J. Adv. Appl. Geol.* **2019**, *9*, 480–497. [CrossRef]
22. Sergeeva, N.D. Deposits of the Lower Karatau subseries of the Karatau series of the Alatau anticlinorium in the section along the Zilim river in the area of Tolparovo (Southern Urals). *Geol. Bull.* **2021**, *1*, 3–13. [CrossRef]
23. Larionov, N.N.; Tsvetkova, A.A. *State Geological Map of the Russian Federation, Scale of 1:200,000*; Series South Ural. Sheet N-40-XXII (Toucan); Ministry of Natural Resources and Environment of the Russian Federation: Saint Petersburg, Russia, 2003.
24. Rykus, M.I.; Snachev, V.I.; Nasibullin, R.A.; Rykus, N.G.; Savelyev, D.E. *Sedimentation, Magmatism and Ore Content of the Northern Part of the Uraltau Zone*; Bashkir State Univ.: Ufa, Russia, 2002; 268p.
25. Dolgov, V.S.; Sereda, M.S.; Kozlov, A.V. *Minerals of the Zlatoust Urals*; FotoMir: Zlatoust, Russia, 2007; 208p.
26. Michurin, S.V.; Kovalev, S.G.; Gorozhanin, V.M. *The Genesis of Sulfates and Sulfides in the Lower Riphean Deposits of the Kamsko-Belsk Avlakogen and Bashkir Meganticlinorium*; Designpoligrafservice: Ufa, Russia, 2009; 192p.
27. Kuznetsov, A.B.; Bekker, A.; Ovchinnikova, G.V.; Gorokhov, I.M.; Vasilyeva, I.M. Unradiogenic strontium and moderate-amplitude carbon isotope variations in early Tonian seawater after the assembly of Rodinia and before the Bitter Springs Excursion. *Precam. Res.* **2017**, *298*, 157–173. [CrossRef]
28. Maslov, A.V.; Erokhin, Y.V.; Gerdes, A.; Ronkin, Y.L.; Ivanov, K.S. First results of U-Pb LA–ICP–MS isotope dating of detrital zircons from arkose sandstone of the Biryan subformation of Zilmerdak formation (Upper Riphean, South Urals). *Dokl. Earth Sc.* **2018**, *482*, 1275–1277. [CrossRef]
29. Golovanova, I.V.; Danukalov, K.N.; Salmanova, R.Y.; Levashova, N.M.; Parfiriev, N.P.; Sergeeva, N.D.; Meert, J.G. Magnetic field hyperactivity during the early Neoproterozoic: A paleomagnetic and cyclostratigraphic study of the Katav formation, southern Urals, Russia. *Geosci. Front.* **2023**, *14*, 101558. [CrossRef]
30. Pavlov, V.; Gallet, Y. Variations in geomagnetic reversal frequency during the Earth's middle age. *Geochem. Geophys. Geosys.* **2010**, *11*, Q01Z10. [CrossRef]
31. Danukalov, K.N.; Golovanova, I.V.; Salmanova, R.Y.; Parfiriev, N.P. Paleomagnetic studies of the Upper Riphean redbeds on the Southern Urals. *Geol. Bull.* **2020**, *3*, 47–54. [CrossRef]
32. Sergeeva, N.D.; Dyakova, S.A. Vend of the eastern flank of the Avdyrdak anticlinal of the Alatau anticlinorium (Southern Urals). *Geol. Bull.* **2023**, *1*, 24–34. [CrossRef]
33. Levashova, N.M.; Bazhenov, M.L.; Meert, J.G.; Kuznetsov, N.B.; Golovanova, I.V.; Danukalov, K.N.; Fedorova, N.M. Paleogeography of Baltica in the Ediacaran: Paleomagnetic and geochronological data from the clastic Zigan Formation, South Urals. *Precam. Res.* **2013**, *236*, 16–30. [CrossRef]
34. Zaitseva, T.S.; Kuznetsov, A.B.; Gorozhanin, V.M.; Gorokhov, I.M.; Ivanovskaya, T.A.; Konstantinova, G.V. The lower boundary of the Vendian in the Southern Urals as evidenced by the Rb–Sr age of glauconites of the Bakeevo formation. *Stratigr. Geol. Correl.* **2019**, *27*, 82–96. [CrossRef]
35. Jowett, E.C. The evolution of ideas about the genesis of stratiform copper-silver deposits. *Econ. Geol.* **1991**, *8*, 117–132.
36. Oummouch, A.; Essaifi, A.; Zayane, R.; Maddi, O.; Zouhair, M.; Maacha, L. Geology and metallogenesis of the sediment-hosted Cu-Ag deposit of Tizert (Igherm inlier. Anti-atlas Copperbelt. Morocco). *Geofluids* **2017**, *2017*, 7508484. [CrossRef]

37. Hitzman, M.W.; Broughton, D.W. Discussion: «Age of the Zambian Copperbelt» by Sillitoe et al. *Miner. Depos.* **2017**, *52*, 1273–1275. [CrossRef]
38. Sillitoe, R.H.; Perello, J.; Creaser, R.A.; Wilton, J.; Wilson, A.J.; Dawborn, T. Reply to discussions of «Age of the Zambian Copperbelt» by Hitzman and Broughton and Muchez et al. *Miner. Depos.* **2017**, *52*, 1277–1281. [CrossRef]
39. Shubin, Y.P. Features of the material composition, genesis and control of copper mineralization of the Bakhmut basin of Donbass. *Collect. Sci. Pap. SEI HE «DonSTI» (LPR)* **2020**, *20*, 23–30.
40. Chen, G.; Chen, M.; Ke, C.; Tang, Y. Paleozoic VMS-type stratiform mineralization overprinted by Mesozoic vein-type mineralization in the Yushui copper deposit. Eastern Guangdong. South China. *Ore Geol. Rev.* **2023**, *158*, 105498. [CrossRef]
41. Barra, F.; Broughton, D.; Ruiz, J.; Hitzman, M. Multi-stage mineralization in the Zambian Copperbelt based on Re-Os constraints. *Geol. Soc. Am. Abstr. Programs* **2004**, *36*, 516.
42. Muchez, P.; Andre-Mayer, A.S.; Desouky, H.A.E.; Reisberg, L. Diagenetic origin of the stratiform Cu-Co deposit at Kamoto in the Central African Copperbelt. *Miner. Depos.* **2015**, *50*, 437–447. [CrossRef]
43. Sillitoe, R.H.; Perello, J.; Creaser, R.A.; Wilton, J.; Wilson, A.J.; Dawborn, T. Age of the Zambian Copperbelt. *Miner. Depos.* **2017**, *52*, 1245–1268. [CrossRef]
44. Saintilan, N.J.; Selby, D.; Creaser, R.A.; Dewaele, S. Sulphide Re-Os geochronology links orogenesis, salt and Cu-Co ores in the Central African Copperbelt. *Sci. Rep.* **2018**, *8*, 14946. [CrossRef]
45. Qiu, H.N.; Zhu, B.Q.; Sun, D.Z. Age significance interpreted from 40Ar/39Ar dating of quartz samples from the Dongchuan Copper Deposits (Yunnan, SW China) by crushing and heating. *Geochem. J.* **2002**, *36*, 475–491. [CrossRef]
46. Ye, L.; Liu, Y.P.; Li, C.Y.; Liu, J.J. The Ar-Ar isotopic age in Dongchuan Taoyuan type copper deposit. Yunnan Province and its significance. *Miner. Petrol.* **2004**, *24*, 57–60.
47. Zhao, X.F.; Zhou, M.F.; Li, J.W.; Qi, L. Late Paleoproterozoic sedimentary rock-hosted stratiform copper deposits in South China: Their possible link to the supercontinent cycle. *Miner. Depos.* **2013**, *48*, 129–136. [CrossRef]
48. Cailteux, J.L.H.; Kampunzu, A.B.; Lerouge, C.; Kaputo, A.K.; Milesi, J.P. Genesis of sediment-hosted stratiform copper–cobalt deposits, central African Copperbelt. *J. Afr. Earth Sci.* **2005**, *42*, 134–158. [CrossRef]
49. Trubachev, A.I.; Korolkov, A.T.; Radomskaya, T.A. Parageneses of minerals and forms of their recognition reflecting the stages of formation of cupriferous sandstones and shales deposit. *Bull. Tomsk Polytech. Univ. Geo Assets Eng.* **2019**, *330*, 70–89. [CrossRef]
50. Larionov, N.N.; Bergazov, I.R.; Granovskaya, N.V.; Nigmatullina, A.M. *State Geological Map of the Russian Federation, Scale of 1:200,000*, 2nd. ed.; Series South Ural. Sheet N-40-XXII (Toucan). Explanatory not; The Moscow branch–A.P. Karpinsky Russian Geological Research Institute: Moscow, Russia, 2015.
51. Knyazev, Y.G.; Knyazeva, Y.G.; Snachev, V.I.; Karimov, T.R.; Masagutov, R.K.H.; Bykova, L.S.; Utaev, M.A.; Aidarov, E.M.; Alslanova, E.R. *Report on the Results of Work on the Object: "Creation of a set of the State Geological Map, Scale of 1:1,000,000. Sheet N-40-Ufa"*; A.P. Karpinsky Russian Geological Research Institute: St. Petersburg, Russia, 2013; 512p.
52. Kozlov, V.I. *The Upper Riphean and the Vend of the Southern Urals*; Nauka: Moscow, Russia, 1982; 128p.
53. Karamova, A.M.; Kazbulatova, G.M.; Michurin, S.V.; Zakharova, A.G. Photometric method of determination of fluorine in rocks by formation of alizarin complexone–lanthanide(III)–fluoride. *Bashkir Chem. J.* **2019**, *26*, 42–47.
54. Gromet, L.P.; Dymek, R.F.; Haskin, L.A.; Korotev, R.L. The «North American shale composite»: Its compilation, major and trace element characteristics. *Geochim. Cosmochim. Acta* **1984**, *48*, 2469–2482. [CrossRef]
55. Cathelineau, M. Cation site occupancy in chlorites and illites as a function of temperature. *Clay Miner.* **1988**, *23*, 471–485. [CrossRef]
56. Hey, M.H. A new review of chlorites. *Miner. Mag.* **1954**, *30*, 277–292. [CrossRef]
57. Talantsev, A.S. *Geothermobarometer by Dolomite-Calcite Paragenesis*; Nauka: Moscow, Russia, 1981; 136p.
58. Sharipova, A.A. Mineralogy and Geochemistry of Gold Deposits and Ore Occurrences of the Avzyan Ore Region (Southern Urals). Ph.D. Thesis, Kazan Federal University, Kazan, Russia, 2021.
59. Chang, Y.A.; Goldberg, D.; Neumann, J.P. Phase diagrams and thermodynamic properties of ternary gold-copper-silver systems. *J. Phys. Chem. Ref. Data* **1977**, *6*, 621–674. [CrossRef]
60. Murzin, V.V.; Malyugin, A.A. New data on the instability of natural solid solutions of the Au–Ag–Cu system in the temperature range below 350 °C. *Rep. USSR Acad. Sci.* **1983**, *269*, 723–724.
61. Knight, J.; Leitch, C.H.B. Phase relations in the system Au–Cu–Ag at low temperatures based on natural assemblages. *Can. Mineral. J.* **2001**, *39*, 889–905. [CrossRef]
62. Chudnenko, K.V.; Palyanova, G.A. Thermodynamic modeling of native formation Cu–Ag–Au–Hg solid solutions. *Appl. Geochem.* **2016**, *66*, 88–100. [CrossRef]
63. Prince, A.; Velikanova, T.; Turchanin, M.A. *Silver–Gold–Copper. Ternary Alloy Systems. Phase Diagrams, Crystallographic and Thermodynamic Data. Landolt Boernstein, Numerical Data and Functional Relationships in Science and Technology (New Series)*; Group IV: Physical Chemistry; Martiensen, W., Effenberg, G., Ilyenko, S., Eds.; Library of Congress Cataloging: Germany, Berlin, 2006; pp. 10–41.
64. Spiridonov, E.M.; Pletnev, P.A. *The Deposit of Copper Gold Zolotaya Gora (about the «Gold-Rodingite» Formation)*; Nauchny Mir: Moscow, Russia, 2002; 216p.
65. Moles, N.; Chapman, R.J.; Warner, R. The significance of copper concentrations in natural gold alloy for reconnaissance exploration and understanding gold-depositing hydrothermal systems. *Geochem Explor. Environ. Anal.* **2013**, *13*, 115–130. [CrossRef]

66. Palyanova, G.A. Gold and silver minerals in sulfide ore. *Geol. Ore Dep.* **2020**, *62*, 383–406. [CrossRef]
67. Shaybekov, R.I.; Sokerina, N.V.; Isaenko, S.I.; Zykin, N.N.; Shanina, S.N. Gold–telluride–palladium mineralization, a new type of mineralization in gabbro-dolerites of the Pai-Khoi Ridge (Yugor Peninsula, Russia). *Russ. Geol. Geophys.* **2020**, *61*, 268–285. [CrossRef]
68. Yurichev, A.N. Gold and silver accessory minerals in ultramafites of the Kyzyr-Burlyuksky ultramafic massif (Western Sayan). *Ores Met.* **2021**, *4*, 109–120. [CrossRef]
69. Michurin, S.V.; Sharipova, A.A. The first finds of copper gold in the Riphean deposits of the Bashkir meganticlinorium (Southern Urals). *Bull. Voronezh State Tech. University. Ser. Geol.* **2022**, *3*, 52–65.
70. Murzin, V.; Palyanova, G.; Mayorova, T.; Beliaeva, T. The gold–palladium Ozernoe occurrence (Polar Urals, Russia): Mineralogy, conditions of formation, sources of ore matter and fluid. *Minerals* **2022**, *12*, 765. [CrossRef]
71. Tauson, V.L.; Makeyev, A.V.; Akimov, V.V.; Paradina, L.F. Copper distribution in minerals of zinc sulfide. *Geochemistry* **1988**, *4*, 492–505.
72. Hardardottir, V.; Hannington, M.; Hedenquist, J.; Kjarsgaard, I.; Hoal, K. Cu-rich scales in the reykjanes geothermal system, Iceland. *Econ. Geol.* **2010**, *105*, 1143–1155. [CrossRef]
73. Raabe, K.C.; Sack, R.O. Growth zoning in tetrahedrite-tennantite from the Hock Hocking mine, Alma, Colorado. *Can. Miner.* **1984**, *22*, 577–582.
74. Lyubimtseva, N.G.; Prokofieva, V.; Bortnikov, N.S. Coexisting tetrahedrite–(Zn) and sphalerite at the Teremki Gold-Ore Deposit (East Transbaikal): Chemical composition and formation conditions. *Geol. Ore Dep.* **2021**, *63*, 454–464. [CrossRef]
75. Lusk, J.; Calder, B.O.E. The composition of sphalerite and associated sulfides in reactions of the Cu–Fe–Zn–S, Fe–Zn–S and Cu–Fe–S systems at 1 bar and temperatures between 250 and 535 °C. *Chem. Geol.* **2004**, *203*, 319–345. [CrossRef]
76. Kojima, S.; Sugaki, A. Phase relations in the Cu–Fe–Zn–S system between 500 and 300 °C under hydrothermal conditions. *Econ. Geol.* **1985**, *80*, 158–171. [CrossRef]
77. Barton, P.B., Jr.; Skinner, B.J. Sulfide mineral stabilities. In *Geochemistry of Hydrothermal Ore Deposits*, 2nd ed.; Barnes, H.L., Ed.; Wiley: New York, NY, USA, 1979; pp. 278–403.
78. Afifi, A.M.; Kelly, W.C.; Essene, E.J. Phase relations among tellurides, sulfides, and oxides; Pt. II. Applications to telluride-bearing ore deposits. *Econ. Geol.* **1988**, *83*, 395–404. [CrossRef]
79. Karakaya, I.; Thompson, W.T. The Ag–Te (silver–tellurium) system. *J. Phase Equilibria* **1991**, *12*, 56–63. [CrossRef]
80. Voronin, M.V.; Osadchii, E.G.; Brichkina, E.A. Thermochemical properties of silver tellurides including empressite (AgTe) and phase diagrams for Ag–Te and Ag–Te–O. *Phys. Chem. Miner.* **2017**, *44*, 639–653. [CrossRef]
81. Zhai, D.; Liu, J. Gold-telluride-sulfide association in the Sandaowanzi epithermal Au-Ag-Te deposit, NE China: Implications for phase equilibrium and physicochemical conditions. *Miner. Petrol.* **2014**, *108*, 853–871. [CrossRef]
82. Liu, J.; Wang, Y.; Huang, S.; Wei, R.; Sun, Z.; Hu, Q.; Hao, J. The gold occurrence in pyrite and Te–Bi mineralogy of the Fancha gold deposit, Xiaoqinling gold field, southern margin of the North China Craton: Implication for ore genesis. *Geol. J.* **2019**, *55*, 5791–5811. [CrossRef]
83. Alfonso, P.; Ccolqque, E.; Garcia-Valles, M.; Martinez, A.; Yubero, M.T.; Anticoi, H.; Sidki-Rius, N. Mineralogy and mineral chemistry of the Au-Ag-Te-(Bi-Se) San Luis Alta Deposit, Mid-South Peru. *Minerals* **2023**, *13*, 568. [CrossRef]
84. Vikentyev, I.V. Precious metal and telluride mineralogy of large volcanic-hosted massive sulfide deposits in the Urals. *Miner. Petrol.* **2006**, *87*, 305–326. [CrossRef]
85. Fleet, M.E. Phase equilibria at high temperatures. *Rev. Miner. Geochem.* **2006**, *61*, 365–419. [CrossRef]
86. Craig, J.R.; Scott, S.D. Sulfide phase equilibria: Reviews in Mineralogy. *Miner. Soc. Am.* **1976**, *1*, CS1–CS110.
87. Whiteside, L.S.; Goble, R.Y. Structural and compositional changes in copper sulfides during leaching and dissolution. *Can. Miner.* **1986**, *24*, 247–258.
88. Brugger, J.; Ogierman, J.; Pring, A.; Waldron, H.; Kolitsch, U. Origin of the secondary REE-minerals at the Paratoo copper deposit near Yunta. *S. Aus. Mag. Miner.* **2006**, *70*, 609–627. [CrossRef]
89. Sawlowicz, Z. REE and their relevance to the development of the Kupferschiefer copper deposit in Poland. *Ore Geol. Rev.* **2013**, *55*, 176–186. [CrossRef]
90. Sadati, S.N.; Mao, J.; Yazdi, M.; Benzadi, M.; Adabi, M.H.; Xu, L.; Mokhtari, M.A.A. REE distribution during diagenetic processes of sedimentary copper deposit in Nahand-Ivand Area. NW Iran. *Acta Geol. Sin. Engl.* **2014**, *88*, 460–462. [CrossRef]
91. Kuznetsov, A.B.; Semikhatov, M.A.; Maslov, A.V.; Gorokhov, I.M.; Prasolov, E.M.; Krupenin, M.T.; Kislova, I.V. New data on Sr-and C-isotopic chemostratigraphy of the Upper Riphean type section (Southern Urals). *Stratigr. Geol. Correl.* **2006**, *14*, 25–53. [CrossRef]
92. Turekian, K.K.; Wedepohl, K.H. Distribution of the elements in some major units of the earth's crust. *Geol. Soc. Am. Bull.* **1961**, *72*, 175–192. [CrossRef]
93. Wedepohl, K.H. The composition of the continental crust. *Geochim. Cosmochim. Acta* **1995**, *59*, 1217–1232. [CrossRef]
94. Chapman, R.J.; Banks, D.A.; Styles, M.T.; Walshaw, R.D.; Piazolo, S.; Morgan, D.J.; Grimshaw, M.R.; Spence-Jones, C.P.; Matthews, T.J.; Borovinskaya, O. Chemical and physical heterogeneity within native gold: Implications for the design of gold particle studies. *Miner. Depos.* **2021**, *56*, 1563–1588. [CrossRef]

95. Zaccarini, F.; Pushkarev, E.; Garuti, G.; Krause, J.; Dvornik, G.P.; Stanley, C.; Bindi, L. Platinum-group minerals (PGM) nuggets from alluvial-eluvial placer deposits in the concentrically zoned mafic-ultramafic Uktus complex (Central Urals. Russia). *Eur. J. Miner.* **2013**, *25*, 519–531. [CrossRef]
96. Pals, D.W.; Spry, P.G. Telluride mineralogy of the low-sulfidation epithermal Emperor gold deposit, Vatukoula, Fiji. *Miner. Petrol.* **2003**, *79*, 285–307. [CrossRef]
97. Spry, P.G.; Scherbath, N.L. The gold-vanadium-tellurium association at the Tuvatu gold-silver prospect, Fiji: Conditions of ore deposition. *Miner. Petrol.* **2006**, *87*, 171–176. [CrossRef]
98. Bortnikov, N.S.; Volkov, A.V.; Savva, N.E.; Prokofiev, V.Y.; Kolova, E.E.; Dolomanova-Topol, A.A.; Galyamov, A.L.; Murashov, K.Y. Epithermal Au–Ag–Se–Te deposits of the Chukchi Peninsula (Arctic zone of Russia): Metallogeny, mineral assemblages and fluid regime. *Russ. Geol. Geophys.* **2022**, *63*, 435–457. [CrossRef]
99. Maslennikov, V.V.; Ayupova, N.R.; Safina, N.P.; Tseluyko, A.S.; Melekestseva, I.Y.; Large, R.R.; Herrington, R.J.; Kotlyarov, V.A.; Blinov, I.A.; Maslennikova, S.P.; et al. Mineralogical features of ore diagenites in the Urals massive sulfide deposits, Russia. *Minerals* **2019**, *9*, 150. [CrossRef]
100. Novakov, R.M.; Kungurova, V.E.; Moskaleva, S.V. Formation conditions of noble metal mineralization in sulfide cobalt-copper-nickel ores of Kamchatka (on the example of Annabergitovaya Schel ore occurrence. *J Min. Inst.* **2021**, *248*, 209–222. [CrossRef]
101. Cook, N.J.; Ciobanu, C.L.; Danyushevsky, L.V.; Gilbert, S. Minor and trace elements in bornite and associated Cu–(Fe)-sulfides: A LA-ICP-MS study. *Geochim. Cosmochim. Acta* **2011**, *75*, 6473–6496. [CrossRef]
102. Maslov, A.V.; Krupenin, M.T.; Gareev, E.Z. Lithological, lithochemical and geochemical indicators of paleoclimate: Evidence from Riphean of the Southern Urals. *Lith. Miner. Res.* **2003**, *38*, 427–446. [CrossRef]
103. Parnachev, V.P. Fluorine and chlorine in Late Precambrian sedimentary rocks of the Bashkir uplift in connection with the issues of their sedimentogenesis. In *Geochemistry of Volcanic and Sedimentary Rocks of the Southern Urals*; UB RAS: Sverdlovsk, Russia, 1987; pp. 35–47.
104. Gareev, E.Z. Geochemistry of Sedimentary Rocks of the Stratotypic Riphean Section, Academic Degree of Candidate of Geological and Mineralogical Sciences. Ph.D. Thesis, V.I. Vernadsky Institute of Geochemistry and Analytical Chemistry, Moscow, Russia, 1989.
105. Strakhov, N.M. Principles of lithogenesis. *Geol. Mag.* **1962**, *3*, 577. [CrossRef]
106. Brown, A.C. A process-based approach to estimating the copper derived from redbeds in the sediment-hosted stratiform copper model. *Econ. Geol.* **2009**, *104*, 857–868. [CrossRef]
107. Brown, A.C. Constraints on conceptual and quantitative modeling of early diagenetic sediment-hosted stratiform copper mineralization. *Minerals* **2017**, *7*, 192. [CrossRef]
108. Kovalev, S.G.; Salikhov, D.N.; Puchkov, V.N. *Mineral Resources of the Republic of Bashkortostan (Metals)*; Alfa-Reklama: Ufa, Russia, 2016; 554p.
109. Krasnobaev, A.A.; Puchkov, V.N.; Sergeeva, N.D.; Lepekhina, E.N. Zirconology of pyroxenites from the Kiryabinka pyroxenite-gabbro complex (Southern Urals). *Dokl. Earth Sc.* **2013**, *450*, 531–535. [CrossRef]
110. Puchkov, V.N. General features relating to the occurrence of mineral deposits in the Urals: What, where, when and why. *Ore Geol. Rev.* **2016**, *85*, 4–29. [CrossRef]
111. Puchkov, V.N. The Plumes–A new word in geology of the Urals. *Litosfera* **2018**, *18*, 483–499. [CrossRef]
112. Puchkov, V.N.; Ernst, R.E.; Ivanov, K.S. The importance and difficulties of identifying mantle plumes in orogenic belts: An example based on the fragmented large igneous province (LIP) record in the Ural fold belt. *Precam. Res.* **2021**, *361*, 106–186. [CrossRef]

minerals MDPI

Article

Evolution of the Hydrothermal Fluids Inferred from the Occurrence and Isotope Characteristics of the Carbonate Minerals at the Pogo Gold Deposit, Alaska, USA

Yuichi Morishita [1,2,3,*] and Jamie R. Rogers [4]

1 Center for Integrated Research and Education of Natural Hazards, Shizuoka University, Shizuoka 422-8529, Japan
2 Geological Survey of Japan (GSJ), National Institute of Advanced Industrial Science and Technology (AIST), Tsukuba 305-8567, Japan
3 Museum of Natural and Environmental History, Shizuoka 422-8017, Japan
4 Northern Star Resources Ltd., Perth, WA 6008, Australia; jrogers@nsrltd.com
* Correspondence: morishita.yuichi@shizuoka.ac.jp

Abstract: Pogo is identified as a deep-seated, intrusion-related gold deposit. Carbonate minerals have a close spatial relationship to hydrothermal gold mineralization in all of its principal ore zones. The carbon and oxygen isotopic ratios of carbonate minerals (siderite, ankerite, and calcite) present within the deposit illustrate the isotopic evolution of the ore-forming fluid. The initial hydrothermal fluid phase is interpreted to be magmatic in origin. The fluid evolution was characterized by a gradual decrease in $\delta^{18}O$ and a slight increase in $\delta^{13}C$ with decreasing temperature. The dominant carbon-bearing species was CO_2, with methane introduced sporadically. Siderite is associated with early-stage mineralization and occurs with ankerite in main-stage ore assemblages. Calcite is recognized in the later stages of mineralization. Gold in the Pogo deposit occurs as native gold, Au-Bi-Te minerals, inclusions in sulfide minerals, or as "invisible gold". The latter is found in pyrite, chalcopyrite, arsenopyrite, and quartz, based on ion microprobe analysis. The presence of invisible gold in these minerals has significant metallurgical implications for gold processing at the Pogo mine.

Keywords: Pogo deposit; siderite; ankerite; calcite; pyrite; carbon and oxygen isotope ratios; genesis and evolution of hydrothermal fluid; IRMS; SIMS

Academic Editor: Huan Li

Received: 19 November 2024
Revised: 24 December 2024
Accepted: 27 December 2024
Published: 12 January 2025

Citation: Morishita, Y.; Rogers, J.R. Evolution of the Hydrothermal Fluids Inferred from the Occurrence and Isotope Characteristics of the Carbonate Minerals at the Pogo Gold Deposit, Alaska, USA. *Minerals* **2025**, *15*, 67. https://doi.org/10.3390/min15010067

1. Introduction

The Pogo deposit is a high-grade gold deposit located in the arc-shaped Tintina Gold Province (TGP, Figure 1). The TGP extends over 1200 km across Alaska (USA) and the Yukon (Canada) and hosts a series of world-class gold deposits that are predominantly Early Cretaceous to Eocene in age (e.g., [1–4]). Pogo has been classified as an intrusion-related gold deposit [4–8], although some argue that its characteristics are more typical of orogenic gold systems [9–12].

The Pogo gold mine has operated continuously since 2006. Gold production for the financial year 2024 (ending 30 June 2024: FY24) totaled 8.7 tons [13]. Total (measured, indicated, and inferred) mineral resources and (proved and probable) ore reserves amount to 2.05×10^7 tons at 10.1 g/t Au and 5.9×10^6 tons at 8.0 g/t Au, respectively [13].

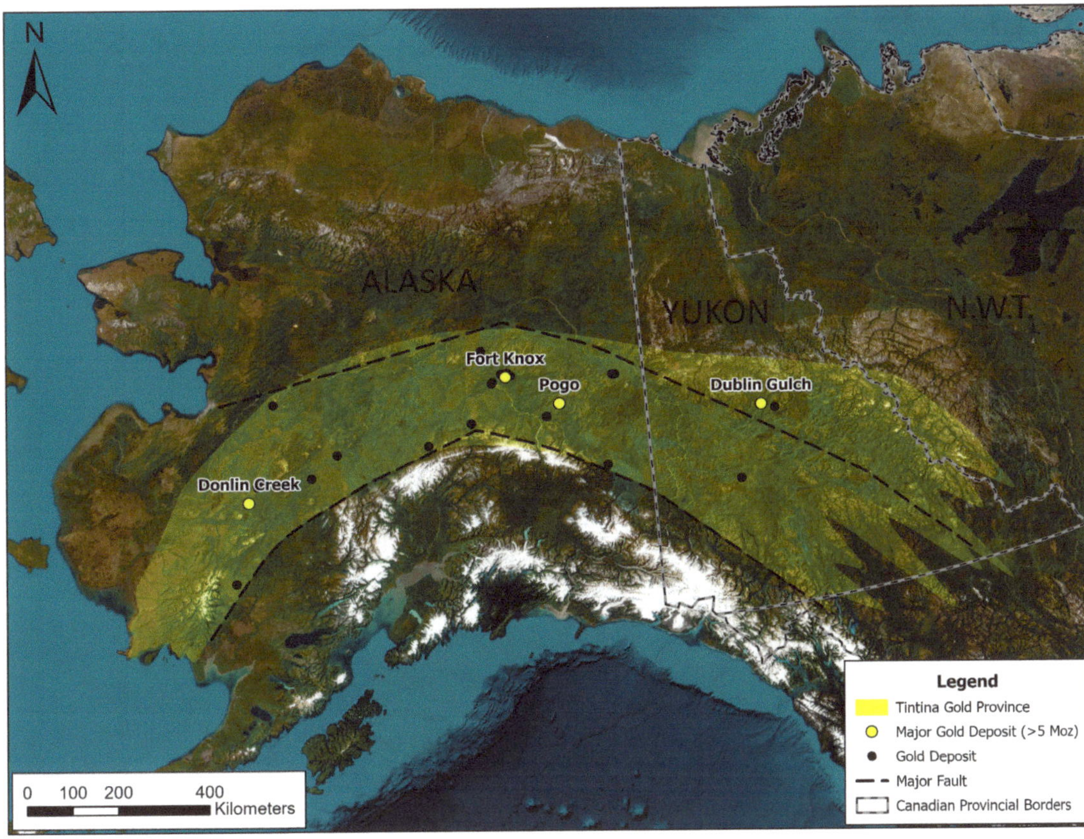

Figure 1. Gold deposits (indicated by circles) distributed in the Tintina Gold Province (pale yellow colored area in the figure) in interior Alaska, USA, and adjacent Yukon, Canada. Major gold deposits (>5 Moz), including the Pogo deposit, are represented by yellow symbols.

The low salinity and high CO_2 contents of intrusion-related gold systems differ from the characteristics of most other magmatic gold environments [12]. Analysis of ore-related carbonates can provide valuable insights into the nature of the hydrothermal mineralizing system. Since the carbon and oxygen isotopic ratios of carbonates are independently arranged, the trend of the two-isotope set can provide valuable information regarding potential ore precipitation mechanisms (e.g., [14]).

The Pogo ore assemblage includes carbonate minerals such as calcite, the ferroan dolomite–ankerite series, and siderite. Stable isotopes of the carbonate minerals can predict the genesis and evolution of the gold mineralizing fluids. The carbon and oxygen isotope ratios of carbonate minerals in this study were determined by isotope ratio mass spectrometry (IRMS). Understanding the nature of the Pogo hydrothermal system on the basis of isotope data can help to map the extent of the mineralization footprint and guide the discovery of new mining areas.

Gold in the Pogo deposit occurs as native gold, Au-Bi-Te minerals, inclusions in sulfide minerals, or as "invisible gold". The invisible gold is further divided into solid solution in sulfides and gold nanoparticles (NPs; <100 nm grains, [15,16]). Microbeam techniques have been applied across a variety of deposits to investigate the presence of gold in sulfides (e.g., [17–23]). The application of the secondary ion mass spectrometry (SIMS) or ion microprobe technique to the Pogo deposit allows for high sensitivity microanalysis, making it possible to elucidate the nature of the gold in the ore-related minerals. These determinations have significant metallurgical implications for gold processing at the Pogo mine.

2. Geology and Mineralogy

The TGP is bound by the Tintina fault system to the north and the Denali fault system to the south (Figure 1). The Pogo deposit is located along the Goodpaster River, 145 km southeast of Fairbanks (inset in Figure 2). Gold-bearing quartz veins of the Pogo deposit are hosted by gneisses of the late Proterozoic to mid-Paleozoic Yukon–Tanana Terrane (e.g., [6,24]) and a suite of mid-Cretaceous felsic to intermediate plutons and dikes [25]. A simplified geologic map of the Pogo mine area is shown in Figure 2.

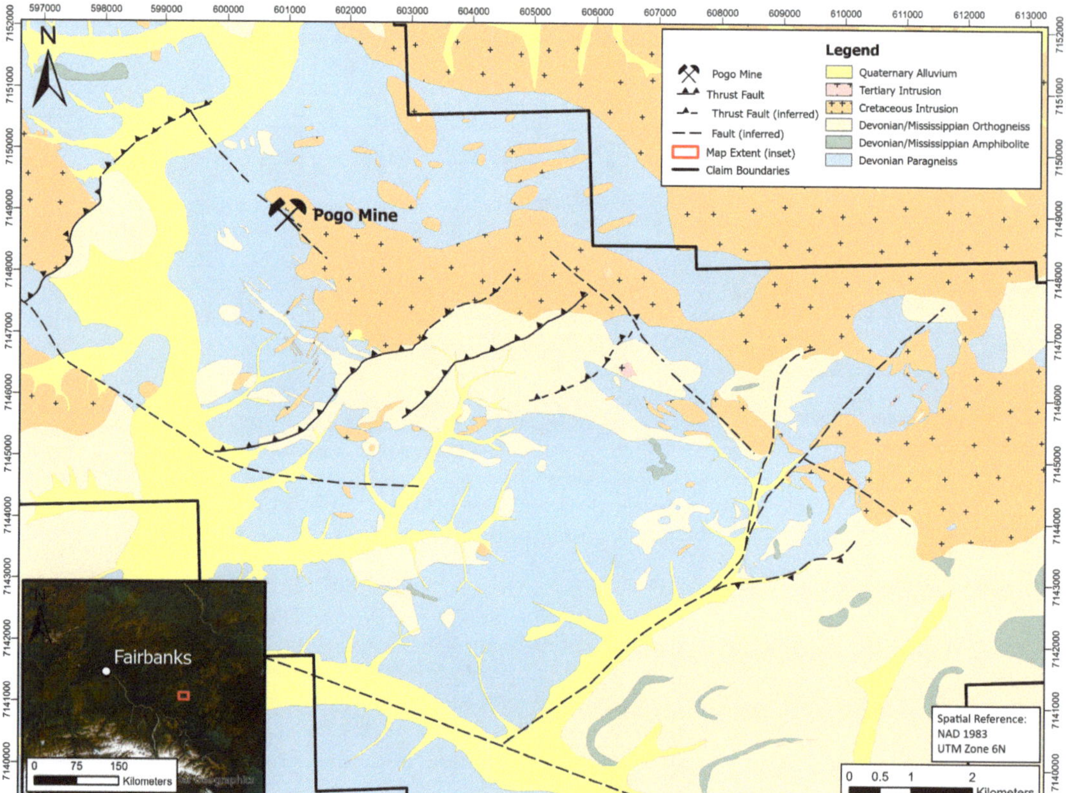

Figure 2. Simplified geologic map of the Pogo mine area. The inset is a small-scale map showing the Pogo mine and Fairbanks.

The Pogo deposit can be subdivided into five main mining areas: Liese, South Pogo, East Deep, North Zone, and Central Lodes (Figure 3a). Several peripheral mineral resources have also been identified at the Goodpaster and 4021 prospects. The Liese, and South Pogo vein systems are located south of the Liese Creek intrusive complex. Each area comprises a series of stacked ore bodies that dip shallowly (25° to 30°) to the northwest (Figure 3b, [6]). A mean ^{187}Re-^{187}Os molybdenite age of 104.2 ± 1.1 Ma was obtained for the Liese veins [26]. An ^{40}Ar/^{39}Ar age of 92.7 ± 0.3 Ma, obtained for a mafic dike that cuts the uppermost (L1) vein structure, constrains the minimum age of gold mineralization [7]. This age is comparable to the 91.2 ± 0.4 Ma and 91.7±0.4 Ma $^{40}Ar/^{39}Ar$ ages obtained for Liese Zone alteration assemblages [6].

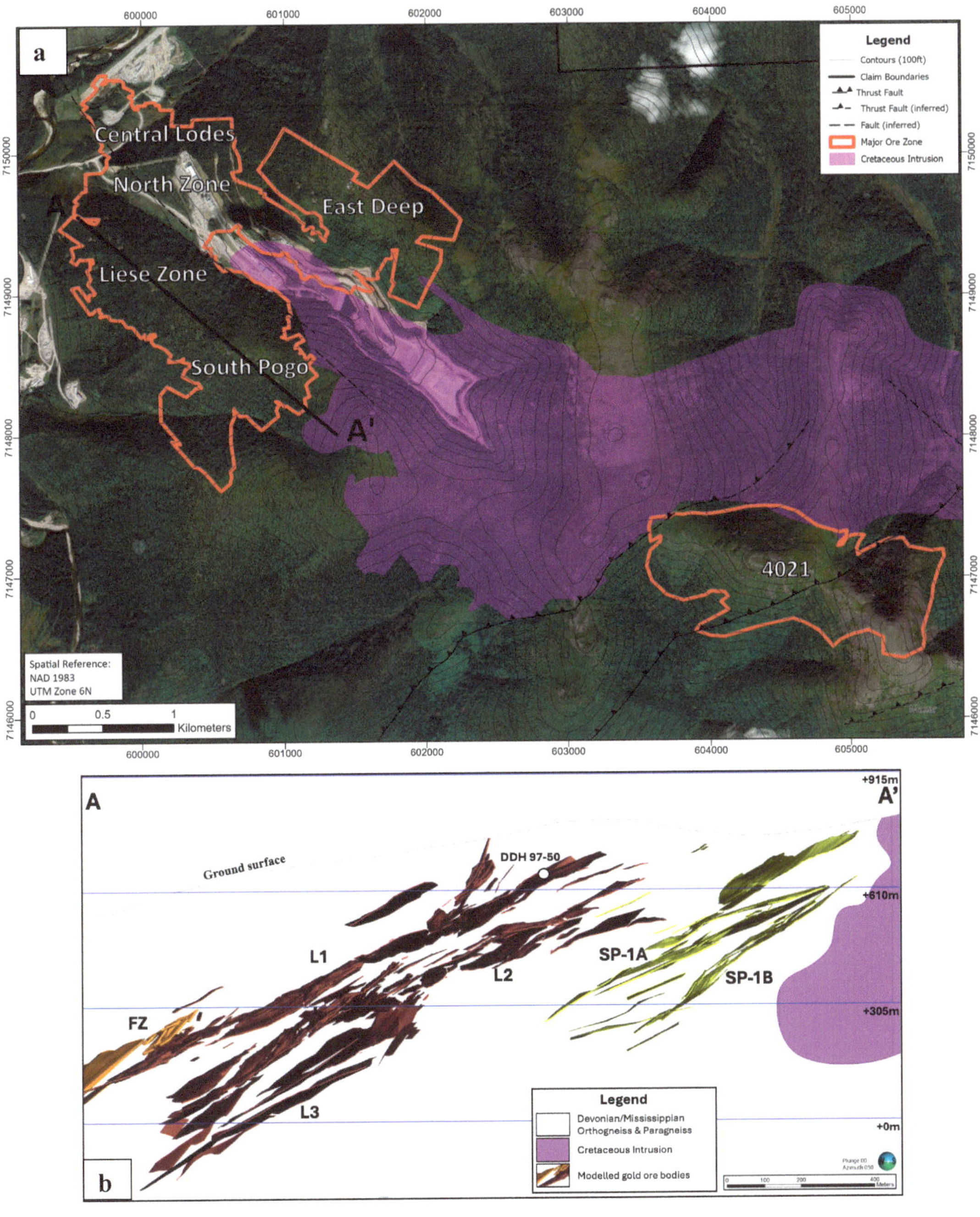

Figure 3. (**a**) The major ore zones (Liese, South Pogo, East Deep, North Zone, and Central Lodes) of the Pogo mining area. The cross-section line A–A′ for (**b**) is shown. (**b**) Cross-section showing the Liese Zone and South Pogo ore bodies. The elevation above mean sea level (AMSL) is indicated.

The East Deep mining area contains two stacked quartz veins (E0.5 and E1) with similar orientations and mineral assemblages to those of the Liese veins. The East Deep veins are interpreted as extensions of the Liese vein system, now separated by the post-mineral Liese Creek intrusive complex [25]. Three principal NNW-striking, subvertical mineralized veins are mined at the North Zone (NZ1, NZ2, and NZ3 veins from west to east). The Central Lodes area is characterized by an extensive domain of sheeted vein arrays hosted in granitic orthogneiss.

Gold mineralization in all mining areas is associated with varying proportions of Fe-As-S minerals including pyrite, arsenopyrite, loellingite, and pyrrhotite, accompanied by minor bismuthinite and chalcopyrite [6]. Aggregates of pyrite or arsenopyrite occur in a matrix of quartz locally intergrown with Mg–Fe carbonate [7]. The Au–Bi assemblage, associated pyrite, arsenopyrite, and pyrrhotite, is typical of plutonic-related gold deposits [5]. Thirty-five intrusive rocks from the vicinity of the Liese zone were collected for major and trace element analyses [6]. They fall within the "I-type" field of Chappell and White [27] and belong to the "ilmenite series" of Ishihara [28]. These compositions are characteristic of intrusion-related gold mineralization in Central Alaska [6].

A detailed study of Pogo hydrothermal alteration assemblages by Smith et al. [6] included descriptions of carbonates relevant to this study. Very fine-grained silica with fine-grained, disseminated pyrite and/or arsenopyrite is typically associated with sericite + ferroan dolomite + chlorite ± quartz alteration, and the rocks are commonly fractured and sheared [6].

3. Materials and Methods

3.1. Samples

The carbonate-bearing samples used in this study were collected from the Liese, East Deep, and North Zone vein systems and from mineralized drill core samples in the general Pogo mining area (Table 1, Figure 4). The principal carbonate mineral in the Liese Zone is siderite ($FeCO_3$), with ferroan dolomite–ankerite series carbonates (Ca (Fe^{2+}, Mg, Mn) $(CO_3)_2$) and calcite ($CaCO_3$) distributed more widely across the deposit.

Table 1. Summary table of the Pogo samples used in this study.

No.	Sample Name	Cal	Ank	Sid	Vein or Area, Location (or Drill Depth), Description	Coordinate X	Coordinate Y	Coordinate Z	REE	$\delta^{13}C$ of Graphite
1	2011032201				L1 Blk 10, 1760C6E, graphite schist	1813630	3820717	1777	•	•
2	2011032202		•		L1 Blk 10, 1760C6E, qtz vein with ankerite and sulfide	1813630	3820717	1777		
3	2011032205		•		L1 Blk 08, 1558C9W, qtz vein with ankerite and sulfide	1811272	3820623	1579		
4	2012080709	•			East Deep, 1300EXP, qtz vein with calcite and sulfide	1814935	3821196	1257		
5	2012080711		•	○	North Zone, NZ02-1331, carbonate dissemination in altered intrusive	1812013	3822688	1318		

Table 1. *Cont.*

No.	Sample Name	Cal	Ank	Sid	Vein or Area, Location (or Drill Depth), Description	Coordinate			REE	$\delta^{13}C$ of Graphite
						X	Y	Z		
6	2012081001	•	○		New Portal Bench, carbonate vein in weathered diorite (ground)	1814453	3821485	2090		
7	2012081003		•		North Zone, drill core 12-685, 635′, qtz–ank vein	1814677	3823024	2081	•	
8	2012081004	•			Drill core 12u130, 305.5′, thin qtz–cal vein with sulfide	1811122	3821419	887		
9	2012081005	•			North Zone, drill core 12-685, 697′, qtz vein with calcite and sulfide	1814670	3823017	2020		
10	2012081006	•			North Zone, drill core 12-598, (Loc 4021), 66′, calcite dissemination in altered rock	1823121	3814731	3108	•	
11	2012081011		•		L1, Ankerite dissemination in altered intrusive	1811995	3822640	1318		
12	2012081108		•		L1 BLK10, 1320, S3, qtz vein with ankerite and sulfide	1812948	3821164	1336	•	
13	2014081806				X-Vein, North Zone, graphite schist with pyrite crystal	1812801	3822967	895	•	•
14	2014082206	•			East Deep, 1020Cut#2E, E1, qtz vein with calcite and sulfide	1815072	3821642	1050		
15	2014082207	•			Estimated L1 extension, qtz–cal vein (ground)	1818057	3817256	822		
16	2014082208	•	○		40 cm above 2014082207, qtz–cal vein (ground)	1818057	3817256	822		
17	2014082210	○	•		Estimated L1 extension, qtz vein with brown carbonate (ground)	1818057	3817256	822		
18	2014082211	•	○		Estimated L1 extension, qtz–cal vein (ground)	1818057	3817256	822		
19	2014082212	•	○		Estimated L1 extension, carbonate veinlet (ground)	1818057	3817256	822		
20	2014082213	•			Estimated L1 extension, just below 2014082207, qtz–cal vein (ground)	1818057	3817256	822		
21	2014082215	•			Qtz–Cal vein (ground)	1816789	3818489	822		
22	2014082216	•			Qtz–Cal vein with sulfide (ground)	1816789	3818489	822		
23	2014082304-2				Burn, fracture-filling qtz vein with arsenopyrite and visible Au	1802710	3818930	511	•	

Table 1. *Cont.*

No.	Sample Name	Cal	Ank	Sid	Vein or Area, Location (or Drill Depth), Description	Coordinate			REE	$\delta^{13}C$ of Graphite
						X	Y	Z		
24	M000002a	•			Hill 4021 prospecting area, drill core 13-610, qtz–cal vein in altered diorite	1823453	3815885	2807	•	
25	M000004a	•			Hill 4021 prospecting area, drill core 13-610, qtz–cal vein in altered diorite	1823445	3815896	2784		
26	M000013	•			East Deep East, drill core 13-678, calcite dissemination in EDE	1817768	3820455	155		
27	M000015a	•			South Pogo, drill core 13-759, qtz–cal vein	1814301	3819357	1745		
28	M000034a	•			NZ03, drill core 13EXP007, qtz–cal vein between NZ04-NZ03	1812626	3823140	134	•	
29	M000068c			•	L1 NorthWest, drill core 12-691, qtz–sid vein with sulfide	1808261	3822132	101		
30	10DL007A	•			East Deep East, 10-507, 841′, cal vein	1817712	3821658	1395		
31	10DL007B	•	•		East Deep East, 10-507, 841′, qtz–cal–ank vein	1817712	3821658	1395		
32	10DL008	•			East Deep, 10-508, 560′, crystalline calcite with rhodochrosite in a drusy cavity	1815203	3821563	1442		
33	10DL012A	•			L1, 10-517, 371′, cal vein	1809548	3822370	1475		
34	10DL012B	•	•		L1, 10-517, 371′, cal–ank vein	1809548	3822370	1475		
35	11DL011A	•	•		Under the L1, 10u129, 108′, qtz–cal–ank vein with muscovite	1811935	3822586	1019		
36	11DL011B		•		Under the L1, 10u129, 108′, qtz–ank vein	1811935	3822586	1019		
37	11DL011DOS		•		Under the L1, 10u129, 108′, qtz–ank vein with DOS alterataion	1811935	3822586	1019		
38	11DL012A	•	•		Under the L1, 10u129, 144′, cal–ank vein	1811935	3822596	1168		
39	11DL012B	•	•		Under the L1, 10u129, 144′, qtz–cal–ank vein	1811935	3822596	1168		
40	11DL012C	•	•		Under the L1, 10u129, 144′, cal–ank vein	1811935	3822596	1168		
41	11DL025	•			Under the L3, 11u133, 120′, qtz–cal vein	1812390	3820875	-92		

Table 1. *Cont.*

No.	Sample Name	Cal	Ank	Sid	Vein or Area, Location (or Drill Depth), Description	Coordinate			REE	$\delta^{13}C$ of
						X	Y	Z		Graphite
42	11DL028	•	•		Above the ED01, 11-573, 528′, qtz–cal–ank vein	1815763	3823111	1404		
43	11-584A			•	East Deep area, 11-584, 980′, qtz–sid vein with chlorite and sulfide	1816137	3822155	1644		
44	11-584B	•			East Deep area, 11-584, 548′, qtz–cal vein	1816140	3822098	2109		
45	11-584C		•		East Deep area, 11-584, 430′, qtz–ank vein with sulfide	1816141	3822081	2240		
46	11-584D	•			East Deep area, 11-584, 430′, cal vein	1816141	3822081	2240		
47	11u201A	•	•		Between L1 and NZ02, 11u201, 60′, qtz–cal–ank vein	1811795	3823003	1038		
48	14-716_278.2ft		○	•	Qtz–Sid vein	1813519	3818448	2226		
49	14-742_474.6ft		•		SP-1a, qtz–ank vein with sulfide	1813107	3818234	1850		
50	14-742_1027.2ft	•			SP-1b, qtz–cal vein	1813145	3818381	1319		
51	14-752_1188.6ft			•	SP-1b, qtz–sid vein with muscovite	1812047	3817786	1252	•	
52	14-765_589.9ft			•	SP-1a, qtz–sid vein with sulfide	1811963	3818573	1566		
53	14-765_824.3ft			•	SP-1b, qtz–sid vein with sulfide	1812030	3818527	1346		
54	14-771_876.1ft	•			SP-1b?, qtz–cal vein	1811607	3818969	1488		
55	14-773_875.8ft		•		SP-1a?, qtz–ank vein with sulfide	1811862	3819396	1455		
56	14-795_1242.8ft	•			X-Vein, qtz–cal vein	1812263	3823758	1132	•	
57	14-795_1434.4ft		•		NZ04, qtz–ank vein with sericite	1812186	3823706	964		
58	14-805_857.1ft	•			X-Vein, qtz–cal vein	1811981	3823809	1420		
59	14-805_1008.9ft	•			NZ04, Qtz vein with calcite and sulfide	1811980	3823808	1419		
60	14-805_1020ft	•			NZ04, qtz–cal vein	1811894	3823760	1292		
61	15-882_316.5	•			Star, Silicified rock with calcite	1819343	3812334	2633		
62	15-882_833.5	•			Star, qtz–cal vein	1819148	3812519	2192		
63	15-884_500.5	•			Star, qtz–cal vein	1819394	3811999	2469		
64	15-896_2500.4		•		East Deep, qtz–ank vein with sericite and sulfide	1814790	3823129	487		
65	15-896_2581.3	•			East Deep, qtz–cal vein	1814801	3823084	420		

Table 1. *Cont.*

No.	Sample Name	Cal	Ank	Sid	Vein or Area, Location (or Drill Depth), Description	Coordinate			REE	$\delta^{13}C$ of Graphite
						X	Y	Z		
66	15-900_156	•			East Deep, qtz–cal vein	1814985	3825571	2426		
67	15-900_1116.8	•			East Deep, qtz–cal vein	1814639	3825093	1670		
68	15-901_2500		•		East Deep, qtz–ank vein in altered host rock	1814636	3822901	615		
69	15-926_1388	•			SP-1a, South Pogo, qtz–cal vein with disseminated sulfide	1814092	3819327	1334		
70	15-926_1396		•		SP-1a, South Pogo, qtz–ank vein in altered host rock	1814093	3819328	1326		
71	15-949_833.5	•			Northern extension of the Liese Zone, qtz–cal vein in altered host rock	1810795	3822859	963		
72	15-949_988.4		•		Northern extension of the Liese Zone, qtz–ank vein in altered host rock	1810696	3822866	844		
73	15EXP008_1177	•	•		Liese Zone, qtz–cal–ank vein in alterd host rock	1812500	3821393	464		
74	15EXP043_389.7		•		East Deep, qtz–ank vein	1813646	3822714	404		
75	15EXP043_582.9	•			East Deep, qtz–cal vein	1813773	3822786	278		
76	555_stope access_X-Vein		•		X-Vein, North Zone, qtz–ank vein	1812723	3822854	544		

DOS: dolomite–sericite alteration; qtz: quartz; cal: calcite; ank: ankerite; sid: siderite. The brief description and its X-Y-Z coordinate of the sample are listed in the table. Spatial reference: NAD 1983 UTM Zone 6N. The XRD analyses for carbonate identification are qualitatively indicated by a circle in the table. The solid circles indicate that the analysis described in the respective column was performed. No analysis has been done on open circles.

One drill core sample was taken from the Liese (L1) orebody (Figure 3b) for SIMS analysis (DDH97-50, 486.0–485.3 feet). The sample consists of quartz and trace amounts of arsenopyrite, pyrite, and chalcopyrite. Two graphite-containing samples were taken from underground exposures of the Leise and North Zone veins (Table 1).

Photographs of representative carbonate-bearing samples and photomicrographs of their thin sections are shown in Figure 5.

3.2. Analytical Methods

3.2.1. X-Ray Diffraction (XRD) Analysis

Powder method X-ray diffraction (XRD) analysis was performed to determine the relative abundance of carbonate minerals in the collected samples. A Rotaflex (rotating anode) RINT2500 (Rigaku, Tokyo, Japan) instrument at the Geological Survey of Japan (GSJ), National Institute of Advanced Industrial Science and Technology (AIST), and the SmartLab (3 kW) (Rigaku, Tokyo, Japan) apparatus at Shizuoka University were used to conduct the analyses.

Carbonate minerals can be distinguished as calcite, ferroan dolomite–ankerite series carbonates, and siderite using XRD analysis. Since calcite ($CaCO_3$) and siderite ($FeCO_3$) contain one kind of metal cation, the d-spacing of the crystal lattice is constant, and the mineral can be clearly identified by XRD analysis. On the other hand, ferroan dolomite–ankerite series carbonates express the more complicated chemical formula of Ca (Fe^{2+}, Mg, Mn) $(CO_3)_2$. A dolomite with an Fe/Mg molar ratio > 0.25 is classified as ankerite (e.g., [29]). Further identification of dolomite is made possible when XRD analyses are combined with XRF analysis to determine major element compositions.

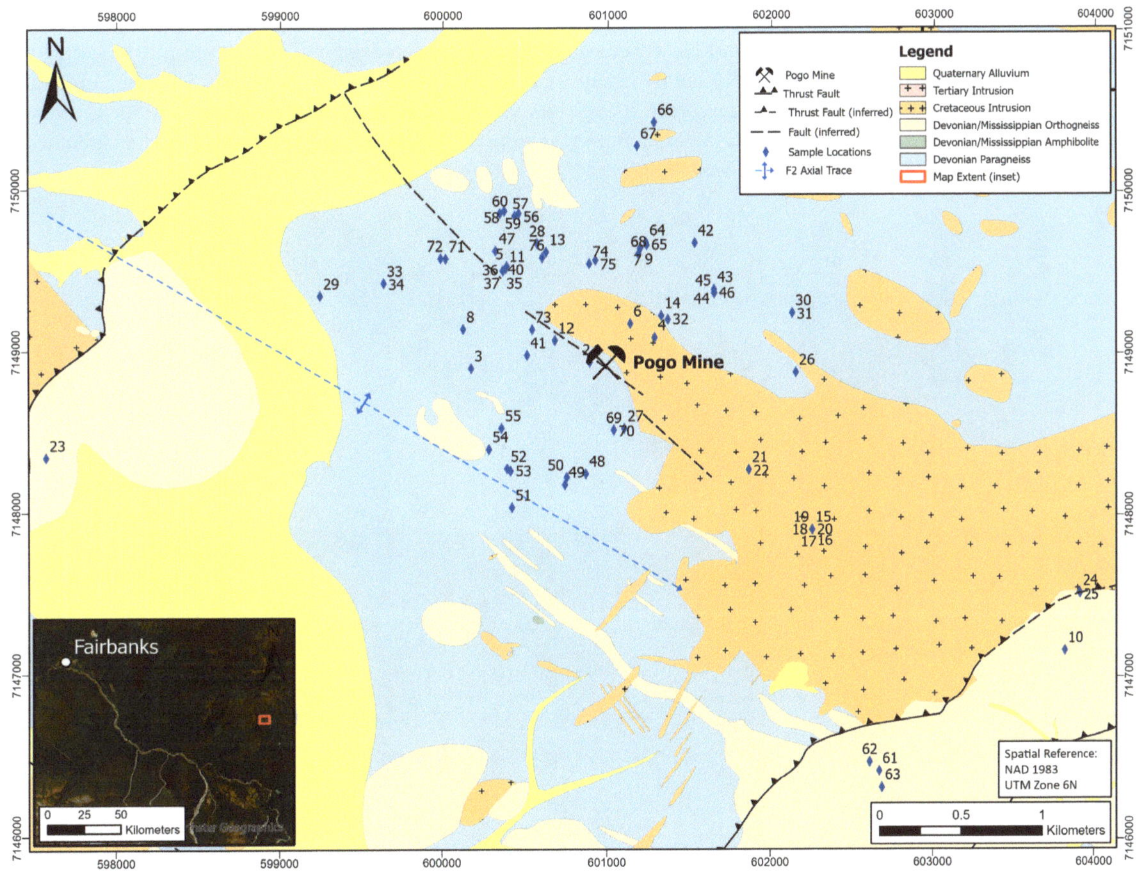

Figure 4. The locality of samples used in this study (Table 1). Note that sample No. 23 (from the Burn exploration area) is located west-southwest of the Pogo deposit across the Goodpaster River.

3.2.2. Chemical Analysis Methods

Several carbonate-bearing rock samples were analyzed for gold by instrumental neutron activation analysis (INAA) and for major element compositions (Ca, Fe, Mg, and Mn) using X-ray fluorescence (XRF) after total fusion, and inductively coupled plasma mass spectrometry (ICP-MS) after near total digestion at Activation Laboratories Ltd. (Actlabs) in Canada. Concentrations of rare earth elements (REE) were obtained by ICP-MS after near total digestion. Samples analyzed for REE are listed in Table 1.

3.2.3. Isotope Ratio Mass Spectrometry (IRMS)

Carbon and oxygen isotopic ratios of carbonate minerals identified by XRD analysis were analyzed using a Finnigan MAT251 mass spectrometer (Thermo Finnigan, Bremen, Germany) at the GSJ, AIST, or using a Finnigan MAT250 mass spectrometer (Thermo Finnigan, Bremen, Germany) at Shizuoka University. At the GSJ (GSJ Method), carbon dioxide was directly liberated from mixtures of calcite and silicate minerals by treatment with 100% phosphoric acid for one hour in an evacuated reaction tube heated to 25 °C in a water bath [30]. Because silicate minerals are inactive with phosphoric acid, we were able to collect CO_2 samples from hydrothermal alteration zones, as well as from veins. Carbonate minerals besides calcite react slowly with phosphoric acid and require higher temperatures for the complete reaction. Ankerite (or ferroan dolomite) is treated at 50 °C for more than 12 h. When calcite and ankerite coexist in the same sample, an initial CO_2 sample is collected by reacting calcite with phosphoric acid at 25 °C for one hour, with a second CO_2 sample recovered by reacting ankerite with phosphoric acid at 50 °C for more than 12 h. CO_2 liberation from siderite is carried out using a silicon oil bath maintained at 100 °C for at least 12 h.

Figure 5. *Cont.*

Figure 5. Photographs of representative carbonate-bearing samples and photomicrographs of their thin sections. (**a**) Calcite-bearing drill core sample 2012081005 from the North Zone, (**A**) hand specimen; (**B**) transmission microscope images of calcite and opaque minerals; (**C**) photomicrograph of pyrite in reflected light. (**b**) Ankerite-bearing drill core sample 11-584C from the East Deep Zone, (**A**) hand specimen; (**B**) ankerite and pyrite in reflected light. (**c**) Siderite-bearing drill core sample 11-584A from the East Deep Zone, (**A**) hand specimen; (**B**) transmission microscope images of siderite and opaque minerals; (**C**) photomicrograph of pyrite (whitish) and chalcopyrite (yellowish) in reflected light.

At Shizuoka University (Shizuoka Method; [31]), 0.15 mg aliquots of the powdered sample are placed into steel thimbles and dropped down one-by-one into phosphoric acid at 60 °C, in an online reaction chamber under vacuum conditions. The evolved gas is cryogenically purified to retain CO_2. The reaction chamber is connected to the inlet system of the mass spectrometer. The ankerite and siderite samples are treated offline at 100 °C for more than 12 h in a silicon oil bath.

Isotope ratios of the evolved CO_2 and machine standard CO_2 were alternately measured and are reported in standard δ notation in per mil (‰) relative to the Vienna Standard Mean Ocean Water (SMOW) for $\delta^{18}O$ and relative to the Vienna Pee Dee Belemnite (PDB) for $\delta^{13}C$. The $\delta^{18}O$ values of calcite, ankerite, and siderite are calculated from the isotope ratios of CO_2 using the oxygen isotope fractionation factors between CO_2 and carbonate minerals (for calcite: [32–34]; for ankerite and siderite: [35]) at the reaction temperatures. Using measurements on a laboratory working standard and the NBS 19, the measured isotope ratios were normalized to the limestone reference material [36]. Reproducibility was better than ±0.1‰ (2σ) for both the $\delta^{13}C$ and $\delta^{18}O$ values of calcite.

The two graphite-containing samples (Table 1) were sent to Actlabs for determination of $\delta^{13}C$ values. XRD analysis confirmed that the rock did not contain any carbon-containing minerals, such as carbonates, other than graphite. The sample was combusted to CO_2, and the carbon isotope ratio of the CO_2 was measured using a mass spectrometer at Actlabs. The reproducibility was ±0.2‰ (2σ).

3.2.4. Secondary Ion Mass Spectrometry (SIMS)

Au and As concentrations were determined on a polished sample from the Leise (L1) ore body using an ims-1270 SIMS (Cameca, France) at the GSJ, AIST. A defocused Cs^+ primary beam was restricted to 15 μm in diameter by a circular aperture to obtain a homogeneous primary beam (Köhler illumination) of about 0.1 nA. A square field aperture is introduced into the secondary ion optics, limiting the analyzed area of the sample surface to a central square measuring 3 × 3 μm to avoid the crater-edge effect.

The $^{197}Au^-$ and $^{75}As^-$ secondary ions, as well as the matrix $^{34}S^-$ as an internal reference, were detected under a total impact energy of 20 kV (a primary accelerating voltage of +10 kV and a secondary extraction voltage of −10 kV) using an electron multiplier (EM) in the analysis of sulfide minerals. A high mass resolving power of 3300 was used to eliminate the isobaric interference of $^{133}Cs^{32}S_2$ on ^{197}Au ($M/\Delta M$ = 1728). The detection limits, which were computed as three times the standard deviation of the background noise, for the Au and As measurements were 15 ppb and 0.6 ppm, respectively, for an ^{34}S count of 3×10^5 cps and assuming that the background noise has a Poisson distribution. The $^{197}Au^-$ and $^{28}Si^-$ secondary ions were measured in the analysis of quartz.

Quantitative SIMS analysis requires the use of matrix-matched external standards for each element because of matrix effects on the secondary ion intensities. Calibration of Au concentrations in pyrite and quartz was conducted using external standard samples of pyrite and quartz. These samples were implanted with ^{197}Au ions at a dosage of 1×10^{15} atoms/cm^2 with an implantation energy of 1.6 MeV by the Mining and Mineral Sciences Laboratories at the Canadian Center for Mineral and Energy Technology (CANMET-MMSL) using the 1.7 MeV Tandetron accelerator located at the interface Science Western laboratory at the University of Western Ontario, Canada. Calibration of Au concentrations in chalcopyrite was conducted using an external chalcopyrite standard that was implanted with ^{197}Au ions at a dosage of 1×10^{14} atoms/cm^2 with an implantation energy of 1.0 MeV at the same laboratory. Calibration of As concentrations in pyrite was performed using an external standard sample implanted with As ions at a dosage of 3×10^{15} atoms/cm^2 with an implantation energy of 0.4 MeV. After analyzing the implanted standard sample for depth profiling, the depth of the sputtered borehole was measured using a surface profiler to determine the relative sensitivity factor (RSF), which is then used to calculate the concentration of the targeted elements. Analytical details, including the RSFs for the Au and As measurements in pyrite and chalcopyrite, are found in the work of Morishita et al. [22,23]. The RSF for Au in quartz is 1.26×10^{21} atoms/cm^3 in our laboratory. The key advantages of SIMS analysis are its high sensitivity and its depth-profiling capability.

4. Results

4.1. XRD Analysis

Qualitative results of the XRD analysis are listed in Table 1, with carbonate minerals classified as calcite, ankerite, or siderite. All of the ferroan dolomite–ankerite series minerals are listed as "ankerite" for the sake of convenience. Note that several samples were found to contain two kinds of carbonate mineral: siderite with ankerite and ankerite with calcite. The coexistence of siderite with calcite was not observed.

4.2. Chemical Analysis

The Ca, Fe, Mg, and Mn concentrations of the analyzed carbonate-bearing rock samples are provided in Table 2. The Au concentration and list of non-carbonate minerals present in each sample are also noted. The results of the REE chemical analysis are shown in Table 3. The locality and description of samples in Tables 2 and 3 are provided in Table 1.

Table 2. The Ca, Fe, Mg, Mn, and Au concentrations in the carbonate-bearing samples.

Sample Name	Ca	Fe	Mg	Mn	Au	Minerals Other than Carbonates
	wt %	wt %	wt %	wt %	ppb	
2014082207	29.60	0.86	0.43	0.43	8	Qtz
M000002a	24.00	1.88	0.83	0.00	35	Qtz
10DL007A	38.13	0.46	0.22	0.28	13	Qtz
10DL008	36.56	1.21	0.17	2.17	<2	Qtz, Rho
11DL025	29.90	0.36	0.18	0.39	7	Qtz
11-584A	0.29	42.21	1.54	2.55	373	Qtz, Chl, Sulfide
11-584B	35.93	0.30	0.16	0.49	425	Qtz
11-584C	18.48	10.23	6.39	0.42	<2	Qtz
11-584D	37.90	0.42	0.17	0.70	<2	Qtz
14-765_589.9	0.40	27.80	1.18	0.75	4670	Qtz, Sulfide
14-765_824.3	0.52	26.20	2.05	2.11	503	Qtz, Sulfide
14-795_1434.4	6.31	5.99	2.66	0.11	503	Qtz, Ser
15-896_2500.4	10.60	17.40	3.43	0.81	154	Qtz, Ser, Sulfide

Qtz: quartz; Rho: rhodochrosite; Chl: chlorite; Ser: sericite. Many samples contain traces of unidentifiable clay minerals.

Table 3. REE and Au concentrations in samples from the Pogo deposit and the Burn deposit (southwest of the Pogo deposit).

Pogo Deposit	La	Ce	Pr	Nd	Sm	Eu	Gd	Tb	Dy	Ho	Er	Tm	Yb	Lu	Au
	ppm	ppm	ppm	ppm	ppm	ppm	ppm	ppm	ppm	ppm	ppm	ppm	ppm	ppm	ppb
2011032201	43.3	79.4	10.2	33.7	5.5	0.91	4.8	0.6	3.3	0.6	1.4	0.2	0.7	0.1	32
2012081003	50.9	101	12.1	41.7	7.6	1.83	7.1	1	5.9	1.1	2.8	0.4	2.3	0.4	167
2012081006	60.3	115	14.2	50.2	9.3	1.84	8.5	1.1	6	1.1	2.7	0.4	1.9	0.3	<2
2012081108	48.2	94.7	11.1	38.2	6.4	1.53	6	0.8	4.7	0.9	2.5	0.4	2	0.3	31
M000002a	14.2	28.3	3.5	14.3	3.1	0.79	3.8	0.6	4	0.9	2.6	0.4	2.1	0.4	35
M000034a	56	110	13.5	47	9	1.95	8.4	1.1	5.9	1	2.2	0.3	1.6	0.2	115
M000068c	7	14.6	1.7	6	1.3	0.75	1.5	0.2	1.8	0.4	1.4	0.3	1.7	0.3	64
11-584A	10.6	20.2	2.3	8.2	1.7	1	2	0.3	2.4	0.5	1.5	0.3	1.7	0.3	373
14-752_1188.6ft	52.3	101	11.9	40.8	6.9	1.21	5.8	0.7	3.5	0.5	1.2	0.1	0.8	0.1	30
14-795_1242.8ft	40.4	78.5	9.3	32.7	5.9	1.35	5.6	0.7	3.9	0.7	1.6	0.2	1	0.1	102
2014081806	48.6	104	13.8	48	8.3	1.49	7.4	0.9	4.9	0.8	1.8	0.2	0.9	0.1	77
Burn															
2014082304	8.8	14.7	1.5	4.6	0.9	1.24	1	0.1	1	0.2	0.7	0.1	1	0.2	>30,000

4.3. IRMS

The $\delta^{13}C$ and $\delta^{18}O$ values of the identified calcite, ankerite, and siderite minerals are listed in Tables 4–6, respectively. The sample locations and descriptions are provided in Table 1. There are 48, 28, and 6 isotope data for calcite, ankerite, and siderite, respectively. The $\delta^{13}C$ values of graphite in samples 2011032201 and 2014081806 (Table 1) are −19.5‰ and −11.0‰, respectively.

Table 4. $\delta^{13}C_{PDB}$ and $\delta^{18}O_{SMOW}$ of calcite.

Sample Name	Description	$\delta^{13}C_{PDB}$	$\delta^{18}O_{SMOW}$
2012080709	Qtz vein with calcite and sulfide	−1.5	−5.3
2012081001	Carbonate vein in weathered diorite (ground)	−1.5	−3.2
2012081004	Thin qtz–cal vein with sulfide	−2.1	−2.8
2012081005	Qtz vein with calcite and sulfide	−0.8	−4.6
2012081006	Calcite dissemination in altered rock	−2.2	−2.6
2014082206	Qtz vein with calcite and sulfide	−0.2	−2.0
2014082207	Qtz–Cal vein (ground)	0.0	−2.3
2014082208	Qtz–Cal vein (ground)	−0.2	−2.1
2014082211	Qtz–Cal vein (ground)	−0.1	−1.8
2014082212	Carbonate veinlet (ground)	−0.3	−0.2
2014082213	Qtz–Cal vein (ground)	0.1	−2.3
2014082215	Qtz–Cal vein (ground)	−0.9	−0.2
2014082216	Qtz–Cal vein with sulfide (ground)	−3.3	2.9
M000002a	Qtz–Cal vein in altered diorite	−2.1	−2.2
M000004a	Qtz–Cal vein in altered diorite	−2.4	−1.5
M000013	Calcite dissemination	−0.9	−2.9
M000015a	Qtz–Cal vein	−1.3	−0.7
M000034a	Qtz–Cal vein	−1.0	−0.3
10DL007A *	Cal vein	−0.7	−5.4
10DL007B *	Qtz–Cal–Ank vein	−0.5	−3.2
10DL008 *	Crystalline calcite in a drusy cavity	1.4	2.8
10DL012A *	Cal vein	−1.1	−0.9
10DL012B *	Cal–Ank vein	−1.3	0.9
11DL011A *	Qtz–Cal–Ank vein with muscovite	−1.8	6.4
11DL012A *	Cal–Ank vein	−1.5	−2.6
11DL012B *	Qtz–Cal–Ank vein	−1.7	−0.4
11DL012C *	Cal–Ank vein	−1.3	−2.8
11DL025 *	Qtz–Cal vein	−1.1	−4.2
11DL028 *	Qtz–Cal–Ank vein	−1.0	−2.7
11-584B *	Qtz–Cal vein	−0.6	−5.0
11-584D *	Cal vein	−0.8	−3.6
11u201A *	Qtz–Cal–Ank vein	−3.4	9.3
14-742_1027.2ft	Qtz–Cal vein	−1.3	−3.2
14-771_876.1ft	Qtz–Cal vein	−1.6	−3.1
14-795_1242.8ft	Qtz–Cal vein	−0.7	1.8
14-805_857.1ft	Qtz–Cal vein	−0.6	−2.8
14-805_1008.9ft	Qtz vein with calcite and sulfide	−1.7	12.5
14-805_1020ft	Qtz–Cal vein	−1.5	6.1
15-882_316.5	Silicified rock with calcite	−2.5	6.9
15-882_833.5	Qtz–Cal vein	−1.3	−1.8
15-884_500.5	Qtz–Cal vein	−0.3	−0.4

Table 4. *Cont.*

Sample Name	Description	$\delta^{13}C_{PDB}$	$\delta^{18}O_{SMOW}$
15-896_2581.3	Qtz–Cal vein	−1.7	−5.6
15-900_156	Qtz–Cal vein	−0.8	−3.3
15-900_1116.8	Qtz–Cal vein	−0.7	0.0
15-926_1388	Qtz–Cal vein with disseminated sulfide	−0.9	−2.4
15-949_833.5	Qtz–Cal vein in altered host rock	−1.3	−1.8
15EXP008_1177	Qtz–Cal–Ank vein in alterd host rock	−1.7	−2.1
15EXP043_582.9	Qtz–Cal vein	−1.3	−4.2

* Analyzed using the GSJ Method. Otherwise, the Shizuoka Method was applied.

Table 5. $\delta^{13}C_{PDB}$ and $\delta^{18}O_{SMOW}$ of ankerite.

Sample Name	Description	$\delta^{13}C_{PDB}$	$\delta^{18}O_{SMOW}$
2011032202	Qtz vein with ankerite and sulfide	0.2	5.6
2011032205	Qtz vein with ankerite and sulfide	0.6	6.1
2012080711	Carbonate dissemination in altered intrusive	−2.9	4.5
2012081003a	12-685, 635′, qtz–ank vein	−1.8	−1.9
2012081011b	Ankerite dissemination in altered intrusive	−3.8	8.8
2012081108a	Qtz vein with ankerite and sulfide	−2.6	7.6
2014082210b	Qtz vein with brown carbonate	0.1	2.7
10DL007B *	Qtz–Cal–Ank vein	0.2	−3.7
10DL012B *	Cal–Ank vein	−0.9	1.9
11DL011A *	Qtz–Cal–Ank vein with muscovite	−1.8	7.3
11DL011B *	Qtz–Ank vein	−1.7	−2.1
11DL011DOS *	Qtz–Ank vein with DOS alteration	−1.5	0.1
11DL012A *	Cal–Ank vein	−0.9	−4.5
11DL012B *	Qtz–Cal–Ank vein	−1.6	−1.3
11DL012C *	Cal–Ank vein	−0.6	−4.9
11DL028 *	Qtz–Cal–Ank vein	−1.2	−3.4
11-584C *	Qtz–Ank vein	−1.3	11.1
11u201A *	Qtz–Cal–Ank vein	−4.0	10.5
14-742_474.6ft	Qtz–Ank vein with sulfide	−3.2	10.9
14-773_875.8ft	Qtz–Ank vein with sulfide	−2.8	10.9
14-795_1434.4ft	Qtz–Ank vein with sericite	−2.3	12.5
15-896_2500.4	Qtz–Ank vein with sericite and sulfide	−2.0	12.7
15-901_2500	Qtz–Ank vein in altered host rock	−2.0	−2.5
15-926_1396	Qtz–Ank vein in altered host rock	−0.1	1.9
15-949_988.4	Qtz–Ank vein in altered host rock	−2.0	1.8
15EXP008_1177	Qtz–Cal–Ank vein in altered host rock	−1.5	−3.7
15EXP043_389.7	Qtz–Ank vein	−1.8	−3.7
555_stope access_X-Vein	Qtz–Ank vein	−2.7	6.7

* Analyzed using the GSJ Method. Otherwise, the Shizuoka Method was applied.

Table 6. $\delta^{13}C_{PDB}$ and $\delta^{18}O_{SMOW}$ of siderite.

Sample Name	Description	$\delta^{13}C_{PDB}$	$\delta^{18}O_{SMOW}$
M000068c	Qtz–Sid vein with sulfide	−4.5	12.6
11-584A *	Qtz–Sid vein with chlorite and sulfide	−4.3	12.0
14-716_278.2ft	Qtz–Sid vein	−4.5	12.7
14-752_1188.6ft	Qtz–Sid vein with muscovite	−4.7	14.0
14-765_589.9ft	Qtz–Sid vein with sulfide	−3.3	14.0
14-765_824.3ft	Qtz–Sid vein with sulfide	−5.2	13.2

* Analyzed using the GSJ Method. Otherwise, the Shizuoka Method was applied.

4.4. SIMS

The results of the Au and As measurements are listed in Table 7. The Au concentrations in pyrite, chalcopyrite, and quartz range from 0.06 to 1.29 ppm, 0.28 to 0.89 ppm, and 0.52 to 9.36 ppm, respectively. The Au concentration profile from each analysis is homogeneous in the depth direction, and Au is considered to be distributed homogeneously. The Au7*282.adp measurement was conducted on a grain boundary between pyrite and chalcopyrite. Although the RSF for Au in arsenopyrite is not available, the Au concentration in arsenopyrite is from 3 to 6 ppm when using the RSF value for Au in pyrite to calibrate the arsenopyrite SIMS data. Arsenopyrite values are shown to one significant digit, given that some errors due to matrix effects are expected. The As concentration in pyrite ranges from 12 to 389 ppm.

Table 7. Au and As measurements by SIMS.

Mineral	Core/Rim	Measurement No.	Locality No.	^{28}Si	^{34}S	As	Au	Au	As
				SIMS	SIMS	SIMS	SIMS	ppm	ppm
Pyrite	core	Au5*262.adp	M262		2.70×10^5	5.83	7.10×10^{-2}	0.06	195
Pyrite	core	Au5*263.adp	M263		2.40×10^5	10.3	2.54×10^{-1}	0.22	389
Chalcopyrite	core	Au5*264.adp	M264		2.70×10^5	9.07	7.73×10^{-1}	0.28	
Chalcopyrite	rim	Au7*281.adp	M281	4.92×10^{-1}	3.39×10^5	0.75	3.08	0.89	
Arsenopyrite	core	Au5*265.adp	M265		8.62×10^4	1.11×10^4	2.64	6	
Pyrite	core	Au6*267.adp	M267		2.47×10^5	8.25×10^{-1}	1.05	0.89	30
Arsenopyrite	core	Au6*271.adp	M271		1.15×10^5	1.78×10^4	2.29	4	
Qtz/Apy	boundary	Au6*272.adp	M272	1.19×10^4	7.03×10^4	1.07×10^4	2.00	16	
Quartz	rim	Au6*273.adp	M273	2.11×10^4	0.53	4.92×10^{-1}	1.87×10^{-1}	1.38	
Quartz	rim	Au6*274.adp	M274	1.63×10^4	1.03×10^3	6.85×10^2	9.82×10^{-1}	9.36	
Arsenopyrite	core	Au7*276.adp	M276	1.59	7.83×10^4	1.03×10^4	9.83×10^{-1}	3	
Quartz	core	Au7*277.adp	M277	5.39×10^4	9.49×10^{-1}	7.31×10^{-1}	1.81×10^{-1}	0.52	
Qtz/Qtz	boundary	Au7*279.adp	M279	9.37×10^4	2.27×10^3	1.36×10^2	1.44	2.38	
Pyrite	core	Au7*280.adp	M280	1.60	2.56×10^5	3.33×10^{-1}	1.57	1.29	12
Py/Cpy	boundary	Au7*282.adp	M282	9.29	3.13×10^5	9.46×10^{-1}	3.05	1.5	27

The measurement Au7*279.adp is on a boundary between two quartz grains. Qtz: quartz; Apy: arsenopyrite; Cpy: chalcopyrite.

5. Discussion

5.1. Chemical Analyses and REE Patterns

The chemical formulae of the carbonate minerals analyzed in this study are shown in Table 8. The presence of quartz does not affect the cation concentrations of the carbonates, but some trace impurities do affect the analytical data (Table 2) and therefore, the chemical formula in Table 8. All of the ferroan dolomites are classified as ankerite, based on their Fe/Mg molar ratios (see Section 3.2.1).

Table 8. Chemical formulas of the analyzed carbonates.

	Chemical Formulas of Carbonates	Carbonate Name
2014082207	$(Ca_{0.95}Fe_{0.02}Mg_{0.02}Mn_{0.01})CO_3$	Calcite
M000002a	$(Ca_{0.90}Fe_{0.05}Mg_{0.05})CO_3$	Calcite
10DL007A	$(Ca_{0.98}Fe_{0.01}Mg_{0.01}Mn_{0.01})CO_3$	Calcite
10DL008	$(Ca_{0.93}Fe_{0.02}Mg_{0.01}Mn_{0.04})CO_3$	Calcite
11DL025	$(Ca_{0.97}Fe_{0.01}Mg_{0.01}Mn_{0.01})CO_3$	Calcite
11-584A	$(Ca_{0.01}Fe_{0.87}Mg_{0.07}Mn_{0.05})CO_3$	Siderite
11-584B	$(Ca_{0.98}Fe_{0.01}Mg_{0.01}Mn_{0.01})CO_3$	Calcite
11-584C	$(Ca_{0.50}Fe_{0.20}Mg_{0.29}Mn_{0.01})CO_3$	Ankerite
11-584D	$(Ca_{0.97}Fe_{0.01}Mg_{0.01}Mn_{0.01})CO_3$	Calcite
14-765_589.9	$(Ca_{0.02}Fe_{0.87}Mg_{0.09}Mn_{0.02})CO_3$	Siderite
14-765_824.3	$(Ca_{0.02}Fe_{0.78}Mg_{0.14}Mn_{0.06})CO_3$	Siderite
14-795_1434.4	$(Ca_{0.42}Fe_{0.29}Mg_{0.28}Mn_{0.01})CO_3$	Ankerite
15-896_2500.4	$(Ca_{0.36}Fe_{0.43}Mg_{0.19}Mn_{0.02})CO_3$	Ankerite

The formula may deviate slightly from that of pure carbonate because. the carbonate-containing whole sample was analyzed (Table 2).

The chondrite-normalized REE patterns of the analyzed Pogo samples are illustrated in Figure 6a. The deviation of the redox-sensitive Eu and Ce elements from a smooth REE pattern can provide important insights into the oxidation state of the hydrothermal mineralizing fluids. The valence of REE elements is usually trivalent, but Eu can take a bivalent state under a reducing environment, whereas Ce can take a tetravalent form in an oxidizing environment. If the hydrothermal fluid is depleted in Eu when bivalent Eu is incorporated into the surrounding rocks before precipitating the carbonate phase, negative Eu anomalies are found in the samples. Negative Eu anomalies are found in the REE patterns of most Pogo samples (Figure 6a), which indicates that Eu was bivalent under a reducing environment at the time of mineralization. The environment is similar to that of the average upper continental crust (Figure 6b). The REE pattern of sample M000002a is different from the others. This sample is from Hill 4021, a peripheral deposit located southeast of the Pogo vein systems (Figure 3a), and the hydrothermal properties may differ from those inferred for the main deposit area. Contrastingly, Figure 7 illustrates REE patterns with positive Eu anomalies. Samples M000068c and 11-584A contain siderite and were collected from the Leise vein system at the center of the deposit (Table 1). Sample 2014082304 was collected from an outcropping mineralized vein in the Burn exploration area, located across the Goodpaster River to the west of the Pogo mining area (Figure 4). These samples with positive Eu anomalies in the REE patterns signify the presence of a high-temperature hydrothermal component [37], and may have been located closer to the center of the magmatic hydrothermal system. Absorption of magmatic vapors, including SO_2, into deeply circulating hydrothermal fluids leads to the formation of highly acidic, oxidizing fluids [38].

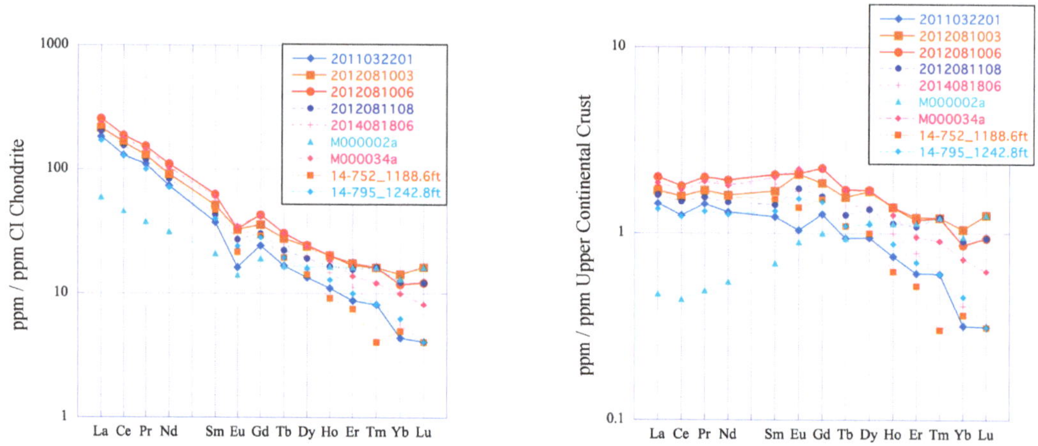

(a) (b)

Figure 6. (**a**) Samples collected from the Pogo deposit with negative Eu anomalies in CI chondrite-normalized REE patterns. (**b**) Upper continental crust-normalized REE patterns for the same samples. The REE concentrations in CI chondrite and the upper continental crust are taken from Refs. [39] and [40], respectively.

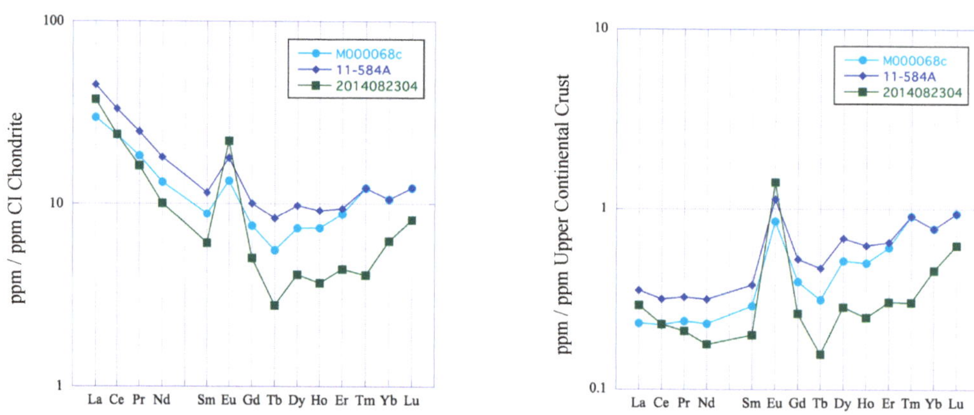

(a) (b)

Figure 7. (**a**) CI chondrite-normalized and (**b**) upper continental crust-normalized REE patterns with positive Eu anomalies of samples from the Pogo deposit and a prospecting area of Burn. The REE concentrations in CI chondrite and the upper continental crust are taken from Refs. [39] and [40], respectively.

5.2. SIMS Analysis

The locations of SIMS microanalyses on a Leise (L1) ore sample are shown on a reflected light photomicrograph in Figure 8. Au concentrations of 1.38 ppm (M273) and 9.36 ppm (M274) measured at a quartz grain margin are higher than those measured in the core of the grain (0.52 ppm, M277). Likewise, an Au concentration of 0.89 ppm measured at the rim of a chalcopyrite grain (M281) is higher than that measured in its core (0.28 ppm,

M264). Au concentrations of 2.38 ppm and 16 ppm were measured at the boundary between two quartz grains (M279) and between quartz and arsenopyrite (M272), respectively. The higher Au concentrations at mineral grain boundaries might be due in part to matrix effects, but Au is generally thought to be concentrated on grain boundaries. The existence of Au located inside the quartz grains implies that the quartz precipitated from a hydrothermal mineralizing fluid.

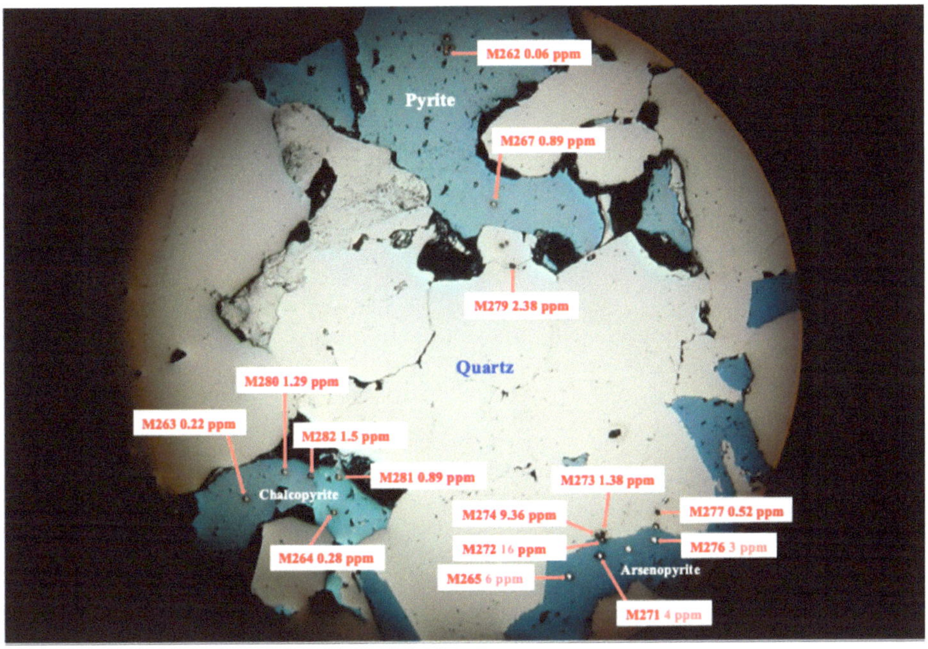

Figure 8. The Au concentrations of the SIMS microanalyses in quartz, pyrite, chalcopyrite, and arsenopyrite (Table 7) are shown on a reflected light photomicrograph.

5.3. Origin of the Ore-Forming Fluid

The Pogo deposit is defined as an intrusive-related gold system formed at a depth of six kilometers [3,4,24]. CO_2 is abundant in fluid inclusions in hydrothermal veins in intrusion-related Au systems [4]. McCoy et al. [2] and Thompson et al. [5] argue that such fluids are of magmatic origin. The homogenization temperatures of fluid inclusions in quartz from the Liese and North Zones of the Pogo deposit range between 290 and 470 °C, with a maximum recorded temperature of 580 °C [41]. Takaoka et al. [41] estimated a fluid $\delta^{18}O$ range of 5.1 to 10.4‰, based on the homogenization temperatures and on $\delta^{18}O$ measurements for vein quartz. The fluid was considered to originate from magmatic water or metamorphic water [41]. The homogenization temperatures of fluid inclusions in Au and sulfide-bearing quartz veins range between 308 and 470 °C, with trapping pressures estimated to be 1.7 to 2 kbar [6,42].

The $\delta^{13}C$ and $\delta^{18}O$ values of the carbonates collected from veins and hydrothermal alteration zones in the Pogo mining area are shown in Figure 9. The distribution of Pogo isotope data is different from that of a typical epithermal gold-bearing vein deposit. For example, the Kushikino deposit was formed from meteoric water ($\delta^{18}O = -7‰$, $\delta^{13}O = -11‰$; [43]), and the hydrothermal fluid of the high-grade Hishikari deposit is almost entirely meteoric in origin, with sporadic injections of magmatic fluid [44]. A fluid of magmatic origin does not indicate that it originates entirely from magma, but rather that it includes fluids in isotopic equilibrium with magma.

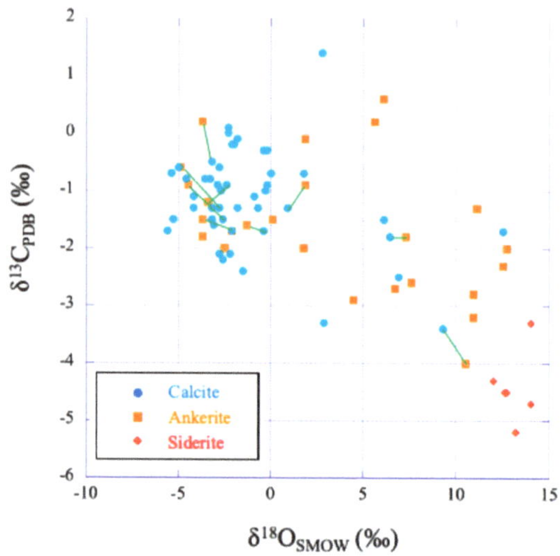

Figure 9. The $\delta^{13}C$ and $\delta^{18}O$ values of calcite, ankerite, and siderite from the veins and hydrothermal alteration zones in the Pogo mining area are shown. The ankerite and calcite connected by a green tie line were separated from the same sample. The isotope ratios of ankerite and calcite in the same sample lie close to each other, indicating that there are no significant differences in their crystallization stages.

All the siderites in this study (Figure 9; Table 6) display high $\delta^{18}O$ and low $\delta^{13}C$ values relative to those of ankerites and calcites, and the $\delta^{13}C$ and $\delta^{18}O$ values fall within a relatively narrow range. The average $\delta^{13}C$ and $\delta^{18}O$ values of five siderites (excluding one outlier) are −4.7‰ and 12.9‰, respectively. We can thus formulate a working hypothesis that the siderite occurred at the earliest stage of the Pogo mineralization, followed by the precipitation of ankerite and calcite, based on their $\delta^{13}C$ and $\delta^{18}O$ values. Assuming that siderite is in equilibrium with a magmatic fluid at 500 °C, the oxygen isotope fractionation factor ($10^3\ln\alpha$) between siderite and water is 2.88‰ [45], and the carbon isotope fractionation factor ($10^3\ln\alpha$) between siderite and the dissolved carbon-bearing species of CO_2 is −0.15‰ [46,47]. The isotope fractionation factors were chosen with an emphasis on comparison between the three carbonate minerals, rather than by using the values from individual papers. While there may be some absolute error in this approach, it allows for a relatively consistent comparison.

The $\delta^{13}C$ and $\delta^{18}O$ values of the early-stage hydrothermal fluid using the isotope fractionation factors and average $\delta^{13}C$ and $\delta^{18}O$ values of the siderites (−4.7‰ and 12.9‰) are calculated as −4.8‰ and 10.0‰, respectively. The estimated $\delta^{18}O$ value of the early-stage hydrothermal fluid is thus near the upper end of the range for primary magmatic water [48,49]. In the ilmenite-series intrusion-related Pogo gold system, the magmatic fluids that interacted with the surrounding rocks are likely to have had high $\delta^{18}O$ values. The $\delta^{13}C$ of magmatic CO_2 varies from −4.2 to −7.5‰ in mid-ocean ridge basalt (MORB) and from −2.8 to −6.7‰ in glasses from Hawaii [50]. The mode of the $\delta^{13}C$ value of the Oka carbonatite lies at −5.0‰ [51]. The estimated $\delta^{13}C$ value of dissolved inorganic carbon (DIC) in the early-stage hydrothermal fluid at Pogo (−4.8‰) is therefore consistent with that of magmatic water.

In summary, the results indicate that a magmatic fluid ($\delta^{18}O$ = 10.0‰, $\delta^{13}C$ = −4.8‰) was the initial ore-forming fluid for Pogo mineralization; however, the other carbonates (ankerite and calcite) exhibit lower $\delta^{18}O$ values (Figure 9). Morishita [52] proposed a fluid evolution model for the 93 Ma Ohtani and Kaneuchi hypothermal tungsten–quartz vein deposits based on oxygen and carbon isotopic evidence. In the ore zones, a relatively high temperature magmatic fluid ($\delta^{18}O$ = 12‰ at 600 °C) produced a greisen in the granodiorite host rock at a depth of 5 km. This alteration phase was later cut by mineralized veins at lower temperatures ($\delta^{18}O$ = 12‰ at 500 °C). The decrease in temperature of the initial magmatic fluid from 600 °C to 500 °C was due to heat conduction into the surrounding rocks. An adiabatic expansion of the fluid and subsequent incorporation of interstitial water with a low $\delta^{18}O$ value can cause a significant decrease in the $\delta^{18}O$ value of the ore-forming fluid [52]. A similar mechanism is invoked to account for the decreasing $\delta^{18}O$ values of the ore-forming fluids at Pogo.

5.4. Fluid Evolution in the Pogo Hydrothermal System

The earliest phase of the Pogo hydrothermal fluid system is assumed to be of magmatic origin. The $\delta^{13}C$ and $\delta^{18}O$ values are calculated as −4.8‰ and 10.0‰, respectively; see Section 5.3. Siderite-bearing samples with positive Eu anomalies in REE patterns (Figure 7) show different REE patterns than do the others (Figure 6), indicating a high-temperature hydrothermal fluid [37].

We assume that the end member of the counterpart fluid for mixing is meteoric water, which at the Pogo deposit has a $\delta^{18}O$ value of −20‰ [53]. This value is consistent with that of meteoric water (−20‰) across southwestern Alaska [8]. The $\delta^{13}C$ value of DIC in meteoric water ($\delta^{13}C$ = 0‰) is taken as the $\delta^{13}C$ value of DIC in sea water, modified locally from equilibration with atmospheric CO_2 in the high latitude ocean (e.g., [54–56]).

Figure 10 shows a mixing line between the magmatic fluid and meteoric water for the ore-forming hydrothermal fluid. The measured $\delta^{13}C$ and $\delta^{18}O$ values of siderite, ankerite, and calcite are superimposed on the simulation results (Figure 10). Most carbonates are in the range of the simulation results, although several ankerites and calcites display higher $\delta^{13}C$ values than those in the simulation range. These anomalies are explained below.

Crustal contamination through assimilation occurs in magmas associated with intrusion-related gold systems [4]. Graphite schist in the Pogo deposit has a $\delta^{13}C$ value of −19.5‰ (Sample 2011032201, Table 1). This ratio can be explained by considering that sedimentary carbon generated by photosynthesis in plants is depleted in the heavier isotope ^{13}C. The carbon source is assumed to be organic matter. Graphitic schist sample 2014081806 has a $\delta^{13}C$ value of −11.0‰. This value is higher than expected for organic carbon (Sample 2011032201, for example). The graphite in this sample might have originated from a partial reaction with hydrothermal carbonate. The CO_2/CH_4 ratios in fluid inclusions from the Liese Zone are interpreted to range between 1 and 49 [41,57]. The methane content in the Pogo fluid was typically only several percentage points, but occasionally increased up to 50%, corroborating a generally reducing environment for Pogo ore formation. The carbon isotope fractionation factors ($10^3 \ln\alpha$) between CO_2 and CH_4 at 500 °C and 400 °C are 15.3‰ and 19.2‰, respectively [58], which are consistent with the research findings of Bottinga [59] and Horita [60]. When CH_4 is produced in the hydrothermal fluid, the $\delta^{13}C$ of CH_4 becomes significantly lower than the average $\delta^{13}C$ of DIC in the fluid. As a result, the $\delta^{13}C$ of CO_2 becomes higher than the average, which is also concluded in Ref. [52]. The carbon isotope fractionation between CO_2 and CH_4 is much larger than that between carbonates and CO_2, so carbonates with $\delta^{13}C$ values higher than those of the simulation range are expected to occur (Figure 10). Therefore, a carbonate precipitated from a fluid containing CH_4 has a higher $\delta^{13}C$ value than that precipitated from a fluid without CH_4. Two ankerite-bearing samples with higher $\delta^{13}C$ values between 0 and 1‰ (Samples 2011032202 and 2011032205, Figure 10) are located near graphite schist, and it is assumed that the CH_4 content of the fluid was relatively high. A large euhedral calcite crystal (10DL008 in Table 1) with the highest $\delta^{13}C$ value (Figure 10) is thought to have slowly crystallized from a low-temperature dilute fluid late in the mineralization history. Large variations in $\delta^{13}C$ values (> 10‰) due to changes in the CO_2/CH_4 ratio have been observed in other intrusion-related mineral systems [52].

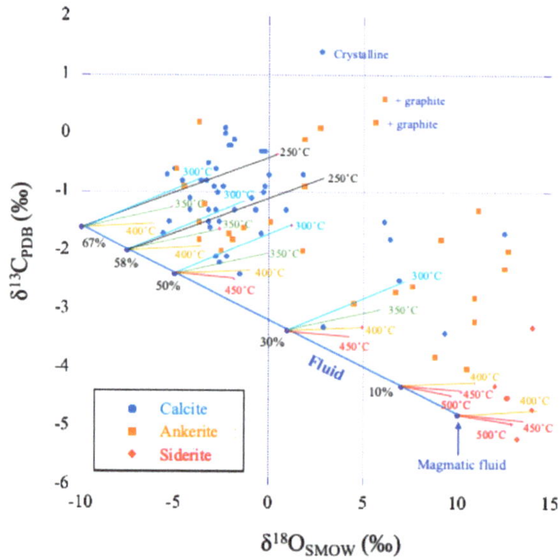

Figure 10. A mixing line (blue in color) between the magmatic fluid and meteoric water for the ore-forming hydrothermal fluid. The calculation of the $\delta^{13}C$ and $\delta^{18}O$ values of carbonate minerals that precipitate at different temperatures and different mixing rate from the mixed fluid are shown. The 10% mark, for example, on the fluid mixing line indicates that 10% meteoric water was mixed into the magmatic fluid. The lower right end of the fluid mixing line corresponds to the initial magmatic fluid ($\delta^{18}O$ = 10‰, $\delta^{13}C$ = −4.8‰. See the Section 5.3). The calculated $\delta^{13}C$ and $\delta^{18}O$ values of siderites that precipitate from the magmatic fluid at 500, 450, and 400 °C are drawn with thin lines. The opposite end of the thin line from the mixing line is the siderite value. The fractionation factors used for the $\delta^{13}C$ and $\delta^{18}O$ calculations of siderite follow those of Mumin et al. [46] and Zheng [45], respectively. Similarly, the $\delta^{13}C$ and $\delta^{18}O$ of ankerites at each temperature are calculated using the methods of Mumin et al. [46] and Zheng [45] for 10%, 30%, 50%, 58%, and 67% mixed fluids. The carbon isotope fractionation factor between ankerite and DIC in fluid is substituted with that between siderite and DIC in fluid. The measured $\delta^{13}C$ and $\delta^{18}O$ values of siderite, ankerite, and calcite are superimposed on them.

The simulation results are consistent with the working hypothesis proposed in Section 5.3. Siderite is first precipitated from a magmatic fluid ($\delta^{18}O$ = 10‰, $\delta^{13}C$ = −4.8‰) at 500 °C. The REE patterns with positive Eu anomalies (Figure 7) confirm the magmatic nature of the early-stage fluid. Ankerite was then precipitated during the gradual incorporation of surrounding fluid mixed with meteoric water-hosted DIC into the magmatic fluid. As a result, the $\delta^{18}O$ and temperature of the fluid gradually decreased, while $\delta^{13}C$ slightly increased. The dominant carbon-bearing species of DIC was CO_2, with CH_4 introduced sporadically. Calcite was precipitated with the further incorporation of surrounding meteoric fluid during the latter stages of mineralization. Pogo is defined as a deep-seated, intrusion-related gold deposit. However, the amount of magmatic fluid input appears unexpectedly small, with most of the Pogo mineralizing fluids interpreted to be of meteoric origin.

6. Concluding Remarks

Carbonate minerals have a close spatial relationship to hydrothermal gold mineralization in the Pogo ore deposit. Most Fe-bearing carbonates (siderite and ankerite) occur in association with hydrothermal alteration and sulfide minerals. Siderite is associated with early-stage gold mineralization and occurs with ankerite in main stage ore assemblages. Calcite is mainly recognized in the later mineralization stages.

The carbon and oxygen isotopic ratios of carbonate minerals (siderite, ankerite, and calcite) present within the deposit illustrate the isotopic evolution of the ore-forming fluid. The initial hydrothermal fluid phase is interpreted to be magmatic in origin. The fluid evolution was characterized by a gradual decrease in $\delta^{18}O$ and a slight increase in $\delta^{13}C$ with decreasing temperature. The general geologic environment of the Pogo deposit is assumed to be reducing at the time of ore formation, based on REE analyses and the presence of graphite. The dominant carbon-bearing species was CO_2, with methane generated sporadically, resulting in the localized crystallization of carbonates with higher $\delta^{13}C$ values.

Invisible gold is found in pyrite, chalcopyrite, arsenopyrite, and quartz, based on SIMS or ion microprobe analysis. The presence of invisible gold in these minerals has significant metallurgical implications for gold processing at the Pogo mine.

Author Contributions: Conceptualization, Y.M.; methodology, Y.M.; resources, Y.M. and J.R.R.; writing—original draft preparation, Y.M.; writing—review and editing, J.R.R.; funding acquisition, Y.M. All authors have read and agreed to the published version of the manuscript.

Funding: This work was supported by JSPS KAKENHI Grant Numbers JP18K03758 and JP22K03736.

Data Availability Statement: Data are contained within the article.

Acknowledgments: The first author is grateful to the Pogo mine, specifically to D. Larimer, K. Puchlik, M. Ishii, E. Fukuda, and N. Ushirone, for facilitating the underground surveys. The authors are indebted to the Pogo Exploration team for helping to draft the site maps and sections. The first author is also indebted to M. Hirano and K. Goto for their help with the XRD analyses. The authors are grateful to Northern Star Resources Limited for permission to publish this paper. Three anonymous reviewers are thanked for their careful and constructive comments, which improved the manuscript.

Conflicts of Interest: The authors declare no conflicts of interest. Jamie R. Rogers is employees of Northern Star Resources Ltd. The paper reflects the views of the scientists and not the company.

References

1. Newberry, R.J.; McCoy, D.T.; Brew, D.A. Plutonic-hosted gold ores in Alaska: Igneous versus metamorphic origins. *Resour. Geol. Spec. Issue* **1995**, *18*, 57–100.
2. McCoy, D.M.; Newberry, R.J.; Layer, P.; Dimarchi, J.J.; Bakke, A.; Masterman, J.S.; Minehane, D.L. Plutonic-related gold deposits of interior Alaska. *Econ. Geol.* **1997**, *9*, 191–241.
3. Lang, J.R.; Baker, T.; Hart, C.J.R.; Mortensen, J.K. An exploration model for Intrusion-Related Gold Systems. *Soc. Econ. Geol.* **2000**, *40*, 1–15. [CrossRef]
4. Lang, J.R.; Baker, T. Intrusion-related gold system: The present level of understanding. *Miner. Depos.* **2001**, *36*, 477–489. [CrossRef]
5. Thompson, J.F.H.; Sillitoe, R.H.; Baker, T.; Lang, J.R.; Mortensen, J.K. Intrusion-related gold deposits associated with tungsten-tin provinces. *Miner. Depos.* **1999**, *34*, 323–334. [CrossRef]
6. Smith, M.T.; Thompson, J.F.H.; Bressler, J.; Layer, P.; Mortensen, J.K.; Abe, I.; Takaoka, H. Geology of the Liese zone, Pogo property, east-central Alaska. *Soc. Econ. Geol.* **1999**, *38*, 11–21. [CrossRef]

7. Rhys, D.; DiMarchi, J.; Smith, M.; Friesen, R.; Rombach, C. Structural setting, style and timing of vein-hosted gold mineralization at the Pogo deposit, east central Alaska. *Miner. Depos.* **2003**, *38*, 863–874. [CrossRef]
8. Goldfarb, R.J.; Ayuso, R.; Miller, M.L.; Ebert, S.W.; Marsh, E.E.; Petsel, S.A.; Miller, L.D.; Bradley, D.; Johnson, C.; McClelland, W. The Late Cretaceous Donlin Creek gold deposit, southwestern Alaska: Controls on epizonal ore formation. *Econ. Geol.* **2004**, *99*, 643–671. [CrossRef]
9. Goldfarb, R.J.; Hart, C.J.R.; Miller, M.L.; Miller, L.D.; Farmer, G.L.; Groves, D. The Tintina gold belt—A global perspective. In *The Tintina Gold Belt: Concepts, Exploration and Discoveries*; Tucker, T., Smith, M., Eds.; British Columbia and Yukon Chamber of Mines: Vancouver, BC, Canada, 2000; Volume 2, pp. 5–34.
10. Hart, C.J.R.; Mccoy, D.T.; Goldfarb, R.J.; Smith, M.; Roberts, P.; Hulstein, R.; Bakke, A.A.; Bundtzen, T.K. Geology, Exploration and Discovery in the Tintina Gold Province, Alaska and Yukon. In *Integrated Methods for Discovery: Global Exploration in the Twenty-First Century*; Society of Economic Geologists: Littleton, CO, USA, 2002; Volume 9, pp. 241–274.
11. Hart, C.J.R. Mid-Cretaceous Magmatic Evolution and Intrusion-related Metallogeny of the Tintina Gold Province, Yukon and Alaska. Unpublished. Ph.D. Thesis, University of Western Australia, Perth, Australia, 2005.
12. Groves, D.I.; Goldfarb, R.J.; Robert, F.; Hart, C.J.R. Gold deposits in metamorphic belts; Overview of current understanding, outstanding problems, future research, and exploration significance. *Econ. Geol.* **2003**, *98*, 1–29.
13. Northern Star Resources Ltd. Annual Report 2024. Available online: https://www.nsrltd.com/media/kmlbwkzn/2-2024-annual-report-double-page-22-08-2024.pdf (accessed on 18 November 2024).
14. Morishita, Y. Calcite as a tracer of ore-forming hydrothermal fluids: Carbon and oxygen isotopic evidence. In *Calcite Formation, Properties and Applications*; Dobrev, J., Markovic, P., Eds.; Nova Science Publishers: New York, NY, USA, 2012; pp. 1–36.
15. Deditius, A.P.; Utsunomiya, S.; Renock, D.; Ewing, R.C.; Ramana, C.V.; Becker, U.; Kesler, S.E. A proposed new type of arsenian pyrite: Composition, nanostructure and geological significance. *Geochim. Cosmochim. Acta* **2008**, *72*, 2919–2933. [CrossRef]
16. Deditius, A.P.; Utsunomiya, S.; Reich, M.; Kesler, S.E.; Ewing, R.C.; Hough, R.; Walshe, J. Trace metal nanoparticles in pyrite. *Ore Geol. Rev.* **2011**, *42*, 32–46. [CrossRef]
17. Cabri, L.J.; McMahon, G. SIMS analysis of sulfide minerals for Pt and Au: Methodology and relative sensitivity factors (RSF). *Can. Mineral.* **1995**, *33*, 349–359.
18. Chryssoulis, S.L.; Weisener, C.G. Secondary ion mass spectrometry relative sensitivity factors for Ru, Rb, Pb, Ag, Os, Ir, Pt and Au in sulfide minerals. In *Secondary Ion Mass Spectrometry: SIMS X*; Benninghoven, A., Hagenhoff, B., Werner, H.W., Eds.; John Wiley & Sons: New York, NY, USA, 1997; pp. 983–986.
19. Larocque, A.C.L.; Cabri, L.J. Ion-microprobe quantification of precious metals in sulfide minerals. *Rev. Econ. Geol.* **1998**, *7*, 155–167.
20. Chouinard, A.; Paquette, J.; Williams-Jones, A.E. Crystallographic controls on trace-element incorporation in auriferous pyrite from the Pascua epithermal high-sulfidation deposit, Chile-Argentina. *Can. Mineral.* **2005**, *43*, 951–963. [CrossRef]
21. Morishita, Y.; Hammond, N.Q. Sub-microscopic gold from the Kalahari Goldridge deposit, Kraaipan Greenstone belt, South Africa. In *Digging Deeper, Proceedings of the ninth Biennial SGA Meeting*; Andrew, C.J., Ed.; Irish Association for Economic Geology: Dublin, Ireland, 2007; Volume 2, pp. 1019–1022.
22. Morishita, Y.; Shimada, N.; Shimada, K. Invisible gold in arsenian pyrite from the high-grade Hishikari gold deposit, Japan: Significance of variation and distribution of Au/As ratios in pyrite. *Ore Geol. Rev.* **2018**, *95*, 79–93. [CrossRef]
23. Morishita, Y.; Hammond, N.Q.; Momii, K.; Konagaya, R.; Sano, Y.; Takahata, N.; Ueno, H. Invisible gold in pyrite from epithermal, banded-iron-formation-hosted, and sedimentary gold deposits: Evidence of hydrothermal influence. *Minerals* **2019**, *9*, 447. [CrossRef]
24. Baker, T.; Pollard, P.J.; Mustard, R.; Mark, G.; Graham, J.L. A comparison of granite-related tin, tungsten, and gold-bismuth deposits: Implications for exploration. *Soc. Econ. Geol.* **2005**, *61*, 5–17. [CrossRef]
25. Larimer, D.; Uesugi, J.; Puchlik, K.; Fukuda, E. Discovery of Deep Ore Body of the pogo Deposit. *Resour. Geol.* **2013**, *63*, 91–104.
26. Selby, D.; Creaser, R.A.; Hart, C.J.R.; Rombach, C.S.; Thompson, J.F.H.; Smith, M.T.; Bakke, A.A.; Goldfarb, R.J. Absolute timing of sulphide and gold mineralization: A comparison of Re-Os molybdenite and Ar-Ar mica methods from the Tintina gold belt, Alaska. *Geology* **2002**, *30*, 791–794. [CrossRef]
27. Chappell, B.W.; White, A.J.R. Two constrasting granite types. *Pac. Geol.* **1974**, *8*, 173–174.
28. Ishihara, S. The magnetite-series and ilmenite-series granitic rocks. *Min. Geol.* **1977**, *27*, 293–305.
29. Deer, W.A.; Howie, R.A.; Zussman, J. *An Introduction to the Rock-Forming Minerals*, 2nd ed.; Longman Group UK Limited: Essex, UK, 1992; 696p.
30. McCrea, J.M. On the isotopic chemistry of carbonates and a paleotemperature scale. *J. Chem. Phys.* **1950**, *18*, 49–857. [CrossRef]
31. Wada, H.; Niitsuma, N.; Saito, T. Carbon and oxygen isotopic measurements of ultra-small samples. *Geosci. Rep. Shizuoka Univ.* **1982**, *7*, 35–50. (In Japanese with English Abstract)

32. Friedman, I.; O'Neil, R. *Compilation of Stable Isotope Fractionation Factors of Geochemical Interest*; USGS Professional Paper 440-KK; USGS: Reston, VA, USA, 1977.
33. Coplen, T.B.; Kendall, C.; Hopple, J. Comparison of stable isotope reference samples. *Nature* **1983**, *302*, 236–238. [CrossRef]
34. Swart, P.K.; Burns, S.J.; Lede, J.J. Fractionation of the stable isotopes of oxygen and carbon in carbon dioxide during the reaction of calcite with phosphoric acid as a function of temperature and technique. *Chem. Geol.* **1991**, *86*, 89–96. [CrossRef]
35. Rosenbaum, J.; Sheppard, S.M.F. An isotopic study of siderites, dolomites and ankerites at high temperatures. *Geochim. Cosmochim. Acta* **1986**, *50*, 1147–1150. [CrossRef]
36. Morishita, Y.; Matsuhisa, Y. Measurement of carbon and oxygen isotope ratios of carbonate reference samples. *Bull. Geol. Surv. Jpn.* **1984**, *35*, 69–79. (In Japanese with English Abstract)
37. Bau, M.; Dulski, P. Distribution of yttrium and rare-earth elements in the Penge and Kuruman iron formations, Transvaal Supergroup, South Africa. *Precambrian Res.* **1996**, *79*, 37–55. [CrossRef]
38. Giggenbach, W.F.; Garcia, N.; Londoño, A.; Rodriguez, L.; Rojas, N.; Calvache, M. The chemistry of fumarolic vapor and termal-spring discharges from the Nevado del Ruiz volcanic-magmatic-hydrothermal system, Colombia. *J. Volcanol. Geother. Res.* **1990**, *42*, 13–29. [CrossRef]
39. McDonough, W.F.; Sun, S.S. The composition of the Earth. *Chem. Geol.* **1995**, *120*, 223–253. [CrossRef]
40. McLennan, S.M. Relationships between the trace element composition of sedimentary rocks and upper continental crust. *Geochem. Geophys. Geosyst.* **2001**, *2*, 2000GC000109. [CrossRef]
41. Takaoka, H.; Abe, I.; Murakami, T. Discovery of gold deposits at Pogo in the Stone Boy area of Alaska. *Resour. Geol.* **1999**, *49*, 125–145. (In Japanese with English Abstract)
42. Rombach, C.S.; Newberry, R.J.; Goldfarb, R.J.; Smith, M. Geochronology and mineralization of the Liese zones, Pogo deposit, Alaska. *Geol. Soc. Am. Abstr.* **2002**, *34*, A114.
43. Morishita, Y.; Nakano, T. Role of basement in epithermal deposits: The Kushikino and Hishikari gold deposits, southwestern Japan. *Ore Geol. Rev.* **2008**, *34*, 597–609. [CrossRef]
44. Morishita, Y.; Yabe, Y. Genesis and evolution of hydrothermal fluids in the formation of the high-grade Hishikari gold deposit: Carbon, oxygen, and sulfur isotopic evidence. *Minerals* **2022**, *12*, 1595. [CrossRef]
45. Zheng, Y.-F. On the theoretical calculations of oxygen isotope fractionation factors for carbonate-water systems. *Geochem. J.* **2011**, *45*, 341–354. [CrossRef]
46. Mumin, A.H.; Fleet, M.E.; Longstaffe, F.J. Evolution of hydrothermal fluids in the Ashanti gold belt, Ghana: Stable isotope geochemistry of carbonates, graphite, and quartz. *Econ. Geol.* **1996**, *91*, 135–148. [CrossRef]
47. Golyshev, S.I.; Padalko, N.L.; Pechenkin, S.A. Fractionation of stable oxygen and carbon isotopes in carbonate systems. *Geochem. Int.* **1981**, *18*, 85–99.
48. Sheppard, S.M.F. Characterization and isotope variations in naturel waters. *Rev. Mineral.* **1986**, *16*, 165–183.
49. Taylor, H.P., Jr. Oxygen and hydrogen isotope relationships in hydrothermal mineral deposits. In *Geochemistry of Hydrothermal Ore Deposits*, 3rd ed.; Barnes, H.L., Ed.; John Wiley & Sons, Inc.: New York, NY, USA, 1997; pp. 229–302.
50. Mattey, D.P.; Carr, R.H.; Wright, I.P.; Pillinger, C.T. Carbon isotopes in submarine basalts. *Earth Planet. Sci. Lett.* **1984**, *70*, 196–206.
51. Deines, P. The carbon and oxygen isotopic composition of carbonates from the Oka carbonatite complex, Quebec, Canada. *Geochim. Cosmochim. Acta* **1970**, *34*, 1199–1225. [CrossRef]
52. Morishita, Y. Fluid evolution and geobarometry on the Ohtani and Kaneuchi tungsten-quartz vein deposits, Japan: Oxygen and carbon isotopic evidence. *Mineral. Depos.* **1991**, *26*, 40–50. [CrossRef]
53. Teck Cominco. *Post-Mining Groundwater Chemistry, Pogo Mine, Alaska*; Water Management Plan Appendix, K; Teck Cominco: Vancouver, BC, Canada, 2002.
54. Kroopnick, P. Correlations between ^{13}C and SCO_2 in surface waters and atmospheric CO_2. *Earth Planet. Sci. Lett.* **1974**, *22*, 397–403. [CrossRef]
55. Kroopnick, P. The distribution of ^{13}C of SCO_2 in the world oceans. *Deep Sea Res. Part A Oceanogr. Res. Pap.* **1985**, *32*, 57–84. [CrossRef]
56. Galbraith, E.D.; Kwon, E.Y.; Bianchi, D.; Hain, M.P.; Sarmiento, J.L. The impact of atmospheric pCO_2 on carbon isotope ratios of the atmosphere and ocean. *Glob. Biogeochem. Cycles AGU* **2015**, *29*, 307–324. [CrossRef]
57. McCoy, D.M.; Olson, I. *Fluid Inclusions and Mineral Equilibria: Implications for P-T-X Conditions and Deposit Type*; Stone Boy Project Annual Report, Appendix 1.1; WGM Inc.: Missoula, MT, USA, 1997.
58. Kuetera, N.; Schmidta, M.W.; Lilley, M.D.; Bernasconi, S.M. Experimental determination of equilibrium CH_4–CO_2-CO carbon isotope fractionation factors (300–1200 °C). *Earth Planet. Sci. Lett.* **2019**, *506*, 64–75.

59. Bottinga, Y. Calculated fractionation factors for carbon and hydrogen isotope exchange in the system calcite–carbon dioxide–graphite–methane–hydrogen–water vapor. *Geochim. Cosmochim. Acta* **1969**, *33*, 49–64. [CrossRef]
60. Horita, J. Carbon isotope exchange in the system CO_2–CH_4 at elevated temperatures. *Geochim. Cosmochim. Acta* **2001**, *65*, 1907–1919. [CrossRef]

minerals

MDPI

Article

A Genetic Model for the Biggenden Gold-Bearing Fe Skarn Deposit, Queensland, Australia: Geology, Mineralogy, Isotope Geochemistry, and Fluid Inclusion Studies

Mansour Edraki [1], Alireza K. Somarin [2,*] and Paul M. Ashley [3]

1. Sustainable Minerals Institute, The University of Queensland, St Lucia, QLD 4072, Australia; m.edraki@uq.edu.au
2. Department of Geology, Brandon University, Brandon, MB R7A 6A9, Canada
3. Earth Sciences, University of New England, Armidale, NSW 2351, Australia; papags47@gmail.com

* Correspondence: somarina@brandonu.ca

Abstract: The Biggenden gold-bearing Fe skarn deposit in southeast Queensland, Australia, is a calcic magnetite skarn that has been mined for Fe and gold (from the upper portion of the deposit). Skarn has replaced volcanic and sedimentary rocks of the Early Permian Gympie Group, which formed in different tectonic settings, including island arc, back arc, and mid-ocean ridge. This group has experienced a hornblende-hornfels grade of contact metamorphism due to the intrusion of the Late Triassic Degilbo Granite. The intrusion is a mildly oxidized I-type monzogranite that has geochemical characteristics intermediate between those of granitoids typically associated with Fe-Cu-Au and Sn-W-Mo skarn deposits. The skarn mineralogy indicates that there was an evolution from prograde to various retrograde assemblages. Prograde garnet ($Adr_{11-99}Grs_{1-78}Alm_{0-8}Sps_{0-11}$), clinopyroxene ($Di_{30-92}Hd_{7-65}Jo_{0-9}$), magnetite, and scapolite formed initially. Epidote and Cl-bearing amphibole (mainly ferropargasite) were the early retrograde minerals, followed by chlorite, calcite, actinolite, quartz, and sulfides. Late-stage retrograde reactions are indicated by the development of nontronite, calcite, and quartz. Gold is mainly associated with sulfide minerals in the retrograde sulfide stage. The fluids in equilibrium with the ore-stage calcites had $\delta^{13}C$ and $\delta^{18}O$ values that indicate deposition from magmatically derived fluids. The calculated $\delta^{18}O$ values of the fluids in equilibrium with the skarn magnetite also suggest a magmatic origin. However, the fluids in equilibrium with epidote were a mixture of magmatic and meteoric water, and the fluids that deposited chlorite were at least partly meteoric. δD values for the retrograde amphibole and epidote fall within the common range for magmatic water. Late-stage chlorite was deposited from metasomatic fluids depleted in deuterium (D), implying a meteoric water origin. Sulfur isotopic compositions of the Biggenden sulfides are similar to other skarn deposits worldwide and indicate that sulfur was most probably derived from a magmatic source. Based on the strontium ($^{87}Sr/^{86}Sr$) and lead ($^{206}Pb/^{204}Pb$, $^{207}Pb/^{204}Pb$ and $^{208}Pb/^{204}Pb$) isotope ratios, the volcanic and sedimentary rocks of the Gympie Group may have contributed part of the metals to the hydrothermal fluids. Lead isotope data are also consistent with a close age relationship between the mineralization at Biggenden and the crystallization of the Degilbo Granite. Microthermometric analysis indicates that there is an overall decrease in fluid temperature and salinity from the prograde skarn to retrograde alterations. Fluid inclusions in prograde skarn calcite and garnet yield homogenization temperatures of 500 to 600 °C and have salinities up to 45 equivalent wt % NaCl. Fluid inclusions in quartz and calcite from the retrograde sulfide-stage homogenized between 280 and 360 °C and have lower salinities (5–15 equivalent wt % NaCl). In a favored genetic model, hydrothermal fluids originated from the Degilbo Granite at depth and migrated through the shear zone,

Academic Editors: Yuichi Morishita and Napoleon Q. Hammond

Received: 23 December 2024
Revised: 16 January 2025
Accepted: 17 January 2025
Published: 20 January 2025

Citation: Edraki, M.; Somarin, A.K.; Ashley, P.M. A Genetic Model for the Biggenden Gold-Bearing Fe Skarn Deposit, Queensland, Australia: Geology, Mineralogy, Isotope Geochemistry, and Fluid Inclusion Studies. *Minerals* **2025**, *15*, 95. https://doi.org/10.3390/min15010095

intrusive contact, and permeable Gympie Group rocks and leached extra Fe and Ca and deposited magnetite upon reaction with the adjacent marble and basalt.

Keywords: Biggenden mine; Gympie Group; gold-bearing skarn; fluid inclusions; Degilbo Granite

1. Introduction

The Biggenden mine in southeast Queensland, Australia (Figure 1) can be classified as an iron skarn, however, it has geochemical and mineralogical characteristics that overlap with worldwide gold and copper skarn deposits (e.g., [1–6]). There are also differences between the Biggenden deposit and other skarn deposits in terms of hydrothermal alteration and geochemical features of the adjacent pluton, Degilbo Granite.

The Biggenden mine was opened in 1888, and it was first operated for Cu, Au, and Bi. Quantities of commodities extracted from the mine are not accurately known, but approximate values provide some indication of the changing focus of mining. Early production of approximately 6000 ounces of gold (1888–1901) was followed by 200 tons of bismuth (1901–1938), with Commercial Minerals commencing open pit mining of magnetite in 1967. An estimated 580,000 tons of magnetite were produced for the Australian coal industry by 1990, during a transition into underground mining. The total measured recoverable reserve at Biggenden was estimated to be 107,000 tons of wet product, from 323,000 tons of ore with an average head grade of 30.7% magnetite [7]. The mine was closed in 1999 due to the exhaustion of economic resources.

The Biggenden skarn has been mentioned in review articles of world skarns [8,9] but has never been described in detail. Despite previous works (e.g., [10–12]), its genesis has long remained an enigma. The spectacular skarn exposures at the Biggenden mine resulting from three decades of open pit and underground mining offer an opportunity to study this deposit more closely and to apply new techniques to investigate the genesis of this deposit.

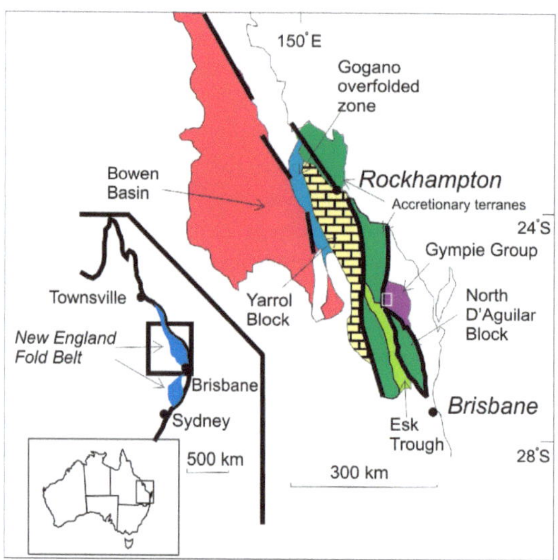

Figure 1. Major tectonic terranes of the Northern New England Orogen and the location of the Biggenden mine in the Gympie Group (also called Gympie Terrane [13]). Location of Figure 2 is shown as white rectangle. Modified after [14].

2. Regional Geological Setting

The Biggenden mine is situated in the Gympie Group, which is part of the New England Orogen, eastern Australia (Figure 1). Various origins have been suggested for the Gympie Group, ranging from an accreted arc to an exotic displaced terrane derived perhaps from New Zealand or New Caledonia [15–17] or, alternatively, that it represents a foreland basin originating not far from the Australian continental crust [13,18]. The mine area is underlain by the Early Permian rocks of the Gympie Group in the east and the Late Triassic Degilbo Granite in the west (Figure 2). The Gympie Group sequence is steeply dipping and locally faulted, sheared, and folded, and affected by the lowermost greenschist facies metamorphism. To the south, the Triassic Mount Marcella Volcanics overlie the Gympie Group, and in some places, Pleistocene Barambah Basalt overlies all the units, especially [15]. The host rocks for mineralization at the Biggenden mine area comprise intercalated basaltic and andesitic flows, tuffaceous and fragmental rocks of sedimentary/volcanic provenance, siltstone, and limestone (Figure 3). The basalts are partly vesicular with local pillow structures. Dolerite dikes also occur to the northeast of the Biggenden mine. Geochemical studies of the volcanic rocks of the Gympie Group suggest that these rocks formed in different tectonic settings, including island arc, back arc, and mid-ocean ridge settings [17,19,20], and were tectonically juxtaposed in the Triassic.

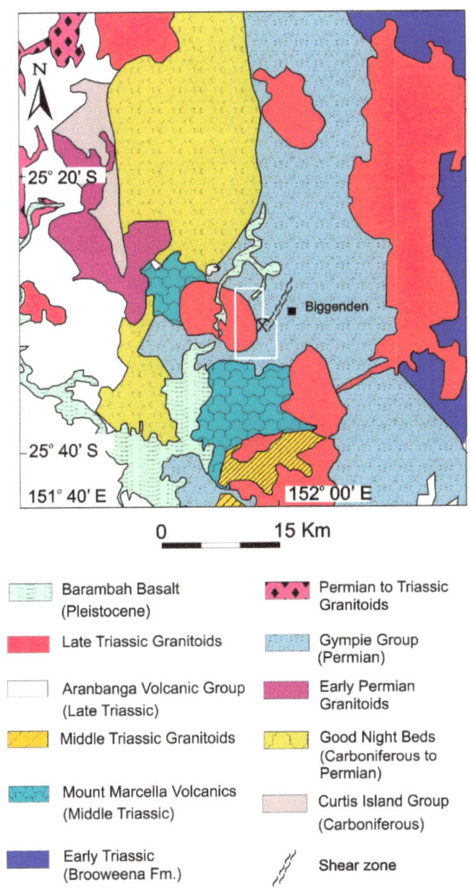

Figure 2. Regional geological map of the Biggenden mine area. Location of Figure 3 is shown as white rectangle. Geology modified after [15].

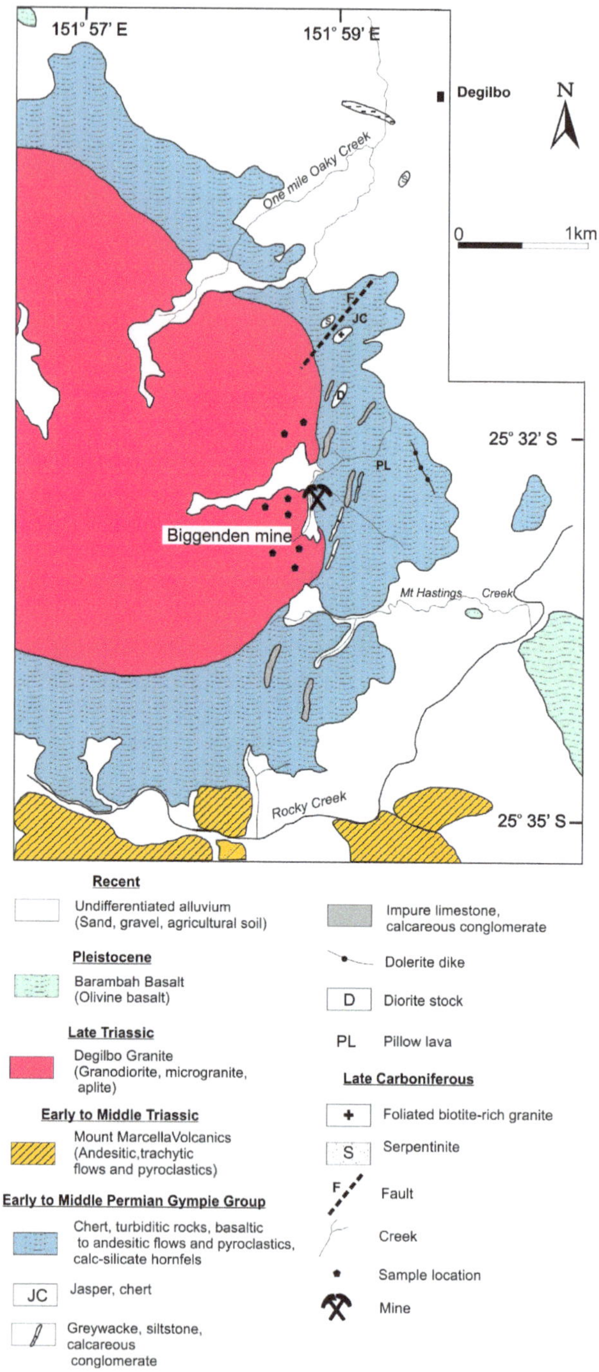

Figure 3. Geological map of the open pit area at the Biggenden mine (after [15]).

The Degilbo Granite is a massive, biotite-hornblende monzogranite of I-type affinity, and except for evidence of minor actinolite endoskarn vein development at the Biggenden mine, it is generally unaltered. It has intruded and is in most parts surrounded by the

Gympie Group, except in the western contact, where it has intruded the andesitic flows and pyroclastics of the Mount Marcella Volcanics. The age of the pluton has been reported to be 220 Ma (a K-Ar biotite age [21]) and 216 ± 6 Ma (a K-Ar age on contact metamorphic hornblende [15]).

Other prospects and abandoned mines in the area are aligned roughly north-south with the Biggenden deposit and are mainly concentrated along the eastern margin of the Degilbo Granite (Figure 3). To the north of the area, in the vicinity of the Commonwealth mine (Figure 3), there are outcrops of small masses of serpentinite and foliated biotite granite that are associated with an NE-trending fault. This biotite granite is distinguished from the Degilbo Granite based on its mineralogy, texture, and deformation. The biotite granite is interpreted to be older and part of a fault melange [19].

Mine Geology

The Gympie Group in the mine area (~2 km^2) consists of five primary rock units:

1. Lava Flows: These amygdaloidal basalts and andesites occur in the eastern part of the open pit and display coherent volcanic flow textures. They feature intergranular, subophitic, or porphyritic textures, with albitized plagioclase laths and augite in a groundmass of altered plagioclase, ferromagnesian minerals, and accessory titanomagnetite and ilmenite. The amygdales, up to 1 cm, are filled mainly with epidote and chlorite.
2. Volcaniclastic Rocks: These rocks are closely associated with the lava flows but differ in their dominant fragmental texture. The outcrops in the open pit transition from a gradational contact with the volcanic flows to a gradational contact with polymictic rocks.
3. Polymictic Rocks: This unit, found between the granite contact and the volcaniclastic unit, includes a tuffaceous lithic wacke and breccia. These rocks are black to greenish-black, due to the presence of metamorphic hornblende, clinopyroxene, and biotite. The key difference from the volcaniclastic unit is the inclusion of non-volcanic clasts such as chert, siltstone, and limestone. Contact metamorphism has replaced some of these fragments with calc-silicate minerals.
4. Limestone: The mine area features several large and small marble bodies, with a granoblastic polygonal texture. Over 98% of the rock is calcite, with disseminated garnet and other common calc-silicate minerals.
5. Metasiltstone: This unit is an orange-brown, intensely jointed horizon in the open pit area. It has a granoblastic texture and locally contains up to 50% metamorphic biotite near the contact zone.

The Degilbo Granite is medium to coarse-grained with a hypidiomorphic granular texture and is locally porphyritic, containing phenocrysts of pink perthitic K-feldspar and white plagioclase. There is no significant mineralogical variation within the granite, and no notable endoskarn has developed. A few aplite and microgranite porphyry dikes cut through the granite and surrounding rocks, suggesting late-stage offshoots of the main granite body. The pink color of the K-feldspar and characteristic minerals, including hornblende, titanite, allanite, and magnetite, are typical of I-type granites [22,23].

The orebody at Biggenden formed through metasomatic replacement (exoskarn) of various protoliths in the Gympie Group. The orebody is 360 m long, up to 35 m wide, and extends to a depth of approximately 135 m. Magnetite occurs in large, irregular masses, lenses, veins, and disseminated grains. The strata strike N-S to N10° E and dip steeply west, possibly indicating subvertical thrust stacks [15]. A near-vertical shear zone occurs close to the granite contact, extending up to 30 m away. Stronger mineralization is found near shear zones. Contact metamorphic minerals overprint the shear fabric, with mineralization observed in pressure shadow tails and boudinage structures. Aeromagnetic data [24] reveal a northeast-striking linear structure, approximately 5 km long and 200 m wide, which includes the shear zone at the mine. Previous studies by Cranfield (1994) [15] also report a regional shear zone extending northeast of the Biggenden mine.

The prominent hanging wall fault, striking parallel to the western margin of the magnetite orebody (Figure 4A), is barren and filled with clay gouge. The most recent movement on this fault post-dates mineralization and is likely post-granite emplacement, as the gouge is not metamorphosed. Broad zones of brecciation, up to several meters wide, occur in the mine, particularly in association with marble (Figure 4B). Breccia fragments, up to 0.5 m in size, are composed of hornfels, mainly formed from polymictic rocks, which are replaced at the margins by garnet and magnetite. No significant structural displacements of the orebody have been observed that could explain its proximity to the unaltered granite (Figure 5).

Figure 4. Photographs showing field features at the Biggenden mine. (**A**) Part of the open pit shows outcrops of dark-brown-gray lenticular magnetite bodies set in a cement of coarse-grained hydrothermal calcite. Lenses of marble occur behind and to the right of the person in the photograph. The upper part of the photograph shows outcrops of hornfels. (**B**) Brecciation showing meter-size fragments of breccia set in a cement of coarse calcite. Fragments comprise magnetite, skarn silicates (predominantly garnet), and remnants of the hornfels protoliths. (**C**) Garnet assemblage formed as reaction skarn at the contact of marble and volcaniclastic rocks. (**D**) Prograde veining and replacement of the volcaniclastic rocks at the open pit by garnet skarn are followed by the deposition of hornblende at a later retrograde stage.

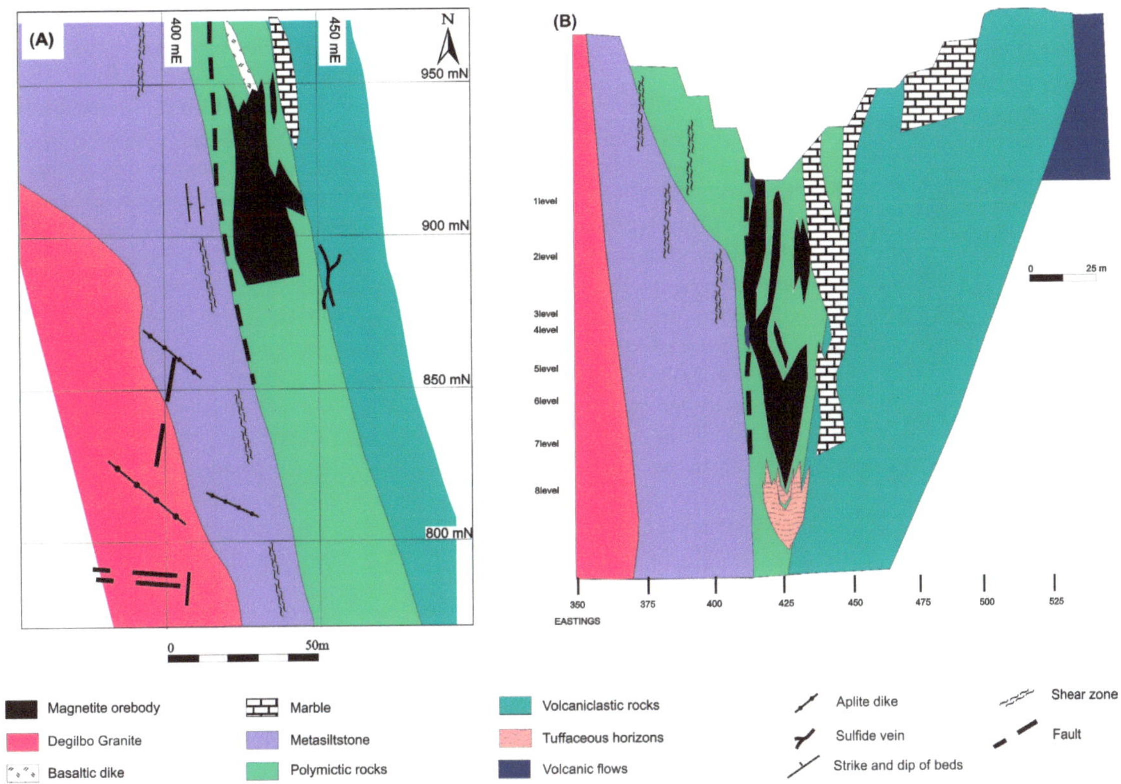

Figure 5. Geology of level 5 at the Biggenden mine (**A**) and cross-section along line 950 mN (**B**) (after [7]; Edraki, 2000 [15]).

3. Materials and Methods

The geochemistry of the intrusive rocks has been investigated in 19 samples, including Degilbo Granite, microgranite, and aplite dikes. Samples were analyzed by XRF at the University of New England (UNE) for major and trace elements. Microanalysis of metamorphic and skarn minerals was carried out at University of New England, Armidale, Australia using a JEOL JSM 5800LV scanning electron microscope equipped with energy dispersive spectrometer (EDS). The acceleration potential, beam current, and analysis time were 20 kv, 20 μA, and 120 s, respectively.

Fluid inclusion samples were mainly collected from the ore and chlorite assemblages. Thermometric analyses were carried out on an SGE heating/freezing stage. Vapor-liquid homogenization temperature (T_h), last ice melting (T_m), and the disappearance temperature of salt (T_d) were recorded. Heating was carried out twice for each fluid inclusion, and an average was recorded as the homogenization temperature. The precision of the temperature measurements was ± 5 °C. The coexistence of vapor-rich and liquid-rich fluid inclusions in the same analyzed grains is considered to be representative of the boiling of hydrothermal fluid [25], and therefore pressure correction was not applied to the acquired homogenization temperatures.

All mineral separations and sample preparations for isotope analyses of carbonates and sulfides were performed at UNE. Sample preparations for silicates and all stable isotope analyses were carried out at the Centre for Isotope Studies at CSIRO, North Ryde, Australia. For oxygen isotope analysis, oxygen in the silicates and magnetite was extracted by the

BrF_5 technique according to the method of Clayton and Mayeda (1963) [26]. For hydrogen isotope analyses, water was extracted from the hydrous silicates according to the method of Friedman (1953) [27].

$\delta^{34}S$ values of the sulfides were determined on SO_2 gas, which was extracted from sulfide minerals by combustion with Cu_2O powder at 1100 °C, according to the method of Robinson and Kusakabe (1975) [28]. Carbonates were prepared following the method of McCrea (1950) [29]. The values are reported here in the usual delta notation (δ) in per mil (‰) with $\delta^{13}C$ relative to V-PDB and $\delta^{18}O$ and δD relative to the V-SMOW standard.

For Sr isotope analysis, pulverized granite and other silicate rock samples were dissolved overnight in a Teflon bomb using HF, HNO_3, and $HClO_4$ at 180 °C until digestion was complete. Carbonates were dissolved in CH_3COOH and HCl. Separation of Sr was conducted using standard cation exchange chromatography. The dissolved samples were centrifuged in 1 mL of 6 M HCl and then loaded on columns of ion-exchange resin. The separated Sr was loaded on W filaments and analyzed by thermal ionization mass spectrometry.

An accelerator mass spectrometer system called AUSTRALIS [30] was used for lead isotope analysis at CSIRO. The system enables in situ microanalysis of samples, and Pb isotope measurements were performed using a 30-micron microbeam in polished thin sections containing galena.

4. Results and Analysis

4.1. Mineralogy and Mineral Compositions

Major minerals in the Degilbo Granite include quartz, plagioclase, K-feldspar, hornblende, and biotite with accessory apatite, zircon, titanite, allanite, and magnetite. The intrusion of this granite has produced a hornblende hornfels grade of contact metamorphism in the rock units of the Gympie Group, extending up to 1.3 km from the intrusive contact. The lowest grade of contact metamorphism is represented by chlorite, epidote, albite, pale-green actinolite, quartz, and calcite in the mafic rocks consistent with the albite-epidote hornfels facies. As the pluton is approached, at approximately 700 m from the contact, green hornblende appears in the basalts and dolerites, and actinolite is partially replaced by hornblende. The highest-grade contact metamorphic rocks occur in the inner part of the aureole at a distance of up to 150 m from the granite contact. Hornblende generally increases in abundance, and biotite appears in association with hornblende. At the mine, garnet ± clinopyroxene-amphibole bands (reaction skarn) have formed at the contact of hornfels with blocks of marble (Figure 4C).

Skarn mineralogy indicates that there was an evolution from prograde through to various retrograde assemblages (Figure 6), as hydrous retrograde minerals overprinted anhydrous prograde assemblages (Figure 4D). The orebody has been formed by metasomatic replacement of several protolith types, however, no significant endoskarn was developed in the granite. The complex nature of the fluid pathways in the host rocks, diverse protoliths, and contrast in permeability among host rocks prevented the development of clear mineralogical zonation patterns such as those documented in other skarn deposits [3,8]. However, close to the granite contact, clinopyroxene and clinopyroxene-plagioclase-titanite ± scapolite are the dominant high-temperature prograde skarn assemblages. They form veins and lenses, predominantly in the polymictic rocks. Clinopyroxene is paragenetically the earliest mineral, and it is found as inclusions in other prograde skarn minerals, commonly garnet.

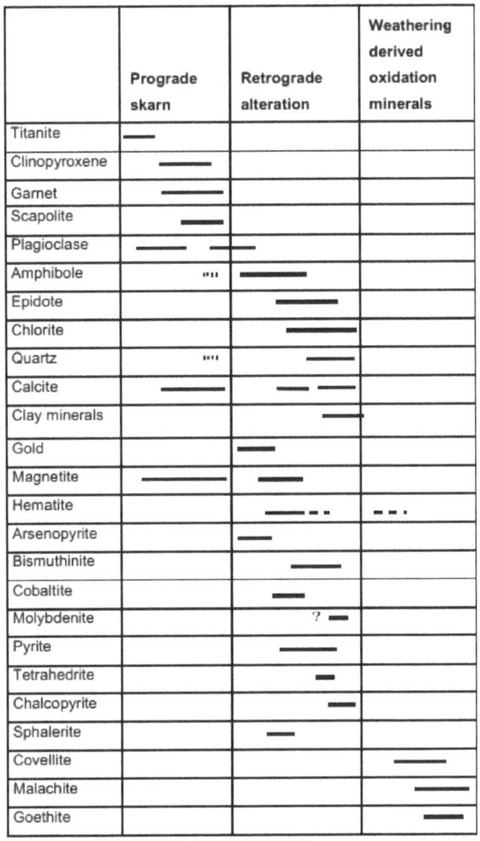

Figure 6. Schematic diagram showing paragenetic stages of skarn formation and mineralization at the Biggenden mine. Short dash lines show discontinuous mineralization. Question mark indicates possible mineralization.

Garnet, magnetite, and calcite were mainly deposited within and close to the orebody, where three prograde mineral assemblages are dominant: (1) garnet, (2) garnet-clinopyroxene, and (3) magnetite-garnet-calcite ± scapolite ± clinopyroxene. Scapolite replaces garnet and earlier-formed plagioclase; veinlets of scapolite commonly cut garnet and magnetite. Calcite was partially contemporaneous with magnetite and garnet based on textural evidence. Large euhedral crystals of calcite, up to a few centimeters in size, fill the fractures and matrix in the ore breccia. Late-stage calcite occurs as open-space-filling transparent crystals, up to a few cm in size, associated with clay and quartz.

The main retrograde assemblages include (1) amphibole, (2) amphibole-chlorite, (3) epidote-fibrous amphibole-calcite, (4) epidote-chlorite, (5) chlorite-quartz ± muscovite, with quartz, calcite, sulfides, and minor amounts of hematite, muscovite, apatite, and albite, and (6) montmorillonite-nontronite-calcite. Retrograde assemblages, particularly epidote-chlorite, are more abundant to the east of the ore zone, where the permeability of pyroclastic rocks may have controlled the distribution of these hydrous minerals.

Sulfides occur as disseminated grains, veins, and fracture fillings and are commonly associated with the retrograde assemblages (Figure 6). The average total sulfide content in the orebody is approximately 3 volume% with pyrite, chalcopyrite, and bismuthinite as the most common phases. Molybdenite, arsenopyrite, tetrahedrite, and trace cobaltite are also found locally as disseminated grains. Gold is dispersed erratically in trace amounts through the orebody. It is mainly associated with bismuthinite and trace native bismuth (Figure 7), and like other gold-bearing iron skarn deposits, its presence is related to the abundance of sulfide minerals [31]. In the past, gold was mined from the "actinolite rock", which is part of the retrograde skarn [10]. A late-stage retrograde clay assemblage locally replaced both the prograde and earlier retrograde skarn minerals and is characterized by the abundance of montmorillonite, nontronite (determined by XRD), and calcite. At shallow depths, supergene oxidation has produced hematite, covellite, malachite, and goethite from the skarn assemblage.

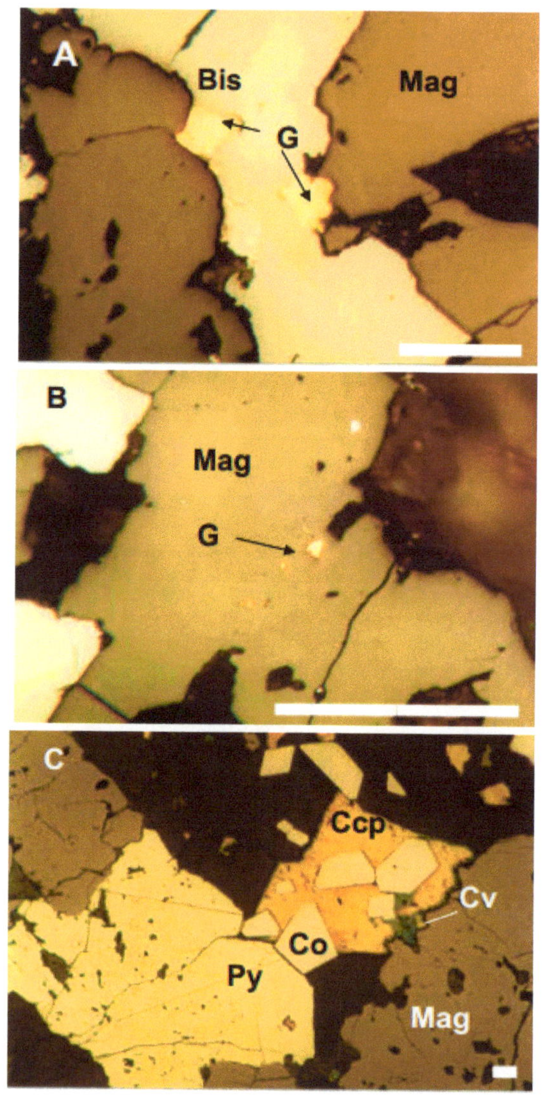

Figure 7. (**A**) Photomicrograph of gold (G) and bismuthinite (Bis) filling a fracture in magnetite (Mag) and/or replacing magnetite along a grain boundary. (**B**) Association of gold and magnetite. The linear arrangement of the small grains of gold may indicate that they follow a grain boundary in the magnetite aggregate. (**C**) Association of magnetite, pyrite (Py), chalcopyrite (Ccp), cobaltite (Co), and covellite (Cv). All are in reflected, plane-polarized light, and the scale bars are 50 μm.

Clinopyroxene is dominated by diopside-hedenbergite solid solution with minor johannsenite ($Di_{30-92}Hd_{7-65}Jo_{0-9}$) (Table 1). Pyroxene compositions fall within the fields of Fe and Au skarns (Figure 8). However, they are diopside-rich and are comparable to Cu skarns with respect to diopside content. The average Mn/Fe ratio for the Biggenden pyroxene is relatively high (0.12) for an iron skarn. Nakano et al. (1994) [32] have reported pyroxene with a high johannsenite component or Mn/Fe ratios in a few iron skarns in Japan. Matsueda (1981) [33] and Einaudi and Burt (1982) [34] also noted that some Fe skarns have high johannsenite-rich pyroxene. Gold skarn pyroxene can be more Al-rich than pyroxene in other skarn types [35]. The Al content of the Biggenden pyroxene is mainly comparable to Au skarns.

Scapolite is marialite-rich (average 76 mole% marialite) and Cl-bearing (Table 2). The proportions of marialite and Cl contents in Biggenden scapolite are similar to those in other gold and iron skarn deposits, e.g., Nickel Plate in British Columbia (Meionite 30–54 mole% and Cl = 1.7%–2.9% [36,37]).

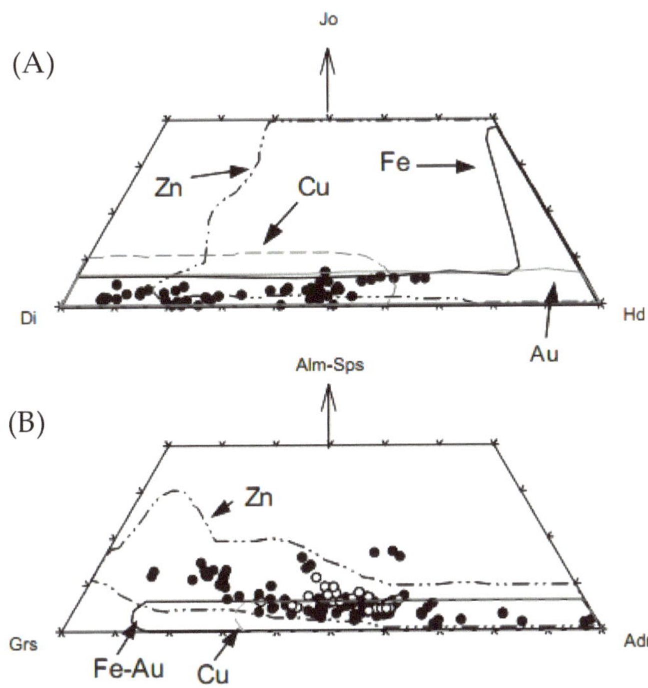

Figure 8. Ternary plots of Biggenden skarn clinopyroxene (**A**) and garnet (**B**) compositions (black circles) compared to Fe, Au, Zn, and Cu skarns [2]. The white circles represent garnets in the ore. Jo = Johannsenite, Hd = Hedenbergite, Di = Diopside, Alm-Sps = Almandine-spessartine, Adr = Andradite, Grs = Grossular.

The composition of plagioclase in the clinopyroxene-plagioclase assemblage is An_{95-99} mole%. Garnet grains are predominantly andradite-grossular solid solutions, containing commonly less than 10 mole% of other components and ranging in composition from Adr_{11} to Adr_{99} (Table 1). The earlier generation of garnet is isotropic and less commonly shows optical zonation. The later generation is anisotropic, grossular-rich, and forms overgrowth rims on isotropic garnets. The composition of Biggenden garnet is similar to the classical examples of iron skarns (Figure 8; e.g., those from western British Columbia [38]). However, the analyzed garnet grains are slightly enriched in Mn and may approach the compositions of garnet in Zn skarns [1,2,32]. Magnetite in the Biggenden skarn is close to stoichiometric Fe_3O_4, with low Ti.

Table 1. Range of chemical composition of garnet, amphibole, epidote, chlorite, and scapolite from the Biggenden mine. The number of analyses is in parentheses.

	Garnet					Clinopyroxene		
Assemblage	**Garnet-Pyroxene (2)**	**Magnetite Ore (2)**	**Garnet Skarn (2)**	**Calc-Silicate (2)**		**Garnet-Pyroxene (4)**	**Magnetite Ore (2)**	**Pyroxene-Plagioclase (4)**
SiO_2	37.46–37.80	37.04–37.26	36.83–37.48	36.9–37.72	SiO_2	49.8–51.95	51.14–51.82	50.88–53.8
TiO_2	1.03–1.79	0.88–21.59	0.79–2.12	0.53–2.07	TiO_2	ND-0.43	ND-0.20	ND
Al_2O_3	8.64–9.23	9.40–11.15	5.93–8.77	9.86–11.73	Al_2O_3	1.22–1.85	1.50–2.87	0.20–0.92
Fe_2O_3 *	16.01–16.24	15.71–15.95	16.19–22.37	11.99–15.76	Fe_2O_3 *	0.85–1.92	ND-1.88	ND-1.30
FeO	0.32–0.72	0.89–2.04	0.28–0.65	1.01–6.79	FeO *	0.14–15.08	0.81–14.89	8.94–14.93
MnO	0.91–1.74	1.19–1.19	0.68–1.07	0.28–0.31	MnO	ND-1.31	1.47–1.55	0.35–1.32
MgO	ND	ND-0.96	ND-0.31	ND	MgO	9.68–14.84	8.49–16.18	9.18–13.08
CaO	32.72–33.6	30.8–33.15	32.79–33.29	29.21–33.69	CaO	22.97–25.77	22.60–24.26	21.84–23.84
Total	98.91–99.30	99.21–100.00	99.57–100	98.49–99.36	Total	99.15–100.91	99.49–100.17	99.39–100.09
Number of ions on the basis of 24 O					Cations on the basis of 6 O			
Si	6.02–6.05	5.94–5.96	5.99–6.03	5.94–6.02	Si	1.87–1.99	1.90–1.97	1.96–2.01
Ti	0.12–0.22	0.11–0.19	0.1–0.26	0.06–0.25	Ti	ND-0.01	ND-0.01	ND
Al^{IV}	1.63–1.75	0.04–0.06	0.01–1.15	1.81–2.21	Al^{IV}	ND-0.07	0.03–0.10	ND-0.04
Al^{VI}	ND	1.71–2.06	ND-1.64	ND-0.06	Al^{VI}	ND-0.01	0.02–0.04	ND-0.01
Fe^{3+}	2.15–2.17	1.85–2.14	2.17–2.76	1.60–2.12	Fe^{3+}	0.03–0.19	ND-0.06	ND-0.04
Fe^{2+}	0.04–0.10	0.12–0.31	0.04–0.09	0.13–0.92	Fe^{2+}	0.03–0.48	0.03–0.48	0.28–0.49
Mn	0.12–0.24	0.16–0.31	0.1–0.15	0.04	Mn	ND-0.04	0.05	0.01–0.04
Mg	ND	ND-0.23	ND-0.08	ND	Mg	0.55–0.83	0.49–0.89	0.53–0.73
Ca	5.64–5.77	5.28–5.69	5.7–5.79	5.04–5.76	Ca	0.93–1.04	0.93–0.96	0.91–0.98
Total	16.01	16–16.01	16–16.03	16.00	Total	4.00	3.99–4.00	3.98–4.00
Pyrope	0.00	0.00–3.82	0.00–1.28	0.00	Johannsenite	0.0–4.5	4.5–5.0	1.1–3.9
Almandine	0.73–1.62	1.99–5.23	0.64–1.46	2.27–15.24	Diopside	51.4–77.8	47.9–87.3	50.1–71.5
Spessartine	2.08–3.97	2.70–2.71	1.59–2.44	0.64–0.70	Hedenbergite	20.5–47.4	8.2–47.1	27.4–45.9
Grossularite	36.75–37.19	36.79–39.14	24.69–34.93	29.35–50.34				
Andradite	57.23–60.44	49.11–58.51	61.17–71.8	46.76–54.71				
	Amphibole			Epidote (4)		Chlorite (5)		Scapolite
	Skarn (6)	Contact metamorphic (2)						
SiO_2	37.76–48.92	45.82–50.18	SiO_2	37.46–38.7	SiO_2	26.79–31.29	SiO_2	54.26–59.73
TiO_2	ND-0.50	ND	TiO_2	ND-0.27	Al_2O_3	16.5–18.79	Al_2O_3	25.09–26.34
Al_2O_3	4.68–15.17	7.42–12.89	Al_2O_3	22.54–24.7	FeO *	26.79–31.29	CaO	5.95–9.90
FeO	19.30–24.70	9.13–12.73	Fe_2O_3	10.44–3.32	MnO	0.75–1.170	Na_2O	7.56–8.92
MnO	0.71–1.03	ND-0.57	MnO	ND-0.33	MgO	6.75–15.01	K_2O	0.21–1.04
MgO	4.84–9.44	13.74–15.97	CaO	23.1–24.61	Total	86.48–89.24	Cl	ND-3.43
CaO	11.43–12.81	10.75–11.69	Na_2O	0.41–0.64			−O=Cl	ND-0.77
Na_2O	0.33–2.74	1.57–3.62	Total	96.34–98.87			Total	99.24–100.82
K_2O	0.28–1.87	0.22–0.26						
Cl	ND-0.72	ND-0.22						
−O=Cl	ND-0.16	ND-0.05						
Total	97.16–99.26	98.33–98.40						
Number of ions on the basis of 23 O			Numbers of ions on the basis of 25 O				Number of ions on the basis of 25 O	
Si	5.84–7.37	6.53–7.25	Si	5.98–6.11	Si	5.73–5.95	Si	7.58–8.31
Al^{IV}	0.63–2.16	0.75–1.47	Ti	ND-0.03	Al^{IV}	2.05–2.27	Al	4.12–4.34
Al^{VI}	0.20–0.65	0.52–0.69	Al	4.22–4.64	Al^{VI}	1.97–2.65	Ca	0.89–1.48
Ti	ND-0.06	ND	Fe^{3+}	1.25–1.59	Fe^{2+}	4.75–6.34	Na	2.06–2.41
Fe^{3+} *	0.05–0.48	0.12–0.36	Mn	ND-0.04	Mn	0.14–0.21	K	0.04–0.18
Fe^{2+}	1.97–3.09	0.56–1.41	Ca	3.93–4.18	Mg	2.44–4.74	Cl	ND-0.82
Mn	0.09–0.13	ND-0.07	Na	0.13–0.20			Total	15.65–16.45
Mg	1.12–2.12	2.96–3.39	Total	16.06–16.16	Total cations	19.24–19.71		
Ca	1.91–2.07	1.64–1.81						
Na	0.10–0.83	0.44–1.00	Clinozoisite	73.50–80.20				
K	0.05–0.37	0.04–0.05	Pistacite	19.80–26.50				
Cl	0.00–0.19	ND-0.05	Piemontite	0–0.80				
Total cations	15.17–16.12	15.44–15.68						

* Recalculated based on stoichiometry. ND = Not detected.

Table 2. Representative chemical analyses of the Degilbo Granite and related dikes. Oxides and trace elements are in weight% and ppm, respectively.

Sample	R76342	R76340	R76343	R76344	R76345	R76346	R76410	R76366	R76458	R76325	R76440	R76368	R76337
	Granite											Microgranite Dike	Aplite Dike
Distance from Contact	600 m	500 m	200 m	150 m	100 m	80 m	60 m	50 m	3 m	1 m	20 cm		
SiO_2	68.94	68.87	71.09	69.38	69.22	69.23	68.86	70.31	68.3	63.9	63.92	70.06	77.17
TiO_2	0.41	0.45	0.35	0.54	0.43	0.42	0.4	0.37	0.44	0.69	0.67	0.43	0.09
Al_2O_3	15.44	15.37	14.73	14.49	15.2	15.42	15.19	14.59	15.65	16.82	18.16	15.23	12.49
Fe_2O_3	1.49	1.19	1.12	1.34	1.30	1.34	1.30	1.13	1.16	1.42	1.25	0.08	0.35
FeO	1.43	1.77	1.41	2.12	1.72	1.66	1.57	1.55	1.72	2.85	2.20	1.78	0.25
MnO	0.06	0.07	0.05	0.07	0.06	0.05	0.05	0.07	0.06	0.11	0.09	0.04	0.03
MgO	0.79	0.95	0.66	1.03	0.85	0.79	0.89	0.79	0.86	1.31	1.25	0.77	0.11
CaO	2.14	2.03	1.74	1.93	2.06	2.00	1.94	1.69	2.33	3.15	3.12	2.24	0.63
Na_2O	3.66	3.63	3.56	3.52	3.73	3.70	3.76	3.66	3.83	4.13	4.92	3.72	3.39
K_2O	4.46	4.30	4.50	4.26	4.23	4.24	4.28	4.29	4.35	3.72	2.75	3.64	4.78
P_2O_5	0.15	0.15	0.11	0.17	0.14	0.15	0.13	0.13	0.15	0.24	0.09	0.13	0.01
S	0.01	0.02	0.01	0.01	0.01	0.01	0.01	0.01	0.01	0.01	0.02	0.14	0.02
LOI	0.73	0.94	0.58	0.74	0.67	0.71	0.77	0.77	0.74	1.00	1.05	0.85	0.56
Total	99.69	99.74	99.9	99.58	99.58	99.71	99.13	99.34	99.6	99.34	99.46	99.08	99.85
Ba	735	704	607	735	627	669	742	703	610	818	541	691	80
Rb	179	170	181	179	187	182	181	195	184	161	138	135	259
Sr	273	248	213	273	205	250	264	251	216	280	381	248	33
Pb	17	16	19	17	19	16	16	14	17	20	19	17	24
Th	17	20	33	17	22	18	15	18	23	16	21	58	62
U	3	4	4	3	4	4	2	3	5	4	4	6	17
Zr	240	249	218	240	304	246	233	239	215	244	541	211	87
Nb	10	10	13	10	13	10	10	9	11	8	16	12	9
Y	32	34	35	32	43	36	37	34	34	28	22	33	25
La	36	43	35	36	45	41	37	41	33	42	46	28	24
Ce	65	69	63	65	83	75	72	67	55	72	78	52	41
Nd	31	32	30	31	44	35	32	32	25	30	37	25	18
Sc	7	7	7	7	16	7	6	9	4	13	14	6	3
V	30	34	24	30	38	32	33	34	30	33	39	29	5
Cr	11	7	6	11	13	10	11	5	12	7	13	7	2
Ni	6	5	3	6	5	6	6	3	3	5	6	4	4
Cu	9	13	45	9	25	14	13	13	8	14	16	95	2
Zn	36	43	37	36	48	40	37	36	50	46	61	43	38
Ga	17	16	16	17	17	18	17	18	17	17	19	16	15
Sn	<3	<3	9	<3	8	4	<3	10	<3	<3	<3	<3	<3
As	5	4	4	5	4	5	4	6	11	6	9	352	58
Mo	2	11	2	2	1	<1	6	3	1	2	15	1	2
Bi	4	73	3	4	2	5	<1	<1	1	2	5	5	2

The Biggenden skarn amphibole belongs to the ferropargasite and hastingsite groups (Table 1) and shows a continuous variation from $Fe^{3+} < Al^{VI}$ to $Fe^{3+} > Al^{VI}$ with up to 1.29% Cl. These amphibole grains resemble those in both Au and Fe skarns with regard to Fe, Mg, Mn, Ca, Al, Na, and K contents [2]. Epidote ranges in composition from 19.8 to 29.3 mole% pistacite with low piemontite content (Table 1). Chlorite is dominantly chamosite with $\Sigma Fe/Fe + Mg$ ratio ranging from 0.50 to 0.72 (Table 1). The range of calculated temperature for the retrograde skarn, using the chlorite geothermometer of Cathelineau (1988) [39], is between 252 °C and 375 °C, with an average of 310 °C.

4.2. Geochemistry of the Degilbo Granite

Based on modal mineralogy and using the classification scheme of Le Bas and Streckeisen (1991) [40], the Degilbo Granite is a monzogranite in composition. The whole rock chemical analyses of samples from the Degilbo Granite show a restricted compositional range, and the variations among samples are not related to the distance from the margin of the pluton (Table 2, Figure 9). There is a range of SiO_2 content from 68 to 71%, except for samples R76325 and R76440 which may be affected by incipient actinolite endoskarn development and consequently have lower SiO_2 values. The granite contains high

K_2O, K/(K + Na), Rb, Pb, and Th and is a high-K granitoid according to the classification of Peccerillo and Taylor (1976) [41]. It is mildly peraluminous (0%–1.63% normative corundum), and the Alumina Saturation Indices (ASI = molar $Al_2O_3/CaO + Na_2O + K_2O$) in all analyzed samples are less than 1.1, which is considered to be typical in I-type granites [22].

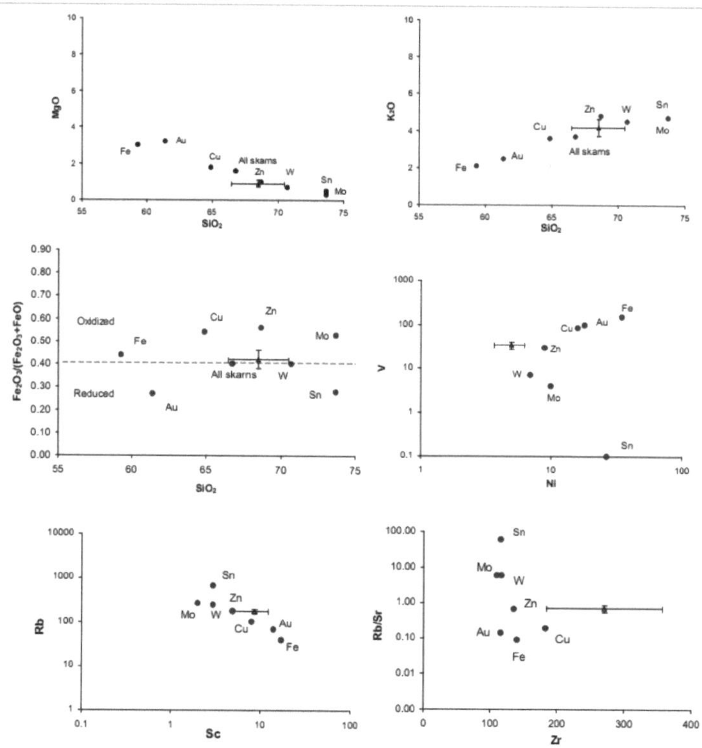

Figure 9. Average major and trace element content of plutons associated with different types of skarn deposits (filled circles after [3,8]), compared to the average composition of the Degilbo Granite (triangle, data from Table 2). Bars show the range of compositions of analyzed samples.

4.3. Microthermometric Studies

Fluid inclusions of measurable size (>5 microns) are present in garnet, quartz, and calcite from the Biggenden skarn assemblage. They are divided into three main types based on the major phases present. Type I fluid inclusions contain an aqueous liquid phase (L), a vapor phase (V), and one or several daughter crystals (S) and are commonly primary (using the criteria of Roedder, 1984 [26]). They are typically found in calcite grains in ore assemblages. Cubic, isotropic, and colorless crystals of halite are the main solid daughter phase in type I inclusions. The second isotropic daughter phase with a round shape was identified as sylvite based on its lower melting temperature. In some of the type I fluid inclusions, prismatic crystals, which are commonly smaller than halite, are also present. These daughter crystals did not change upon heating and are identified as anhydrites. Two types of opaque solid phases are present in these fluid inclusions; cubic grains possibly entrapped in magnetite and an unknown irregular solid phase attached to halite crystals.

Type II fluid inclusions are the most abundant and are characterized by a high proportion of liquid, with vapor content ranging from 10 to 40 volume% at room temperature. They are both primary and secondary or pseudosecondary, and the vapor homogenizes into a liquid phase upon heating

Type III inclusions are less common and are characterized by a high proportion of gas ranging from 70 to 100 volume% and homogenize into a gas phase. Some of the inclusions of this type contain gas phase only, which may indicate that the fluids were trapped near the two-phase curve [26].

Fluid inclusions in calcite occur in various sizes and shapes and include all three types, with sizes ranging from less than 10 to 30 μm. Fluid inclusions in quartz are type II (V + L) inclusions, have irregular shapes, and range in size from 10 to 50 μm. Fluid inclusions in garnet are less abundant; they are two-phase (type II), have irregular shapes, and are commonly less than 10 μm in size, although in one sample, larger fluid inclusions up to 30 μm were found. Only a few measurements of T_h were obtained from fluid inclusions in garnet, as on heating above 500 °C the host mineral commonly became dark brown and fluid inclusions were obscured. The majority of analyzed fluid inclusions homogenized to a liquid phase. However, a few calcite samples contained scattered vapor-rich, type III inclusions that commonly homogenized into a vapor phase and are interpreted to reflect trapping under boiling conditions. The scarcity of this type of inclusion may imply that boiling occurred only locally or intermittently (cf. [42–44]).

4.3.1. Homogenization Temperatures

Two types of primary fluid inclusions are recognized based on homogenization temperature (Figure 10): (1) high-temperature (500–560 °C) fluid inclusions that are found in garnet and calcite from the magnetite ore assemblage, (2) moderate-temperature (280 °C to 360 °C) fluid inclusions in calcite and quartz from the retrograde (mainly chlorite) assemblage. A few inclusions in calcite (outside the ore assemblage) homogenize between 380 °C and 440 °C. Homogenization temperature in the secondary and pseudosecondary fluid inclusions ranges from 200 °C to 280 °C.

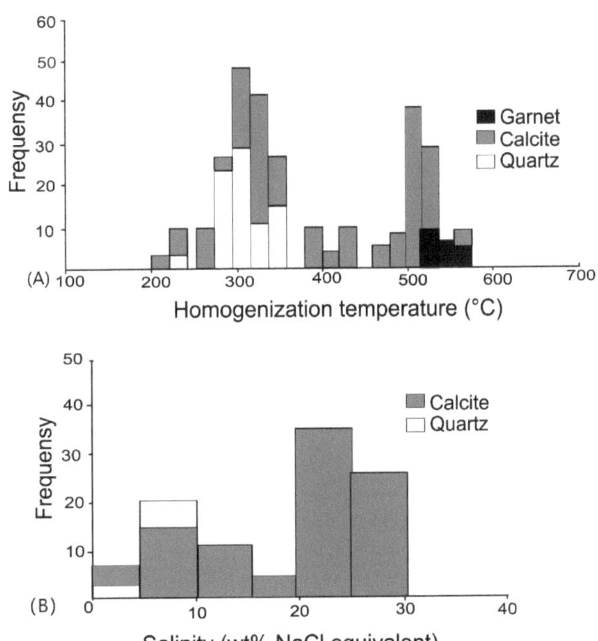

Figure 10. (**A**) Frequency distribution of the homogenization temperatures of primary fluid inclusions in garnet, calcite, and quartz samples. (**B**) Frequency distribution of the salinity (wt % equiv NaCl) based on the last ice melting and halite dissolution temperature measurements.

4.3.2. Salinities

Salinities of the fluid inclusions were measured from either the last ice melting temperature [45] or the halite dissolution temperature [46]. Halite in type I fluid inclusions was dissolved at temperatures between 240 °C and 386 °C, corresponding to 31 and 43 wt % NaCl equivalent. Type II (L + V) fluid inclusions show a bimodal distribution of salinity; one group between 5 and 15 wt % NaCl equivalent and the other group between 20 and 30 wt % NaCl equivalent (Figure 10). The latter is mainly from fluid inclusions with higher homogenization temperatures.

4.4. Isotope Geochemistry

$\delta^{18}O$ and $\delta^{13}C$ in carbonates: Samples of various generations of calcite, as well as marble, limestone, and calc-silicate hornfels, were analyzed for oxygen and carbon isotopes (Table 3). Skarn calcite, which occurs interstitially between skarn silicates and magnetite, has $\delta^{18}O$ values ranging from 9.6 to 12.2% and $\delta^{13}C$ values ranging from −0.7 to −7.1%. Average $\delta^{18}O$ and $\delta^{13}C$ in marble samples are 20.9% and 4.9%, respectively. Skarn calcite samples have oxygen and carbon isotope values only slightly different from those of fine-grained calcite samples (Figure 11). Although limestone has a eutectic melting temperature of 600–675 °C [47], evidence of limestone melting (e.g., melt inclusions; [48]) was not found in the Biggenden skarn. One calcite sample plot in the igneous field on the $\delta^{18}O$ and $\delta^{13}C$ diagram (Figure 11), which may be due to the intensive reaction of the limestone with magmatic fluid. Calcite formed in the latest stage of retrograde hydrothermal alteration has $\delta^{13}C$ values that reflect the skarn calcite compositions but have heavier $\delta^{18}O$ values, which are comparable to those of marine carbonates [49]. In marbles, despite recrystallization, the marine limestone isotopic signature has been retained. The carbon isotopic values for the calc-silicate hornfels are lower than those in the limestones (Table 3).

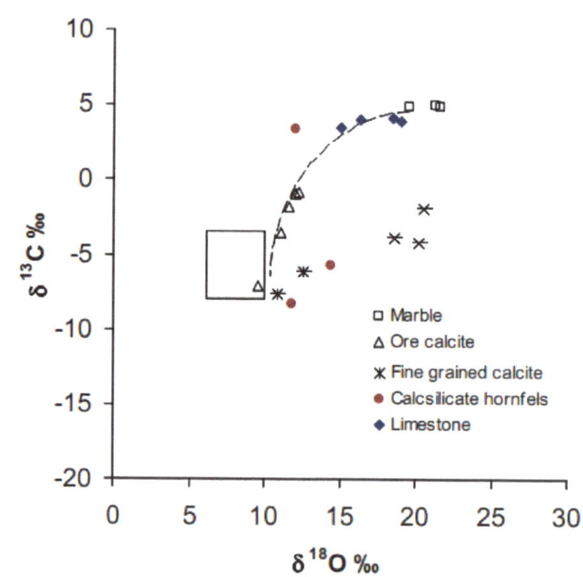

Figure 11. Carbon versus oxygen isotope diagram showing the isotopic compositions of the marble, limestone, ore calcite associated with magnetite skarn, fine-grained interstitial calcite, calcite from the late-stage retrograde alteration, and calcite from calcsilicate hornfels. The curve defines an inferred evolutionary trend from heavier oxygen and carbon isotope values in limestone and marble to lighter values in skarn, as the result of decarbonation and interaction of the infiltrating fluid. The box represents igneous carbonate (after [50])

Table 3. Carbon and oxygen isotope data for calcite samples.

Sample	Description	$\delta^{13}C$ (% PDB)	$\delta^{18}O$ (% SMOW)
R76483	Limestone	3.9	19.0
R76477	Limestone	4.2	18.5
R76477	Limestone	3.5	15.1
R76467	Limestone	4.0	16.3
R76329	Marble	4.9	21.6
R76330	Marble	5.0	21.3
R76318	Marble	4.9	19.6
R76347	Ore assemblage calcite	−3.5	11.1
R76335	Ore assemblage calcite	−0.8	12.3
R76436	Ore assemblage calcite	−7.1	9.6
R76379	Ore assemblage calcite	−0.9	12.0
R76389	Ore assemblage calcite	−1.8	11.6
R76377	Late-stage alteration calcite	−3.9	18.6
R76363	Late-stage alteration calcite	−4.2	20.2
R76441	Late-stage alteration calcite	−1.9	20.5
R76364	Fine-grained calcite	−7.7	10.9
R76365	Fine-grained calcite	−6.1	12.6
R76468	Calc-silicate hornfels	−8.3	11.8
R76466	Calc-silicate hornfels	−5.7	14.4
R76488	Calc-silicate hornfels	3.4	12.0

$\delta^{18}O$ in silicates: The $\delta^{18}O$ values for magnetite and garnet are in the range of 7.6–10.4% and 2.4–4.4%, respectively (Table 4). The $\delta^{18}O$ values of the hydrous retrograde minerals (amphibole, epidote, and chlorite) range from 7.8 to 9.7%, 8 to 8.2%, and 0.2 to 4.1%, respectively. The oxygen isotope composition of the granite sample (9.5%) and the narrow range of granite quartz separates lie well within the range for I-type igneous rocks (e.g., [51]).

Using fractionation factors of Bottinga and Javoy (1973) for coexisting calcite and magnetite, which show equilibrium textures in the magnetite ore assemblage, temperatures of 572 °C and 604 °C were calculated for samples R76335 and R76379, respectively [52]. These temperatures are close to those estimated from the metamorphic assemblages [19] and measured in fluid inclusions.

δD in hydrous silicates: Hydrous silicate minerals from the retrograde skarn assemblage and also biotite separates from the Degilbo Granite were analyzed both for oxygen and hydrogen isotopes (Table 4). For the retrograde skarn minerals, δD values cluster in a narrow range for each mineral; amphibole –117 to –132%, epidote –97 to –100%, chlorite –120 to –128%.

$\delta^{34}S$ in sulfides: The $\delta^{34}S$ values of all analyzed sulfides (molybdenite, bismuthinite, chalcopyrite, arsenopyrite, and pyrite) fall in the range of –5.6–2.3% and are comparable to the $\delta^{34}S$ values of sulfides in other skarn deposits (Table 5, Figure 12). The six pyrite samples have $\delta^{34}S$ values ranging from −5 to +2.3% with the maximum $\delta^{34}S$ value from pyrite in metamorphosed sedimentary rocks approximately 300 m outside the mine area.

Table 4. Oxygen and deuterium isotope data in silicates.

Sample	Mineral/Rock	$\delta^{18}O$ (%)	δD (%)	T °C [1]	$\delta^{18}O_{H2O}$ (%) [2]	δD_{H2O} (%)
R76296	Amphibole	8.6	−133	300–400	9.1 to 9.9	−63 to −81
R76331	Amphibole	7.8	−128	300–400	8.3 to 9.1	−58 to −77
R76388	Amphibole	9.7	−126	300–400	10.2 to 11	−56 to −75
R76430	Amphibole	8.1	−130	300–400	8.6 to 9.4	−60 to −79
R76435	Amphibole	8.7	−117	300–400	9.2 to 10	−48 to −66
R76463	Epidote	8.0	−100	300–400	5.9 to 7.4	−50
R76406	Epidote	8.2	−97	300–400	6.1 to 7.6	−47
R76302	Epidote	8.0	−98	300–400	5.9 to 7.4	−48
R76410	Biotite	6.2	−133	650	8.7	−99
R76344	Biotite	6.3	−128	650	8.8	−93
R76342	Biotite	5.7	−121	650	8.2	−87
R76338	Biotite	6.0	−114	650	8.5	−80
R76344	Granite	9.5	-	-	-	-
R76307	Chlorite	4.1	−120	310	4.2	−80
R76447	Chlorite	0.3	−122	310	0.4	−82
R76441	Chlorite	0.2	−128	310	0.3	−88
R76297	Garnet	9.2	-	550	11.1	-
R76335	Garnet	8.1	-	550	10.0	-
R76352	Garnet	8.2	-	550	10.1	-
R76358	Garnet	7.6	-	550	9.5	-
R76444	Garnet	10.4	-	550	12.3	-
R76379	Magnetite	4.4	-	550	10.3	-
R76427	Magnetite	3.6	-	550	9.5	-
R76335	Magnetite	2.8	-	550	8.7	-
R76353	Magnetite	2.4	-	550	8.3	-
R76410	Quartz	10.6	-	650	8.7	-
R76344	Quartz	10.0	-	650	8.1	-
R76342	Quartz	10.3	-	650	8.4	-
R76338	Quartz	10.4	-	650	8.5	-

[1] Based on fluid inclusions, mineral paragenesis, and chlorite thermometry. [2] Mineral-water fractionations of oxygen are calculated from Bottinga and Javoy (1973) [52] for amphibole and magnetite; Graham et al. (1980) [53] for epidote; Bottinga and Javoy (1973) [52] for biotite; Bottinga and Javoy (1975) [54] for garnet; Matsushisa et al. (1979) [55] for quartz; Wenner and Taylor (1971) [56] for chlorite. Mineral-water fractionations of hydrogen are calculated from Graham et al. (1984) [57] for amphibole; Graham et al. (1980) [53] for epidote; Suzuoki and Epstein (1976) [58] for biotite; and Kyser (1987 [59]: after Suzuoki and Epstein, 1976 [58]) for chlorite.

Table 5. Sulfur isotope data for sulfides from the Biggenden mine.

Sample	Mineral	$\delta^{34}S$ (% CDT)
R76375	Molybdenite	−1.6
R76378	Bismuthinite	−1.1
R76380	Bismuthinite	−1.1
R76524	Chalcopyrite	−1.1
R76430	Chalcopyrite	0.5
R76383	Arsenopyrite	−5.6
R76384	Arsenopyrite	1.4
R76385	Pyrite	−1.2
R76447	Pyrite	−5.0
R76376	Pyrite	−2.6
R76486	Pyrite	2.3
R76382	Pyrite	−1.0
R76381	Pyrite	−1.8

Strontium isotopes: In order to use the Sr data (Table 6) as hydrothermal tracers and compare the results in different rock types, the initial $^{87}Sr/^{86}Sr$ ratios have been calculated for the age of the Degilbo Granite (215 Ma). A comparison of the Sr isotope values of ore-stage calcite samples to the nearby rocks (Figure 13) shows that the $^{87}Sr/^{86}Sr$ ratio in calcite has a narrow range and is lower than that in limestone and metasiltstone and higher than that in the Degilbo Granite. One sample from metavolcanic rock with a higher

$^{87}Sr/^{86}Sr$ ratio is from the mine (R76402, Table 6), and its strontium isotope composition may have been modified during fluid/rock interaction. The other sample (R76501, Table 6) collected outside the mine area is a metabasalt.

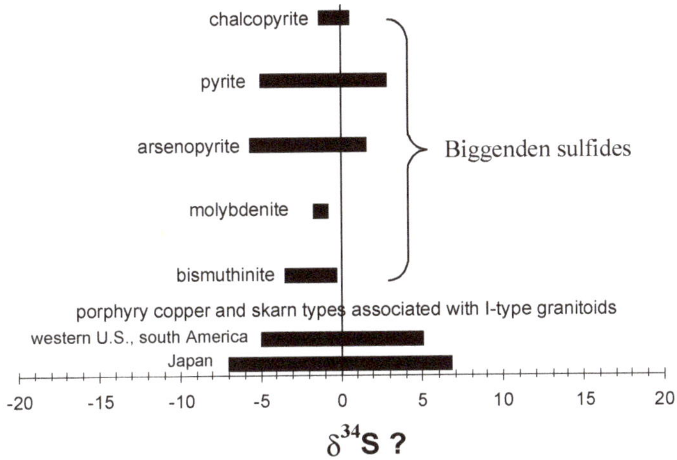

Figure 12. Sulfur isotope composition of the Biggenden sulfides compared to those of other skarn deposits in the world (data from [60]).

Table 6. Strontium isotope data for rocks from the Biggenden mine.

Sample	Description	$^{87}Rb/^{86}Sr$	$^{87}Sr/^{86}Sr$	$^{87}Sr/^{86}Sr_{initial}$	Sr ppm	Rb ppm
R76472	Metasiltstone	1.9483	0.7128	0.70684	169	114
R76402	Metavolcanic	0.2352	0.7062	0.70548	442	36
R76501	Metavolcanic	0.0593	0.7034	0.70322	341	7
R76410	Granite	2.2000	0.7106	0.70387	256	195
R76344	Granite	2.6346	0.7121	0.70404	205	187
R76483	Limestone	0.0543	0.7065	0.70633	319	6
R76502	Ore assemblage calcite	0.0245	0.7047	0.70463	118	<1
R76335	Ore assemblage calcite	0.0162	0.7046	0.7046	178	<1
R76357	Ore assemblage calcite	0.0222	0.7048	0.70473	130	<1

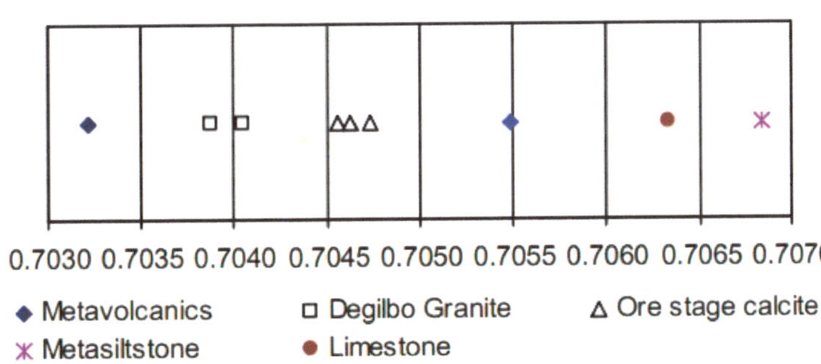

Figure 13. Plot of $^{87}Sr/^{86}Sr$ isotope ratios for different rocks in the Biggenden area compared to those of three ore-stage calcite samples from the Biggenden mine.

Lead isotopes: Lead isotope data for two galena samples were measured on polished thin sections by in situ analysis. The analyzed galena grains belong to sulfide-bearing retrograde assemblages that formed at lower temperatures (300–400 °C) than the prograde skarn minerals. The Biggenden galena samples have a narrow range of Pb isotope values and have $^{206}Pb/^{204}Pb$, $^{207}Pb/^{204}Pb$, and $^{208}Pb/^{204}Pb$ ratios that are similar to those of other hydrothermal deposits associated with Triassic intrusions in southeast Queensland (Figure 14; Table 7). The measured values are initial ratios because the U/Pb and Th/Pb ratios of galena are so low that isotopic ratios do not change significantly with time [61].

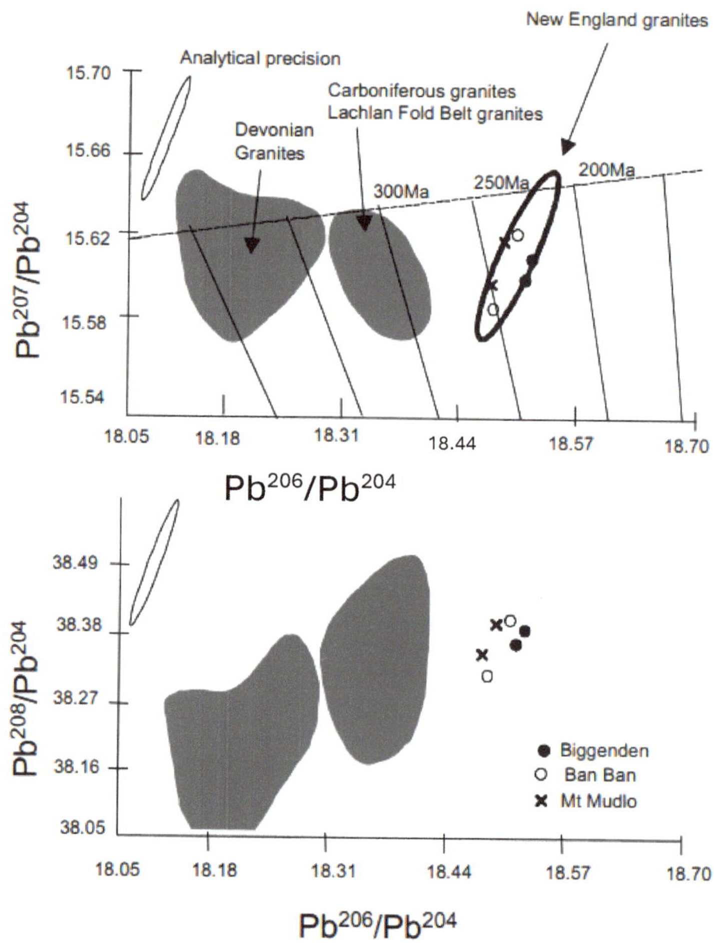

Figure 14. Lead isotope ratios for galena samples from the Biggenden mine and two other deposits (Ban Ban Zn skarn and Mt Mudlo mesothermal Cu-Mo-W vein) related to the Triassic granitoids in southeast Queensland. Fields for Devonian and Carboniferous granites from the Lachlan Fold Belt of eastern Australia are from [62]. New England Orogen granite ellipse from G. Carr (unpublished data).

Table 7. Pb isotope data for Biggenden galena and two other hydrothermal deposits associated with the Triassic granitoids in the area [15].

Location	Sample	$^{206}Pb/^{204}Pb$	$^{207}Pb/^{204}Pb$	$^{208}Pb/^{204}Pb$
Biggenden	R76448	18.505	15.602	38.369
	R76441	18.514	15.613	38.391
Ban Ban	BB1	18.472	15.589	38.315
	BB3	18.495	15.625	38.406
Mt Mudlo	MM22	18.466	15.602	38.350
	MM44A	18.480	15.620	38.401

5. Discussion

5.1. Isotope Interpretations

The $\delta^{13}C$ and $\delta^{18}O$ values are relatively high in Biggenden marbles (Figure 11) and are similar to those of marbles elsewhere (e.g., [63]). Skarn calcite samples isotopically resemble calcite in other skarn deposits [64–66] and display a trend towards lighter $\delta^{13}C$ and $\delta^{18}O$ values compared to those of marble and limestone. Similar depletion trends have been reported for other skarn deposits worldwide [50], with initial limestone in the range of $\delta^{18}O$ = 20 to 25‰ and $\delta^{13}C$ = −2 to 5‰, and final skarn calcite values near igneous values (Figure 11). This depletion can result either from the evolution of CO_2 during decarbonation or from the interaction with infiltrating fluid (e.g., [67]). Metamorphic decarbonation preferentially removes heavy isotopes through the loss of CO_2 (e.g., [68]), although the maximum decrease in $\delta^{18}O$ by this mechanism in marbles is 2‰. A large decrease in $\delta^{18}O$ values in marbles and low $\delta^{13}C$ values in skarn calcite has been interpreted in other skarns to be the result of water-rock exchange or infiltration metasomatism from multiple fluid sources mainly involving magmatic-hydrothermal fluids (e.g., [69]).

The isotopic composition of late-stage calcite is similar to that in some late-stage open space-filling calcite in other skarn deposits (e.g., [70]). Heavy $\delta^{18}O$ values for late-stage calcite may indicate the interaction of magmatic fluids with more ^{18}O-enriched sediments during skarn formation. The $\delta^{13}C$ values for the calc-silicate hornfels are lower than those in the limestone samples and likely reflect the involvement of metasomatic fluids.

A good correlation between the $\delta^{18}O$ values of calcite from skarns and the whole rock values of the associated plutons has been cited as evidence for the likely magmatic origin of most skarn-forming fluids (e.g., [71]). The $\delta^{18}O$ value of Degilbo Granite and the mean value for skarn calcite are aligned with the correlation trend of Nabelek (1991) [71]. Moreover, the calculated $\delta^{18}O$ of the metasomatic fluid in equilibrium with the skarn calcite at 500 °C (8.4–11‰) (fractionation factors from Friedman and O'Neil, 1977 [72]) is close to that calculated for the fluid in equilibrium with the granite and anhydrous skarn silicates (Table 4).

The oxygen isotope composition of magnetite at Biggenden (2.4–4.4‰) is similar to that in other Fe skarn deposits (e.g., Iron Hat skarn [73] and Peña Colorada deposit [74]). The range of oxygen isotope composition (8.1 to 10.4‰) in the Biggenden garnet is similar to that in some gold skarns (e.g., Fortitude deposit [8]).

The $\delta^{18}O$ values for the fluid in equilibrium with skarn minerals as well as mineral separates from the granite have been calculated for the related temperature ranges (Figure 15). In these calculations, a temperature of 550 °C has been used for prograde garnet and magnetite. This temperature estimate is based on fluid inclusion studies, isotope thermometry, and contact metamorphic assemblages of the hornblende-hornfels facies. A temperature of 650 °C was used in the calculations for the quartz and biotite separates from the Degilbo Granite, assuming that these minerals formed at this temperature during the crystallization of the granitic melt. The 300–400 °C temperature used for amphibole and epidote is based on the paragenetic relationship of these minerals in the retrograde assemblages, which formed earlier than chlorite and probably at slightly higher temperatures. Fluid inclusion studies in quartz associated with chlorite and thermometry based on the chemical composition of chlorite [39] indicate that chlorite formed at average temperatures of 310 °C.

The $\delta^{18}O$ values of fluid in equilibrium with garnet and amphibole are slightly heavier than the $\delta^{18}O$ of fluids in many skarn deposits associated with felsic magmatic rocks [75]. This could be due to the effect of fluid/rock interaction; granitoid-derived fluids interacted with more ^{18}O-enriched sediments during skarn formation. However, the oxygen isotopic values of the fluid in equilibrium with chlorite in the retrograde assemblage are lower, and meteoric water is a more likely source for this fluid. The range of $\delta^{18}O$ values for epidote could also indicate the involvement of meteoric water. The general decreasing trend of the $\delta^{18}O$ in fluid from 8.0 to 3.0% has been attributed in other skarn deposits to the mixing of magmatic fluid with meteoric water (e.g., [67]).

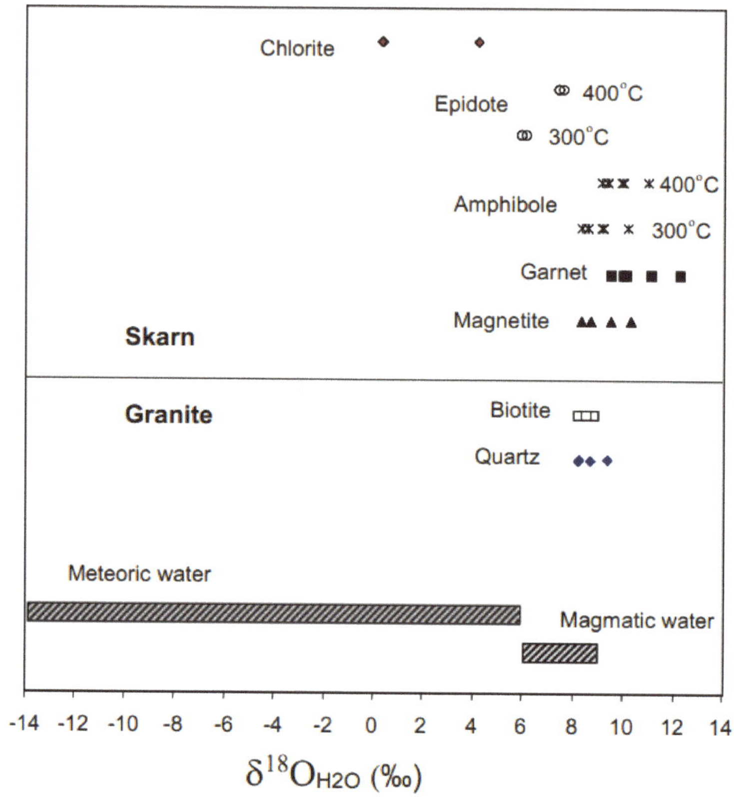

Figure 15. Calculated $\delta^{18}O$ values for the fluids in equilibrium with the skarn minerals and mineral separates from the Degilbo Granite (see Table 4). The range for magmatic and meteoric waters is from [68].

Biotite from the Degilbo Granite has δD values (−114 to −133%, Figure 16, Table 4) that suggest equilibrium with magmatic water of δD = −80 to −99% (using the fractionation factors of Suzuoki and Epstein 1976 [58]). The δD values of biotite do not show systematic changes with distance from the igneous contact that could infer the mixing of magmatic waters with the meteoric waters. This is also supported by field and petrological evidence that the Degilbo Granite has not interacted significantly with the hydrothermal fluids. An effect of imposed weathering on the δD values of biotite is unlikely as the analyzed samples are quite fresh. The involvement of meteoric water that would be highly depleted in D, however, could be explained by the geographic position of this part of eastern Australia in Triassic time at a latitude of 50–55° S [76,77].

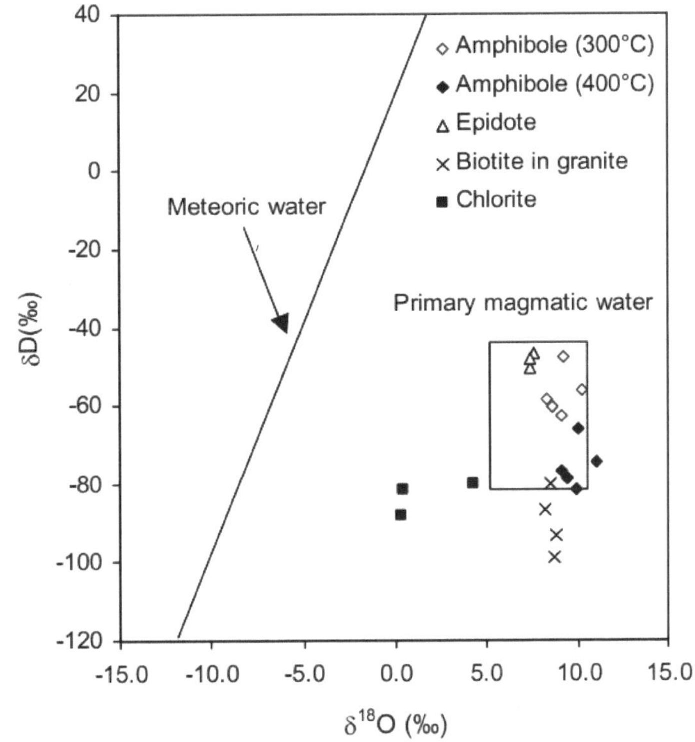

Figure 16. Calculated $\delta^{18}O$ versus δD diagram for fluids at temperatures in the range of 300–550 °C in equilibrium with skarn minerals and biotite samples from granite (see Table 4). The field for magmatic water and the line for meteoric waters are from [78].

Amphibole samples have δD values of −133 to −126%, which is consistent with the contribution of meteoric water during the retrograde stage. Similarly, the isotopic composition of water in equilibrium with chlorite samples at 310 °C (fractionation factors from Taylor and O'Neil, 1977 [79]) is lighter and may suggest the presence of meteoric water during the retrograde skarn formation. In contrast, the δD values of water in equilibrium with the amphibole samples at 300–400 °C range between −48 and −81%, which is consistent with δD for magmatic water. Although the $\delta^{18}O$ values of fluid in equilibrium with epidote show a slight shift towards the meteoric water composition, δD values for these fluids (−47 to −50% based on fractionation factors from Graham et al., 1980 [53]) are close to magmatic values. It is not clear to what extent the early prograde magmatic fluid was mixed with highly depleted meteoric water during the retrograde alteration, but the chlorite oxygen values show a significant contribution of meteoric water at the retrograde stage.

The sulfur isotope values of Biggenden sulfides are in the range of those for other magnetite-bearing skarn deposits and have been attributed to a magmatic origin (e.g., [2,60,80]). In terms of the Sr isotope values (Figure 12), the intermediate position of the ore-stage calcite between igneous rocks (Degilbo Granite and metabasalt) and sedimentary units (limestone and metasiltstone) may indicate that both the Degilbo Granite and the Gympie Group rocks supplied Sr to the hydrothermal fluid. The low initial $^{87}Rb/^{86}Sr$ ratios of the Degilbo Granite (0.70387–0.70404) are comparable to those of I-type granites (e.g., [22,23]). The lead isotope ratio suggests a model age of ~225 Ma for mineralization (Figure 14), which is consistent with the age proposed for the Degilbo Granite [15]. The lead isotope data also lie below the Cumming and Richards (1975) [81] growth curve, suggesting a mantle contribution to the ore Pb. Mantle Pb could have been derived from the Degilbo Granite, which is an I-type pluton, or by leaching of Pb from the Gympie Group basalts.

5.2. Classification of the Biggenden Skarn Deposit

Based on mineralogy and dominant metals present, the Biggenden deposit is a gold-bearing iron skarn [3,25]. Skarn deposits containing economic concentrations of magnetite are found in post-orogenic terrains similar to Biggenden and are associated with Fe-rich plutons [1]. However, Biggenden is different from typical island arc calcic magnetite skarn deposits, as they are usually associated with relatively primitive plutons and show widespread sodium metasomatism [1,3].

Biggenden has been mined for gold in the past, and it has common features with gold skarns. These common features include the presence of biotite hornfels, garnet-pyroxene alteration, clastic or volcaniclastic-rich protoliths, As and Bi minerals, and association with a shallow-level, silicic, magnetite-bearing pluton. Characteristics such as high ratios of garnet/pyroxene, low total sulfides, high abundance of pyrite, lack of pyrrhotite, minor occurrence of chalcopyrite and galena, and the presence of a pyrite and magnetite assemblage are all consistent with gold skarns, except that garnets are Fe-rich in Biggenden [82].

The average composition of the Degilbo Granite is higher in SiO_2 (68.5%) and K_2O (4.18%) and lower in MgO (0.9%) compared to plutons typically associated with Fe skarns (Figure 9). It is a weakly oxidized pluton and has an average Fe_2O_3/Fe_2O_3 + FeO ratio similar to plutons typically associated with Fe skarns (Figure 9). The Degilbo Granite has lower average values of Ni (5 ppm), Cr (9 ppm), V (34 ppm), and Sc (9 ppm) and also contains on average more Zr (270 ppm), Rb (176 ppm), Ba (694 ppm), and Sc (9 ppm) than plutons commonly associated with Fe skarns (Figure 9), which suggests that the Degilbo Granite crystallized from more fractionated magmas. It also has higher Rb/Sr (0.7) ratios than plutons associated with Fe-skarns (Figure 9). The major and trace element content of the Degilbo Granite is intermediate between the Fe-Cu-Au group and the Sn-W-Mo group of skarn-forming granitoids [3,82,83].

Garnet and pyroxene compositions from the Biggenden mine are analogous to those in several gold skarn deposits, particularly Nambija, Ecuador, and McCoy, Nevada [8]. The presence of Cl-rich scapolite in Biggenden is similar to both iron and gold skarn deposits (e.g., [36,84–86]).

Distribution of homogenization temperature data in the prograde-high temperature and retrograde-low temperature assemblages is common in iron skarns (e.g., Iron Hat [73]; Vyhne-Klokoc [87]). Moderate to high salinities and high homogenization temperatures in Biggenden are also similar to some gold skarn deposits [8].

In summary, in terms of ore mineralization, Biggenden is a gold-bearing iron skarn deposit. However, the associated granitic body, Degilbo Granite, at the exposed level is more fractionated, felsic, and oxidized than granites associated with common iron deposits. As a result, in terms of associated pluton, this deposit shows some similarities with Au and even the Sn-W-Mo group of skarn-forming granitoids.

5.3. Genetic Models

The spatial association of the Biggenden deposit with the Degilbo Granite may suggest that the ore-bearing fluids were derived directly from the adjacent pluton (model 1, Figure 17). Field evidence such as the scarcity of fracture systems and veins, lack of any significant structures that could act as fluid pathways, lack of greisen, and the general absence of endoskarn replacement in the granite may imply that there has been little infiltration across the contact into the pluton. The geochemical characteristics of the Degilbo Granite are analogous to those in intrusions associated with Fe-Cu-Au skarns as well as those associated with other skarn types. The presence of Cu and Au in the orebody suggests that there may be a genetic link between skarn mineralization and a deep-seated intermediate intrusion (perhaps of porphyry Cu-Au type) that provided some of the ore constituents, including Fe, Cu, and Au (model 2, Figure 17). It is noteworthy that a porphyry Cu system (Coalstoun deposit) is hosted by an early Triassic tonalitic intrusion in the district [88]. The formation of skarns from hidden intrusions has been suggested by other workers (e.g., [89–91]). There is, however, no field evidence (such as near-surface dikes) that could imply the presence of another pluton at depth.

Field relations, such as the proximity of ore deposition to a shear zone, numerous faults, and steeply dipping volcanic and sedimentary host rocks, suggest that the hydrothermal fluids responsible for skarn formation have most likely risen from depth along structural zones and sedimentary layer boundaries, rather than passing directly across the intrusive contact.

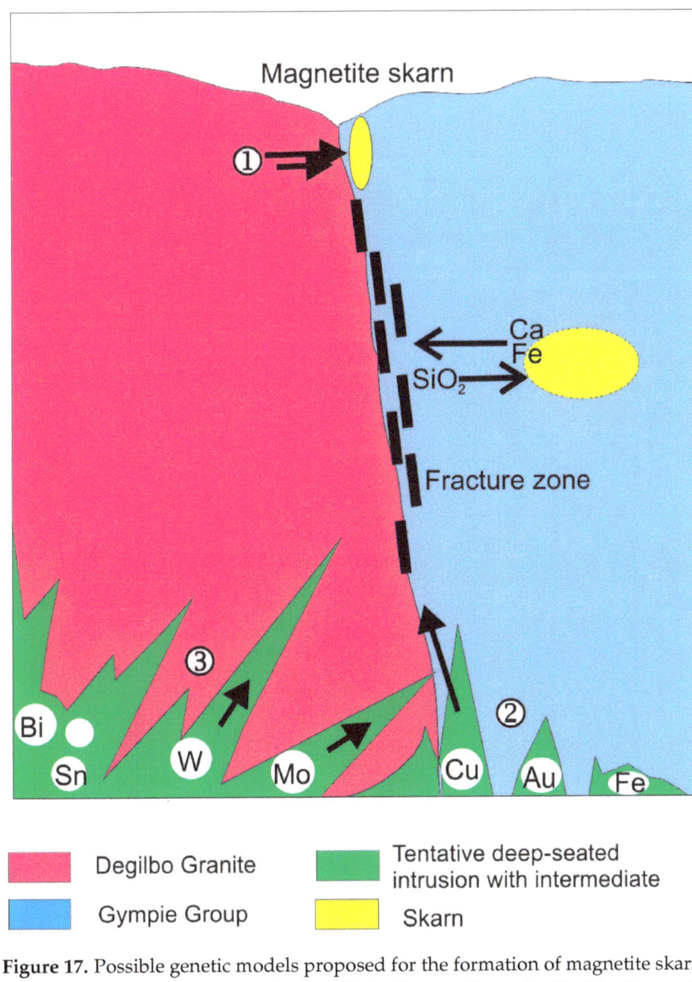

Figure 17. Possible genetic models proposed for the formation of magnetite skarn at the Biggenden mine. Metasomatism of SiO_2, Ca, and Fe is shown schematically. Numbers refer to the potential sources of metals.

Geochemical studies of the orebody [19] show that "granitophile" elements Sn, W, Mo, and Bi are enriched. In addition, a genetic link between skarn mineralization and the emplacement of the Degilbo Granite is, to some extent, supported by isotopic evidence. Skarn calcite has $\delta^{13}C$ and $\delta^{18}O$ values, which indicate deposition from magmatically derived fluids with isotopic compositions similar to that calculated for the fluid in equilibrium with the Degilbo Granite. Moreover, the calculated $\delta^{18}O$ values of the fluid in equilibrium with the skarn magnetite and amphibole are in the range of those for biotite and quartz from the Degilbo Granite. Lead and strontium isotope data also support a third genetic model, which involves both the Degilbo Granite and the Gympie Group.

The inconsistencies between the geochemical evidence (which suggests the role of the Degilbo Granite in the ore genesis) and the field evidence (that does not clearly support a direct genetic link between granite and skarn) can be explained by assuming that the Degibo Granite is in fact the source of hydrothermal fluids and metals but not at the current level of exposure. In our preferred model, hydrothermal fluids emanating from the Degilbo Granite at depth may have been channeled by the shear zone, intrusive contact, and through permeable Gympie Group rocks, reacted with adjacent marble and basalt, and deposited magnetite both as replacement bodies and as vein and breccia fillings. High-temperature and high-salinity fluids at Biggenden, similar to other Au-skarns [3,8], can transport significant amounts of gold as chloride complexes. This would allow a long-distance (and large-volume) metasomatic equilibration with low-Si and high-Fe rocks of the Gympie Group, leading to desilication of originally magmatic fluids and their enrichment in Fe and Ca (model 3, Figure 17).

6. Conclusions

- The Biggenden Gold-Bearing Fe Skarn is a calcic magnetite skarn type that is hosted by the Early Permian volcanic and sedimentary rocks of the Gympie Group due to the intrusion of the Late Triassic Degilbo Granite.
- The skarn mineralogy varies from the prograde garnet, clinopyroxene, magnetite, and scapolite to retrograde epidote and ferropargasite, followed by chlorite, calcite, actinolite, quartz, sulfides, gold, nontronite, and quartz.
- The prograde mineralization occurred at a high temperature (500–600 °C) from high salinity (up to 45 equivalent wt % NaCl) hydrothermal fluids. Lower temperatures (280–360 °C) and lower salinity (5–15 equivalent wt % NaCl) solutions precipitated the retrograde sulfide minerals.
- The δ18O values of the fluids indicate that the prograde minerals were precipitated from magmatic water, whereas the retrograde minerals were mainly deposited from a meteoric fluid. Sulfur was likely derived from a magmatic source.
- Sr and Pb isotopes suggest that the volcanic and sedimentary rocks of the Gympie Group may have partially provided the metals to the hydrothermal solution.

Author Contributions: M.E. did fieldwork, analysis, and initial interpretation of data. A.K.S. helped in the mineral analysis, mineral calculations, and interpretation. P.M.A. helped with project initiation, methodology, and interpretation. All authors have read and agreed to the published version of the manuscript.

Funding: This work was funded by research scholarship from the University of New England, Australia.

Data Availability Statement: Data are contained within the article.

Acknowledgments: The study has benefited from discussions with Nick Stephenson, Peter Flood, Barrie McKelvey, Warwick Sivell, and Bernd Lottermoser. The authors would like to thank the Centre for Isotope Studies at the CSIRO, especially Anita Andrew, for their help with the isotope analyses. Commercial Minerals and Geoff Weekes are thanked for allowing access to the mine and the drill cores. The manuscript benefitted greatly from critical reviews by anonymous reviewers

Conflicts of Interest: The authors declare no conflicts of interest.

References

1. Einaudi, M.T.; Meinert, L.D.; Newberry, R.J. Skarn deposits. *Econ. Geol.* **1981**, *75*, 317–391.
2. Meinert, L.D. Skarns and skarn deposits. *Geosci. Can.* **1992**, *19*, 145–162.
3. Meinert, L.D.; Dipple, G.M.; Nicolescu, S. World Skarn Deposits. In *Economic Geology*; Hedenquist, J.W., Thompson, J.F.H., Goldfarb, R.J., Richards, J.P., Eds.; Society of Economic Geologists: Littleton, CO, USA, 2005; 100th Anniversary Volume, pp. 299–336.
4. Ren, L.; Huang, J.; Wang, X.; Yang, S.; Yang, C.; Zhao, C.; Wang, L.; Mei, W.; Deng, M.; Zhou, Y. Magmatic control on orebody distribution of porphyry-skarn gold-copper deposit: A case study of Beiya deposit from Sanjiang metallogenic belt in the southwest China. *Ore Geol. Rev.* **2024**, *175*, 106350. [CrossRef]
5. Sui, J.X.; Li, J.W.; Wen, G.; Jin, X.Y. The Dewulu reduced Au-Cu skarn deposit in the Xiahe-Hezuo district, West Qinling orogen, China: Implications for an intrusion-related gold system. *Ore Geol. Rev.* **2017**, *80*, 1230–1244. [CrossRef]
6. Zhou, H.; Sun, X.; Wu, Z.; Huang, Q. Timing of skarn gold deposition in the giant Beiya polymetallic gold deposit, southwest China: Constraints from in situ monazite SIMS U-Th-Pb geochronology. *Ore Geol. Rev.* **2019**, *106*, 226–237. [CrossRef]
7. Weekes, G. Biggenden magnetite mine, assessment of reserves March 1992 and preliminary proposal for final development, Commercial Minerals. 1992; Unpublished report No. 5/92. Brisbane, Queensland, Australia.
8. Meinert, L.D. A review of skarns that contain gold. *Mineral. Assoc. Can. Short Course Ser.* **1988**, *26*, 359–414.
9. Seccombe, P.K.; Downes, P.M.; Ashley, P.M.; Brathwaite, R.L.; Green, G.R.; Murray, C.G.; Rubenach, M.J. World Skarn Deposits: Skarns of the Southwest Pacific. In *Economic Geology*; Hedenquist, J.W., Thompson, J.F.H., Goldfarb, R.J., Richards, J.P., Eds.; Society of Economic Geologists: Littleton, CO, USA, 2005; 100th Anniversary Volume, pp. 299–336.
10. Clarke, D.E. The Geology of the Mount Biggenden Gold and Bismuth Mine and Environs. Unpublished Bachelor's Thesis, University of Queensland, Brisbane, Australia, 1963; p. 137.
11. Clarke, D.E. *Geology of the Mount Biggenden Gold and Bismuth Mine and Environs*; Report nr. 32; Geological Survey of Queensland: Brisbane, Australia, 1 January 1969; p. 16.
12. Tellam, L. Geology of the Mt. Biggenden Magnetite Mine Environs, Southeast Queensland, Geochemical and Mineralogical Aspects. Unpublished Bachelor's Thesis, University of Queensland, Brisbane, Australia, 1980; p. 71.
13. Li, P.; Rosenbaum, G.; Yang, J.; Hoy, D. Australian-derived detrital zircons in the Permian-Triassic Gympie terrane (eastern Australia): Evidence for an autochthonous origin. *Tectonics* **2015**, *34*, 858–874. [CrossRef]
14. Holcombe, R.J.; Stephens, C.J.; Fielding, C.R.; Gust, D.; Little, A.; Sliwa, R.; Kassan, J.; McPhie, J.; Ewart, A. Tectonic evolution of the northern New England fold belt: The Permian-Triassic Hunter-Bowen event. In *Tectonics and Metallogenesis of the New England Orogen*; Ashley, P.M., Flood, P.G., Eds.; Geological Society Special Publication: London, UK, 1997; Volume 19, pp. 52–65.
15. Cranfield, L.C. *1: 250,000 Geological Series-Explanatory Notes, Maryborough*; Queensland Sheet SG56-6; Department of Minerals and Energy: Brisbane, Australia, 1994; p. 120.
16. Jell, P.A.; Cranfield, L.C. Gympie Province. In *Geology of Queensland*; Jell, P.A., Ed.; Geological Survey of Queensland: Brisbane, Australia, 2013; pp. 369–371.
17. Sivell, W.J.; Waterhouse, J.B. Petrogenesis of Gympie Group volcanics; evidence for remnants of an Early Permian volcanic arc in eastern Australia. *Lithos* **1988**, *21*, 81–95. [CrossRef]
18. Hoy, D.; Rosenbaum, G. Episodic behavior of Gondwanide deformation in eastern Australia: Insights from the Gympie Terrane. *Tectonics* **2017**, *36*, 1497–1520. [CrossRef]
19. Edraki, M. Geochemistry, Mineralogy and Genesis of the Biggenden Au-Bi-Fe Skarn Deposit, Southeast Queensland, Australia. Unpublished Ph.D. Thesis, University of New England, Armidale, Australia, 2000; p. 318.
20. Cranfield, L.C.; Murray, C.G. Geochemistry and tectonic setting of granitic rocks in the Maryborough 1:250 000 sheet area, Southeast Queensland—Caractéristiques géochimiques et tectoniques des roches granitiques sur la feuille au 1:250 000 de Maryborough, SE de Queensland. *Queensl. Gov. Min. J.* **1989**, *90*, 408–415.
21. Webb, A.W.; McDougall, I. Isotopic dating evidence on the age of the Upper Permian and Middle Triassic. *Earth and Planetary Sci. Lett.* **1967**, *2*, 483–488. [CrossRef]
22. Chappell, B.W.; White, A.J.R. I- and S-type granites in the Lachlan Fold Belt. *Trans. R. Soc. Edinb. Earth Sci.* **1992**, *83*, 1–26.
23. Chappell, B.W.; White, A.J.R. Two contrasting granite types: 25 years later. *Aust. J. Earth Sci.* **2001**, *48*, 489–499. [CrossRef]
24. Cobine, T.J. The Analytic Signal and Cross-Correlation Applied in Rapid, Detailed Analysis of Aeromagnetic and Radiometric Data. Unpublished Master's Thesis, University of New England, Armidale, Australia, 1997; p. 75.
25. Roedder, E. Fluid inclusions. *Rev. Mineral.* **1984**, *12*, 644.
26. Clayton, R.N.; Mayeda, T.K. The use of bromine pentafluoride in the extraction of oxygen from oxides and silicates for isotopic analysis. *Geochem. Cosmochim. Acta* **1963**, *27*, 43–52. [CrossRef]
27. Friedman, I. Deuterium content of natural waters. *Geochim. Cosmochim. Acta* **1953**, *4*, 89–103. [CrossRef]

28. Robinson, B.W.; Kusakabe, M. Quantitative preparation of sulfur dioxide for $^{34}S/^{32}S$ analysis from sulfides by combustion with cuprous oxide. *Anal. Chem.* **1975**, *47*, 1179–1181. [CrossRef]
29. McCrea, J.M. The isotope geochemistry of carbonates and a paleotemperature scale. *J. Chem. Phys.* **1950**, *18*, 849–857. [CrossRef]
30. Sies, S.H.; Niklaus, T.R.; Sims, D.A.; Bruhn, F.; Suter, G.; Cripps, G. AUSTRALIS: A new tool for the study of isotopic systems and geochronology in mineral systems. *Aust. J. Earth Sci.* **2002**, *49*, 601–611.
31. Theodore, T.G.; Orris, G.J.; Hammarstrom, J.M.; Bliss, J.D. *Gold-Bearing Skarns*, 1930th ed.; U.S. Government Printing Office: Washington, DC, USA, 1991; pp. 1–35.
32. Nakano, T.; Yoshino, T.; Shimazaki, H.; Shimizu, M. Pyroxene composition as an indicator in the classification of skarn deposits. *Econ. Geol.* **1994**, *89*, 567–1580. [CrossRef]
33. Matsueda, H. Pyrometasomatic iron-copper ore deposits of the Sampo Mine, Okayama Prefecture; II, The modes of occurrences, mineral paragenesis and chemical compositions of skarn and ore. *Min. Geol.* **1981**, *6*, 1–43.
34. Einaudi, M.T.; Burt, D.M. Introduction—Terminology, classification, and composition of skarn deposits. *Econ. Geol.* **1982**, *77*, 745–754. [CrossRef]
35. Meinert, L.D. Gold skarn deposits–geology and exploration criteria. *Econ. Geol.* **1989**, *65*, 537–552.
36. Pan, Y. Scapolite in skarn deposits; petrogenetic and geochemical significance. *Mineral. Assoc. Can. Short Course Handb.* **1998**, *26*, 169–210.
37. Pan, Y.; Fleet, M.E.; Ray, G.E. Scapolite in two Canadian gold deposits: Nickel Plate British Columbia and Hemlo. *Ontario Can. Mineral.* **1994**, *32*, 825–837.
38. Meinert, L.D. Mineralogy and petrology of iron skarns in western British Columbia, Canada. *Econ. Geol.* **1984**, *79*, 869–882. [CrossRef]
39. Cathelineau, M. The chlorite and illite geothermometers. *Chem. Geol.* **1988**, *70*, 182–183. [CrossRef]
40. Le Bas, M.J.; Streckeisen, A.L. The IUGS systematics of igneous rocks. *J. Geol. Soc. Lond.* **1991**, *148*, 825–833. [CrossRef]
41. Peccerillo, A.; Taylor, S.R. Geochemistry of Eocene calc-alkaline volcanic rocks from the Castamoun area, northern Turkey. *Contrib. Mineral. Petrol.* **1976**, *58*, 63–81. [CrossRef]
42. Kwak, T.A.P. Fluid inclusions in skarns (carbonate replacement deposits). *J. Metamorph. Geol.* **1986**, *4*, 363–384. [CrossRef]
43. Kwak, T.A.P.; Tan, T.H. The geochemistry of zoning in skarn minerals at the King Island (Dolphin) Mine. *Econ. Geol.* **1981**, *76*, 468–497. [CrossRef]
44. Meinert, L.D. Skarn zonation and fluid evolution in the Groundhog Mine, Central mining district, New Mexico. *Econ. Geol.* **1987**, *82*, 523–545. [CrossRef]
45. Potter, R.W.; Clynne, M.A.; Brown, D.L. Freezing point depression of aqueous sodium chloride solutions. *Econ. Geol.* **1978**, *73*, 284–285. [CrossRef]
46. Bodnar, R. Revised equation and table for determining the freezing point depression of H_2O-Nacl solutions. *Geochim. Cosmochim. Acta* **1993**, *57*, 683–684. [CrossRef]
47. Lentz, D.R. Carbonatite genesis: A reexamination of the role of intrusion-related pneumatolytic skarn processes in limestone melting. *Geology* **1999**, *27*, 335–338. [CrossRef]
48. Xu, X.; Xu, X.; Szmihelsky, M.; Yan, J.; Xie, Q.; Steele-MacInnis, M. Melt inclusion evidence for limestone assimilation, calc-silicate melts, and "magmatic skarn". *Geology* **2023**, *51*, 491–495. [CrossRef]
49. Ohmoto, H. Stable isotope geochemistry of ore deposits. In *Stable Isotopes in High Temperature Geological Processes*; Valley, J.W., Taylor, H.P., O'Neil, J.R., Eds.; Mineralogical Society of America: Chantilly, VA, USA, 1986; Volume 16, pp. 491–559.
50. Valley, J.W. Stable isotope geochemistry of metamorphic rocks. *Rev. Mineral.* **1986**, *16*, 445–489.
51. Taylor, H.P.; Sheppard, S.M.F. Igneous rocks, I. Processes of isotopic fractionation in isotope systematics. *Rev. Mineral.* **1986**, *16*, 227–271.
52. Bottinga, Y.; Javoy, M. Comments on oxygen isotope geothermometry. *Earth Planet. Sci. Lett.* **1973**, *20*, 250–265. [CrossRef]
53. Graham, C.M.; Sheppard, S.M.F.; Heaton, T.H.E. Experimental hydrogen isotope studies—I. Systematics of hydrogen isotope fractionation in the systems epidote-H_2O, zoisite-H_2O and AlO(OH)-H_2O. *Geochimica Cosmochimica Acta* **1980**, *44*, 353–364. [CrossRef]
54. Bottinga, Y.; Javoy, M. Oxygen isotope partitioning among the minerals in igneous and metamorphic rocks. *Rev. Geophys.* **1975**, *13*, 401–418. [CrossRef]
55. Matsushisa, Y.; Goldsmith, N.; Tanaka, T. Oxygen isotope fractionation in the system quartz-albite-anorthite-water. *Geochim. Cosmochim. Acta* **1979**, *43*, 1131–1140. [CrossRef]
56. Wenner, D.B.; Taylor, H.P., Jr. Temperatures of serpentinization of ultramafic rocks based on $^{18}O/^{16}O$ fractionation between coexisting serpentine and magnetite. *Contrib. Mineral. Petrol.* **1971**, *32*, 165–185. [CrossRef]
57. Graham, C.M.; Harmon, R.S.; Sheppard, S.M.F. Experimental hydrogen isotope studies, hydrogen isotope exchange between amphibole and water. *Am. Mineral.* **1984**, *69*, 128–138.

58. Suzuoki, T.; Epstein, S. Hydrogen isotope fractionation between OH-bearing minerals and water. *Geochim. Cosmochim. Acta* **1976**, *40*, 1229–1240. [CrossRef]
59. Kyser, T.K. Equilibrium fractionation factors for stable isotopes. In *Stable Isotope Geochemistry of Low Temperature Processes*; Mineralogical Association of Canada: Montreal, QC, Canada, 1987; Volume 13, pp. 1–84.
60. Ohmoto, H.; Goldhaber, M.B. Sulfur and carbon isotopes. In *Geochemistry of Hydrothermal Ore Deposits*, 3rd ed.; Barnes, H.L., Ed.; John Wiley and Sons: New York, NY, USA, 1997; pp. 517–612.
61. Faure, G. *Principles of Isotope Geology*, 2nd ed.; John Wiley and Sons: New York, NY, USA, 1986; p. 589.
62. Carr, G.R.; Dean, J.A.; Suppel, D.W.; Heithersay, P.S. Precise lead isotope fingerprinting of hydrothermal activity associated with Ordovician to Carboniferous metallogenic events in the Lachlan Fold Belt of New South Wales. *Econ. Geol.* **1995**, *90*, 1467–1505. [CrossRef]
63. Cartwright, I.; Weaver, T.R. Fluid-rock interaction between syenites and marbles at Stephen Cross Quarry, Québec, Canada: Petrological and stable isotope data. *Contrib. Mineral. Petrol.* **1993**, *113*, 533–544. [CrossRef]
64. Baker, T.; Lang, J.R. Reconciling fluid inclusions, fluid processes and fluid source in skarns: An example from the Bismarck skarn deposit, Mexico. *Miner. Depos.* **2003**, *38*, 474–495. [CrossRef]
65. Bowman, J.R. Stable-isotope systematics of skarns. *Mineral. Assoc. Can. Short Course Ser.* **1998**, *26*, 99–145.
66. Bowman, J.R.; O'Neil, J.R.; Essene, E.J. Contact skarn formation at Elkhorn, Montana. *II: Origin and evolution of C-O-H skarn fluids. Am. J. Sci.* **1985**, *285*, 621–660.
67. Vander-Auwera, J.; Andre, L. Trace elements (REE) and isotopes (O, C, Sr) to characterize the metasomatic fluid sources: Evidence from the skarn deposit (Fe, W, Cu) of Traversella (Ivrea, Italy). *Contrib. Mineral. Petrol.* **1991**, *106*, 325–339. [CrossRef]
68. Taylor, B.E. Stable isotope geochemistry of ore-forming fluids. *Mineral. Assoc. Can. Short Course Handb.* **1987**, *13*, 377–445.
69. Zaw, K.; Singoyi, B. Formation of Magnetite-Scheelite Skarn Mineralization at Kara, Northwestern Tasmania: Evidence from Mineral Chemistry and Stable Isotopes. *Econ. Geol.* **2000**, *95*, 1215–1230. [CrossRef]
70. So, C.-S.; Rye, D.M.; Shelton, K.L. Carbon, hydrogen, oxygen, and sulfur isotope and fluid inclusion study of the Weolag tungsten-molybdenum deposit, Republic of Korea; fluid histories of metamorphic and ore-forming events. *Econ. Geol.* **1983**, *78*, 1551–1573. [CrossRef]
71. Nabelek, P.I. Stable isotope monitors. In *Contact Metamorphism*; Kerrick, D.M., Ed.; Reviews in Mineralogy: Chantilly, VA, USA, 1991; Volume 26, pp. 395–435.
72. Friedman, I.; O'Neil, J.R. Compilation of stable isotope fractionation factors of geochemical interest. In *Data of Geochemistry*; Fleischer, M., Ed.; United States Government Printing Office: Washington, DC, USA, 1977.
73. Hall, D.L.; Cohen, L.H.; Schiffman, P. Hydrothermal alteration associated with Iron Hat skarn deposit, San Bernardino County, California. *Econ. Geol.* **1988**, *83*, 568–587. [CrossRef]
74. Zürcher, L.; Ruiz, J.; Barton, M.D. Paragenesis, elemental distribution, and stable isotopes at the Peña Colorada iron skarn, Colima, Mexico. *Econ. Geol.* **2001**, *96*, 535–557. [CrossRef]
75. Taylor, H.P. Comparison of hydrothermal systems in layered gabbros and granites and the origin of low ^{18}O magmas. *Geochem. Soc. Spec. Publ.* **1987**, *1*, 337–357.
76. Ashley, P.M.; Andrew, A.S. The Mt Ninderry acid sulphate alteration zone and its relation to epithermal mineralization in the North Arm Volcanics, southeast Queensland. *Aust. J. Earth Sci.* **1992**, *39*, 79–98. [CrossRef]
77. Sun, S.S.; Eadington, P.J. Oxygen isotope evidence for the mixing of magmatic and meteoric waters during tin mineralization in the Mole Granite, New South Wales, Australia. *Econ. Geol.* **1987**, *82*, 45–52. [CrossRef]
78. Taylor, H.P. Oxygen and hydrogen isotope studies of plutonic granite rocks. *Earth Planet. Sci. Lett.* **1974**, *38*, 177–210. [CrossRef]
79. Taylor, B.E.; O'Neil, J.R. Stable isotope studies of metasomatic Ca-Fe-Al-Si skarns and associated metamorphic and igneous rocks, Osgood Mountains, Nevada. *Contrib. Mineral. Petrol.* **1977**, *63*, 1–49. [CrossRef]
80. Baranov, E.N.; Grinenko, L.N.; Pavlov, G.P. Pyrite sulfur isotope compositions in Ural skarn-magnetite deposits. *Geochem. Int.* **1986**, *23*, 81–90.
81. Cumming, G.L.; Richards, J.R. Ore lead isotope ratios in a continuously changing earth. *Earth Planet. Sci. Lett.* **1975**, *28*, 155–171. [CrossRef]
82. Meinert, L.D. Compositional variation of igneous rocks associated with skarn deposits—Chemical evidence for a genetic connection between petrogenesis and mineralization. *Mineral. Assoc. Can. Short Course* **1995**, *23*, 401–418.
83. Newberry, R.J. The formation of sub-calcic garnet in scheelite-bearing skarns. *Can. Mineral.* **1983**, *21*, 529–544.
84. Frietsch, R.; Tuisku, P.; Martinsson, O.; Perdahl, J.-A. Early proterozoic Cu, Au, and Fe ore deposits associated with regional NaCl metasomatism in northern Fennoscandia. *Ore Geol. Rev.* **1998**, *12*, 1–34. [CrossRef]
85. Steven, N.M.; Moore, J.M. Pan-African tungsten skarn mineralization at the Otjua Prospect, central Namibia. *Econ. Geol.* **1994**, *89*, 1431–1453. [CrossRef]
86. Zhao, Y.; Lin, W.; Bi, C.; Li, D.; Jiang, C. *Skarn Deposits of China*; Chinese Academy Geological Sciences: Beijing, China, 1994.

87. Kodera, P.; Rankin, A.H.; Lexa, J. Evolution of fluids responsible for iron skarn mineralization: An example from the Vyhne-Klokoc deposit, Western Carpathians, Slovakia. *Mineral. Petrol.* **1998**, *64*, 119–147. [CrossRef]
88. Ashley, P.M.; Billington, W.G.; Graham, R.L.; Neale, R.C. Geology of the Coalstoun porphyry copper prospect, southeast Queensland, Australia. *Econ. Geol.* **1978**, *73*, 945–965. [CrossRef]
89. Lee, C.H.; Lee, H.K.; Kim, S.J. Geochemistry and mineralization age of magnesian skarn-type iron deposits of the Janggun mine, Republic of Korea. *Miner. Depos.* **1998**, *33*, 379–390. [CrossRef]
90. Meinert, L.D. Skarn, manto, and breccia pipe formation in sedimentary rocks of the Cananea mining district, Sonora, Mexico. *Econ. Geol.* **1982**, *77*, 919–949. [CrossRef]
91. Mueller, A.G. The Savage Lode magnesian skarn in the Marvel Loch gold-silver mine, Southern Cross greenstone belt, Western Australia, Part 1; Structural setting, petrography, and geochemistry. *Can. J. Earth Sci.* **1991**, *28*, 659–685. [CrossRef]

minerals

MDPI

Article

Magnetite Texture and Geochemistry in the Takab Ore Deposit (NW Iran): Implications for a Complex Hydrothermal Evolution

Christiane Wagner [1,*], Omar Boudouma [1], Nicolas Rividi [2], Beate Orberger [3], Ghasem Nabatian [4], Maryam Honarmand [5] and Iman Monsef [5]

1 Sorbonne Université, CNRS-INSU, Institut des Sciences de la Terre de Paris, ISTeP, F-75005 Paris, France; omar.boudouma@sorbonne-universite.fr
2 Sorbonne Université, CNRS-INSU, Institut de Physique du Globe de Paris, IPGP, F-75005 Paris, France; nicolas.rividi@sorbonne-universite.fr
3 Laboratoire Géosciences Paris-Saclay, GEOPS, Université Paris-Saclay, CNRS, F-91405 Orsay, France; beate.orberger@universite-paris-saclay.fr
4 Department of Geology, Faculty of Science, University of Zanjan, University Blvd., Zanjan 45371-38791, Iran; gh.nabatian@znu.ac.ir
5 Department of Earth Sciences, Institute for Advanced Studies in Basic Sciences (IASBS), Zanjan 45137-66731, Iran; m.honarmand@iasbs.ac.ir (M.H.); iman.monsef@iasbs.ac.ir (I.M.)
* Correspondence: christiane.wagner_raffin@sorbonne-universite.fr

Abstract: The massive magnetite deposit from Takab (NW Iran) is hosted in amphibolite layers intercalated with the chemical and terrigenous sediments of the Takab BIF. A comprehensive textural and chemical study allowed three types of magnetite (Mt) to be distinguished. Mt1 forms large (≤1 mm) inhomogeneous grains surrounded and locally invaded by magnetite Mt2. Oscillatory zoning is present in Mt1 and Mt2. Mt3 forms bands aligned along fracture planes. Mt3 may contain hematite relicts and is porous in proximity to hematite. Mt1 shows variable and higher Si (up to 1.4 wt. %), Al, Ca, and Mg and lower Fe content (68 wt. %) than Mt2. Mt3 has the lowest Si (<0.3 wt. %) and highest Fe (71 wt. %) contents. The temperature of formation decreases from Mt1 (600 °C) to Mt2 (500–550 °C) and Mt3 (380–440 °C). Mt1 likely formed in a reducing Si-rich environment. The close spatial relationship, sharp compositional boundaries, similar crystallographic structure of Mt1 and Mt2, and porosity in Mt2 suggest a fluid-assisted coupled dissolution of Mt1 and precipitation of Mt2 (CDR process). Microfracturing allowed the penetration of oxidizing fluid and the formation of platy hematite bands. Mt3 (mushketovite) formed after hematite by interaction with a reducing lower temperature fluid through a redox transformation.

Keywords: magnetite; textural evolution; trace elements; CDR process; hydrothermal fluid; EMPA; SEM; EBSD; Takab; Iran

Academic Editors: Yuichi Morishita and Napoleon Q. Hammond

Received: 24 September 2024
Revised: 24 January 2025
Accepted: 25 January 2025
Published: 29 January 2025

Citation: Wagner, C.; Boudouma, O.; Rividi, N.; Orberger, B.; Nabatian, G.; Honarmand, M.; Monsef, I. Magnetite Texture and Geochemistry in the Takab Ore Deposit (NW Iran): Implications for a Complex Hydrothermal Evolution. *Minerals* **2025**, *15*, 137. https://doi.org/10.3390/min15020137

1. Introduction

Magnetite, a common mineral in many ore deposits, can contain minor and trace elements which have been widely used to constrain the environment of formation [1–3]. However, both the texture and trace-element composition of magnetite can be significantly modified or re-equilibrated by metasomatic fluids, e.g., [4–7]. Therefore, detailed studies of texture and chemical composition must be carried out to defining the processes of magnetite formation.

Iran is the ninth most important iron ore producer in the world with about 4 billion tons of iron ore annually produced from more than 200 iron deposits [8–11]. Iron ore deposits were formed during several metallogenic stages: (1) Neoproterozoic–early Cambrian,

(2) late Paleozoic–early Mesozoic, and (3) Cenozoic. The largest iron ore deposits formed during stage 1, mainly as Kiruna-type (IOA, iron oxide apatite) deposits, and during stage 3, mainly as Fe-skarn deposits [12–15]. Their spatial distribution is related to the main suture zones of the fragmented Iranian continental blocks resulting from the complex evolution of the Tethyan oceans [10,16,17]. The most important iron deposits are in the Central Iranian Zone (CIZ), in the Sanandaj–Sirjan zone (SSZ), and in the Alborz (AMB)–Azerbaijan belts (Figure 1; e.g., [18–20] and references therein). The SSZ extends 1500 km from northwest to southeast parallel to the Zagros belt, on the east side of the Main Zagros Fault. It hosts numerous iron and iron–manganese deposits of various origins (Figure 1; [10,21]). For example, in the south of the SSZ, the major Gol-e-Gohar iron deposit and the Heneshk iron–manganese deposit are related to the opening of the Neo-Tethys Ocean and are interpreted to be of volcano–sedimentary or mixed hydrothermal and volcano–sedimentary origin [18]. Further north, the Fe-skarn-type Hamekasi iron deposits (e.g., Baba Ali) are related to the subduction and closure of the Neo-Tethys Ocean (Figure 1; [10] and references therein). In northwestern Iran, the major deposit of the Zanjan-Soltaniyeh area, which belongs to the near structural Central Iranian Zone, comprises both skarn-type and volcano–sedimentary iron deposits [10,20]. A hundred km west of Zanjan in the northwestern part of the SSZ, several iron mineralizations have been reported in the Takab area. In this area, some outcrops were initially explored in the 1990s and studied for their geology and geochemistry by a few authors: at 60 km east of Takab town (Figure 1), the Shahrak mining complex represents reserves of a total of 60 million tons of iron ore with an average Fe grade of 50% [22,23]. It comprises 10 ore deposits, among them the Korkora skarn deposit [23] and the Kosaj volcano–sedimentary-type iron deposit [24]. The iron deposit of Mianaj, near Halab (Figure 2), was also reported as a metamorphosed and folded volcano-sedimentary-type iron deposit [25]. Different petrogenetic models have been proposed, such as the superior-type BIF [26] and Algoma-type BIF [27], for the Halab and Ouzijan iron mineralizations.

In a recent study on the Takab magnetite mineralizations near Halab, Honarmand et al. [8] investigated the banded, nodular, and disseminated magnetite layers in the folded succession of micaschist and quartzite (the studied outcrops are represented by yellow circles in Figure 2). Based on geochemical, Nd isotopic, and geochronological data, they proposed that this iron ore corresponds to a Rapitan-type banded iron formation. These data indicate that it originated in a back-arc basin through iron precipitation and the incorporation of ca. 20% of the terrigenous materials in the chemical precipitates. The Neoproterozoic felsic–intermediate crystalline basement was the main sediment source [8]. Furthermore, trace-element and Fe and O isotopic investigations of the banded, nodular, and disseminated magnetite [28] support their genesis from a mixture of seawater and hydrothermal fluid environments. Isotopic data point out the complex hydrothermal history from seafloor high-T° fluid to later re-equilibration with lower-T° fluid produced during the decarbonatization of the intercalated carbonate-rich schists.

Close to the nodular and disseminated magnetite mineralization, massive magnetite layers are intercalated in amphibolite. These layers were never studied in detail for petrography, iron oxide chemistry, or textural features.

This study presents for the first time major- and trace-element electron microprobe data and EBSD analyses of the massive magnetite layers in the vicinity of banded, nodular, and disseminated magnetite layers that were previously studied.

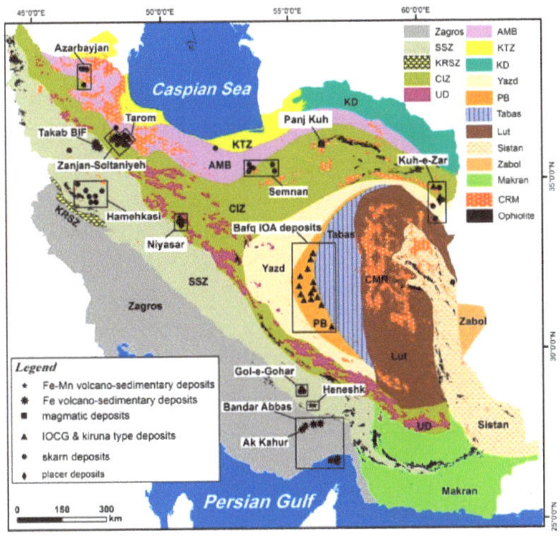

Figure 1. Tectonic and structural map of Iran (after [10]). The red star indicates the location of the Takab study area (Takab town GPS coordinates are 36°24′2″ E and 47°6′42″ E). The black boxes correspond to the locations of different types of iron ore deposits in Iran. Zagros, Zagros ranges; KRSZ, Kermanshah Radiolarites subzone; SSZ, Sanandaj–Sirjan zone; UD, Urumieh–Dokhtar magmatic arc; the Central Iranian microcontinent includes Yadz, Posht-e-Badam block (PB), Tabas, and Lut blocks; AMB, Alborz magmatic arc; KTZ, Khazar–Talesh–Ziveh structural zone; CIZ, Central Iranian zone; Sistan, East Iran ranges; Makran, Makran zone; KD, Kopeh–Dagh ranges; Zabol, Zabol area; CMR, Cenozoic magmatic rocks [10].

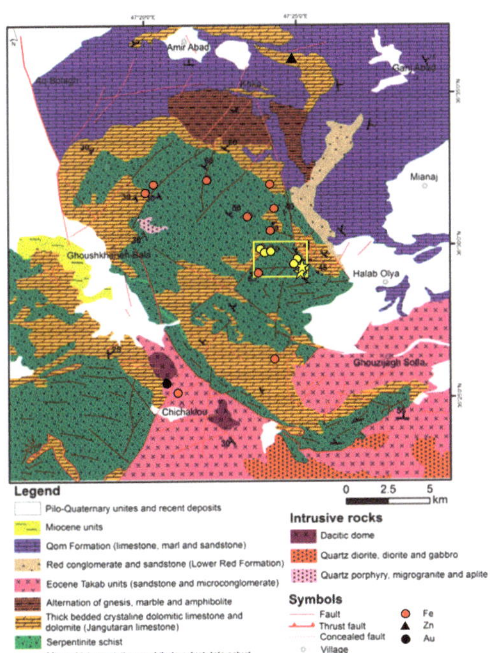

Figure 2. Simplified geological map of the Takab area modified after [8]. The marked area (47°23′51″ E to 47°25′39″ E longitude and 36°29′13″ N to 36°30′03″ N latitude) limits the location of the Takab iron ore samples from previous studies (yellow circles: banded, nodular, and disseminated iron ore, [8,28]) and from this study (yellow stars: massive iron ore).

2. Materials and Methods

2.1. Geological Settings

The Takab area is located in the northwestern part of the Sanandaj–Sirjan magmatic-metamorphic structural zone (Figure 1). The Takab area consists of metamorphic rocks including gneisses, amphibolites, mica- and calc-schists, quartzite, jaspilite, marble, and rare granulite. These rocks were intruded by granitoids, diorite, gabbros, and rare rhyolite [8,29,30]. The area experienced several metamorphic and deformation events during the Precambrian and the Tertiary. The latter event is related to the Alpine orogeny, leading to the faulting and folding of the Precambrian metamorphic units. The Tertiary metamorphic event produced the migmatites, at ca. 25 Ma [31], and the granitoids [10,29,31].

The metamorphic units of the late Neoproterozoic–early Cambrian (548–568 Ma) age [29,31] were overlain unconformably by Tertiary sedimentary units. The lack of Paleozoic and Mesozoic rocks indicates that this region was exhumated at that time [32]. Exhumation started during the Early to Middle Jurassic back-arc extension and Tertiary crustal extension related to compressional tectonics [29]. The different rock units experienced upper greenschist to amphibolite facies metamorphism (450° to 700 °C and 0.6–1 GPa) in the Middle Jurassic. After exhumation and weathering, the ore was covered by Quaternary sediments [18].

2.2. Ore Deposit and Sample Material

The Takab iron ore outcrops are located west of the village of Halab (marked area in Figure 2). Recently, Honarmand et al. [8] and Wagner et al. [28] studied outcrops of banded and nodular magnetite layers up to 2 m in thickness in the folded succession of micaschist and quartzite from the Takab BIF, 3 to 6 km west of Halab (47°23′51″ E to 47°25′39″ E longitude and 36°29′13″ N to 36°30′03″ N latitude, Figure 2, yellow circles). In this study, we investigate the massive magnetite layers intercalated with micaschist and amphibolite (former meta-basalt) in the Takab BIF. The samples were collected ca. 3 km west of Halab along an exposure 200 m in length situated at 47°25′28.1″ E and 36°29′25.5″ N (Figure 2, yellow stars).

The iron ore is hosted in a metamorphosed volcano–sedimentary sequence composed of mica- and calc-schists, quartzite, and jaspilite. Some marble and amphibolite (meta-basalt) bands are intercalated with the micaschists (Figure 3). U-Pb dating on zircons from micaschist lower and upper layers of the ore body indicates a depositional age of ca. 560 Ma [8]. The banded, nodular, and disseminated iron ore consist of magnetite, which may be partially hematized and surrounded by goethite in some samples [8,28]. The massive magnetite layers are hosted in amphibolite interlayered with the micaschists and quartz sequence. The host amphibolite is a massive, dark green rock with schistose texture (Figures 4a and 5). At the microscale, transparent patches of albite (Ab_{92}, Supplementary Table S1) and microquartz occur in a light to dark green matrix mainly composed of amphibole and minor biotite (Figure 4a). The amphibole forms elongate crystals of actinolite and a few acicular crystals of Mg-hornblende (Table 1). Actinolite hosts relicts of a more Mg-rich amphibole (Figure 4b and Table 1). Euhedral crystals of titanite, interstitial calcite, and large patches of epidote and chlorite are also present (Figure 4b–e). Chlorite is mostly developed along amphibole and mica cleavages (Figure 4d) or as a few rosette-shaped interstitial crystals. Accessories are barite (Figure 4f), rare Cl-free apatite and allanite, and late pyrite. (Figure 4g,h).

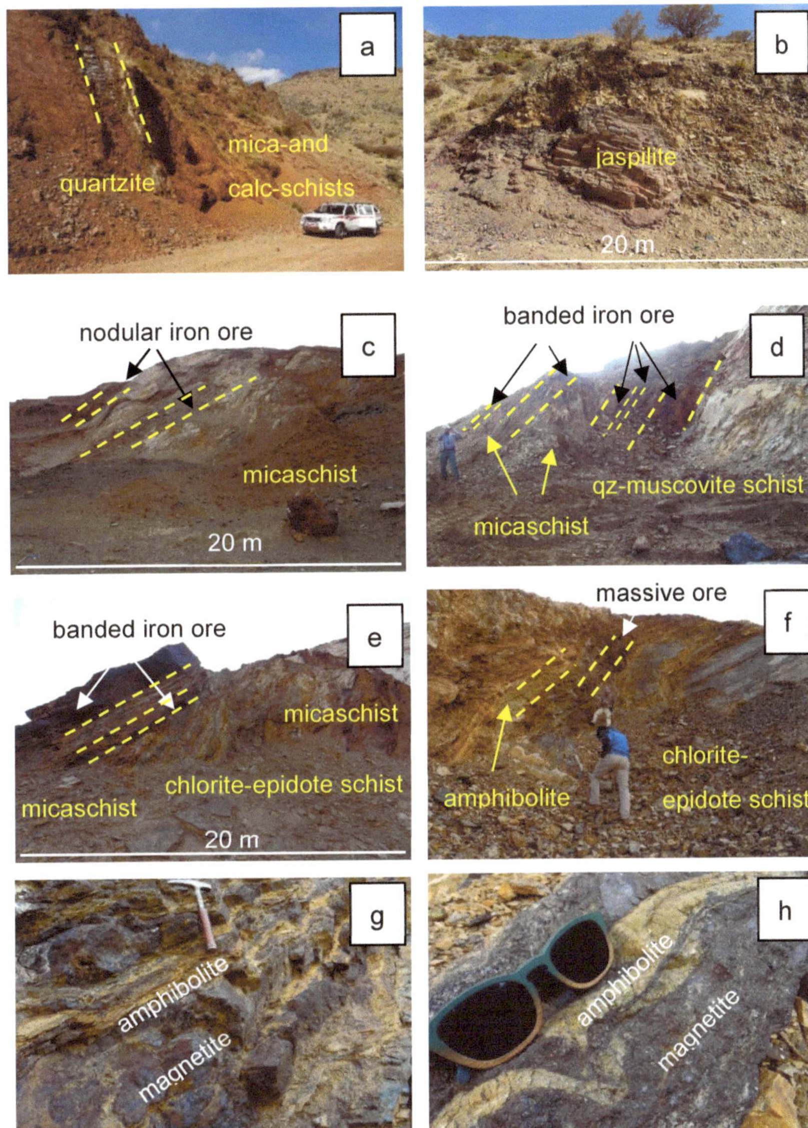

Figure 3. (**a–f**) Field photographs of the outcrops of the Takab iron ore deposit. (**g**,**h**) Close-up view of the amphibolite hosting massive magnetite ore.

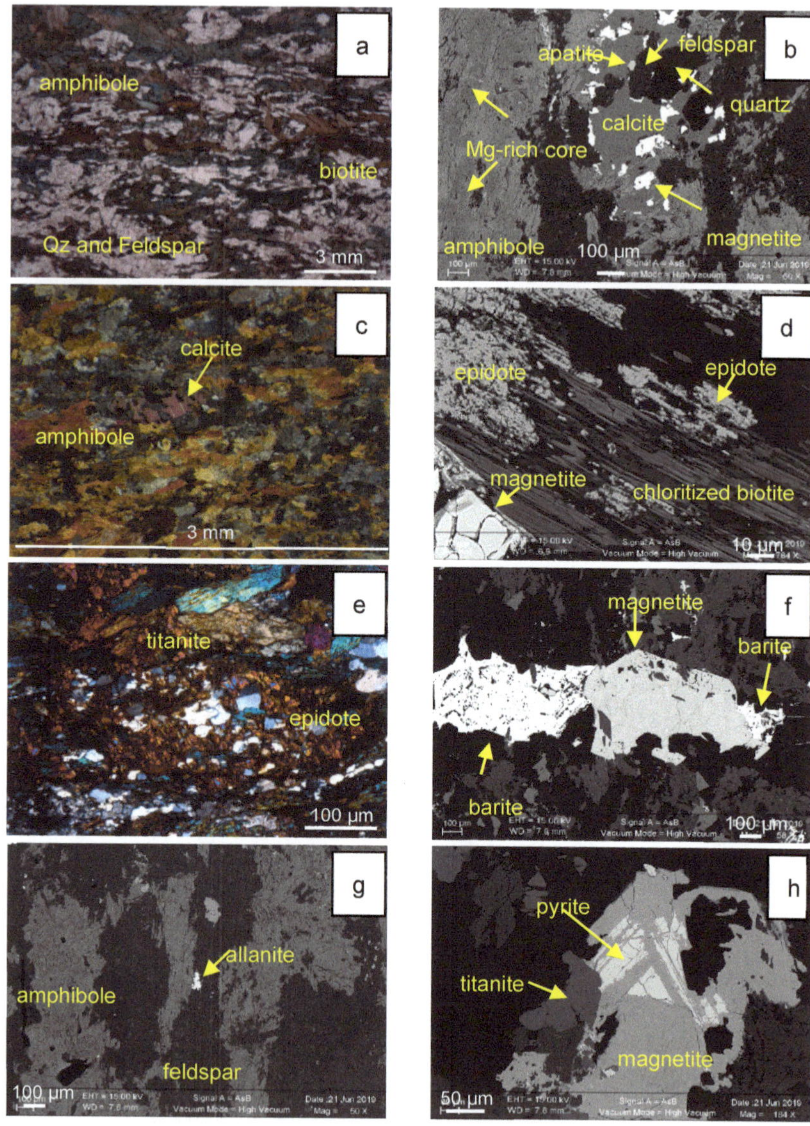

Figure 4. Amphibolite mineralogy: optical photomicrographs (**a**) in plane-polarized light and (**c**,**e**) in cross-polarized light; backscattered images (**b**,**d**,**f**–**h**). (**a**) Actinolite and minor biotite alternating with quartz and feldspar transparent patches; (**b**) actinolite with a Mg-rich core; interstitial calcite crystal also shown in (**c**) and accessory apatite. (**d**) Epidote and chlorite developed in biotite. (**e**–**h**) Accessory titanite, barite, allanite, and pyrite.

Figure 5. Hand specimen (**a**) and enlarged views of the massive magnetite ore from Takab. (**b**) Thin section (3.8 cm x 2.5 cm), (**c**) close-up view of the contact between the amphibolite and the magnetite ore, and (**d**) BSE image of the area marked in (**c**) showing the amphibole Mg-rich core. (**e**,**f**) Optical photomicrographs in cross-polarized light of prismatic, fibrous, and acicular amphibole crystals. (**f**) Rosette-shaped interstitial chlorite.

Table 1. Composition (in wt. %) of amphibole from Takab host amphibolite and iron ore.

	Host										Iron Ore		
	Actinolite						Tr	Mhb			Actinolite		
						core	core	acicular crystals					core
SiO_2	54.44	51.83	50.8	53.9	55.97	55.97	57.32	49.36	47.34	48.45	52.55	51.75	55.93
TiO_2	0.12	0.11	0.08	0.1	0	0	0.05	0.37	0.41	0.43	0.05	0.06	0.03
Al_2O_3	1.51	2.97	3.61	1.27	2.85	2.85	3.02	7.42	7.63	8.05	2.1	2.83	3.81
$Cr2O_3$	0.04	0.11	0.05	0.06	0.03	0.03	0.01	0.13	0.13	0.14	0.03	0	0.02
FeO	15.96	18.01	18.34	15.6	5.9	5.9	1.97	13.93	16.36	13.9	16.81	17.02	3.82
MnO	0.19	0.12	0.23	0.16	0.1	0.1	0.1	0.5	0.38	0.57	0.12	0.12	0
MgO	13.95	12.53	12.31	14.04	20.21	20.21	22.53	13.79	12.4	13.49	13.46	12.97	20.83
CaO	12.02	11.65	11.87	12.33	12	12	12.63	11.28	11.57	11.33	12.11	12.36	12.46
Na_2O	0.45	0.67	0.84	0.38	1.18	1.18	1.13	1.5	1.28	1.48	0.49	0.64	1.38
K_2O	0.08	0.17	0.16	0.1	0.26	0.26	0.08	0.16	0.34	0.15	0.15	0.13	0.17
Total	98.76	98.16	98.3	97.95	98.5	98.5	98.85	98.44	97.84	98	97.87	97.88	98.45
mg#	61.1	55.6	54.7	61.8	86.1	86.1	95.4	64.1	57.7	63.6	59	57.8	90.7

Tr = tremolite, Mhb = magnesio hornblende. mg#= 100*Mg/(Mg+Fe).

The amphibolite hosts massive magnetite layers of several tens to about one meter in thickness, which are slightly folded and/or boudinized (Figure 3g,h). The contact between the magnetite layers and the amphibolite is either abrupt or gradual (Figure 5b,c). In the magnetite layers, albite, amphibole, and minor biotite are interstitial to the magnetite crystals (Figure 5) and have the same composition as in the amphibolite (Table 1). Amphibole and mica can be locally altered to epidote, calcite, and chlorite. Rare interstitial calcite and rosette-like chlorite crystals are also observed (Figure 5f).

2.3. Analytical Methods

Three representative samples of the massive iron ore were studied. The samples were prepared as standard polished thin sections for study by optical microscopy, high-resolution scanning electron microscopy (HR-SEM), electron backscatter diffraction (EBSD), and electron microprobe analysis (EMPA).

2.3.1. Electron Microprobe Analysis

EMPA was performed with a Cameca SX100 electron microprobe equipped with five wavelength dispersive spectrometers at CAMPARIS, Sorbonne-Université, Paris, France. Fe and O were analyzed under the following operating conditions: monochromators LLIF for the measure of Fe and LPC1 for oxygen; accelerating voltage 15 kV; beam current 10 nA; counting times, 10 s over the peak and 10 s for background measured on each side of the peak; Fe_2O_3 as standard. WDS multi-element X-ray mapping was obtained from the same EMP.

For trace elements, monochromators were LTAP (Si, Al, and Mg) and PET (for Ca). To decrease the detection limits Ti, V, Mn, Ni, Cr, and Cu were simultaneously analyzed with LIF and LLIF monochromators on three different spectrometers. The operating conditions were 25 keV, 300 nA, 40 s to 70 s, and a beam size of 1 μm. Any damage effects on the samples were checked during analyses. The concentration of V using the Kα lines was not affected by the adjacent Ti Kβ line at the observed low Ti concentration. Standards were synthetic metal oxides and natural minerals: Ti, Mn: $MnTiO_3$; Fe, O: Fe_2O_3; Al; orthoclase; Cr: Cr_2O_3; Ni: NiO; Mg, Si, Ca: diopside; V: vanadinite; Cu: Cu native. Carbon coating was performed simultaneously for standards and samples to ensure the same coating thickness. These settings yielded minimum detection limits, with average concentrations as low as ~3–5 for Ca, Ti, and Mn and ~11–19 ppm for Mg, Al, and Si. The concentrations in V, Ni, Cr, and Cu were below or near the detection limit (~8–15 ppm) and could not be accurately determined.

2.3.2. Scanning Electron Microscopy and Electron Backscattered Diffraction Analysis

Backscattered imaging by scanning electron microscopy (SEM) and electron backscattered diffraction (EBSD) were carried out on a ZEISS Supra 55 VP at ISTeP (Sorbonne Université, Paris, France). The EBSD analysis was performed using an ARGUS^{TM} FSE/BSE imaging system mounted on a Bruker e$^-$Flash^{FS} detector. The data were processed with the ESPRIT 2.2 software package from the QUANTAX EDS/EBSD system from the Bruker Company. EBSD is detailed hereafter. Carefully polished thin sections from three samples were prepared and then analyzed uncoated by SEM. Representative areas were selected. The initial conditions were as follows: angular deviation lower than 5°, minimum pixel size of 5, and spike correction of zero-solution measurements relative to the surrounding points. EBSD measurements were used to acquire mineral phases, crystal orientations, and microstructural information. Crystal orientations are described using the set of Euler angles, and each orientation is represented by a combination of RGB colors in an Euler map. Pole figure (PF) and inverse pole figure (IPF) are also used in this study. The produced pole figures are displayed in terms of points or in contoured versions indicating the strength of the clustering of poles relative to that of a random distribution. The IPF maps were generated for the three selected viewing directions parallel to the X, Y, and Z axes, quoted as IPFX, IPFY, and IPFZ.

3. Results

3.1. Morphology and Textures of Magnetite

Magnetite grains are enclosed in a matrix mainly composed of actinolite and albite (Figure 4). Based on BSE imaging and WDS X-ray mappings, three types of magnetite have been identified (Figures 6–8).The first type (Mt1) occurs as sub- to anhedral grains of an inhomogeneous dark gray magnetite in BSE images, ranging up to 1 mm in dimension. Mt1 is surrounded by a light gray type 2 magnetite (Mt2). The boundary between Mt1 and Mt2 is usually sharp (Figure 6a, bottom right, Figure 7a) but may also locally invade dark Mt1, forming lighter gray patches (Figure 6a, upper left). Three different subtypes of Mt2 have been locally recognized in sample TAK-Z1: Mt2-A and Mt2-C are light gray on BSE images, while Mt2-B is darker gray (Figure 8). Both Mt1 and Mt2-B show an oscillatory growth zoning of dark and light zones, suggesting a higher concentration of elements with a low atomic number in the dark zones (Figure 8). Mt3 appears as bands (10–80 µm width) with needle-like textures of bright magnetite aligned along microfracture planes that crosscut the Mt1–Mt2 grains (Figures 6a and 7; Supplementary Figure S1). Mt3 bands replace platy hematite, which remains as rare relict-aligned patches (Figure 6). In some bands, Mt3 appears to be porous only close to the hematite relicts. Aggregates of euhedral magnetite grains (100–200 µm) from sample TAK-Z4 also display well-developed chemical oscillatory zoning (Figure 9).

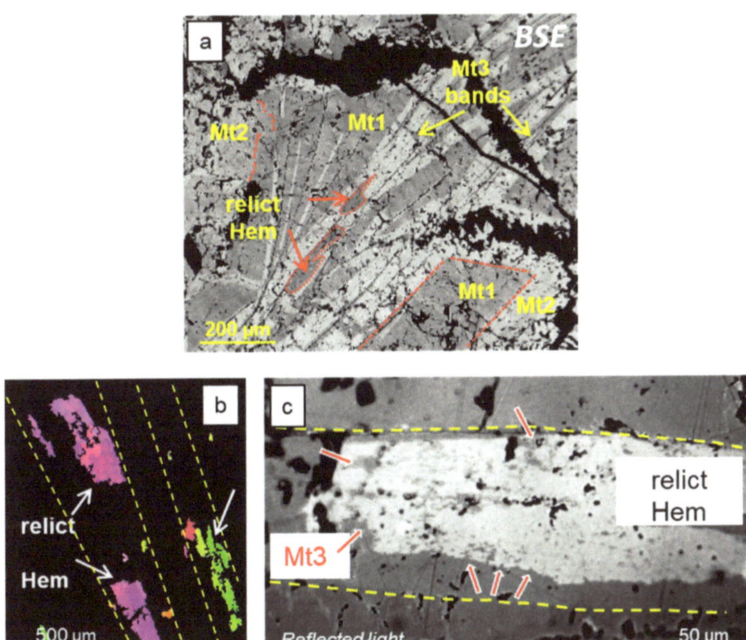

Figure 6. Morphology and textures of magnetite. BSE image of magnetite grains from sample TAK-Z3. (**a**) Dark gray magnetite Mt1 is surrounded by lighter gray magnetite Mt2 (bottom right), which locally invaded Mt1 (upper left). Magnetite Mt3 forms bright bands, which may host relict hematite (Hem). (**b**) Inverse pole figure (IPF) of relict hematite phases (white arrows) in two Mt3 bands outlined by the yellow dashed lines (sample TAK-Z1). Different relict hematite patches within one single Mt3 band display similar colors, indicating that hematite patches have similar crystallographic orientation and, thus, represent a unique hematite band. (**c**) Reflected light. Enlarged view of an Mt3 band (outlined by the yellow dashed lines) with relict hematite (bright patches) undergoing transformation in Mt3 (red arrows).

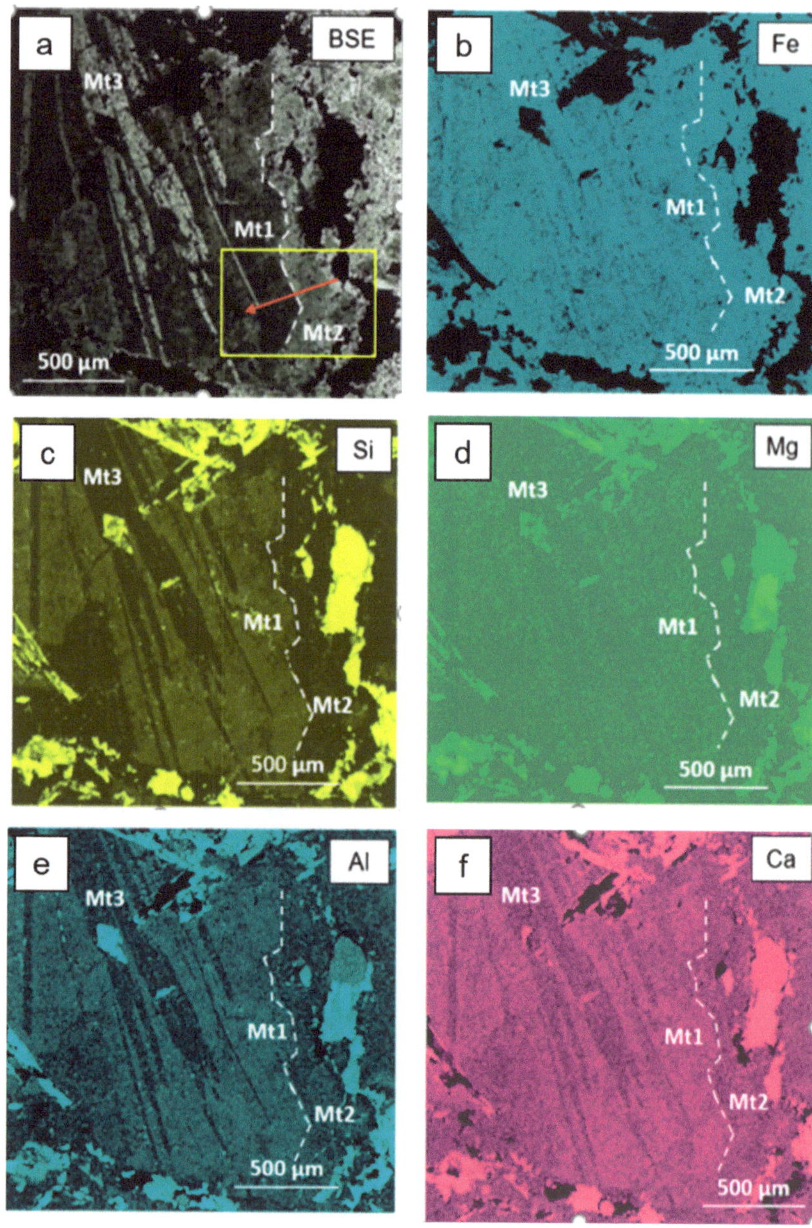

Figure 7. BSE image (**a**) and WDS X-ray maps of selected elements in a magnetite grain from sample TAK-Z1 (**b–f**). Magnetite Mt1 contains higher Si, Al, Mg, and Ca than magnetite Mt2. Mt3 bands have the highest Fe and the lowest Si, Al, Mg, and Ca contents. The yellow rectangle corresponds to the zone detailed in Figure 8. The red arrow is precised in Figure 8.

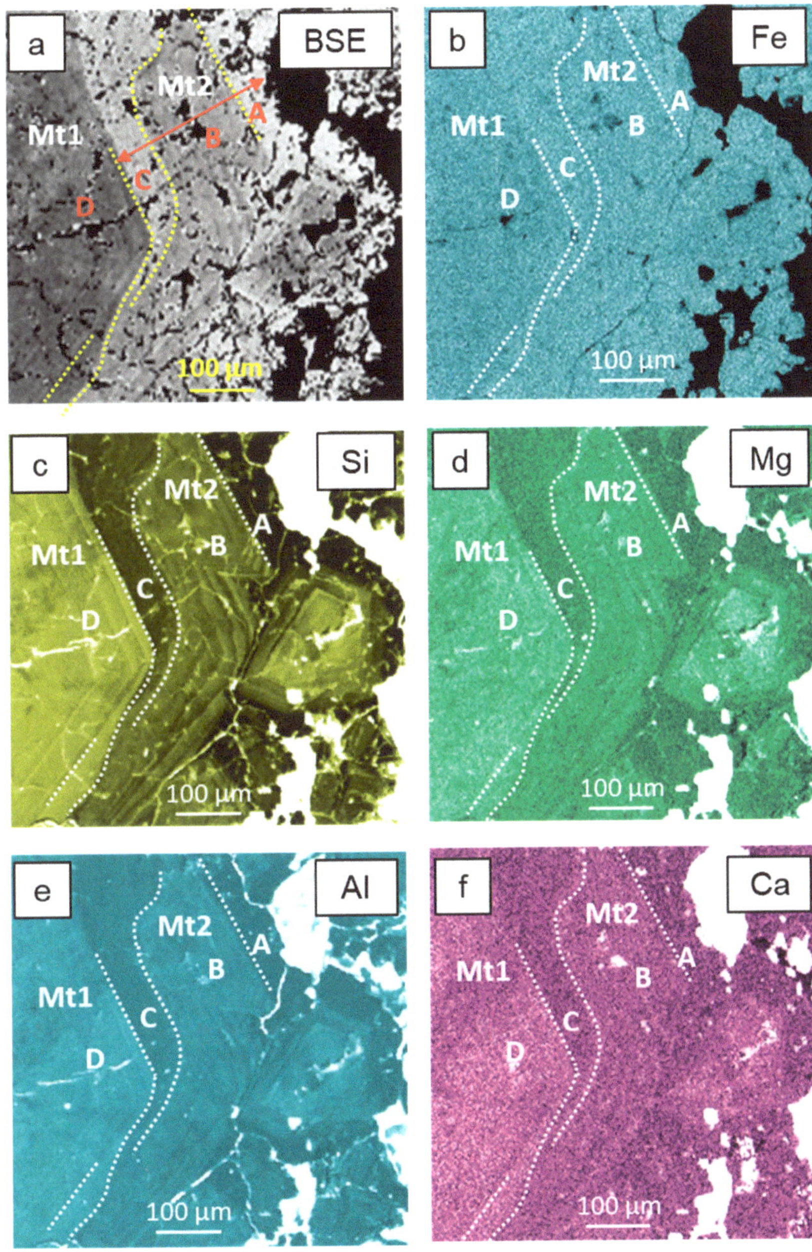

Figure 8. Detailed BSE image (**a**) and WDS X-ray maps of selected elements (**b**–**f**) in a magnetite grain from sample TAK-Z1 showing dark magnetite Mt1 (D) and the different subzones of gray magnetite Mt2 (zones A, B, C). Oscillatory zoning is visible in Mt1 and Mt2-B. The total grain is shown in Figure 7. The red arrow limits the different zones of magnetite Mt2, see text for details.

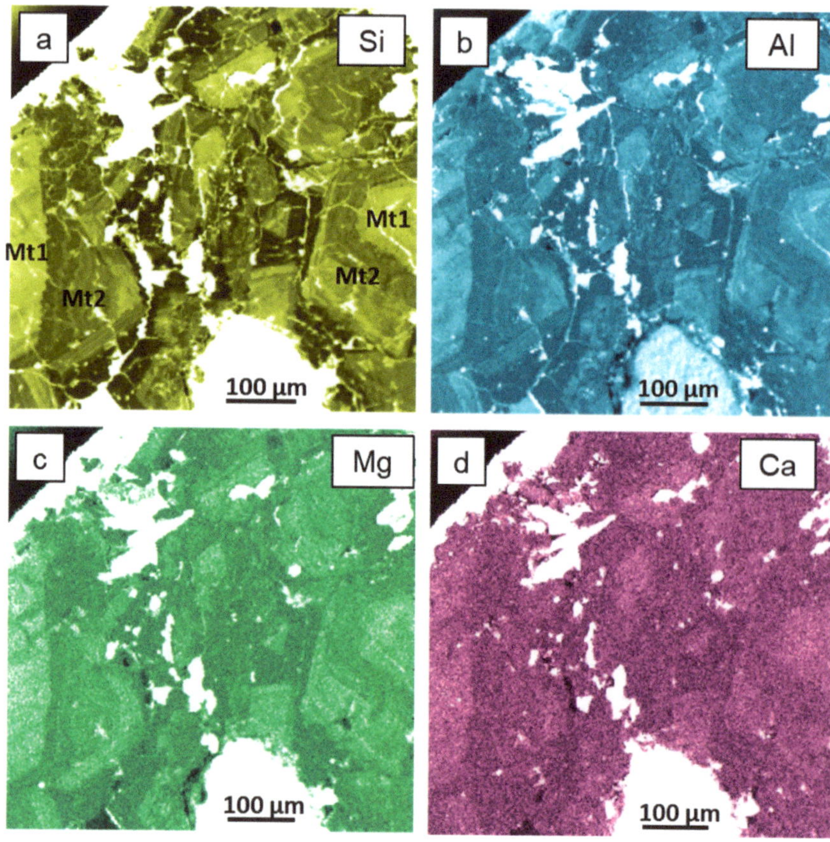

Figure 9. (**a**–**d**) WDS X-ray maps of selected elements in an aggregate of smaller magnetite grains from sample TAK-Z4 with compositional and zoning characteristics like to those of large magnetite grains (Figure 8).

3.2. Electron Backscatter Diffraction Mapping

EBSD misorientation maps of magnetite show both sudden orientation changes across grain boundaries and gradual or wavy misorientation unrelated to grain boundaries. For example, Figure 10 shows wavy misorientation in Mt1 (lower left) as well as in the Mt3 band, delimited by the dotted lines.

A small area of the different magnetite grains was selected to generate pole figures (PFs) and inverse pole figures (IPFs). Magnetite grains with a uniform color in IPF maps show discrete point clusters in PFs, reflecting uniform intragrain orientation (Figure 10, subset 1). Magnetite grains with wavy colors show less-clustered points in PFs, and the crystal orientations may disperse along small circle girdles in stereoplots of <100>, <110>, and <110> directions in individual magnetite grains (Figure 10, subset 2), which is related to the wavy intragrain misorientation. This indicates that the intragrain microstructures formed through deformation by dislocation movement without recovery or rearrangement in subgrains [33]. Magnetite thus shows at the microscale the effect of deformation, which is observed at the macroscale by the folded magnetite layers in the host amphibolite (Figure 3g,h).

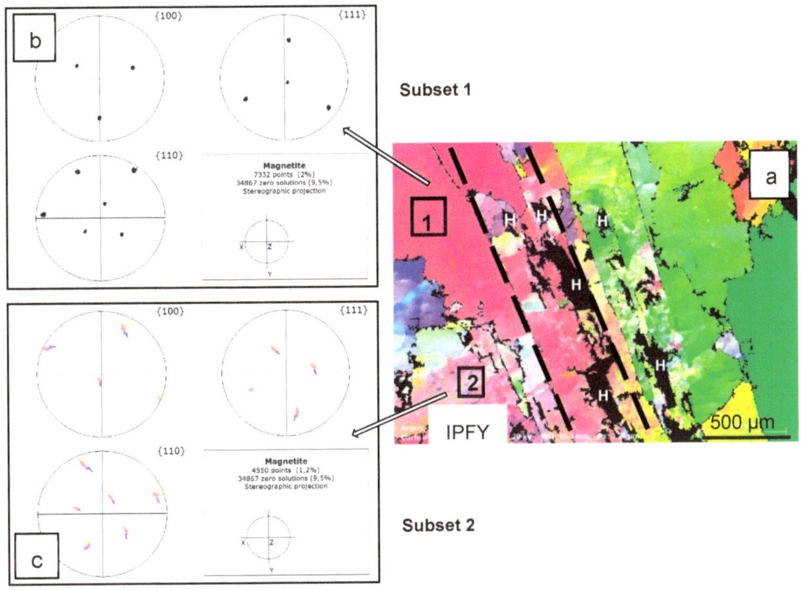

Figure 10. Inverse pole figure IPF Y (**a**) and pole figures (**b**,**c**) of magnetite from sample TAK-Z1. The dotted lines delimit the larger Mt3 band of Figure 6b. Abrupt changes in color reflect changes in orientation across grain boundaries. Wavy misorientation reflects intragrain deformation. In the latter case, the pole figure shows dispersed points along circle girdles (subset 2) compared to the clustered points of uniform intragrain orientation (subset 1). H: hematite.

3.3. Chemical Composition of Magnetite

WDS X-ray elemental mapping was performed to determine the distribution of minor and trace elements. Si, Al, Mg, and Ca maps clearly distinguish the different magnetite types (Figures 6–8; Supplementary Figure S1). Mn has a uniform distribution throughout the different magnetite types, whereas Ni, Cr, and Cu are below or close to the detection limits and thus give no information. Therefore, these elements are not further discussed here. Of the three texturally identified magnetite types, Mt1 is rich in Si, Al, Mg, and Ca (Figures 6–8). Mt2 shows variable enrichment or depletion of Si, Al, Mg, and Ca in the different zones Mt2-A, -B, and -C (Figure 8). Magnetite from zone Mt2-B presents oscillatory enrichment or depletion, which is also observed in magnetite Mt1 (Figure 8) and magnetite from the aggregates (Figure 9). Mt3 is characterized by the lowest contents of these elements and by the highest Fe content (Figure 7).

The average concentrations, standard deviations, and ranges of values are listed in Table 2. The complete analytical results are given in Supplementary Table S2.

Table 2. Fe and O (wt. %) minor- and trace-element (ppm) contents of magnetite from Takab.

Sample		TAK-Z1	TAK-Z1	TAK-Z1	TAK-Z4	TAK-Z4
Mt type		**Mt1**	**Mt2**	**Mt3**	**Mt1**	**Mt2**
n analyses		12	4	10	8	6
Fe (wt. %)	Average	68.0	69.8	70.8	67.6	69.6
	Range	66.1–69.4	69.3–70.4	69.9–71.3	67.2–68.7	69.0–70.1
O (wt. %)	Average	29.4	28.6	28.5	29.1	28.7
	Range	29.0–29.7	28.2–28.8	28.0–29.1	28.0–29.6	27.8–29.1
Si	Average	11,017	4336	214	9942	3550
	sd	*1107*	*409*	*195*	*1444*	*397*
DL = 11	Range	8899–12779	3846–4840	11–426	8274–11,941	3149–4045

Table 2. *Cont.*

Sample		TAK-Z1	TAK-Z1	TAK-Z1	TAK-Z4	TAK-Z4
Mt type		**Mt1**	**Mt2**	**Mt3**	**Mt1**	**Mt2**
Ti	Average	130	86	26	293	104
	sd	*61*	*35*	*20*	*120*	*18*
DL = 4	Range	51–276	37–126	BDL-63 *(2)*	149–474	73–106
Al	Average	2489	1031	275	2321	985
	sd	*479*	*76*	*94*	*364*	*173*
DL = 12	Range	1428–3103	947–1120	187–434	1617–2773	818–1301
Mn	Average	262	227	261	279	223
	sd	*26*	*59*	*107*	*54*	*27*
DL= 5	Range	198–289	161–305	166–519	198–368	193–273
Mg	Average	1263	410	34	1225	382
	sd	*320*	*134*	*26*	*378*	*131*
DL = 19	Range	807–1882	234–560	BDL-71 *(4)*	825–1794	205–553
Ca	Average	2677	1289	313	2799	1192
	sd	*585*	*109*	*95*	*720*	*294*
DL = 3	Range	1606–3529	1127–1349	*194–484*	1760–3821	1018–1155
V	Average	22	19	34	nd	nd
	sd	10	5	6		
DL = 9	Range	13–43	12–24	22–42		
Sample		**TAK-Z3**	**TAK-Z3**	**TAK-Z3**		**TAK-Z3**
Mt type		**Mt1**	**Mt2**	**Mt3 porous**		**Mt3**
n analyses		23	10	8		13
Fe (wt. %)	Average	68.2	69.7	71		71.1
	Range	67.5–69.2	69.5–70.2	70.1–71.5		70.4–71.8
O (wt. %)	Average	29.4	29.0	28.7		28.8
	Range	28.9–29.9	28.6–29.2	28.3–29.7		28.4–29.5
Si	Average	11806	5430	587		432
	sd	*1231*	*936*	*445*		*274*
DL = 11	Range	9769–13,785	3949–7170	12–870		55–884
Ti	Average	143	56	9		38
	sd	*61*	*13*	*5*		*22*
DL = 4	Range	38–246	37–82	BDL-17 *(2)*		8–65
Al	Average	2781	1190	278		239
	sd	*500*	*252*	*95*		*104*
DL = 12	Range	1745–3480	759–1621	175–468		105–519
Mn	Average	222	240	198		181
	sd	23	32	*45*		*45*
DL= 5	Range	181–262	196–276	148–292		101–261
Mg	Average	1272	775	42		62
	sd	*308*	*220*	*24*		*28*
DL = 19	Range	614–1738	342–1064	BDL-70 *(2)*		18–89
Ca	Average	2866	1661	473		416
	sd	*521*	*353*	*213*		*179*
DL = 3	Range	1966–3575	1150–2226	*249–824*		228–797
V	Average	24	21	nd		27
	sd	*10*	*8*			*7*
DL = 9	Range	11–39	11–30			22–42

Ni and Cu are not reported as their contents are below or close to the detection limits (DL = 12 and 8 ppm, respectively). BDL = below detection limit. (n) = number of analyses below the detection limit. nd = not determined. sd = standard deviation.

The magnetite composition is presented in Figure 11 for sample TAK-Z3, as it is representative of magnetite from the three studied samples. In the three samples of magnetite, Mt1 has the most variable composition and the lowest Fe content (66–68 wt. %), while both Mt2 and Mt3 show a restricted range of composition in minor and trace elements (Figure 11a, b) and higher Fe contents (70 and 71 wt. %, respectively). In the three samples, Mt1 magnetite has high contents of Si (0.83–1.38 wt. %) and moderately high Al (0.14–0.35 wt. %), Ca (0.16–0.38 wt. %), and Mg (0.06–0.19 wt. %). Mt2 magnetite has lower Si (0.32–0.72 wt. %), Ca (0.10–0.22 wt. %), Al (0.08–0.16 wt. %), and Mg (0.02–0.11 wt. %). Mt3 (porous and non-porous) magnetite has the lowest minor- and trace-element contents, with Si ($\ll$0.01–0.08 wt. %), Al (0.02–0.05 wt. %), Ca (0.02–0.08 wt. %), and Mg ($\ll$0.01 wt. %). The three types of magnetite contain similar low Mn (0.02–0.05 wt. %), while Ti is $\ll$0.01 wt. %

in Mt2 and Mt3 but can reach 0.04 wt. % in Mt1. Mt1 and Mt2 have low similar V contents (19–24 ppm), while Mt3 is slightly richer in V (27–34) ppm. The composition of the thin oscillatory dark and lighter gray bands in Mt1 and Mt2-B magnetite grains could not be resolved with the microprobe.

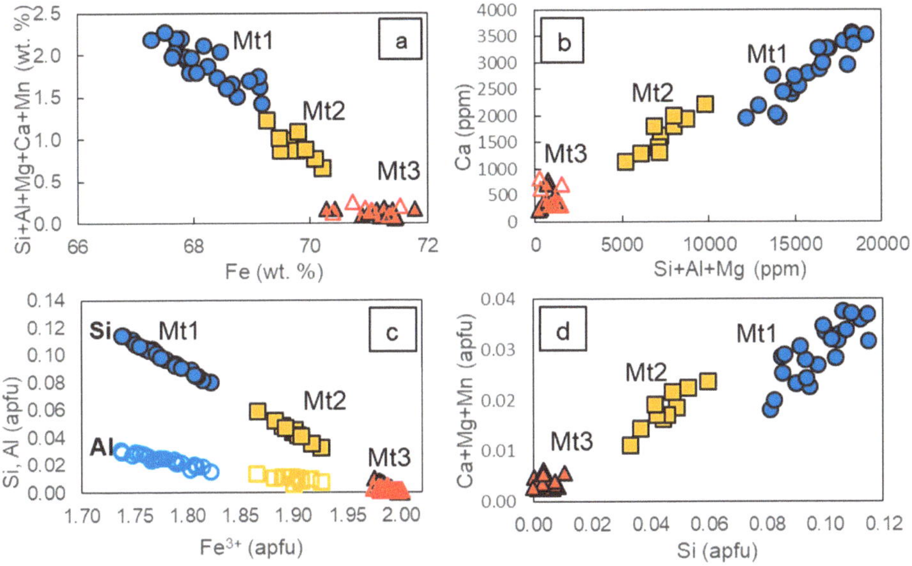

Figure 11. Binary plots of selected elements in magnetite from Takab, sample TAK-Z3. (**a**) [Si + Al + Mg + Ca + Mn] vs. Fe (wt. %) and (**b**) Ca vs. [Si + Al + Mg] (ppm), showing high trace-element and low Fe contents in magnetite Mt1 compared to magnetite Mt2. Magnetite Mt3 has the lowest trace-element and the highest Fe contents. (**c**) Si and Al vs. Fe^{3+} and (**d**) [Ca + Mg + Mn] vs. Si diagrams (in apfu) show the negative correlation of Fe^{3+} with Si and Al, possibly indicating the incorporation of Si in the magnetite structure by substitution for Fe^{3+}, while Si is positively correlated with Ca, Mg, and Mn as valence-state species related to the substitution. Blue circles, Mt1; yellow squares, Mt2; filled red triangles, non-porous Mt3; and red open triangles, porous Mt3. Calculation of Fe^{2+}/Fe^{3+} partitioning explained in Supplementary Table S2.

4. Discussion

4.1. Magnetite Formation Conditions

The minor- and trace-element composition of magnetite has been used in numerous studies to fingerprint its environment of formation based on a series of discrimination diagrams, e.g., [1,3,34]. However, coexisting phases may affect the composition of magnetite, for example, the co-precipitation of amphibole and epidote could decrease the incorporation of Mg and Al, respectively, in magnetite. In this study, as the different magnetite types Mt1, Mt2, and Mt3 co-crystallized with the same coexisting assemblage, any major control of the co-precipitated phases on compositional differences between the three magnetite types is unlikely.

The contents of Mg, Mn, Ti, V, Ni, and Cr in magnetite, for example, can be used to discriminate between formation in a hydrothermal or a magmatic environment. In this study, the different types of magnetite plot in the field of hydrothermal magnetite characterized by low Ti content on a Fe/(Mg + Al + Mn) discrimination diagram (Figure 12). On the V vs. Ti diagram, most of the different types of magnetite plot outside the hydrothermal field due to a very low V content (Supplementary Figure S2), which does not give here any reliable information. On the other hand, the extremely low Ni and Cr contents of all three types of magnetite (Table 2) further support a hydrothermal origin. This agrees with a previous study of the layered, nodular, and disseminated magnetite from the Takab BIF, in which it was shown that both the trace-element and Fe and O isotopic compositions provided evidence for a complex hydrothermal history [8,28].

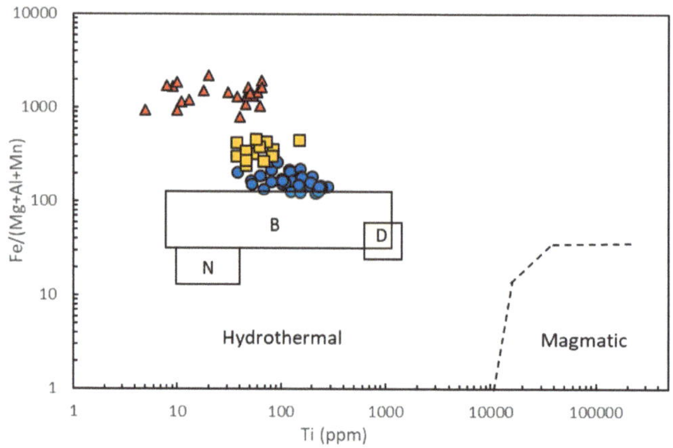

Figure 12. Discrimination diagram of Fe/(Mg + Mn + Al) vs. Ti. The different magnetite types from samples TAK-Z1 and TAK-Z3 plot in the hydrothermal domain. The boxes outline the domains of banded (B), nodular (N), and disseminated (D) magnetite previously studied [28]. Hydrothermal and magmatic domains are after [34].

Element partition coefficients are temperature-dependent and can therefore be used for constraining the temperature of formation of magnetite. The incorporation of Ti increases with temperature in magmatic and in high-temperature hydrothermal environments, whereas it is negligible in low-temperature hydrothermal fluids [35,36]. Discriminating diagrams, such as (Al + Mn)/(Ti + V), have been proposed to outline different fields of temperature [3,37]. As can be seen in Table 2 and Figure 11, magnetite Mt1 has a variable composition with up to 1.4 wt. % Si and up to 0.4 wt. % Al. This raises the question of whether Si is really incorporated in the magnetite structure or present as Si-bearing nanoparticles [37–39]. In the absence of a definitive argument (as discussed below in Section 4.1), we tentatively estimated the possible contamination of the analyses of magnetite by nanoinclusions of amphibole at 5–6% (Supplementary Figure S3). The appropriate correction was applied before using the (Al + Mn)/(Ti + V) diagram and temperature calculations.

The decreasing trend shown in the (Al + Mn)/(Ti + V) diagram (Figure 13) indicates a decreasing temperature from Mt1 to M2 and Mt3. Overall, the different magnetite types formed at medium temperature ca. 300 °C or 200–300 °C. These temperature ranges are similar although slightly lower than that those recorded by the banded, nodular, and disseminated magnetite previously studied [28], shown by the rectangles in Figure 13.

Canil and Lacourse [40] proposed a $T^{\circ}_{Mg\text{-}mag}$ empirical geothermometer, $T^{\circ}_{Mg\text{-}mag} = -8344 (\pm 320)/[\ln X_{Mg} - 4.1 (\pm 0.28)] - 273$, that was applied to magnetite from various settings: igneous rocks, hydrothermally altered rocks, and ore deposits. Figure 14 presents the results of the $T^{\circ}_{Mg\text{-}mag}$ calculations of the three magnetite types excluding the analytical points from the areas showing oscillatory zoning in magnetite Mt1 and Mt2b. The data are presented in Supplementary Table S3.

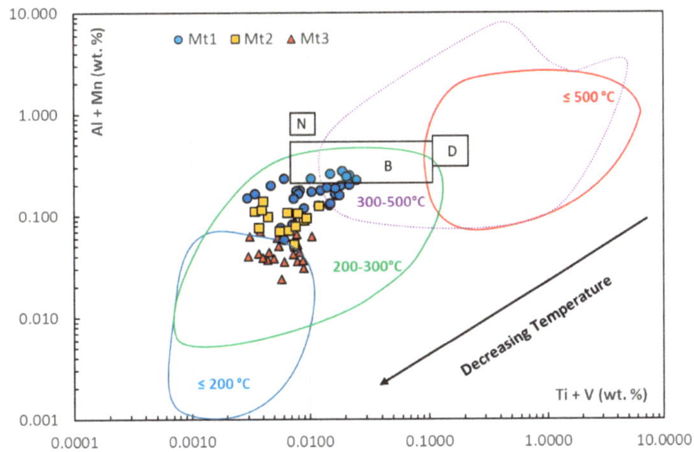

Figure 13. Approximate temperatures of formation of magnetite. Temperature reference fields are after [3,38]. The boxes outline the domains of banded (B), nodular (N), and disseminated (D) iron ore samples previously studied [28].

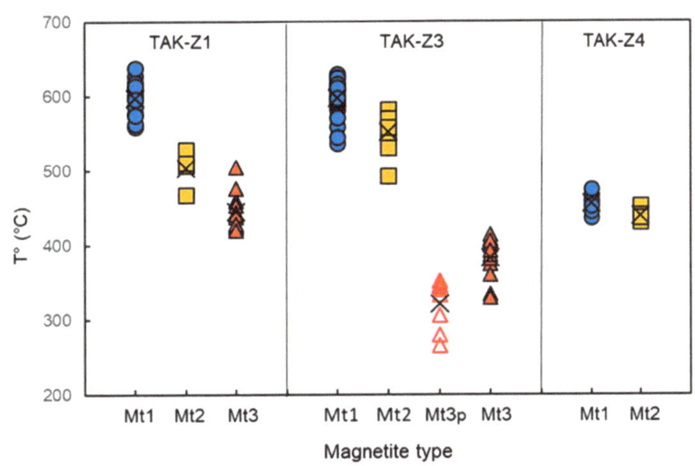

Figure 14. Calculated formation temperatures for Takab magnetite show a decreasing trend from magnetite Mt1 and Mt2 to magnetite Mt3. The temperatures were calculated using the $T°_{Mg\text{-}mag}$ thermometer of [40]. The calculated T° values have an uncertainty of about ±60 °C. In sample TAK-Z3, Mt3p corresponds to porous Mt3 adjacent to relict hematite. Mt3 was not observed in sample TAK-Z4. The points correspond to the total analyses in Supplementary Table S2 and calculated temperatures in Supplementary Table S3. (X): mean values.

A decreasing trend is evidenced from magnetite Mt1 to Mt2 and Mt3 in samples TAK-Z1 and TAK-Z3, whereas in sample TAK-Z4, magnetite Mt1 and Mt2 from the agglomerates formed at a similar temperature. In sample TAK-Z3, the porous magnetite Mt3 records lower temperature than the non-porous magnetite Mt3. However, the calculated temperatures are much higher than those estimated from the (Al + Mn)/(Ti + V) diagram (Figure 13): e.g., average 600 °C for Mt1, 500 °C for Mt2, and 440 °C for Mt3. This difference is difficult to explain.

Oxygen fugacity can also influence the composition of magnetite. For example, V has three valence states (V^{3+}, V^{4+}, and V^{5+}), and as only V^{3+} can enter the magnetite structure in high abundance, its incorporation is linked to fO_2 and increases with deceasing fO_2. In contrast, Ti has only one valence state (Ti^{4+}) and a rather constant partition coefficient. Thus, Ti/V ratios decreasing in magnetite indicates a reduced environment. The Ti/V ratios are low in magnetite Mt1 (6.6), Mt2 (3.8), and Mt3 (1.2), thus suggesting a rather reduced environment of formation. Relatively constant V (Table 2) content in the different magnetite types suggests quite constant fO_2.

In summary, magnetite from the massive magnetite ore formed in a rather reduced hydrothermal environment at medium temperature for Mt1 and Mt2, which slightly decreased for Mt3.

4.2. Occurrence of Si and Elemental Substitution Mechanisms in Magnetite Mt1

Si-bearing magnetite with $SiO_2 > 1$ wt. % have been reported in numerous studies in a variety of rocks from igneous rocks to mid-ocean serpentinites, banded iron formations, porphyry deposits, skarns, and others (see review by [41] and so-called "silician magnetites"). Si can be either structurally incorporated or forms mineral micro- to nanoparticles.

Si and Al can enter in the structure of magnetite due to their small ionic radii (Si^{4+}, 0.26 Å; Al^{3+}, 0.39 Å), preferentially substituting for $^{IV}Fe^{3+}$ (0.57 Å) in tetrahedral sites rather than for $^{VI}Fe^{3+}$ (0.63 Å) or $^{VI}Fe^{2+}$ (0.69 Å) in octahedral sites [42]. Their incorporation in tetrahedral coordination requires coupled substitutions with divalent cations in octahedral sites for valence state balancing, with or without octahedral vacancy [41,43]. In this case, well-defined elemental trends are observed. For example, Fe^{3+} is negatively correlated with Si and Al, while Si is positively correlated with divalent elements, Fe^{2+}, Ca, and Mg. Structural Si has been evidenced by combining high-resolution analytical techniques (X-ray microdiffraction, XRMD, and TEM) [37,41]. Xu et al. [38] reported the presence of Si in magnetite lattice as discrete nanoprecipitates, in which Si replaces $^{IV}Fe^{3+}$ with vacancies in the octahedral sites, using combined Z-contrast imaging and ab initio calculations. Deditius et al. [37] reported two types of trace-element incorporation in magnetite by both structural incorporation and the formation of mineral nanoparticles (NPs). Moreover, Ciobanu et al. [39] reported the presence of submicroscopic Si-Fe precipitates and calc–silicate inclusions in magnetite from the Olympic Dam Iron Oxide Copper Gold deposit, south Australia.

Here, EMPA spot analyses were carefully located to avoid the rare inclusions, which were detected by high-resolution SEM analysis. Magnetite Mt1 contains up to 1.4 wt. % Si and 0.4 wt. % average Al and Ca. EMPA results show the anti-correlations of Fe^{3+} with Si and Al, while Si is positively correlated with divalent elements, Fe^{2+}, Ca, and Mg as well as Al (Figure 11c,d). These trends would be consistent with the structural incorporation of Si. However, in the absence of further analysis by high-resolution techniques, the presence of nanoparticles in magnetite cannot be excluded. To examine the possible presence of amphibole inclusions in magnetite, we checked whether the amphibole plots on the continuation trend of the magnetite compositions. The slopes of the compositions of magnetite and the line connecting pure magnetite and amphibole average composition are similar in terms of Ca/(Si + Al), for example (Supplementary Figure S3). This allowed the approximation of up to 5–6% of the amphibole component in the magnetite composition. However, this admixture fails to reproduce the observed Ca content of magnetite for a given (Si + Al) content. On the other hand, there is no agreement between the slopes when adding Mg to the check in terms of Ca/(Si + Al + Mg).

Nevertheless, regardless of the mechanisms involved, the presence of these elements in magnetite Mt1 suggests its precipitation in a Si (Al, Mg)-rich and reducing environment, as reported for silician magnetite from hydrothermal, volcano–sedimentary, and metamorphic environments [41,43–46].

4.3. Dissolution–Reprecipitation Replacement of Mt1 by Mt2

Magnetite Mt1 was replaced by magnetite Mt2 mainly along grain margins. Mt1 magnetite is rich in minor elements (Si, Ca, and Al) compared to magnetite Mt2 (Table 2 and Figure 11). The relations between Mt1 and Mt2 are the following: (1) A close spatial relationship between Mt1 and Mt2 with a sharp boundary between Mt1 and Mt2 visible on BSE images and WDS X-ray maps (Figures 6–8). IPF maps of composite Mt1 and Mt2 magnetite grains indicate a single-crystal domain shown by the same color, indicating the same orientation (Figure 15a,b). Similarly, the pole figures of selected areas in Mt1 and Mt2 show the coincidence of the crystallographic orientations of Mt1 and Mt2 (Figure 15c,d). This supports the inheritance of the crystallographic structure during the epitaxial growth of Mt2. (2) Porosity is locally developed in Mt2. (3) There is a sharp reaction front between Mt1 and Mt2 without diffusion, as shown in EMPA profiles (Figure 16c). Such compositional and textural characteristics suggest a fluid-induced replacement of Mt1 by Mt2 by a coupled dissolution–reprecipitation process (CDR) [5,47–50]. The CDR process allows the dissolution of early trace-element-rich Mt1 magnetite in the core and the precipitation of secondary Si-poor Mt2 magnetite at the rim. The presence of structural Si or nano-Si-rich inclusions in magnetite Mt1 would have caused a crystal lattice defect and/or deformation that would have facilitated its dissolution. Moreover, porosity would have facilitated the infiltration of hydrothermal fluid that, in turn, may have boosted the dissolution process. Coupled dissolution–reprecipitation evidenced here removed Si, Ca, and Mg from Mt1 magnetite near the reaction front and formed secondary Mt2 magnetite with lower trace-element content (type 1-CDR process in [5]).

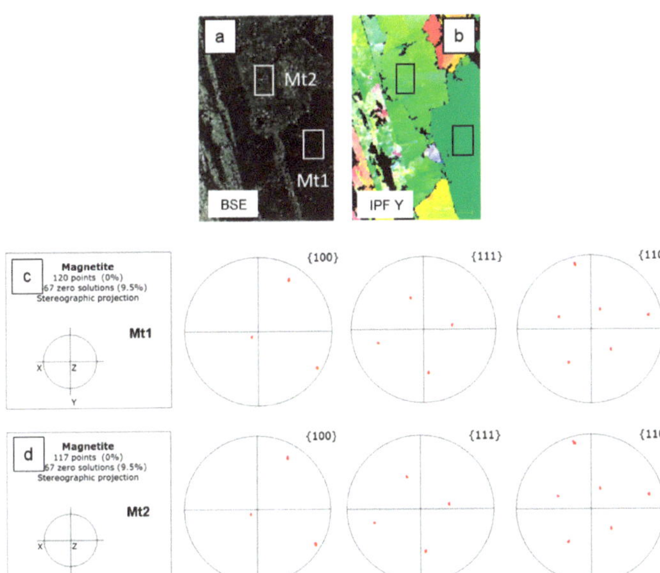

Figure 15. Enlarged BSE image (**a**) and inverse pole figure IPF Y (**b**) of Mt1 and Mt2 adjacent zones in a magnetite grain from sample TAK-Z1, and pole figures (**c**,**d**) of selected areas corresponding to Mt1 and Mt2 showing a similar orientation of Mt1 and Mt2 in favor of a CDR process (see text).

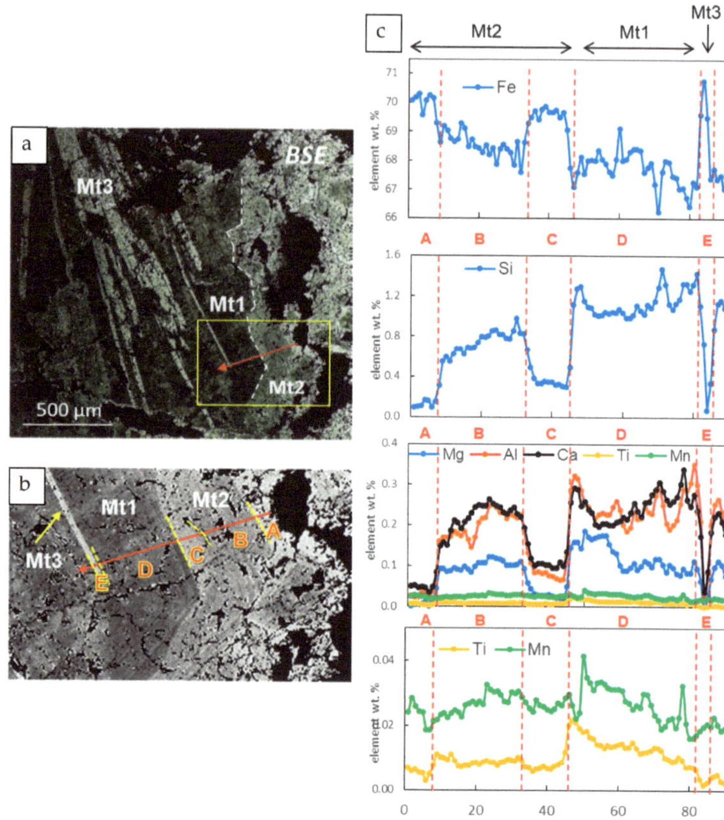

Figure 16. BSE images of the magnetite from sample TAK-Z1 (**a**) and enlarged view of the yellow contoured zone (**b**). In (**b**), the red arrow indicates the EMPA element profile (**c**) through the different magnetite zones Mt1, Mt2, and Mt3. The multi-element profiles show the difference in composition and the sharp contact between the different magnetite types. WDS X-ray maps of the same area are given in Figures 7 and 8. See text for explanations of the different zones A-E.

Among the different factors that may control a dissolution–reprecipitation process, such as temperature, oxygen fugacity, and fluid composition [5,51–53], a change in fluid composition played here a major role in inducing the CDR process of magnetite Mt1/Mt2. The slight temperature decrease from Mt1 to Mt2 (Figure 13) may also have influenced the process. In contrast, the V content, similarly low (19–24 ppm, Table 2) in both Mt1 and Mt2, argues against a possible role of varying oxygen fugacity in the process.

4.4. *Origin of Oscillatory Zoning in Magnetite Mt1*

Oscillatory zoning of thin dark and light gray bands (in BSE) is observed in magnetite Mt1 and Mt-B large grains (Figure 8) and in the magnetite agglomerates (Figure 9). The decrease in brightness in BSE images indicates a lower concentration of trace elements with lower atomic numbers than Fe, such as Si, Al, and Ca. Oscillatory zoning is too fine to be analyzed, and no clear compositional variations could be demonstrated in EMPA profiles (Figure 16, zones B and D). Oscillatory zoning is generally interpreted as growth zoning under cyclical variation in fluid composition, temperature, or oxygen fugacity, which changes trace-element partitioning between magnetite and fluid [2,5,53]. The oscillatory zoning is parallel to the interface between Mt1 and Mt2, which likely results from a growth zoning under fluctuating fluid composition (Figures 8 and 9). The preservation of oscillatory zoning requires crystal growth faster than the intracrystalline diffusion. Moreover, assuming that the thin dark and light gray bands may be similar in composition to Mt1 and Mt2 magnetite, respectively, we suggest that the oscillatory zoning would likely result from a fluctuating fluid composition rather than temperature or oxygen fugacity variation (see previous paragraph).

4.5. Origin of Magnetite Mt3 (Mushketovite)

Magnetite Mt3 has replaced the former platy hematite, now present as rare relicts, and is developed along cleavages and microfracture planes or crosscutting Mt1 and Mt2 magnetite grains (Figure 6). It is characterized by low concentrations of Si, Al, Mg, Ca, Mn, and Ti (Table 2, Figures 7, 11 and 16). The similarity between the crystal structures of magnetite and hematite controls the precipitation of magnetite and leads to the coincidence of orientation between the two minerals (Supplementary Figure S4). The pseudomorphic replacement of hematite by magnetite (mushketovitization) has been reported in many ore deposits in which mushketovite shows a platy morphology (e.g., [4,34] and references therein). It may result from two different processes: redox reactions by the reduction of Fe^{3+} to Fe^{2+} with the removal of oxygen and non-redox reactions by the addition of Fe^{2+} ions [54,55]. In the latter case, the number of Fe atoms increases, whereas in the redox process, the replacement of Fe^{3+} with Fe^{2+} causes a volume decrease calculated at 1.64% [4]. Magnetite Mt3 immediately adjacent to relict hematite shows microporosity, but porosity decreases with distance from hematite or in the bands without relict hematite. Thus, we propose that magnetite Mt3 likely formed via the redox transformation mechanism in a reducing environment compared to that prevailing for the formation of hematite.

5. Conclusions

This first study on the texture and chemistry of massive magnetite layers in amphibolites and micaschists confirms findings from previous studies on the disseminated and layered iron ore in micaschists from the same area, i.e., a Rapitan-type BIF mineralization formed in a back-arc basin, through iron precipitation and incorporation of terrigenous materials in the chemical precipitates of the original BIF. The protolith of the amphibolites was a basalt, typically interlayered with chemical and terrigenous marine sediments. Hydrothermal fluids formed and modified the iron ore.

Our findings support a formation in four steps (Figure 17): Step 1: The silician magnetite Mt1 likely formed in a rather reduced Si-rich environment, a hydrothermal volcano–sedimentary setting typical for BIF. The incorporation of structural or nano-Si-rich inclusions may have caused lattice defects or deformation promoting fluid circulation and alteration of the magnetite grains. Step 2: A fluid-induced coupled dissolution of Mt1 and precipitation of Mt2 (CDR process) is supported by a close spatial relationship, sharp compositional boundaries, the similar crystallographic structure of Mt1 and Mt2, and porosity in Mt2. Porosity promotes the infiltration of hydrothermal fluids, accelerating the CDR process. Step 3: Microfracturing increasing the permeability led to the penetration of a more oxidized fluid along cleavage planes and the formation of platy hematite. By increasing the contact between the fluid and the magnetite grain, microfracturing allows an acceleration of the oxidation process and dissolution of magnetite. Step 4: Finally, after progressive sealing of the fractures, mushketovite Mt3 formed after hematite from a more reduced fluid composition by a redox transformation supported by the volume decrease. Deformation can be attributed to the two latter steps.

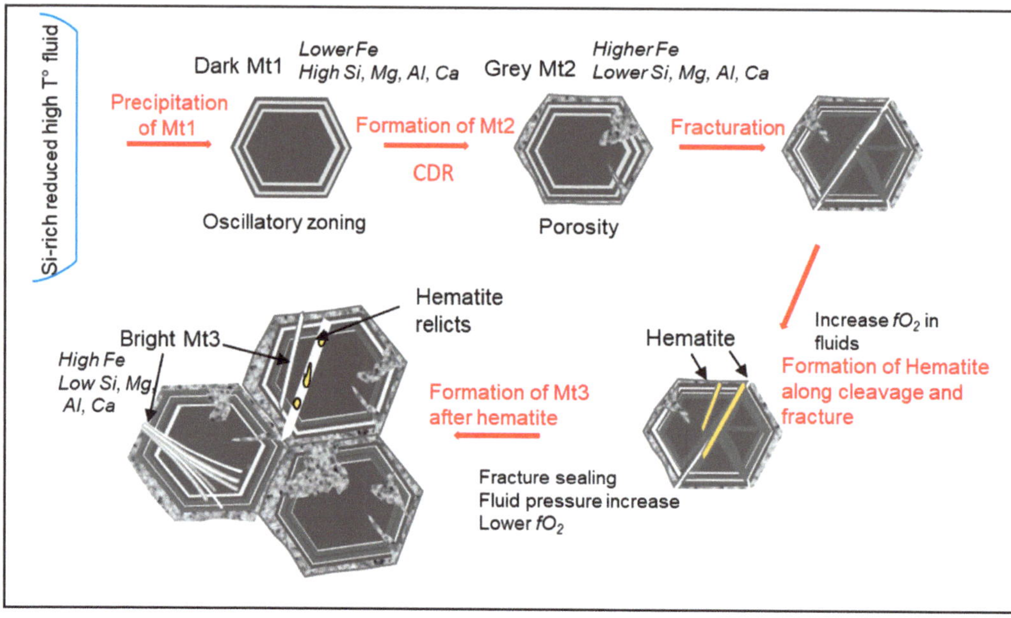

Figure 17. Schematic illustration of the multiple processes involved in the formation of magnetite from the Takab iron ore deposit. CDR: coupled dissolution–reprecipitation process. Dark, gray, and bright magnetite as defined on BSE images.

Supplementary Materials: The following supporting information can be downloaded at: https://www.mdpi.com/article/10.3390/min15020137/s1, Figure S1: WDS X-ray maps of selected elements in a magnetite grain from sample TAK-Z3. Figure S2: Discrimination diagram V vs Ti for the Takab magnetite. Figure S3. Ca vs. (Si + Al) plot of magnetite Mt1 and amphibole from sample TAK-Z1. Figure S4: Pole figures of magnetite Mt3 and relict hematite. Table S1: Composition of feldspar in amphibolite host rock and iron ore from Takab. Table S2: Fe, O, and trace-element composition of magnetite from the Takab iron ore samples. Table S3: Average formation temperature for the Takab magnetite.

Author Contributions: Conceptualization, C.W. and B.O.; investigation, C.W., O.B., N.R. and B.O.; project administration, C.W. and B.O.; resources, C.W., B.O., G.N., M.H. and I.M.; supervision, C.W.; validation, C.W., O.B. and N.R.; visualization, C.W., G.N. and M.H.; writing—original draft, C.W.; writing—review and editing, C.W., B.O., G.N. and M.H. All authors have read and agreed to the published version of the manuscript.

Funding: We thank the Institute of Advanced Sciences of Basic Science and University of Zanjan, Zanjan, for logistic support during fieldwork and the stay of BO in Zanjan. Campus France and the French embassy (Gundishapur project No. 40624TK) are thanked for financing the stay of BO at IASBS, Zanjan. The institute ISTeP of Sorbonne Université is thanked for financial support of the analyses through the French National TRIGGER project, French–Iranian cooperation.

Data Availability Statement: The original contributions presented in this study are included in the article and Supplementary Materials.

Acknowledgments: L. Reisberg (CRPG, Vandoeuvre-lès-Nanvy, France) is kindly acknowledged for a thorough review and English revision of the manuscript. We thank three anonymous reviewers for their helpful comments.

Conflicts of Interest: The authors declare no conflicts of interest.

References

1. Dupuis, C.; Beaudoin, G. Discriminant diagrams for iron oxide trace element finger printing of mineral deposit types. *Miner. Deposita* **2011**, *46*, 319–335. [CrossRef]
2. Dare, S.A.S.; Barnes, S.-J.; Beaudoin, G.; Méric, J.; Boutroy, E.; Potvin-Doucet, C. Trace elements in magnetite as petrogenetic indicators. *Miner. Deposita* **2014**, *49*, 785–796. [CrossRef]
3. Nadoll, P.; Angerer, T.; Mauk, J.L.; French, D.; Walshe, J. The chemistry of hydrothermal magnetite: A review. *Ore Geol. Rev.* **2014**, *16*, 1–32. [CrossRef]
4. Hu, X.; Chen, H.; Beaudouin, G.; Zhang, Y. Textural and compositional evolution of iron oxides at Mina Justa (Peru): Implications for mushketovite and formation of IOGC deposits. *Am. Mineral.* **2020**, *105*, 397–408. [CrossRef]
5. Huang, X.W.; Beaudoin, G. Textures and chemical compositions of magnetite from iron oxide copper-gold (IOCG) and Kiruna-type iron oxide-apatite deposits and their implications for ore genesis and magnetite classification schemes. *Econ. Geol.* **2019**, *114*, 953–979. [CrossRef]
6. Zhang, Y.; Hollings, P.; Shao, Y.; Li, D.; Chen, H.; Li, H. Magnetite texture and trace-element geochemistry fingerprint of pulsed mineralization in the Xinqiao Cu-Fe-Au deposit, Eastern China. *Am. Mineral.* **2020**, *105*, 1712–1723. [CrossRef]
7. Liang, P.; Wu, C.; Hu, X.; Xie, Y. Textures and geochemistry of magnetite: Indications for genesis of the Late Paleozoic Laoshankou Fe-Cu-Au deposit, NW China. *Ore Geol. Rev.* **2020**, *124*, 103632. [CrossRef]
8. Honarmand, M.; Nabatian, N.; Wagner, C.; Monsef, I.; Delpech, G.; Bayon, G.; Orberger, B. Late Ediacaran iron formation, NW Iran: Origin, depositional age and tectonic/climate significance. *Precambrian Res.* **2024**, *406*, 107382. [CrossRef]
9. Karimpour, M. *Applied Economic Geology*; Javid Publication: Mashhad, Iran, 1989; 404p.
10. Nabatian, G.; Rastad, E.; Neubauer, F.; Honarmand, M.; Ghaderi, M. Iron and Fe-Mn mineralisation in Iran: Implications for Tethyan metallogeny. *Aust. J. Earth Sci.* **2015**, *62*, 211–241. [CrossRef]
11. Tuck, C.C. Iron ore. In *Mineral Commodity Summaries*; USGS Reports; USGS Reports; National Minerals Information Center: Reston, VA, USA, 2023.
12. Foster, H.; Jafarzadeh, A. The Bafq mining district in Central Iran—A highly mineralized Infracambrian volcanic field. *Econ. Geol.* **1994**, *89*, 1697–1721. [CrossRef]
13. Mazaheri, S.A.; Andrew, A.S.; Chenhall, B.E. *Petrological Studies of Sangan Iron Ore Deposit*; Center for Isotope Studies Research Report: Sydney, NSW, Australia, 1994; pp. 48–52.
14. Daliran, F. Kiruna-Type Iron Oxide-Apatite Ores and Apatitites of the Bafq District, Iran, with an Emphasis on the REE Geochemistry of Their Apatites. In *Hydrothermal Iron Oxide Copper Gold and Related Deposits. A Global Perspective*; Porter, T.M., Ed.; PGC Publishing: Adelaide, SA, Australia, 2002; Volume 2, pp. 303–320.
15. Daliran, F.; Stosch, H.G.; Williams, P.; Jamli, H.; Dorri, M.B. Lower Cambrian Iron Oxide-Apatite-REE (U) Deposits of the Bafq District, East-Central Iran. In *Exploring for Iron Oxide Copper-Gold Deposits: Canada and Global Analogues*; Corriveau, L., Mumin, H., Eds.; Geological Association of Canada Short Course Notes: St John's, NL, Canada, 2010; Volume 20, pp. 143–155.
16. Alavi, M. Sedimentary and structural characteristics of the Paleo-Tethys remnants in northeastern Iran. *GSA Bull.* **1991**, *103*, 983–992. [CrossRef]
17. Berberian, M.; King, G.C.P. Towards a paleogeography and tectonic evolution of Iran. *Can. J. Earth Sci.* **1981**, *18*, 210–265. [CrossRef]
18. Jafari, A.; Karimpour, M.H.; Mazaheri, S.A.; Shafaroudi, A.M.; Ren, M. Geochemistry of metamorphic rocks and mineralization in the Gol-Gohar iron ore deposit (No. 1), Sirjan, SE Iran: Implications for Paleotectonic setting and ore genesis. *J. Geochem. Explor.* **2019**, *205*, 106330. [CrossRef]
19. Nabatian, G.; Ghaderi, M.; Daliran, F.; Rashidnejad-Omran, N. Sorkhe-Dizaj iron oxide-apatite ore deposit in the Cenozoic Alborz-Azarbaijan magmatic belt, NW Iran. *Resour. Geol.* **2013**, *63*, 42–56. [CrossRef]
20. Nabatian, G.; Li, X.-H.; Honarmand, M.; Melgarejo, J.C. Geology, mineralogy and evolution of iron skarn deposits in the Zanjan district, NW Iran: Constraints from U-Pb dating, Hf and O isotope analyses of zircons and stable isotope geochemistry. *Ore Geol. Rev.* **2017**, *84*, 42–66. [CrossRef]
21. Sarjoughian, F.; Habibi, H.; Lentz, D.R.; Azizi, H.; Esna-Ashari, A. Magnetite compositions from the Baba Ail iron deposit in the Sanandaj-Sirjan zone, western Iran: Implications for ore genesis. *Ore Geol. Rev.* **2020**, *126*, 103728. [CrossRef]
22. Maanijou, M.; Khodaei, L. Mineralogy and electron microprobe studies of magnetite in the Sarab-3 iron Ore deposit, southwest of the Shahrak mining region (east Takab). *J. Econ. Geol.* **2018**, *10*, 267–293. (In Persian)
23. Maanijou, M.; Salemi, R. Mineralogy, chemistry of magnetite and genesis of Korkora-1 iron deposit, east of Takab, NW Iran. *J. Econ. Geol.* **2014**, *6*, 355–374. (In Persian)
24. Ghaderi Piraghoum, Z.; Kouhestani, H.; Tofighi, F. Geology, geochemistry and genesis of the Kosaj Fe occurrence, Takab-Takht-e Soleiman-Angouran metallogenic zone, SW Zanjan. *Adv. Appl. Geol.* **2020**, *10*, 294–313, (In Persian with English abstract).

25. Pourmohammad, F.; Kouhestani, H.; Azimzadeh, A.M.; Nabatian, G.; Mokhtari, M.A.A. Mianaj iron occurrence, southwest of Zanjan: Metamorphosed and deformed volcano-sedimentary type of mineralization in Sanandaj-Sirjan zone. *Sci. Q. J. Geosci.* **2019**, *28*, 161–174, (In Persian with English abstract).
26. Biparva, M. Geochemistry, Mineralogy and Genesis of Fe Deposits, East of Takab (Halab, Mianaj and Quzijan Villages). Master's Thesis, Department of Geology, Shahid Beheshti University, Teheran, Iran, 2013. Unpublished (In Persian with English abstract).
27. Fereiduni, Z. Geology, Mineralogy and Geochemistry of Halab Iron Mineralization, SW Zanjan. Master's Thesis, Department of Geology, University of Zanjan, Zanjan, Iran, 2017. Unpublished (In Persian with English abstract).
28. Wagner, C.; Villeneuve, J.; Boudouma, O.; Rividi, N.; Orberger, B.; Nabatian, G.; Honarmand, M.; Monsef, I. In Situ Trace Element and Fe-O Isotope Studies on Magnetite of the Iron-Oxide Ores from the Takab Region, North Western Iran: Implications for Ore Genesis. *Minerals* **2023**, *13*, 774. [CrossRef]
29. Hassanzadeh, J.; Stockli, D.F.; Horton, B.K.; Axen, G.J.; Stockli, L.D.; Grove, M.; Schmitt, A.K.; Walker, J.D. U-Pb zircon geochronology of late Neoproterozoic-early Cambrian granitoids in Iran: Implications for paleogeography, magmatism, and exhumation history of Iranian basement. *Tectonophysics* **2008**, *451*, 71–96. [CrossRef]
30. Hajialioghli, R.; Moazzen, M.; Jahangiri, A.; Oberhansli, R.; Mocek, B.; Altenberger, U. Petrogenesis and tectonic evolution of metaluminous sub-alkaline granitoids from the Takab complex, NW Iran. *Geol. Mag.* **2011**, *148*, 250–268. [CrossRef]
31. Moazzen, M.; Hajialioggli, R. Zircon SHRIMP Dating of Mafic Migmatites from NW Iran: Reporting the Oldest Rocks from the Iranian Crust. In Proceedings of the 5th Annual Meeting AOGS, Busan, Korea, 16–20 June 2008.
32. Babakhani, A.R.; Ghalamghash, J. Explanatory Text of Takht-e-Soleiman. In *Geological Quadrangle Map 1:100000, No.5463;* Geological Survey of Iran: Tehran, Iran, 2005. (In Persian)
33. Till, J.L.; Moskowitz, B.M. Deformation microstructures and magnetite texture development in synthetic shear zones. *Tectonophysics* **2014**, *629*, 211–223. [CrossRef]
34. Hu, X.; Xiao, B.; Jiang, H.; Huang, J. Magnetite texture and trace element evolution in the Shaquanzi Fe-Cu deposit, Eastern Tianshan, NW China. *Ore Geol. Rev.* **2023**, *154*, 105306. [CrossRef]
35. Nielsen, R.L.; Forsythe, L.M.; Gallahan, W.E.; Fisk, M.R. Major element and trace element magnetite-melt equilibria. *Chem. Geol.* **1994**, *117*, 167–191. [CrossRef]
36. Toplis, M.J.; Caroll, M.R. An experimental study of oxygen fugacity on Fe-Ti oxide stability, phase relations, and mineral-melt equilibria in ferro-basaltic systems. *J. Petrol.* **1995**, *36*, 1137–1170. [CrossRef]
37. Deditius, A.P.; Reich, M.; Simon, A.C.; Suvorova, A.; Knipping, J.; Roberts, M.P.; Rubanov, S.; Dodd, A.; Saunders, M. Nanogeochemistry of hydrothermal magnetite. *Contrib. Mineral. Petrol.* **2018**, *173*, 46. [CrossRef]
38. Xu, H.; Shen, Z.; Konishi, H. Si-magnetite nano-precipitates in silician magnetite from banded iron formation: Z-contrast imaging and ab-initio study. *Amer. Mineral.* **2014**, *99*, 2196–2202. [CrossRef]
39. Ciobanu, C.L.; Verdugo-Ihl, M.R.; Slattery, A.; Cook, N.J.; Ehrig, K.; Courtney-Davies, L.; Wade, B.P. Silician Magnetite: Si–Fe-Nanoprecipitates and Other Mineral Inclusions in Magnetite from the Olympic Dam Deposit, South Australia. *Minerals* **2019**, *9*, 311. [CrossRef]
40. Canil, D.; Lacourse, T. Geothermometry using minor and trace elements in igneous and hydrothermal magnetite. *Chem. Geol.* **2020**, *541*, 119576. [CrossRef]
41. Huberty, J.M.; Konishi, H.; Heck, P.R.; Fournelle, J.H.; Valley, J.W.; Xu, H. Silician magnetite from the Dales Gorge Member of the Brockman Iron Hamersley Group, Western Australia. *Am. Mineral.* **2012**, *97*, 26–37. [CrossRef]
42. Whittaker, E.J.W.; Muntus, R. Ionic radii for use in geochemistry. *Geochim. Cosmochim. Acta* **1970**, *34*, 945–956. [CrossRef]
43. Hu, X.; Chen, H.; Zhang, W. Texture and composition of magnetite in the Duotoushan deposit, NW China: Implications for ore genesis of Fe-Cu deposits. *Min. Mag.* **2020**, *84*, 398–411. [CrossRef]
44. Shimazaki, H. On the occurrence of silician magnetite. *Resour. Geol.* **1998**, *48*, 23–29. [CrossRef]
45. Ohkawa, M.; Miyahara, M.; Ohta, E.; Hoshino, K. Silicon-substituted magnetite and accompanying iron oxides and hydroxides from the Kumano mine, Yamaguchi Prefecture, Japan: Reexamination of the so-called maghemite (g-Fe_2O_3). *J. Mineral. Petrol. Sci.* **2007**, *102*, 182–193. [CrossRef]
46. Sun, S.; Li, Y.-L. Microstructures, crystallography and geochemistry of magnetite in 2500 to 2200 million-year old banded iron formation from South Africa, Western Australian and North China. *Prec. Res.* **2017**, *298*, 292–305. [CrossRef]
47. Putnis, A. Mineral replacement reactions. In *Thermodynamics and Kinetics of Water-Rock Interaction*; Oelkers, E., Schott, J., Eds.; Reviews in Mineralogy & Geochemistry, Mineralogical Society of America: Chantilly, VA, USA, 2009; Volume 70, pp. 87–124.
48. Putnis, A. Transient porosity resulting from fluid-mineral interaction and its consequences. In *Pore-Scale Geochemical Processes*; Steefel, C.I., Emmanuel, S., Anovitz, L.M., Eds.; Reviews in Mineralogy & Geochemistry, Mineralogical Society of America: Chantilly, VA, USA, 2009; Volume 80, pp. 1–23.
49. Altree-Williams, A.; Pring, A.; Ngothai, Y.; Brugger, J. Textural and compositional complexities resulting from coupled dissolution-reprecipitation reactions in geomaterials. *Earth Sci. Rev.* **2015**, *150*, 628–651. [CrossRef]

50. Hu, H.; Li, J.W.; Lentz, D.; Ren, Z.; Zhao, X.F.; Deng, X.D.; Hall, D. Dissolution-re precipitation process of magnetite from the Chengchao iron deposit: Insights into ore genesis and implication for in-situ chemical analysis of magnetite. *Ore Geol. Rev.* **2014**, *57*, 393–405. [CrossRef]
51. Hu, H.; Lentz, D.; Li, J.-W.; McCarron, T.; Zhao, X.-F.; Hall, D. Reequilibration processes in magnetite from iron skarn deposits. *Econ. Geol.* **2015**, *110*, 1–8. [CrossRef]
52. Yin, S.; Ma, C.Q.; Robinson, P.T. Textures and high field strength elements in hydrothermal magnetite from a skarn system: Implications for coupled dissolution-reprecipitation reactions. *Am. Miner.* **2017**, *102*, 1045–1056. [CrossRef]
53. Zhang, S.; Chen, H.; Xiao, B.; Zhao, L.; Hu, X.; Li, J.; Gong, L. Textural and chemical evolution of magnetite from the Paleozoic Shuanglong Fe-Cu deposit: Implications for tracing ore-forming fluids. *Am. Miner.* **2023**, *108*, 178–191. [CrossRef]
54. Ohmoto, H. Nonredox transformations of magnetite-hematite in hydrothermal systems. *Econ. Geol.* **2003**, *98*, 157–161. [CrossRef]
55. Mücke, A.; Cabral, A.R. Redox and non-redox reactions of magnetite and hematite in rocks. *Geochemistry* **2005**, *65*, 271–278. [CrossRef]

minerals

MDPI

Review

Evolution of Auriferous Fluids in the Kraaipan-Amalia Greenstone Belts: Evidence from Mineralogical and Isotopic Constraints

Kofi Adomako-Ansah [1], Napoleon Q. Hammond [2,*], Yuichi Morishita [3] and Daizo Ishiyama [4]

[1] Department of Geological Engineering, University of Mines and Technology, Tarkwa P.O. Box 237, Ghana; kadomakoansah@umat.edu.gh
[2] Department of Geology and Mining, School of Physical and Mineral Sciences, University of Limpopo, Private Bag X1106, Sovenga 0727, South Africa
[3] Faculty of Science, Centre for Integrated Research and Education of Natural Hazards, Shizuoka University, 836 Ohya, Suruga-ku, Shizuoka 422-8529, Japan; morishita.yuichi@shizuoka.ac.jp
[4] Graduate School of International Resource Sciences, Akita University, 1-1 Tegata-gakuen machi, Akita 010-8502, Japan; ishiyama@gipc.akita-u.ac.jp
* Correspondence: quaye.hammond@msn.com

Abstract: The Kraaipan and Amalia greenstone belts in South Africa occur in the western part of the Kaapvaal Craton. The two belts stretch discontinuously in an approximately north–south orientation over a distance of about 250 km from southern Botswana in the north to the Vaal River near Christiana in the south and are separated by a distance of about 90 km. Gold mineralization is hosted in banded iron formation at both the Kalahari Goldridge deposit (Kalgold) in the Kraaipan greenstone belt in the north and the Amalia deposit in the Amalia greenstone belt in the south, with the mineralization associated with quartz–carbonate veins. The footwalls of these deposits are generally composed of mafic volcanic schist and the hanging walls consisting of graywackes, schist and shale units. The Kalgold and Amalia gold deposits show some variation in the redox condition of the mineralizing system and fluid chemistry. The ore mineral assemblage is characterized by magnetite–pyrrhotite–pyrite at Kalgold, which is indicative of reducing conditions, and a magnetite–hematite–pyrite assemblage at Amalia that suggests a relatively oxidizing environment. Average mineralizing temperatures determined from chlorite geothermometry were relatively higher at the Kalahari Goldridge deposit ranging from 350 to 400 °C compared to the slightly cooler range of 330 to 390 °C at Amalia. The composition of the fluids derived from fluid inclusions is indicative of low salinity H_2O--CO_2±CH_4-rich fluids at Kalgold against relatively H_2O-CO_2-rich fluids at Amalia. Evidence from strontium–carbon–oxygen isotopic ratios from carbonates suggests that differences in redox conditions in the deposits could be attributed to different flow pathways by an evolving fluid from a common source (with minimum $^{87}Sr/^{86}Sr$ = 0.70354) to the sites of gold deposition, with a significant ore fluid interaction with a thick sequence of carbonaceous meta-pelitic rock units at the Kalahari Goldridge deposit that is absent in the Amalia deposit.

Keywords: Kraaipan; Amalia; greenstone belt; gold; oxide BIF; strontium isotopes; South Africa; Kaapvaal Craton

Citation: Adomako-Ansah, K.; Hammond, N.Q.; Morishita, Y.; Ishiyama, D. Evolution of Auriferous Fluids in the Kraaipan-Amalia Greenstone Belts: Evidence from Mineralogical and Isotopic Constraints. *Minerals* **2024**, *14*, 1171. https://doi.org/10.3390/min14111171

Academic Editor: Kunfeng Qiu

Received: 25 August 2024
Revised: 11 November 2024
Accepted: 12 November 2024
Published: 18 November 2024

1. Introduction

Several studies on Archaean orogenic gold deposits (e.g., [1–5]) are documented to have been formed from ore-forming fluids that originated from deep within the crust and migrated along crustal-scale faults or fissures to their final depositional sites. The stable isotope data associated with most of the previous studies were unable to clearly define the origin of ore-forming fluids associated with such deposits due to the overlap of isotopic signatures among fluid sources. In recent years, the application of radiogenic isotopes, particularly strontium (Sr), have been widely used as tracers by analyzing strontium-rich gangue minerals to define the source reservoir and flow pathways of these fluids (e.g., [6,7]), because the $^{87}Sr/^{86}Sr$ ratios can identify the origin of Sr from continental, marine or mantle reservoirs as these have characteristic signatures (e.g., [8–14]). The flow pathways can provide significant insight into the nature of water–rock interactions between the fluid and the rocks encountered during hydrothermal fluid migration, thus providing important information on the nature and evolution of the hydrothermal fluid in ore deposit systems.

The Kraaipan and Amalia greenstone belts in South Africa are N-S trending, laterally discontinuous structures that are spatially associated with granitoids of similar petrological characteristics and age (Figure 1). The basement rocks of these greenstone belts are Archaean-aged tonalite–trondhjemite–granodiorite (TTG) gneisses [15]. Epigenetic banded iron formation (BIF)-hosted gold deposits within the Kraaipan-Amalia terrane include the Kalahari Goldridge deposit in the Kraaipan greenstone belt (to the north) and the Amalia deposit in the Amalia greenstone belt (to the south) (Figure 2) and are separated about 90 km from each other. The hanging wall units in both deposits are made up of Archean supracrustal metasedimentary rocks of shales and schists, and the footwall comprises mafic volcanic schists (Figures 2 and 3). However, local lithological variations occur between the BIF and footwall at the Kalahari Goldridge deposit [16]. Thin carbonaceous- and pyrite-bearing metamorphosed pelites (meta-pelites) ranging between 1 and 2 m in thickness occur discontinuously between the contacts of the orebody and footwall and also between the orebody and immediate hanging wall units at the Kalahari Goldridge deposit. The meta-pelites have been subjected to intense deformation as evidenced by the tight to isoclinal folding associated with the unit.

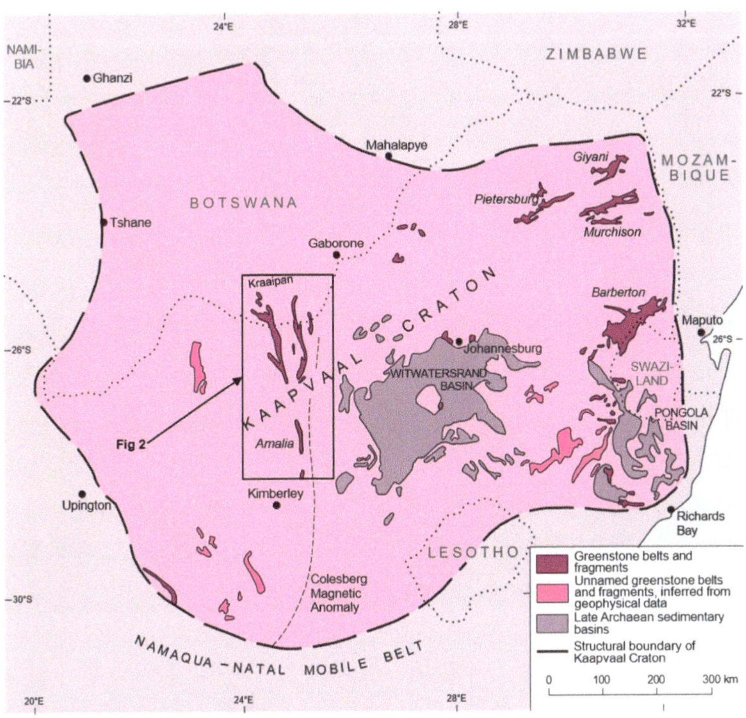

Figure 1. Geological map of the Kaapvaal Craton showing the location of the Kraaipan and Amalia greenstone belt in the western part of the craton. (Modified after [17]).

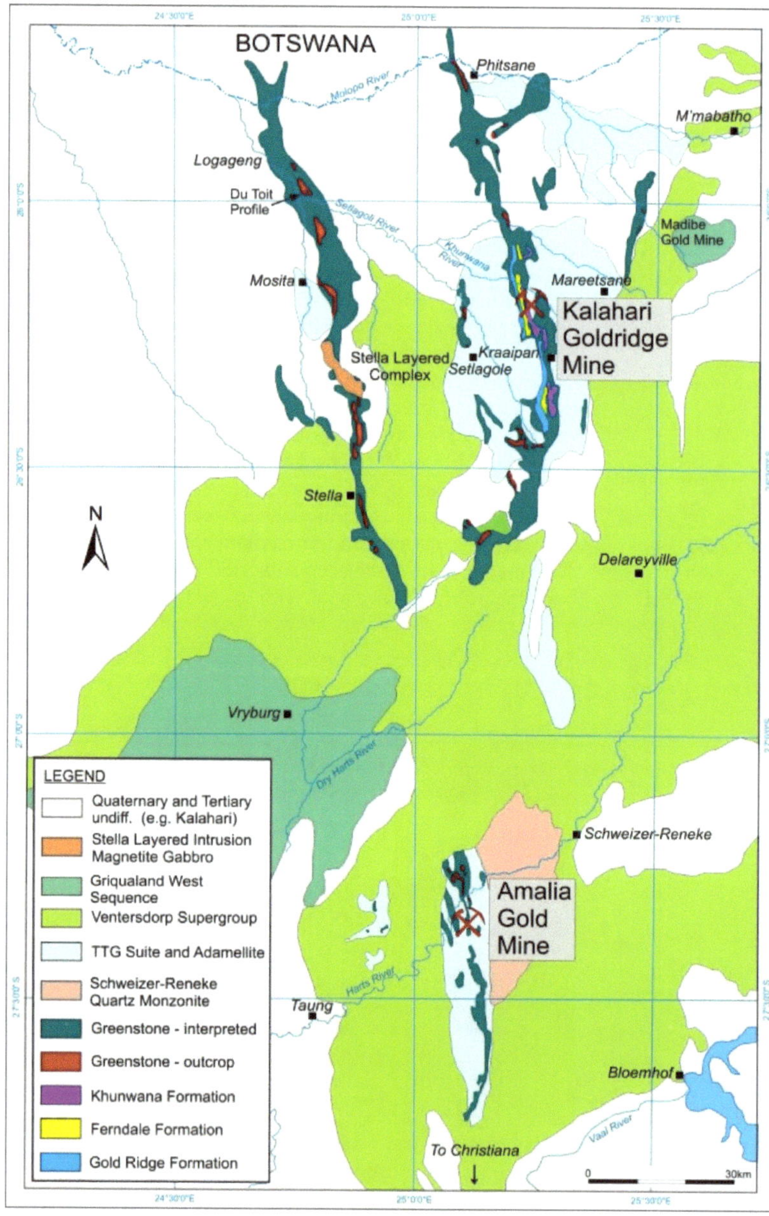

Figure 2. Regional geological map of the Kraaipan and Amalia greenstone belts with the location of the Kalahari Goldridge and Amalia gold deposits.

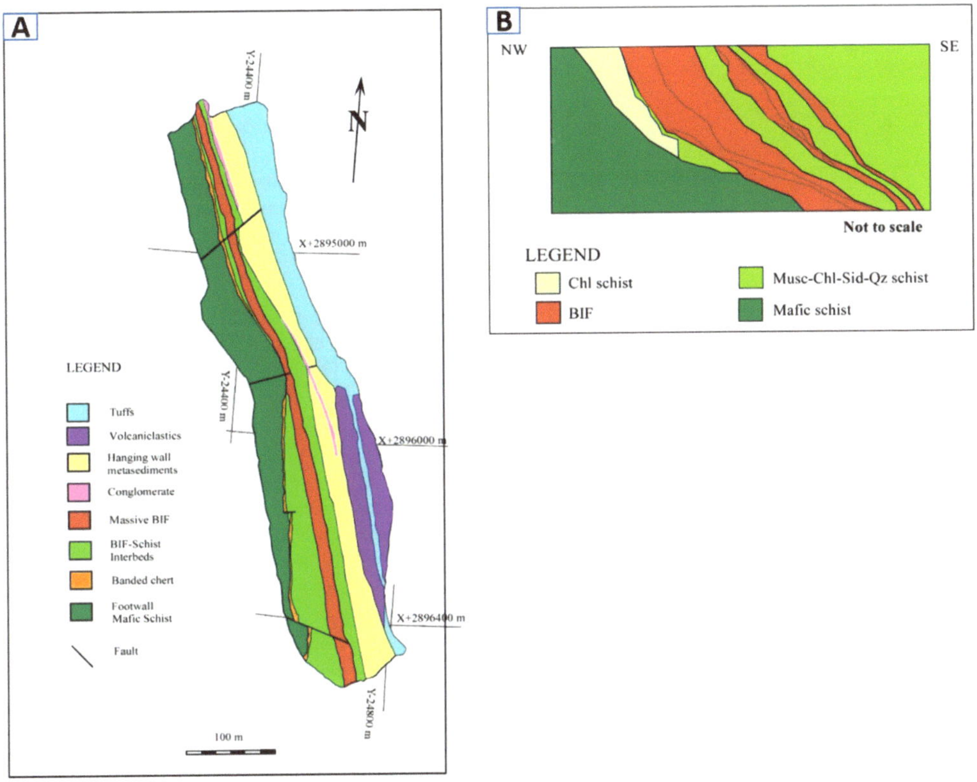

Figure 3. Local geological maps of the Kalahari Goldridge, D Zone orebody (**A**) and the Amalia gold deposit (not to scale) (**B**) as shown in Figure 2. Chl = chlorite; Musc = muscovite; Sid = siderite; Qz = quartz; and BIF = banded iron formation.

Despite similar geological settings, the geochemistry of the mineralization in these deposits exhibits some local variations [18,19]. It is therefore not clear if the mineralizing events in these deposits were spatially linked to a unique fluid source that were modified along different flow pathways to the depositional sites, or if they were discrete mineralizing events with distinct sources. A combination of Sr-C-O isotopic ratios from previous studies on the Kalahari Goldridge [18,20], the C-O isotope on Amalia [19] and our new Sr isotope data for the Amalia deposit (this paper) help to unravel and evaluate the nature and evolution of the fluids associated with gold mineralization in the Kraaipan-Amalia terrane.

2. Geological Setting

The evolutional history of the Kaapvaal craton was initiated by the amalgamation of the ~3.7-Ga-old Witwatersrand block (to the east) and the 3.2-Ga-old Kimberly block (to the west) along the Colesberg Lineament (CL) suture (Figure 1) followed by subduction-accretion and continent-continent collision processes at about 3.0–2.9 Ga ago [21–27]. The Kraaipan-Amalia greenstone belts are located in the Kimberly block of the Kaapvaal craton and aligned parallel to the north-trending CL. The regional thermo-tectonic processes are believed to have resulted in the formation of highly deformed, north-trending subvertical volcano-sedimentary rock units in the Kraaipan-Amalia greenstone belts that are generally fault-bounded and partially engulfed by abundant intrusive syn- to post-tectonic granitoids [22,25]. The Kraaipan-Amalia greenstone belt is flanked to the west by 3.08–2.93 Ga-old reworked TTGs and to the east by the 2.93–2.85 Ga-old Kraaipan group of post-collisional and intrusive granodioritic and magnetite-bearing quartz monzonitic plutons (Figure 1; [15,27,28]). The spatial association between the Kraaipan-Amalia greenstone belts and the granodioritic-monzonitic rocks aroused speculations by earlier researchers on a genetic link between the gold deposits within the greenstone belts and felsic igneous activity e.g., [15,27,28]. The greenstone belts are poorly exposed due to the limited outcrop in the region. A few BIF deposits were reported to be associated with a suite of agglomerates and accretionary lapilli tuffs [15,29]. A geophysical investigation conducted across the CL by [30] revealed that the Kraaipan-Amalia greenstone belt terrane was obducted unto the Kaapvaal craton from the east as an allochthonous unit by accretionary tectonics [30].

3. Hydrothermal Alteration and Gold Mineralization

The host rocks to gold mineralization in the Kalahari Goldridge and Amalia deposits are oxide facies BIFs with variable amounts of chert and magnetite (Figure 3) [16,31]. The BIFs are intercalated with a variety of schists [16,31].

Gold ore in both deposits is associated with shallow-dipping quartz-carbonate veins that cut across the oxide facies BIF layers (Figure 4), with the mineralization concentrated in the altered oxide-rich units. Carbonate minerals in the veins are dominated by ankerite and siderite. At Kalahari Goldridge, the mineralizing quartz–carbonate veins were further classified as Group II (A and B) based on their sizes and cross-cutting relation on the magnetite mesobands (Figure 4B). Field and textural evidence in both deposits indicate pervasive carbonate alteration of the BIF with intense sulfidation. Sulfidation haloes commonly occur at the contacts and selvages between the quartz–carbonate veins and the host BIF units. The replacement of magnetite by pyrite and pyrrhotite is prominent at Kalahari Goldridge (Figure 5A–C), indicating a paragenetically later episode of sulfidation. However, there is an extensive replacement of magnetite by hematite and pyrite at the Amalia deposit with minor chalcopyrite and arsenopyrite (Figure 5D–F). The hematization of magnetite was generally absent at Kalahari Goldridge. Field and geochemical observations showed that gold mineralization in both deposits is closely associated with the altered host rocks; typically, the carbonate-altered and highly sulfidized BIF. These observations are clearly consistent with contrasting ambient redox conditions in the ore-forming fluids during mineralization in these deposits: a more reduced fluid system at Kalahari Goldridge deposit and more oxidized conditions at Amalia.

Figure 4. Photos of quartz–carbonate veins in the BIF at the Kalahari Goldridge deposit (**A**) and Amalia gold deposit (**B**).

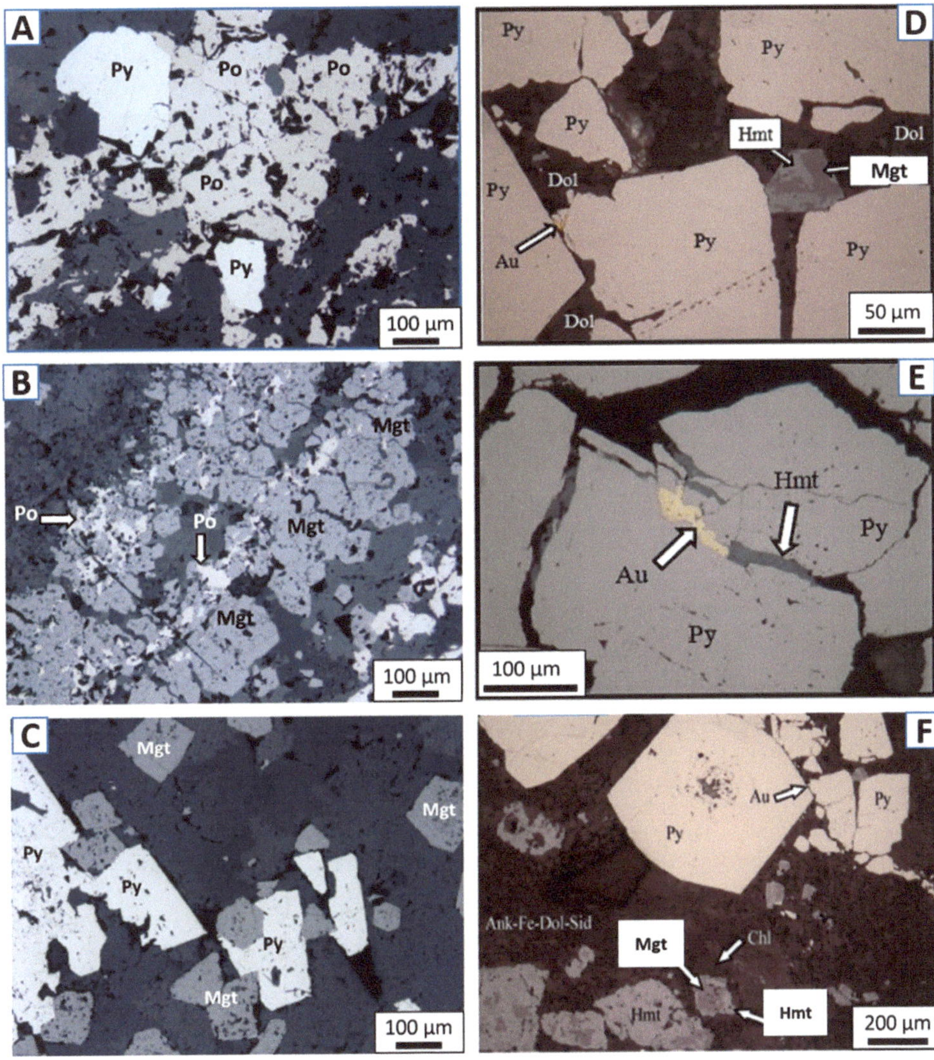

Figure 5. Textural relationships between ore minerals in the BIF at the Kalahari Goldridge (**A–C**) and Amalia gold deposits (**D–F**). Po = pyrrhotite; Py = pyrite; Mgt = magnetite; Hmt = hematite; Au = gold.

The varied redox state of these deposits is further evidenced from microthermometric and Raman analyses of fluid inclusions in quartz veins that indicated low salinity H_2O-NaCl-CO_2 ± CH_4 compositions at the Kalahari Goldridge [18] and H_2O-NaCl-CO_2 at Amalia [19] (Figure 6). Thus, a relatively 'pure' CO_2-rich composition characterizes the ore-forming fluid at Amalia compared to the Kalahari Goldridge, which contains significant CH_4 in some inclusions.

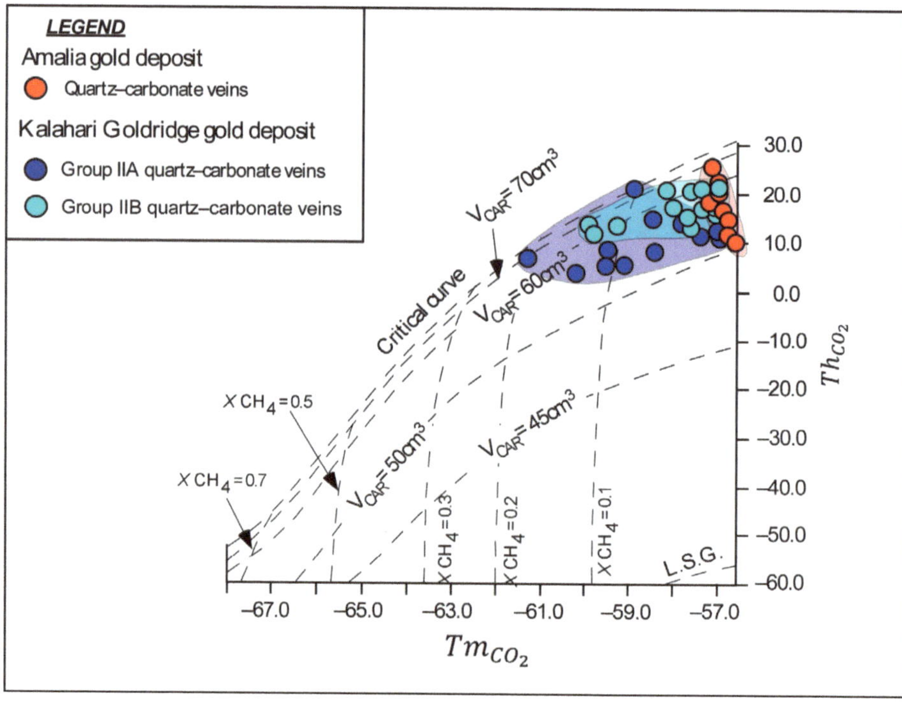

Figure 6. Tm_{CO_2} versus Th_{CO_2} of fluid inclusions in quartz veins from the Kalahari Goldridge and Amalia gold deposits. The inclusions show relatively pure CO_2-rich inclusions at Amalia compared to the Kalahari Goldridge, which have mixed $CO_2 \pm CH_4$-rich inclusions.

Chlorite occurs in all the lithological units at the Kalahari Goldridge deposit. In the pelitic sediments, chlorite typically occurs as fine- to medium-grained schistose or felt-like mass. In the altered BIF units, the chlorite occurs as tabular to equigranular grains intergrown with magnetite and other alteration minerals. Chlorite is rare in the unaltered to less altered BIF units. Chlorite associated with the mineralized quartz veins occurs as sheaves, rosettes or felt-like equigranular aggregates, which may be intergrown with carbonate and quartz in a mosaic texture. Chlorite occurrence at Amalia exhibits similar characteristic features as the Kalahari Goldridge deposits, occurring in all the lithologies. They occur as fine-to-tabular and fibrous grains showing intergrowth with magnetite, siderite, ankerite and pyrite in the highly altered BIF. In the quartz–carbonate veins, it occurs as fine, equigranular grains with a felt-like mass appearance, texturally associated with quartz, ankerite and pyrite. Chlorite also occurs as veinlets cutting magnetite-rich bands in the Amalia BIF.

4. Analytical Procedure

Two independent studies using the same analytical techniques were undertaken for the Kalahari Goldridge and Amalia deposits. These studies are summarized below.

4.1. Mineral Chemistry of Chlorite

The mineral composition of chlorites from the Kalahari Goldridge deposit were obtained by WDS at a 15 kV operating voltage and 20 nA beam current using a JEOL CXA-733 Superprobe (JEOL, Tokyo, Japan) at the Department of Geology, Rhodes University. The counting time was maintained at 20 s. The composition of chlorite from Amalia was acquired by a wavelength dispersive spectrometer (WDS at a 15 kV operating voltage and 20 nA beam current using a JEOL JXA-8800R Superprobe (JEOL, Tokyo, Japan) at the Department of Earth Science and Technology, Akita University, Japan. The peak acquisition time was maintained at 20 s for all elements. In both cases, calibrations were performed using international standards, and data were corrected using ZAF corrections.

4.2. Carbon and Oxygen Isotope Analysis of Carbonates

Carbon and oxygen isotopes analyses for Kalahari Goldridge and Amalia deposits were conducted following the conventional method of McRea (1950). Pulverized carbonate samples were decomposed in orthophosphoric acid (H_3PO_4). The sequential extraction of CO_2 was performed for samples containing more than one carbonate phase. CO_2 gas from carbonates samples from Kalahari Goldridge was liberated from ankerite and siderite. The CO_2 initially evolved at 25° and 50 °C for calcite and ankerite/dolomite, respectively, for at least 4 h. Subsequently, CO_2 was liberated from siderite in a paraffin bath at 100 °C for at least 6 h. The evolved CO_2 was analyzed for C- and O-isotopic ratios using a Finnigan MAT 252 mass spectrometer (Thermo Fisher Scientific, Waltham, MA, USA) at the Department of Geological Sciences, University of Cape Town. The C- and O-isotopic data are presented in the standard notation of $\delta^{13}C$ and $\delta^{18}O$, respectively. In samples from Amalia, CO_2 gas was liberated from ankerite. This step was carried out in a water bath maintained at 100 °C for at least 12 h. The released CO_2 gas was collected in a liquid N_2 trap and then separated from water vapor by the substitution of a dry-ice-acetone trap. The pure CO_2 was analyzed for carbon and oxygen isotopic ratios using a mass spectrometer (Iso-Prime EA-IRMS) at the Research Center for Engineering Sciences, Akita University, Japan. In both cases, carbon and oxygen isotopic ratios are reported as per mil (%) with respect to the Pee Dee Belemnite (PDB) for carbon and Standard Mean Ocean Water (SMOW), respectively.

4.3. Strontium Isotopic Measurements

Nine carbonate samples from mineralized quartz–carbonate veins from the Kalahari Goldridge deposit were selected for Sr isotope studies. Selected samples were further analyzed using X-ray diffraction for high-purity carbonate (siderite and ankerite) separates for Sr isotopic measurements. About 50 to 100 mg of the carbonates were dissolved for 2 h using 0.8 mol^{-1} hydrochloric acid in a Teflon vessel. Strontium was separated by a standard ion exchange technique. $^{87}Sr/^{86}Sr$ ratios were measured on a Finigan MAT 261 mass spectrometer at the Research School of Earth Sciences, Australian National University. Strontium isotopic ratios were normalized to a value of 0.1194 for $^{86}S/^{88}Sr$. During the analytical run, the mean value of nine analyses of the NBS SRM-987 standard was 0.71022 ± 10 (±2σ) for $^{87}Sr/^{86}Sr$. The nine samples were also analyzed for Sr and Rb concentrations using a Quadrupole-ICP-MS at the Research School of Earth Sciences, Australian National University.

From the Amalia gold deposit, strontium isotopic signatures ($^{87}Sr/^{86}Sr$ ratios) were determined from six vein carbonates (ankerite) using a standard ion exchange technique in accordance with the procedure described by [32]. The $^{87}Sr/^{86}Sr$ isotopic measurements were performed on these separates, using a Finnigan MAT261 mass spectrometer (Thermo Fisher Scientific, Waltham, MA, USA) at the Department of Earth Science and Technology, Akita University, Japan. $^{86}Sr/^{88}Sr$ results were normalized to $^{86}Sr/^{88}Sr$ = 0.1194. A mean value of the NBS SRM-987 standard of 0.710244 ± 5 (±2σ) for $^{87}Sr/^{86}Sr$ was used during analytical runs. The concentrations of Sr and Rb were determined from the same vein carbonate samples using an Agilent 7500i Quadrupole ICP-MS instrument (Agilent, Santa Clara, CA, USA) at the Mining Museum of Akita University, Japan.

4.4. Fluid Inclusion Analysis

Microthermometric measurements of fluid inclusions were conducted on primary or pseudo-secondary inclusions in mineralized quartz veins from Kalahari Goldridge and Amalia deposits employing standard procedures. In these studies, five phase transition temperatures were measured during heating runs: (i) the final melting temperature of CO_2

(Tm_{CO_2}), (ii) the final melting temperature of ice (Tm_{ice}), (iii) the dissociation temperature of clathrate ($Tm_{clathrate}$), (iv) the homogenization temperature of CO_2 and CH_4 (Th_{CO_2}/Th_{CH4}) and (v) the final homogenization of mixed aqueous–carbonic (H_2O–NaCl–$CO_2 \pm CH_4$) and aqueous (H_2O–NaCl) inclusions (Th_{total}).

Microthermometric measurements on fluid inclusions in quartz from the Kalahari Goldridge deposit was performed in the Department of Geology at Rhodes University, South Africa using a Linkam THMS 600 heating-freezing (Linkam, Redhill, UK) stage attached to a Nikon microscope. Freezing runs were performed using liquid nitrogen aided by an LNP cooling pump connected to the stage, while heating runs were performed by a thermal resistor with a TM-93 control unit. Microthermometric measurements of fluid inclusions in quartz veins from Amalia were also conducted on a Linkam THMSG 600 heating-freezing stage at the Department of Earth Science and Technology, Akita University. In both cases, mixed aqueous–carbonic inclusions were initially frozen below 100°C and then heated slowly at a rate of 0.5 °C to 1 °C/min for temperatures below 31°C (critical temperature of CO_2) and increased to 0.5 °C to 10 °C/min until total homogenization temperatures. Tm_{CO_2} and Th_{CO_2} measurements were used to evaluate the purity of the CO_2 phase and to ascertain the density of CO_2, respectively.

5. Mineralization Temperatures Based on Chlorite Geothermometry

Chlorite occurs in a wide range of geological environments such as in diagenetic, low- to medium-grade metamorphism and hydrothermally altered rocks. It displays a wide range of non-stoichiometric compositional variations depending on the bulk composition of host rocks, particularly the Fe/(Fe + Mg) ratio, prevailing physico-chemical conditions such as temperature, pressure and the redox state during formation [33–35] and fluid chemistry in systems during the mineralization. The variation in chemical composition in chlorite serve as a useful tool to obtain information on the physico-chemical conditions during the evolution of mineralizing hydrothermal fluids. The geothermometer is generally applied to chlorite in diagenetic, hydrothermal and metamorphic settings (e.g., [36–38]).

The mineralization temperatures for the Amalia and Kalahari Goldridge deposits were determined using the chlorite geothermometry method of [39,40]. Chlorite in mineralizing veins and hydrothermally altered BIFs was analyzed from both deposits. Figure 7 illustrates overlapping temperatures determined from the chlorite geothermometry for the Kalahari Goldridge and Amalia deposits ranging from 350° to 400 °C at the Kalahari Goldridge [18,20] and Amalia from 330° to 390 °C at Amalia [31]. However, there was an overall decrease in temperature with a corresponding decreasing Fe/(Fe+Mg) ratio, with the lowest Fe/(Fe + Mg) ratios associated with chlorites in late fractures in Amalia BIF and the highest Fe/(Fe + Mg) ratio occurring in the sulfidized altered BIF at Kalahari Goldridge. Reference [36] demonstrated factors that cause $Fe^{2+}/(Fe^{2+} + Mg)$ compositional variations in chlorites in hydrothermal ore deposits. They documented in their study that the exchange of Fe^{2+} and Mg^{2+} between the chlorite and hydrothermal solution is a function of physicochemical parameters (e.g., temperature, oxygen fugacity, pH, total dissolved sulfur and activity of Mg^{2+} and Fe^{2+} ions in aqueous solution), as well as the extent of water–rock interactions. The observed positive correlation between the temperature and Fe/(Fe + Mg) ratios at the Kalahari Goldridge and Amalia deposits is inferred to be a regional trend in the Kraaipan-Amalia greenstone terranes, which could possibly be associated with fluid movement along the flow path of mineralizing fluids from the Kalahari Goldridge in the north to the Amalia deposit in the south.

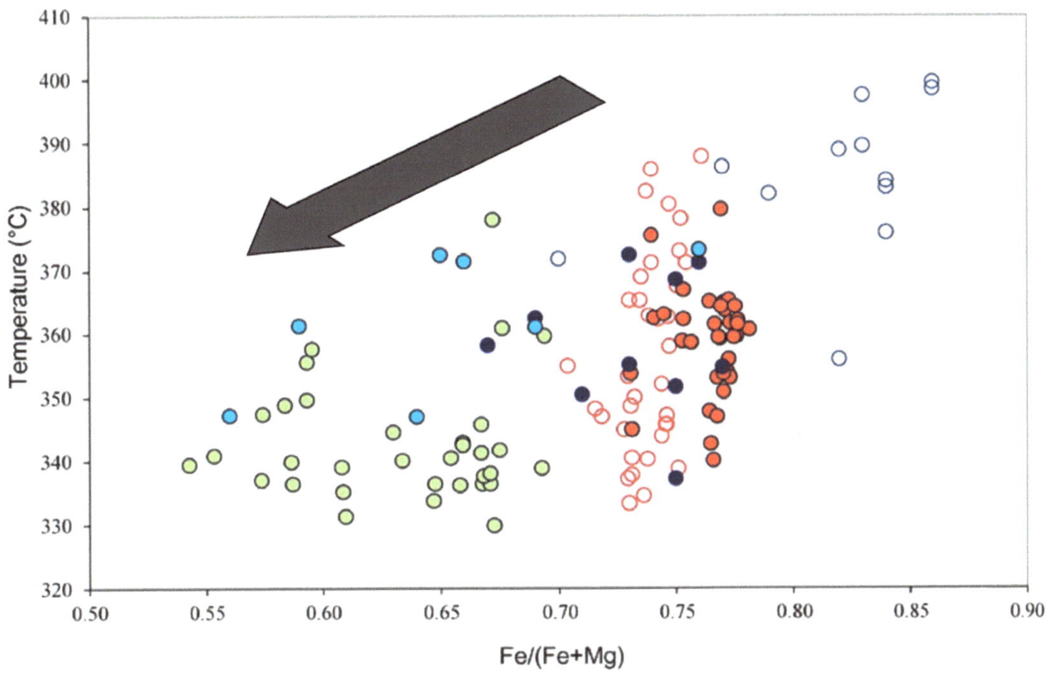

Figure 7. Temperature distribution based on chlorite geothermometry in quartz–carbonate veins and altered BIF at the Kalahari Goldridge and Amalia gold deposits showing a relatively increasing temperature variation from Amalia to the Kalahari Goldridge deposit.

6. Isotopic (Sr, C and O) Signatures and Rb Concentrations

Rubidium–strontium, carbon and oxygen isotopic systematics in carbonates from the alteration carbonates at the Amalia and Kalahari Goldridge deposits were undertaken to deduce the nature and evolution of fluids in these deposits in an attempt to define the fluid origin in the Kraaipan-Amalia terrane. The Rb-Sr isotopic technique was based on the premise that minerals with relatively low concentrations of Rb (<1 ppm) but high concentrations of Sr, such as in carbonates are commonly subjected to Sr-Ca substitution due to similar ionic radii of the two elements. Hence, the $^{87}Sr/^{86}Sr$ ratio can remain relatively unchanged with time due to limited or no influence of radiogenic Sr from the in situ radioactive decay of Rb. $^{87}Sr/^{86}Sr$ ratios represent the approximate initial compositions of fluids at the time of carbonate crystallization. On the basis of this premise, diagnostic fluid–rock interaction trends between evolving fluid systems and crustal components can be constructed to deduce this assertion. The Rb-Sr isotope data for Kalahari Goldridge deposit reported earlier by [20] and compared with new strontium isotopic signatures ($^{87}Sr/^{86}Sr$ ratios) for the Amalia gold deposit are shown in Table 1. Results of the carbonate separates yielded very radiogenic $^{87}Sr/^{86}Sr$ ratios that range from 0.70354 to 0.73914 for the carbonates from Kalahari Goldridge. The Sr and Rb concentrations vary from 1.12 to 168 ppm and 0.04 to 0.4 ppm, respectively. In the Amalia gold deposit, the $^{87}Sr/^{86}Sr$ ratios of the vein carbonate is from 0.703023 to 0.706643, while the Sr and Rb concentrations vary from 3.4 to 157.2 ppm and 0.6 to 2.1 ppm, respectively. The $\delta^{13}C$-$\delta^{18}O$ data of carbonates from quartz–carbonate veins from the Amalia and Kalahari Goldridge deposits are derived from [18,19], respectively, and are also summarized in Table 1.

Table 1. (**A**). Oxygen, carbon, rubidium and strontium isotopic data of carbonates from the Amalia gold deposit, Amalia Greenstone Belt, South Africa (this study). (**B**). Oxygen, carbon, rubidium and strontium isotopic data for carbonates from the Kalahari Goldridge, Kraaipan Greenstone Belt, South Africa [20].

	Mineral	$\delta^{18}O_{SMOW}$ (per mil)	$\delta^{13}C_{PDB}$ (per mil)	$^{87}Rb/^{86}Sr$	Rb (ppm)	Sr (ppm)	Rb/Sr	1/Sr (ppm^{-1})	$^{87}Sr/^{86}Sr$	Std
(A)										
C17-20	Ank	17.2	−4.0	0.010	0.55	157.20	0.004	0.006	0.703023	(±7; 2σ)
C11-5A-3	Ank	16.3	−4.5	0.076	1.00	38.15	0.026	0.026	0.703747	(±6; 2σ)
C17-15B	Ank	13.5	−5.6	1.23	1.46	3.43	0.426	0.292	0.706643	(±6; 2σ)
C17-23	Ank	16.3	−3.9	0.371	0.89	6.94	0.128	0.144	0.704895	(±6; 2σ)
N23-7A	Ank	17.4	−3.5	0.382	1.88	14.22	0.132	0.070	0.705781	(±12; 2σ)
C17-6	Ank	16.3	−4.6	0.078	2.11	78.45	0.027	0.013	0.704318	(±7; 2σ)
(B)										
	Mineral	$\delta^{18}O_{SMOW}$ (per mil)	$\delta^{13}C_{PDB}$ (per mil)	$^{87}Rb/^{86}Sr$	Rb (ppm)	Sr (ppm)	Rb/Sr	1/Sr (ppm^{-1})	$^{87}Sr/^{86}Sr$	Std
ARC 236/11A	Ank	11.3	−5.6	0.034	0.1	23.3	0.004	0.043	0.71042	(±1; 2σ)
ARC 236/11X	Sid	10.1	−5.8	0.188	0.3	4.5	0.067	0.222	0.72325	(±1; 2σ)
ARC 236/16	Sid	15.1	−5.4	0.04	0.04	3.5	0.011	0.286	0.71138	(±1; 2σ)
MSH/W-3	Ank	9.8	−7.6	0.005	0.3	168	0.002	0.006	0.70354	(±1; 2σ)
GDP 531/9C	Sid	11.5	−6.9	0.513	0.2	1.1	0.182	0.909	0.72907	(±2; 2σ)
DZ 40/1	Sid	10.3	−6.7	0.389	0.4	2.9	0.138	0.345	0.73914	(±2; 2σ)
DZ 40/2	Sid	10.6	−6.7	0.047	0.2	12.1	0.017	0.083	0.71583	(±1; 2σ)
DZ 40/3	Sid	10.8	−6.6	0.047	0.3	18.2	0.016	0.055	0.71235	(±1; 2σ)
GDP 531/16B	Sid	10.8	−6.1	0.006	0.1	18.1	0.006	0.055	0.70564	(±2; 2σ)

Std = standard deviation; Ank = ankerite; Sid = siderite.

7. Discussion

The analytical results indicate very low Rb/Sr ratios (<1) for the vein carbonates from both deposits, suggesting that the corresponding $^{87}Sr/^{86}Sr$ ratios remained unchanged through time, making the measured values a good approximation for the initial $^{87}Sr/^{86}Sr$ ratios. Consequently, the $^{87}Sr/^{86}Sr$ ratios are suitable for monitoring the nature and evolution of the ore-forming fluid(s) in the Kraaipan-Amalia terrane.

The basement rocks to the BIF-hosted Kalahari Goldridge and Amalia gold deposits are Archean TTGs of similar age and petrographic characteristics. If these ore-forming fluids interacted with only the basement TTG rocks and/or BIF units, it is expected that the deposits will exhibit similar variations or tendencies in the radiogenic Sr isotopic ratios of their respective vein carbonates. However, this is not the case. Inconsistency in the trend to high Sr isotopic ratios between the two deposits is illustrated in Figure 8, by comparing the Sr isotopic data with the $\delta^{13}C$ values from both deposits (e.g., [41,42]).

The vein carbonates from the Amalia gold deposit are characterized by relatively homogeneous and less radiogenic $^{87}Sr/^{86}Sr$ ratios. On the other hand, the $^{87}Sr/^{86}Sr$ ratios in the vein carbonates from Kalahari Goldridge are relatively heterogenous, more radiogenic and widespread: they range from low values that are similar to that of the Amalia gold deposit to much higher radiogenic values not recorded in the Amalia gold deposit. Given that the basement rocks are the same for both deposits, we postulate that the wider spread to relatively higher radiogenic Sr isotopic values in the precipitating vein carbonates at the Kalahari Goldridge deposit resulted from the fluid–rock interaction between ore-forming fluids and the graphite-bearing metasedimentary rock surrounding the host BIF unit. In addition, the heterogeneous nature of the $^{87}Sr/^{86}Sr$ ratios at Kalahari Goldridge can be attributed to the larger volume of fluid interaction with more crustal rocks and relatively more intensive fluid–rock interaction. The large fluid volume is reflected by the large quartz veins (Group IIB veins) and the multiple ladder veins (Group IIA) [20]. Conversely, the homogeneous $^{87}Sr/^{86}Sr$ ratios in the vein carbonates from Amalia, as illustrated in Figure 8, can be attributed to the interaction of ore-forming fluids with the limited or restricted variety of crustal sedimentary host rocks at Amalia in comparison with

Kalahari Goldridge. The observations at the Kraaipan-Amalia deposits are consistent with several orogenic gold deposits of variable geological settings worldwide (e.g., [4,41]). For example, a review by [4] on fluid and metal sources for orogenic gold deposits reported that strontium was derived from basement rocks below Archean greenstone belts that host these deposits or reflect a significant mantle component, with values altered along the flow path and at the site of gold deposition by host metasedimentary rock sequences. Their study also documented local wall rock sources for Sr or multiple strontium sources in host rocks distal to the gold deposits. In a study by [41], it was noted that the uniformity of initial $^{87}Sr/^{86}Sr$ ratios of mafic volcanic rocks in gold deposit-hosted terranes in the Archean Abitibi subprovince of Canada was consistent with a homogenous upper mantle reservoir. In a similar study in an epithermal gold deposits in the Kyushu region of Japan by [42], they noted that variations in $^{87}Sr/^{86}Sr$ ratios hosted in both the basement carbon-rich metasedimentary rocks of the Shimanto Group and the overlying andesitic volcanic rocks showed relatively high $^{87}Sr/^{86}Sr$ ratios in ore-related calcite, which were inferred to indicate hydrothermal ore-forming fluid interactions with the carbon-rich metasedimentary rocks that contained a much higher $^{87}Sr/^{86}Sr$ composition than the surrounding shallower volcanic rocks hosting the deposits.

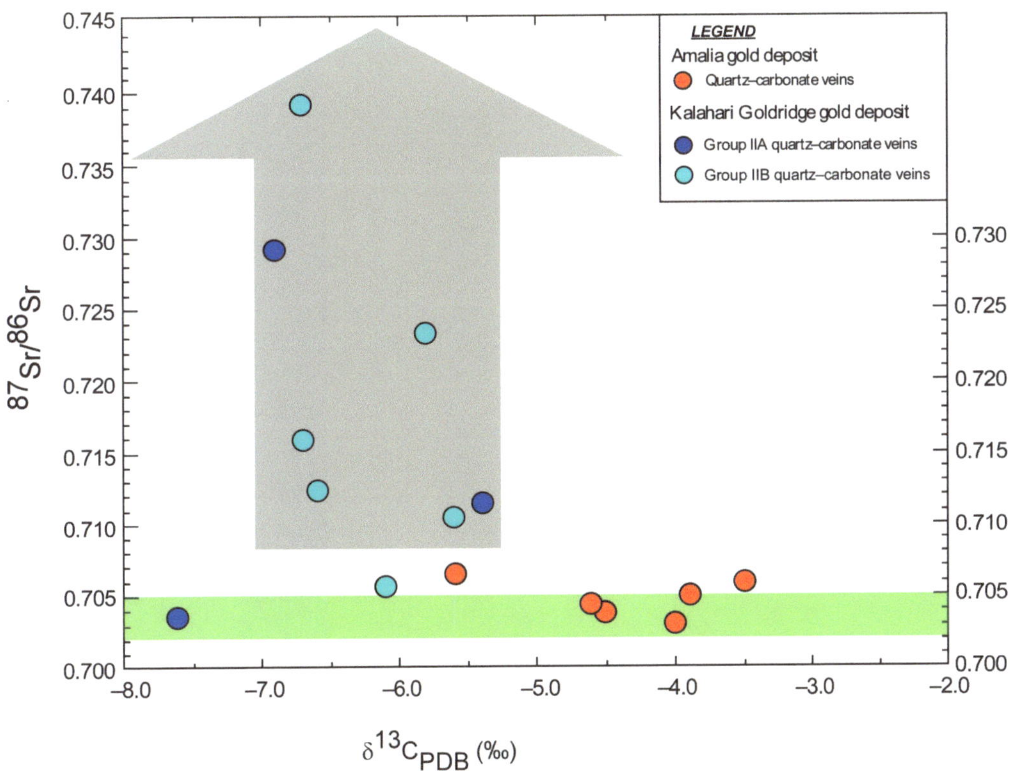

Figure 8. Plot of $^{87}Sr/^{86}S$ versus $\delta^{13}C$ of carbonates in mineralizing quartz–carbonate veins at the Kalahari Goldridge and Amalia gold deposits. The green shaded area depicts the range for the mantle source [8,9,11,13,14]. The gray shaded area depicts the range for supracrustal rocks with high $^{87}Sr/^{86}Sr$ ratios [8,10,12]. The arrow depicts increasing $^{87}Sr/^{86}Sr$ values during the fluid–rock interaction in the Kalahari Goldridge gold deposit.

Like most Archean orogenic gold deposits, the deposits in the Kraaipan-Amalia terrane exhibit similarities in $\delta^{18}O_{water}$ values in quartz from the quartz–carbonate veins, in addition to the pervasive carbonatization and sulfidation mentioned earlier [18,19]. However, variations exist in the $\delta^{13}C$ values of the associated carbonates from both deposits (Figure 9). The $\delta^{13}C$ values of ore-related carbonates from the Kalahari Goldridge deposit (−7.6 to −5.4%; [18]) are lower than the $\delta^{13}C$ values of the Amalia gold deposit (−5.8 to −3.5%; [19]). On the basis of geochemical mass balance calculations, the variation in $\delta^{13}C$ values above or below average mantle/igneous values of $-5 \pm 3\%$; [43–45] has been attributed to basement rocks of heterogeneous TTG compositions [41] or fluid–rock interactions between ore-forming fluids and carbon-rich sedimentary rocks (e.g., [42,46]). In addition, the CO_2-bearing ore-forming fluid at the Amalia contains a negligible CH_4 concentration compared to the appreciable concentration in the ore-forming fluid at Kalahari Goldridge (Figure 6), which can be attributed to the lack of host carbon-rich metasedimentary rocks at Amalia [47] and reasonably explains the absence of CH_4 in the Amalia ore-forming fluid. Therefore, the occurrence of carbon-rich metasedimentary rocks surrounding the Kalahari Goldridge deposit could have partly buffered the ore-forming fluid to reduced conditions, thereby resulting in lower $\delta^{13}C$ values and, consequently, the observed difference in redox conditions at the deposits.

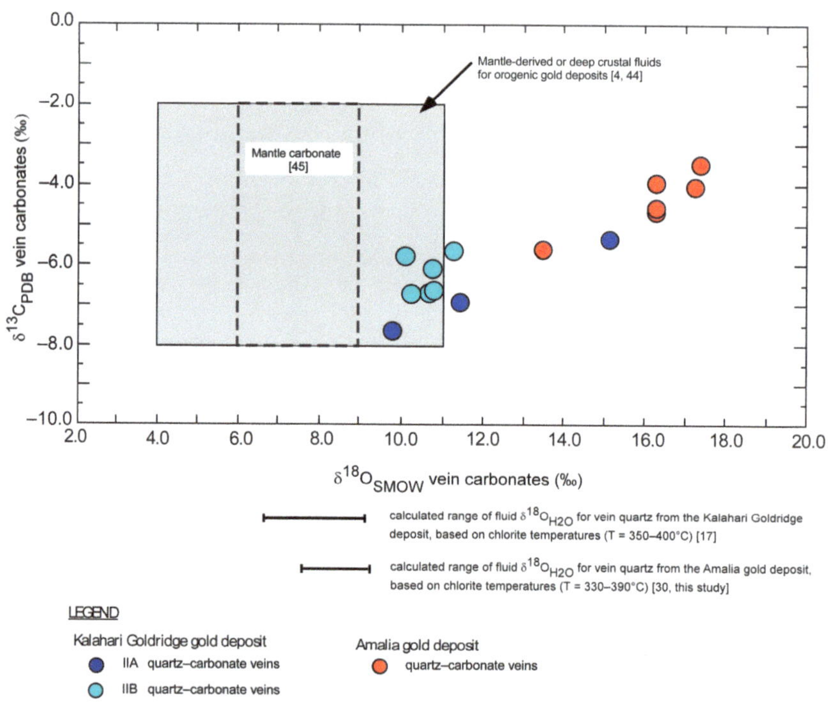

Figure 9. Variation of carbon and oxygen isotope composition in carbonates from mineralizing quartz–carbonate veins at the Kalahari Goldridge and Amalia gold deposits (From [4,17–19,30,44,45]).

The $\delta^{13}C$ and $\delta^{18}O$ of vein carbonates from the Kalahari Goldridge and Amalia gold deposits show a generally positive correlation (Figure 9) with the Amalia deposit exhibiting relatively higher $\delta^{13}C$ and $\delta^{18}O$ values between the oxygen and carbon isotope ratios. References [43,48] documented such positive variation in several studied individual deposits that showed increasing trend of enrichment of ^{18}O and ^{13}C in carbonates in the paragenetic sequence of the mineralization. Reference [43] attributed this trend to (i) decreasing temperature, (ii) increasing CO_2/CH_4 ratios in an evolving fluid system resulting from the local fluid interaction with graphitic rocks or immiscible separation of CO_2 + CH_4 in the hydrothermal fluid and/or (iii) the contribution of CO_2 from other sources, resulting in a progressive increase in the $^{13}C/^{12}C$ and $^{18}O/^{16}O$ ratios. Therefore, the general positive variation between $\delta^{13}C$ and $\delta^{18}O$ in the two deposits is attributed to the increasing oxidation state in an evolving fluid system from a unique homogenous origin that is conformable with the observed $^{87}Sr/^{86}Sr$-Sr variation illustrated in Figure 10. Additionally, this finding is also consistent with the observed regional temperature trend illustrated in Figure 7 where fluid migration is inferred from the Kalahari Goldridge deposit in the north to the Amalia deposit in south. Figure 10 shows a binary plot of $^{87}Sr/^{86}Sr$ ratios and their corresponding Sr concentrations in carbonates veins from the Kalahari Goldridge and Amalia gold deposits, which demonstrate that the carbonates veins from both deposits have a common minimum value characterized by the $^{87}Sr/^{86}Sr$ isotopic ratio of 0.70354. The minimum value indicates that the ore-forming fluids for both deposits possibly originated from a common fluid reservoir source of uniform isotopic composition ($^{87}Sr/^{86}Sr$ ratio = 0.70354), which is consistent with a mantle or mafic igneous signature [8,9,11,13,14]. The plot also illustrates a diverging trend towards increasing radiogenic Sr values along different evolutionary flow paths for each of the quartz–carbonate veining events in these deposits. The increasing radiogenic Sr trend can be attributed to the isotopic exchange resulting from the mixing of multiple fluids of isotopically different signatures or by the fluid–rock interaction between ore-forming fluids and supracrustal wall rocks with higher radiogenic Sr concentrations [8,10,12,49]. However, fluid inclusion and stable isotope data [17,18] are inconsistent with the involvement of multiple fluids in both deposits. Hence, the effect of fluid mixing is consequently ruled out. Thus, the trend to higher $^{87}Sr/^{86}Sr$ ratios illustrated in Figure 8 can be related to isotopic exchange between ore-forming fluids and the supracrustal rocks that surround the deposits. This finding is also consistent with the observed variation of CH_4 and CO_2 concentrations (Figure 6), which suggests a progressive enrichment of ^{13}C and ^{18}O in the fluids during the fluid interaction with host rocks from Kalahari Goldridge to Amalia (Figure 9), as well as the decreasing regional trend in the mineralization temperatures from Kalahari Goldridge to Amalia (Figure 7).

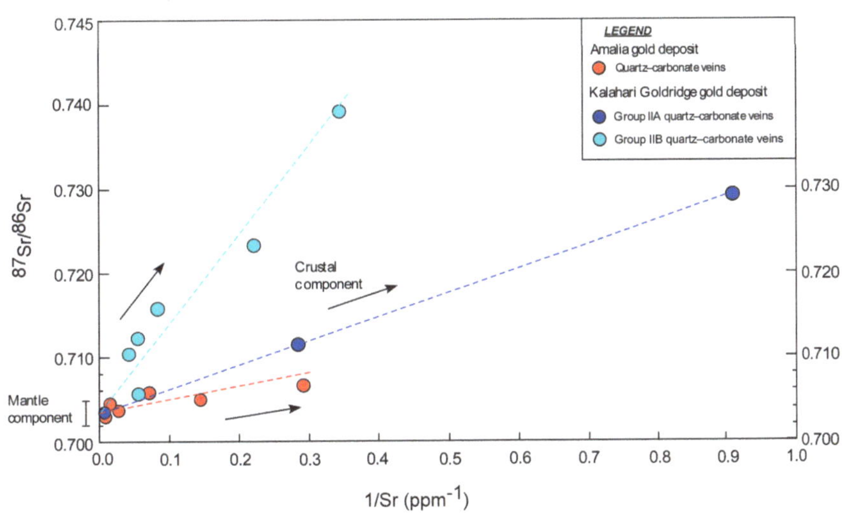

Figure 10. A plot of $^{87}S/^{86}S$ versus 1/Sr depicting the evolutionary trend of ore-forming fluids at the Kalahari Goldridge and Amalia gold deposits.

8. Conclusions

The combination of Sr, C and O isotopic data from the Kalahari Goldridge and Amalia BIF-hosted gold deposits were used to evaluate the nature and evolution of the ore-forming fluids associated with orogenic gold mineralizing systems in the Kraaipan-Amalia region of South Africa. A schematic model of the Kraaipan-Amalia gold mineralization system is illustrated in Figure 11. The two gold deposits had a common source for the ore-forming fluids on the basis of Sr isotopic data acquired on carbonates associated with gold mineralization, irrespective of their contrasting ambient redox conditions, fluid chemistry and temperatures of gold formation, which are best explained as local variations within similar mineralizing systems. The vein-forming fluids interacted with sedimentary rocks having organic materials at Kalahari Goldridge in comparison with the host rocks at Amalia, which lacked or had limited organic materials. This finding suggests that upward-migrating deep crustal ore-forming fluids having mantle-like signatures were associated with the gold mineralization in the deposits. However, the ore-forming fluids interacted with the supracrustal rocks of high isotopic composition en route to the depositional sites in the Kraaipan-Amalia terrane.

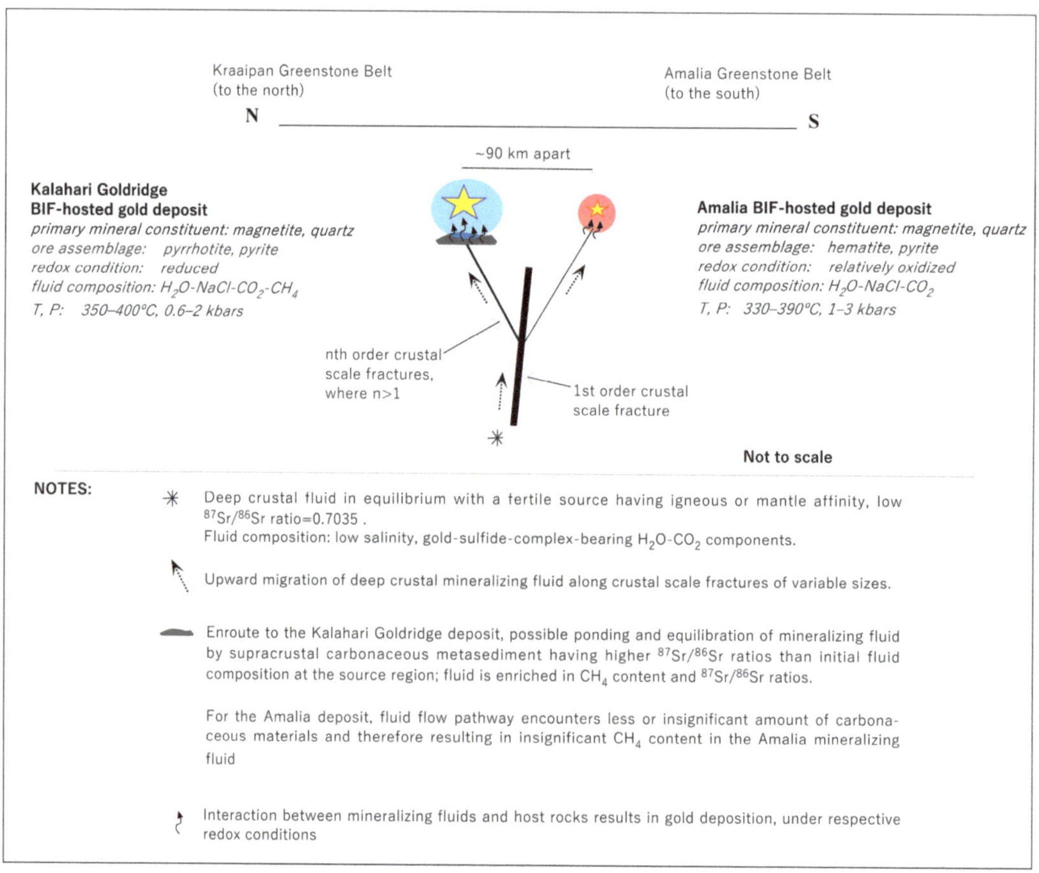

Figure 11. Schematic diagram showing BIF-hosted gold mineralization model in the Kraaipan-Amalia greenstone belts.

The absolute timing of the gold mineralization in these greenstone belts is not constrained; however, evidence from fluid composition data and Sr-C-O isotopic ratios suggests that observed differences in redox conditions in these deposits could be attributed to different local flow pathways by similar evolving fluids from a common source (with minimum $^{87}Sr/^{86}Sr$ = 0.70354) to the sites of the final gold deposition. Fluid–rock interactions between ore-forming fluid and carbonaceous meta-pelitic rock units partly resulted in reducing conditions and heterogeneity in the observed Sr-C isotopic distribution at the Kalahari Goldridge deposit. The oxidized character and homogeneous Sr-C isotopic distribution observed at the Amalia gold deposit is attributed to the lesser fluid–rock interaction between the ore-forming fluid and limited amount of (carbonaceous) supracrustal rocks. The results of this study amplify the fact that, although Archean orogenic gold deposits formed from fluids of similar composition in similar tectonic environments, minor differences in the deposits could be linked to variable host rock composition, redox conditions of gold formation and/or other physico-chemical parameters at individual deposits.

Funding: Funding for the strontium isotope research was provided by Akita University and Japan Society for the Promotion of Science (JSPS).

Acknowledgments: Masatsugu Yamamoto, formerly of Akita University, Japan, is specially thanked for his immense support during the Sr isotope analytical runs. We also thank Toshio Mizuta for his advice during the course of this research.

Conflicts of Interest: The authors declare no conflicts of interest.

References

1. Kerrich, R. Geochemical evidences of the sources of fluids and solutes for shear zone hosted mesothermal gold deposits. In *Mineralization and Shear Zones*; Bursnall, J.T., Ed.; Short Course; Geological Association of Canada: St. John's, NL, Canada, 1989; Volume 6.
2. Ho, S.E.; Groves, D.I.; McNaughton, N.J.; Mikucki, E.J. The source of ore fluids and solutes in Archaean lode-gold deposits of Western Australia. *J. Volcanol. Geotherm. Res.* **1992**, *50*, 173–196. [CrossRef]
3. Groves, D.I.; Goldfarb, R.J.; Gebre-Mariam, M.; Hagemann, S.G.; Robert, F. Orogenic gold deposits—A proposed classification in the context of their crustal distribution and relationship to other gold deposit types. *Ore Geol. Rev.* **1998**, *13*, 7–27. [CrossRef]
4. Goldfarb, R.J.; Groves, D.I. Orogenic gold. Common or evolving fluid and metal sources through time. *Lithos* **2015**, *233*, 2–26. [CrossRef]
5. Groves, D.I.; Santosh, M.; Deng, J.; Wang, Q.; Yang, L.; Zhang, L. A holistic model for the origin of orogenic gold deposits and its implications for exploration. *Miner. Depos.* **2020**, *55*, 275–292. [CrossRef]
6. Glodny, J.; Grauert, B. Evolution of a hydrothermal fluid-rock interaction system as recorded by Sr isotopes: A case study from the Schwarzwald, SW Germany. *Mineral. Petrol.* **2009**, *95*, 163–178. [CrossRef]
7. Toki, T.; Nohara, T.; Urata, Y.; Shinjo, R.; Hokakubo-Watanabe, S.; Ishibashi, J.; Kawagucci, S. Sr isotopic ratios of hydrothermal fluids from the Okinawa Trough and the implications of variation in fluid–sediment interactions. *Prog. Earth Planet Sci.* **2022**, *9*, 59. [CrossRef]
8. Armstong, R.L. A model for the evolution of strontium and lead isotopes in a dynamic earth. *Rev. Geophys.* **1968**, *6*, 175–199. [CrossRef]
9. Smith, C.B. Pb, Sr and Nd isotopic evidence for sources of southern African Cretaceous kimberlites. *Nature* **1983**, *304*, 51–54. [CrossRef]
10. Jacobsen, S.B.; Pimentel-Klose, M.R. Nd isotopic variations in Precambrian banded iron formations. *Geophys. Res. Lett.* **1988**, *15*, 393–396. [CrossRef]
11. Mitchell, R.H. *Kimberlites, Orangeites, and Related Rocks*; Plenum Press: New York, NY, USA, 1995.
12. Shields, G.A.; Veizer, J. Precambrian marine carbonate isotope database: Version 1.1. *Geochem. Geophys. Geosyst.* **2002**, *6*, 1–12. [CrossRef]
13. Donnelly, C.L.; Griffin, W.L.; O'Reilly, S.L.; Pearson, N.J.; Shee, S.R. The Kimberlites and related rocks of the Kuruman Kimberlite Province, Kaapvaal Craton, South Africa. *Contrib. Mineral. Petrol.* **2011**, *161*, 351–371. [CrossRef]
14. Hochscheid, F.; Coltat, R.; Ulrich, M.; Munoz, M.; Manatschal, G.; Boulvais, P. The Sr isotope geochemistry of oceanic ultramafic-hosted mineralization. *Ore Geol. Rev.* **2022**, *144*, 104824. [CrossRef]
15. Anhaeusser, C.R.; Walraven, F. Episodic granitoid emplacement in the western Kaapvaal Craton. Evidence from the Archean-Kraaipan granite-greenstone terrane, South Africa. *J. Afr. Earth Sci.* **1999**, *28*, 289–309. [CrossRef]
16. Hammond, N.Q.; Moore, J.M. Archean lode gold mineralization in banded iron formation at the Kalahari Goldridge deposit, Kraaipan greenstone belt, South Africa. *Miner. Depos.* **2006**, *41*, 483–503. [CrossRef]
17. Brandl, G.; Cloete, M.; Anhaeusser, C.R. Archaean greenstone belts. In *The Geology of South Africa*; Johnson, M.R., Anhaeusser, C.R., Thomas, R.J., Eds.; Geological Society of South Africa/Council for Geoscience Johannesburg/Pretoria: Pretoria, South Africa, 2006; pp. 9–56.

18. Hammond, N.Q.; Moore, J.M.; Sheets, R.W. Physico-chemical conditions of ore-bearing fluids associated with the genesis of the Kalahari Goldridge deposit, Kraaipan greenstone belt, South Africa. *Ore Geol. Rev.* **2007**, *30*, 106–134. [CrossRef]
19. Adomako-Ansah, K.; Mizuta, T.; Ishiyama, D.; Hammond, N.Q. Nature of ore-forming fluid and formation conditions of BIF-hosted gold mineralization in the Archean Amalia greenstone belt, South Africa: Constraints from fluid inclusion and stable isotope studies. *Ore Geol. Rev.* **2017**, *89*, 609–626. [CrossRef]
20. Hammond, N.Q.; Morishita, Y. Source of ore fluids at the Kalahari Goldridge deposit, Kraaipan greenstone belt, South Africa: Evidence from Sr, C and O isotope signatures in carbonates. *Geofluids* **2009**, *9*, 356–364. [CrossRef]
21. Corner, B.; Durrheim, R.J.; Nicolaysen, L.O. Relationships between the Vredefort structure and the Witwatersrand basin within the tectonic framework of the Kaapvaal Craton as interpreted from regional gravity and aeromagnetic data. *Tectonophysics* **1990**, *171*, 49–61. [CrossRef]
22. de Wit, M.J.; Roering, C.; Hart, R.J.; Armstrong, R.A.; de Ronde, C.E.J.; Green, R.W.E.; Tredoux, M.; Peberdy, E.; Hart, R.A. Formation of an Archaean continent. *Nature* **1992**, *357*, 553–562. [CrossRef]
23. McCourt, S. The crustal architecture of the Kaapvaal crustal block South Africa, between 3.5 and 2.0 Ga: A synopsis. *Miner. Depos.* **1995**, *30*, 89–97. [CrossRef]
24. Richardson, S.H.; Shirey, S.B.; Harris, J.W.; Carlson, R.W. Archean subduction recorded by Re-Os isotopes in eclogitic sulfide inclusions in Kimberley diamonds. *Earth Planet. Sci. Lett.* **2001**, *191*, 257–266. [CrossRef]
25. Schmitz, M.D.; Bowring, S.A.; de Wit, M.J.; Gartz, V. Subduction and terrane collision stabilize the western Kaapvaal craton tectosphere 2.9 billion years ago. *Earth Planet. Sci. Lett.* **2004**, *222*, 363–376. [CrossRef]
26. Eriksson, P.G.; Banerjee, S.; Nelson, D.R.; Rigby, M.J.; Catuneanu, O.; Sarkar, S.; Roberts, R.J.; Ruban, D.; Mtimkulu, M.N.; Raju, P.V.S. A Kaapvaal craton debate: Nucleus of an early small supercontinent or affected by an enhanced accretion event? *Gondwana Res.* **2009**, *15*, 354–372. [CrossRef]
27. Poujol, M.; Anhaeusser, C.R.; Armstrong, R.A. Episodic granitoid emplacement in the Archean Amalia-Kraaipan terrane, South Africa: Confirmation from zircon U-Pb geochronology. *J. Afr. Earth Sci.* **2002**, *35*, 147–161. [CrossRef]
28. Robb, L.J.; Meyer, F.M. The Witwatersrand Basin: Geological framework and mineralization processes. *Ore Geol. Rev.* **1995**, *10*, 67–94. [CrossRef]
29. Jones, I.M.; Anhaeusser, C.R. Accretionary lapilli associated with Achaean banded iron formations of the Kraaipan Group, Amalia greenstone belt, South Africa. *Precambrian Res.* **1993**, *61*, 117–136. [CrossRef]
30. de Wit, M.; Tinker, J. Crustal structures across the central Kaapvaal craton from deep-seismic reflection data. *S. Afr. J. Geol.* **2004**, *107*, 185–206. [CrossRef]
31. Adomako-Ansah, K.; Mizuta, T.; Hammond, N.Q.; Ishiyama, D.; Ogata, T.; Chiba, H. Gold mineralization in banded iron formation in the Amalia greenstone belt, South Africa: A mineralogical and sulfur isotope study. *Resour. Geol.* **2013**, *63*, 119–140. [CrossRef]
32. Yamamoto, M.; Maruyama, T. The Sr and Nd isotopic analysis and quantitative analysis of Rb and Sr by using MAT 261. *Rep. Res. Inst. Nat. Resour. Min. Coll. Akita Univ.* **1996**, *61*, 7–30. (In Japanese with English Abstract).
33. Bryndzia, T.; Scott, S.D. The composition of chlorite as a function of sulfur and oxygen fugacity: An experimental study. *Am. J. Sci.* **1987**, *287*, 50–76. [CrossRef]
34. De Caritat, P.; Hutcheon, I.; Walshe, J.L. Chlorite geothermometry: A review. *Clays Clay Miner.* **1993**, *41*, 219–239. [CrossRef]
35. Vidal, O.; Lanari, P.; Munoz, M.; Bourdelle, F.; De Andrade, V. Deciphering temperature, pressure and oxygen-activity conditions of chlorite formation. *Clay Miner.* **2016**, *51*, 615–633. [CrossRef]
36. Shikazono, N.; Kawahata, H. Compositional differences in chlorite from hydrothermally altered rocks and hydrothermal ore deposit. *Can. Miner.* **1987**, *25*, 465–474.
37. Bobos, I.; Noronha, F.; Mateus, A. Fe-, Fe, Mn- and Fe, Mg-chlorite: A genetic linkage to W, (Cu, Mo) mineralization in the magmatic-hydrothermal system at Borralha, northern Portugal. *Mineral. Mag.* **2018**, *82*, 259–279. [CrossRef]
38. Wang, Z.; Chen, B.; Yan, X.; Li, S. Characteristics of hydrothermal chlorite from the Niujuan Ag-Au-Pb-Zn deposit in the north margin of NCC and implications for exploration tools for ore deposits. *Ore Geol. Rev.* **2018**, *101*, 398–412. [CrossRef]
39. Cathelineau, M. Cation site occupancy in chlorites and illites asa function of temperature. *Clay Miner.* **1988**, *23*, 471–485. [CrossRef]
40. Cathelineau, M.; Nieva, D. A chlorite solid solution geothermometer. The Los Aufres geothermal system (Mexico). *Contrib. Mineral. Petrol.* **1985**, *91*, 235–244. [CrossRef]
41. Kerrich, R.; Fryer, B.J.; King, R.W.; Willmore, L.M.; van Hees, E. Crustal outgassing and LILE enrichment in major lithosphere structures, Archean Abitibi greenstone belt: Evidence on the source reservoir from strontium and carbon isotope tracers. *Contrib. Mineral. Petrol.* **1987**, *97*, 156–168. [CrossRef]
42. Morishita, Y.; Nakano, T. Role of basement in epithermal deposits: The Kushikino and Hishikari gold deposits, southwestern Japan. *Ore Geol. Rev.* **2008**, *34*, 597–609. [CrossRef]
43. Ohmoto, H.; Rye, R.O. Isotopes of sulphur and carbon. In *Geochemistry of Hydrothermal Ore Deposits*, 2nd ed.; Barnes, H.L., Ed.; John Wiley and Sons: New York, NY, USA, 1979; pp. 509–562.
44. McCuaig, T.C.; Kerrich, R. P-T-t-deformation-fluid characteristics of lode gold deposits: Evidence from alteration systematics. *Ore Geol. Rev.* **1998**, *12*, 381–453. [CrossRef]
45. Giuliani, A.; Phillips, D.; Kamenetsky, V.S.; Fiorentini, M.L.; Farquhar, J.; Kendrick, M.A. Stable isotope (C, O, S) compositions of volatile-rich minerals in kimberlites: A review. *Chem. Geol.* **2014**, *374–375*, 61–63. [CrossRef]

46. Kerrich, R. The stable isotope geochemistry of Au-Ag vein deposits in metamorphic rocks. In *Stable Isotope Geochemistry of Low Temperature Processes: GAC-MAC Short Course*; Kyser, T.K., Ed.; Mineralogical Association of Canada: Quebec City, QC, Canada, 1987; Volume 11, pp. 318–361.
47. Kiefer, R. Regional Geology, Tectonic Evolution, and Controls of Gold Mineralization in the Archean Amalia Greenstone Belt, South Africa. Ph.D. Thesis, University of the Witwatersrand, Johannesburg, South Africa, 2004; 542p.
48. Rye, R.O.; Ohmoto, H. Sulphur and carbon isotopes and ore genesis: A review. *Econ. Geol.* **1974**, *69*, 642–826. [CrossRef]
49. McLennan, S.M.; Taylor, S.R.; McCulloch, M.T.; Maynard, J.B. Geochemical and Nd-Sr isotopic composition of deep-sea turbidites: Crustal evolution and plate tectonics. *Geochim. Cosmochim. Acta* **1990**, *54*, 2015–2050. [CrossRef]

MDPI AG
Grosspeteranlage 5
4052 Basel
Switzerland
Tel.: +41 61 683 77 34

Minerals Editorial Office
E-mail: minerals@mdpi.com
www.mdpi.com/journal/minerals

www.ingramcontent.com/pod-product-compliance
Lightning Source LLC
Chambersburg PA
CBHW041427210920
51850CB00013B/3359
9783725838288